Notes on Numerical Fluid Mechanics and Multidisciplinary Design

Volume 144

Notes on Numerical Fluid Mechanics and Multidisciplinary Design publishes state-of-art methods (including high performance methods) for numerical fluid mechanics, numerical simulation and multidisciplinary design optimization. The series includes proceedings of specialized conferences and workshops, as well as relevant project reports and monographs.

More information about this series at http://www.springer.com/series/4629

Piotr Doerffer · Pawel Flaszynski ·
Jean-Paul Dussauge · Holger Babinsky ·
Patrick Grothe · Anna Petersen · Flavien Billard
Editors

Transition Location Effect on Shock Wave Boundary Layer Interaction

Experimental and Numerical Findings from the TFAST Project

Springer

Editors
Piotr Doerffer
Institute of Fluid Flow Machinery, Polish Academy of Sciences, IMP PAN
Gdansk, Poland

Pawel Flaszynski
Institute of Fluid Flow Machinery, Polish Academy of Sciences, IMP PAN
Gdansk, Poland

Jean-Paul Dussauge
Supersonic Group
IUSTI
Marseille, France

Holger Babinsky
Department of Engineering
University of Cambridge
Cambridge, UK

Patrick Grothe
Rolls-Royce Deutschland Ltd
Blankenfelde-Mahlow, Germany

Anna Petersen
German Aerospace Center (DLR)
Institute of Propulsion Technology
Göttingen, Niedersachsen, Germany

Flavien Billard
Aircraft Engineering and Loads Division
Dassault Aviation
Saint-Cloud, France

ISSN 1612-2909 ISSN 1860-0824 (electronic)
Notes on Numerical Fluid Mechanics and Multidisciplinary Design
ISBN 978-3-030-47460-7 ISBN 978-3-030-47461-4 (eBook)
https://doi.org/10.1007/978-3-030-47461-4

This Springer imprint is published by the registered company Springer Nature Switzerland AG
The registered company address is: Gewerbestrasse 11, 6330 Cham, Switzerland

Acknowledgements

The editors would like to thank the European Commission for funding the research project FP7-AAT-2010-RTD-1 GA-2011-265455 TFAST Transition Location Effect on Shock Wave Boundary Layer Interaction.

The coordinators would like to express special gratitude to Dietrich Knoerzer for stimulating and friendly supervision of the project.

It should be also emphasized the helpful role of Jean Delery, Erick Dick, Claus Sieverding and Werner Haase, as Advisor Board members, whose comments improved the final shape of the project results.

TFAST coordinators would like also to thank the Computational Centre of TASK (Centrum Informatyczne Trójmiejskiej Akademickiej Sieci Komputerowej - CI TASK) and PL-Grid Infrastructure for allowing access to computational resources.

The publication of the book is a result of a collective effort by contributors and authors of the chapters. The editors wish to express warm gratitude to Michal Piotrowicz for the book composition and preparation for printing.

Contents

Introduction—TFAST Overview

Piotr Doerffer

Abstract In recent decades huge effort has been placed on maintaining laminar boundary layers with respect to a number of applications. To a large extent this research concerned laminar wings, especially focusing on maintaining laminar boundary layer on the longest possible distance. The research of laminar boundary layer is inherently coupled with the investigations of instabilities leading to the laminar-turbulent transition. These research directions are still in progress. The existing knowledge, at its present stage of development, has been fully utilised in the project for the analysis of laminar boundary layer development and the following transition. It must be emphasized that TFAST objective was not to improve knowledge of laminar boundary layers and of transition. The general aim of TFAST was to avoid that the laminar boundary layer is penetrated by the shock wave. The benefits of having laminar boundary layer are so important that the transition should occur as late as possible. The sensitivity of the solution is calling for basic studies before the configurations closer to application may be investigated. It is necessary to carry out advanced experiments and accurate CFD work with the most advanced methods as LES and DNS. The topic of laminar/transitional/turbulent interaction with a shock wave is the most challenging problem in aeronautics, even more when unsteady interaction effects are to be treated. For these challenging tasks a consortium of the high quality was employed. It is not only the high scientific level and skills of the partners which is important but also the quality of partner's cooperation and integration in the consortium. Due to the experience obtained from the consortium that was forming the UFAST project, and because the proposed research was based on the experience gained in UFAST, the core of the TFAST consortium consisted mainly of UFAST partners. Due to the industrial requirements and needs those partners being "only" observers in UFAST were employed as full partners in TFAST. Another "new" partner was DLR which carried out the experimental compressor and turbine cascade investigations. In summary one may conclude that the consortium is not only of the high European level, but that it has already a long tradition of successful collaboration. The main objective of the project was to study the effect of transition location

P. Doerffer (✉)
Institute of Fluid Flow Machinery, Polish Academy of Sciences, IMP PAN, Gdansk, Poland
e-mail: doerffer@imp.gda.pl

P. Doerffer et al. (eds.), *Transition Location Effect on Shock Wave Boundary Layer Interaction*, Notes on Numerical Fluid Mechanics and Multidisciplinary Design 144,
https://doi.org/10.1007/978-3-030-47461-4_1

on the structure of interaction between a shock wave and a boundary layer. Main question was how close the induced transition may be to the shock wave while still maintaining a typical turbulent character of the interaction. In other words, how far the laminar boundary layer may extend without changing the turbulent character of interaction. This question have been answered in WP-2 (i.e. in the basic type of the test section), and also in the flow cases characterising different applications in WP-3, WP-4 and WP-5.

1 Project Summary

Vision-2020, whose objectives include the reduction of emissions and more effective transport systems, puts severe demands on aircraft velocity and weight. These require an increased load on wings and aero-engine components. The greening of air transport systems means a reduction of drag and losses, which can be obtained by keeping laminar boundary layers on external and internal airplane parts. Increased loads make supersonic flow velocities more prevalent and are inherently connected to the appearance of shock waves, which in turn may interact with laminar boundary layers. Such an interaction can quickly cause flow separation, which is highly detrimental to aircraft performance, and poses a threat to safety. In order to diminish the shock induced separation, **the boundary layer at the point of interaction should be turbulent**.

The main objective of the TFAST project is to study the effect of transition location on the structure of interaction. The main question is how close to the shock wave the induced transition may be located, while still maintaining a typical turbulent character of the interaction.

The main study cases—shock waves on wings/profiles, turbine and compressor blades and supersonic intakes—have helped to answer open questions posed by the aeronautics industry and to tackle more complex applications. In addition to basic flow configurations, transition control methods were investigated for controlling transition location, interaction induced separation and inherent flow unsteadiness. TFAST for the first time is providing a characterization and selection of appropriate flow control methods for transition induction as well as physical model of these devices.

Emphasis has been placed on close coupling of experiments and numerical investigations to overcome weaknesses in both approaches.

2 Project Objectives

In recent decades huge effort has been placed on maintaining laminar boundary layers with respect to a number of applications. To a large extent this research concerned laminar wings, especially focusing on maintaining laminar boundary layer on the longest possible distance. The research of laminar boundary layer is inherently coupled with the investigations of instabilities leading to the laminar-turbulent transition. These research directions are still in progress. The existing knowledge, at its present stage of development, has been fully utilised in the project for the analysis of laminar boundary layer development and the following transition.

It must be emphasized that TFAST objective is not to improve knowledge on laminar boundary layers and on transition itself. **The general aim of TFAST is to avoid that the laminar boundary layer is penetrated by the shock wave**. The benefits of having laminar boundary layer are so important that the transition should occur as late as possible. The sensitivity of the solution is calling for basic studies before the configurations closer to application may be investigated. It is necessary to carry out advanced experiments and accurate CFD work with the most advanced methods as LES and DNS. The topic of laminar/transitional/turbulent interaction with a shock wave is the most challenging problem in aeronautics, even more when unsteady interaction effects are to be treated.

In all of the investigated test cases there are two limiting conditions which provide a reference frame for the research. The first one concerns flows with natural transition. Here, the location of transition relative to the shock wave is a key result. It is desirable to obtain a laminar boundary layer interacting with the shock wave in order to understand how transition occurs in such a case and how detrimental this can be. The second condition refers to fully turbulent flow which may be obtained using boundary layer tripping close to the leading edge. Of course, these two limiting cases cause large differences in the shock structure, in the pressure distribution, and in the history and state of the boundary layer.

An important aspect of the TFAST project is to investigate the appropriate place for transition. This means to find the transition location closest to the shock wave while ensuring that the interaction occurs with a minimum of adverse effects (such as separation, unsteadiness and large lambda shock foot). Current knowledge suggests that this requires fully turbulent flow at the shock foot, but one of the objectives of TFAST is to also study transitional interactions which are currently poorly understood. This latter aspect of TFAST is linked to the investigation of methods of transition induction. Apart from the widely used tripping with wires or distributed roughness it is planned to also use fluidic jets and cold plasma control in the experiments carried out in TFAST. This choice of methods is based on results of the EU project UFAST. The key factor for the method selection was to minimize the disturbance caused by the control device to the laminar boundary layer while still ensuring successful transition. From this it follows that traditional fixed vortex generators made as a piece of oblique plate rigidly mounted on the wall are not acceptable. Thus a major challenge of the TFAST project is to tackle a completely new research aim: flow control devices in a laminar boundary layer. Hence, one of the main and most important topics in the TFAST project is to understand how transition can be induced by control devices, without penetrating the main stream i.e. overtripping the boundary layer.

For these challenging tasks a consortium of the highest quality has been employed. It is not only the high scientific level and skills of the partners which were important but also the quality of partner's cooperation and integration in the consortium.

Because of the good experience obtained from the consortium that was forming the UFAST project, and because the proposed research is based on the formerly gained experience, the core of the TFAST consortium consists mainly of UFAST partners. However, only those partners have been selected who have a proven research track record in the topics to be investigated in TFAST. Due to the industrial requirements and needs, those partners who were "only" observers in UFAST, have been employed as full partners in TFAST. Another "new" partner was DLR which carried out the experimental compressor and turbine cascade investigations. Some partners of the TFAST consortium have already worked together with DLR in the AITEB and other projects. In summary one may say that the consortium was not only of the highest European level but that it also already had a long tradition of successful collaboration.

Airplane manufacturers are continuously striving to improve the efficiency of their products to reduce fuel burn. This objectives have even been formalised in the ACARE Strategic Research Agenda which aims at a reduction of 20% for fuel consumption and CO_2 and 80% reduction in NOX compared to year 2000 reference airplanes. The main objective of TFAST is to obtain net fuel saving of about 12% compared to year 2000 reference airplanes. This has been achieved by allowing for the maximal use of the laminar technology in respect to the wing as well as to compressor and turbine blades.

Too little effort had been so far devoted to accurately predicting and controlling phenomena dedicated to unsteady shock wave laminar boundary layer interaction problems. The control of transition location has significant effect on the boundary layer state and on the interaction unsteadiness. TFAST project has improved understanding of flow-physics in respect of unsteady SWBLI phenomena with transition and improved approach to appropriate modelling of the flow of interest and—even more important—to control it in order to keep physical risks for aircraft as low as possible.

Improved predictive capability has been obtained of the presently available CFD methodologies such as URANS, (Unsteady Reynolds Averaged Navier-Stokes), LES, (Large-Eddy Simulation), hybrid RANS-LES approaches and even DNS. TFAST project allowed for a deeper insight into the physics governing the laminar—turbulent transition and unsteadiness of the shock, the shock/boundary layer interaction, the development of buffeting, together with a study on efficient methods for controlling these phenomena, allowing reaching for aircraft with better performance and improved safety.

One of important objectives was how to combine the flow control methods with transition modelling. Most of the former transition modelling approaches had been tested for all flow control methods. This has provided indication which model is superior for which flow control device.

As the mentioned physical phenomena that occur in high speed, i.e. transonic and supersonic, flows for both external and internal aerodynamics lead to boundary layer separation which might cause a damage of the structure itself but would at least downgrade the efficiency of the aircraft or aircraft system. The importance and consequence of such a malfunction on the fatigue of structure are well known, with a direct influence on the qualification of the aircraft. As mentioned above,

SWBLI can occur in supersonic air intakes and reduce their efficiency because the induced separation becomes strongly unsteady with back-coupling to the engine itself. Obviously in these cases the performance of the engine itself is endangered.

For these reasons, the TFAST project had conducted—as a project objective—a relevant set of well focused experiments, relevant to the above mentioned flow-physics phenomena, and the data base, both of experimental and numerical results has provided a sound basis, which can easily be exploited by other interested groups in Europe, but primarily of course to the aeronautics industry. Experiments have been designed by pro-active CFD work in order to link experiment and CFD as closely together as possible and to allow for cross-fertilisation, rather than on competition between them.

To summarise, the treatment of shock wave/boundary layer interaction involved a coupling of a number of different physical aspects, which constituted the key-factors to be met in respect of the objectives and are listed as follows:

- Shock motion that is initiated by the incoming boundary layer and most likely by a shock-induced separation bubble.
- Unsteadiness of the separation bubble, which may be due to a pure pulsation or may result from vortex shedding. In the latter case, the vortices being produced in the separated zone are convected downstream, often over large distances.
- Transition induction and turbulence production (possibly including strong compressibility effects) caused by the shock wave itself.
- Formation of a new boundary layer downstream of the separation, more precisely downstream of re-attachment which is characterized by vortex interactions, but also by low-frequency unsteadiness that might be induced by the shock motion.
- Strong coupling through acoustic waves between the different phenomena—depending on the different flow conditions and characteristics.

The overall objective of the TFAST project was to fill the lack of knowledge on transition location effect on SWBLI, by providing knowledge and expertise on three main topics, easily to be identified as particular objectives:

- Reference experiments focused on unsteady effects (designed by CFD).
- Application and improvement of existing numerical modelling methods.
- Enhanced physical understanding of the complicated physical phenomena.

To tackle this highly complex and innovative objectives and to arrive at general conclusions most of the flow configurations in which shocks play a key role have to be addressed. It is evident that these goals and objectives were met by a highly skilled consortium of organisations with expertise on both experimental and theoretical grounds. This inherently means that a group of experimentalists and a group of theoreticians have interacted for the sake of an improved understanding of the problems of interest. Both groups have offered critical mass that allowed reaching the upstream and cutting-edge goals of the TFAST project. This was ensured by basic research carried out in the project, allowing for cross validation, gathering of new results and at the same time categorizing them into a reliable new framework of a new knowledge base. The knowledge gathered from the basic research was utilized in the research on application test cases. Evidently, a smaller consortium would have been of high risk, because important knowledge and expertise would have been missed.

The first objective of the TFAST project was to provide a comprehensive experimental data base, which has documented flow structure for different transition locations. It has provided insight to both the high frequency events and the properties of the large scale coherent structures in the context of SWBLI. It must be mentioned again: almost no experimental information was available for SWBLI, and more important for industrially relevant flows. The low configurations measured correspond to generic geometries that can be easily exploited when more complex geometries need to be treated, as there are airfoils/wings, nozzles, cascades i.e. all important flow cases governed by normal and oblique shocks. This "shock configuration platform" was necessary when looking for the general interaction unsteady features. And again: a realisation of this objective in the short 3-years project time could only be achieved by an involvement of well-skilled laboratories which shared the huge amount of necessary experimental work.

Data obtained were split into so-called "basic" and "control" cases, the latter carried out to provide the means to industry for reducing risks in the domain of flow-physics, in particular by reducing unsteadiness of flow, i.e. stabilising flow, reducing noise and even fatigue. Control devices were used to control large eddies and include e.g. stream-wise vortex generators and electro-hydrodynamic actuators (cold plasma).

As already mentioned TFAST project has put a great deal of emphasis on the fact that experiments were accompanied by theoretical methods—and vice versa—that had been enhanced according to the needs for SWBLI. This might even mean that experiments were designed by CFD and/or were changed according to the geometrical or flow parameters or were repeated if numerical results indicated a need for that. The TFAST structure involved interactive design of experiments with the help of CFD groups.

The second objective dealt with application of theoretical methods for transitional flows, RANS/URANS, hybrid RANS-LES, LES and DNS approaches for both improving the knowledge on SWBLI and modelling of flow physics. This investigation also included advanced numerics, as well as advanced modelling strategies and investigations on the "range of applicability" for the different methods involved. The outcome of TFAST in this respect is a "best-practice guidelines" of transitional SWBLI problems.

There was a strong need for high accuracy numerical approaches, applied to the main three categories of numerical tools (RANS/URANS, RANS-LES and LES). This requirement stems from the necessity to accurately capture and resolve spontaneous unsteadiness, such as shear layer instabilities generated by shock interactions. In addition to the numerical work carried out to develop these high-order schemes, it was obviously necessary that the still existing deficiencies for modelling shock wave/boundary layer had to be treated as well.

Summarising on the above mentioned items, and getting to the third objective, it can be stated that a major outcome of the TFAST project was the improvement in physical understanding of all phenomena governing transitional shock wave/boundary layer interaction. New knowledge was generated concerning the effect of transition location on SWBLI and concerning the unsteady interaction phenomena as coupling between low frequency vortex shedding and shock movement or turbulence amplification/decay at the shock wave.

To conclude, it was evident that the TFAST project was extremely innovative and was clearly dominated by upstream research. Moreover, the TFAST was not solely aiming at an advanced understanding of flow-physics in general, but was aiming at supporting of industrial needs and requirements in the field of interest which was of utmost importance for the design of current and future aircraft—but had never been investigated thoroughly in the past.

3 Research Consortium

The consortium overview is given in Table 1. It shows that 16 organisations from 7 European Member countries and two from ICPC (Russia and Ukraine) were participating originally. Shown is also the role of the organisation in the project. Besides research organisations the consortium includes:

Table 1 List of partners

No	Participant name	Abrev.	Country	EXP	CFD
1	IMP PAN	IMP	Poland	X	X
2	Rolls-Royce Deutschland, Berlin	RRD	Germany		X
3	Dassault-Aviation	DAAV	France		X
4	CNRS Lab. IUSTI, UMR 7343, Marseille	IUSTI	France	X	X
5	ONERA: (DAFE, DMAE, DAAP)	ONERA	France	X	X
6	DLR: turbines-Göttingen, compressors-Köln	DLR	Germany	X	
7	NUMECA, Belgium, SME	NUMECA	Belgium		X
8	Institute of Aviation, Warsaw	IoA	Poland	X	X
9	Khristianovich Institute of Theoretical and Applied Mechanics SB RAS	ITAM	Russia	X	
10	University of Cambridge, Dept. of Engin.	UCAM	England	X	
11	Delft University of Technology, Aero. Lab.	TUD	Holland	X	
12	University of Southampton, (SES)	SOTON	England		X
13	University of Rome "La Sapienza"	URMLS	Italy		X
14	University of Liverpool, Dept. of Aero. Eng.	LIV	England		X
15	Institute Mécanique des Fluides de Toulouse	IMFT	France		X
16	Ukrainian Academy of Sciences	UAN	Ukraine		X

- 2 large industry partners: RRD (Rolls-Royce Germany) and DAAV (Dassault Aviation)
- 1 SME specialised in CFD; NUMECA is a software provider and research institution.

In addition an Advisory Board has been established consisting of four independent experts who were attending TFAST meetings and were taking part in the discussions, giving some advice and guidance to the research carried out. The presence of Advisory Board was obviously contributing to the quality of the research.

The TFAST project contained two main research approaches, at least in size and volume of research. The first one is experimental which delivered a data base of transitional SWBLI and its control, meeting the present needs of industry. The second one was contributing to numerical modelling using URANS, hybrid RANS/LES and LES and DNS simulations, which delivered an assessment of their applicability to the problem. Additional effort was devoted to an analytical work, focused on the physical modelling of transition control devices.

Looking at the general features of transitional SWBLI, most of the flow configurations exhibiting shock waves in internal and external flows had to be included in the investigations. Unfortunately, this implied a high number of flow cases. In order to fit these into a three-year project different experimental facilities must have been engaged, as well as different theoretical methods, e.g. CFD codes. That is the main reason why 16 partners were taking part in the realisation of the ambitious goals of the TFAST project. This number was not exaggerated and amount of overlapping work was minimal, just sufficient to ensure the verification of results. One should also mention here that ONERA consists really of three independent groups: DAFE, DMAE and DAAP. Also DLR includes two independent laboratories: Göttingen (turbines) and Köln (compressors). Therefore the number of involved research labs is greater than the number of listed partners.

3.1 Experiment—CFD Relation

Experimental work in Fluid Mechanics requires facilities. Among the nine wind tunnels used in the project, four of them are run at universities, and five in Research Establishments. All these facilities are known to produce flows of very good quality, with good repeatability and stability of the aerodynamic conditions. Some of them have remarkable quality. The technical staff who are in charge have a large experience which suggests that no major hitches in wind tunnel operations are to be expected. The characteristics of the wind tunnels are different in Mach and Reynolds number values; so that the consortium covered a wide range of conditions, from rather low Reynolds numbers making LES easier, to larger Reynolds numbers closer to "real life" situations. The groups of scientists working on these facilities have a well known expertise on experimental methods. The size of the group of experimentalists is supercritical, so that collaborations and synergies can be expected.

Similarly, the numerical modelling groups consist of well known and highly skilled experts. Many of them have published original and prominent contributions to turbulence modelling for RANS/URANS and LES, Hybric LES/RANS and DNS. Although each individual team is of limited size, they all have sufficient infrastructure for this challenging project. They all have experience and expertise in the area of numerical simulations of compressible flows. For their computations they all have access to up-to-date computational resources, such as PC clusters running under distributed memory, or they have access to supercomputers or supercomputing centres.

Analysing the contribution of CFD and experimental groups (see the Table 1) it may be noticed that the TFAST project is very well balance in the relation between the experimental and the numerical investigations.

3.2 Inclusion of SME

Particular attention was paid to a very skilled **small enterprise** (SME), which is a part of the Consortium. This is a Belgian company NUMECA, working for many years in the area of aerodynamics with particular application to turbo-machinery. NUMECA is developing CFD codes, is involved in mesh generation, turbo-machinery design, as well as development of models for flow-physics analysis. NUMECA is an ideal partner in presented project because it is able to combine the latest developments in CFD modelling with relevant industrial objectives. Therefore NUMECA had been nominated by the consortium as CFD leader which was helping in more structured approach to CFD methods in TFAST project and was strongly involved in the formulation of conclusions and achievements on the CFD side of the project.

3.3 Inclusion of ICPC Countries

Another important element of the consortium is the participation of the International Cooperation Partner Countries (ICPC) from Russia and Ukraine. One partner was an excellent experiment group ITAM, and the other one UAN is a CFD group working on turbo-machinery applications.

The Institute of Theoretical and Applied Mechanics, Academy of Sciences of Russia, Siberian Branch (Novosibirsk) is well known for the quality of its research on stability and turbulence in compressible flows. Its supersonic wind tunnel is proven to be excellent and by all means comparable to international standards in terms of high quality, which has a very low level of background turbulence. Experiments can be performed under well controlled conditions, which was particularly important for investigations of SWBLI phenomena, which can be sensitive to free stream fluctuations. Moreover, ITAM has developed experimental methods for the measurement of fluctuations in high speed flow, with a high degree of expertise, and is experienced in Electro-Hydro-Dynamics and Magneto-Hydro-Dynamics.

The second ICPC partner is the Aerohydrodynamic Department of the Podgorny Institute for Mechanical Engineering Problems of the National Academy of Sciences of Ukraine. Since 1991 they have experience in numerical simulations and investigations of the viscous flows in aerodynamic passages, taking into account three-dimensionality, compressibility, flow separation, unsteadiness, influence of turbulence and other physical effects. Numerical method for computation of 3D viscous flows, were implemented in the aerodynamic solver *FlowER*, developed by Yershov S. and Rusanov A., the leaders of the scientific team.

Both ICPC partners were members of the UFAST consortium therefore their skills and high quality were fully tested. Unfortunately, breakup of Russian-Ukrainian war has dismantled the UAN research group and made further cooperation impossible. Therefore in the middle of the project this successful cooperation had to be terminated.

4 Project Structure

4.1 Overall Strategy of the Work Plan

The TFAST project is dedicated to the creation of a novel data base. It has a clear goal, concerning the best location of laminar/turbulent transition in order to minimise drag or losses induced by the boundary layer interaction with a shock wave. This rather ambitious goal first of all requires basic knowledge regarding the laminar interaction as well as the turbulent interaction in chosen flow cases. Two shock wave types are taken into account. One is a normal shock which induces separation at Mach numbers = 1.2–1.3. The other is the case of oblique shock reflection in which incipient separation is obtained in the range of M = 1.5–1.7.

The methods of transition induction were investigated as one of the main objectives of the TFAST project. For this purpose proposed was the application of flow control techniques such as: tripping wire, roughness patches, Air Jet Vortex Generators (AJVG), Rod Vortex Generators and Cold Plasma Actuators. These devices had been used formerly only in turbulent boundary layers. Our objective was therefore to discover how these techniques work in a laminar environment and how they induce transition.

To realise the above objectives two basic work packages were foreseen: WP-1 and WP-2.

Both of them focus on basic interaction configurations for:

- normal shock wave in a range of M = 1.2–1.3 (incipient separation parameters)
- the oblique shock reflection for M = 1.6 (as in HISAC project), including off-design Mach numbers of M = 1.5 and 1.7.

WP-1 concerned not only limiting cases of fully laminar and fully turbulent interactions and acquired new knowledge for the experimental data bank, but also played an important role in testing the numerical codes used for flow predictions The obtained results served as a reference for further investigations of flows with natural and controlled transition.

While also dealing with the natural transition of boundary layers from a laminar to a turbulent state, WP-2 focused mainly on transitions induced by flow control methods. The experimental results determined the numerical strategies adequate for these complicated flow problems. Finally, the obtained experimental and numerical results became a basis for the formulation of physical models of flow control devices, as requested by the industry, which needs a simplified treatment of transition control in order to implement it into the design process. The task (Task 2.3) was related to all experiments and CFD concerning the mechanisms of flow control in all Work Packages (see Fig. 1). Only one physical model of a flow control device was developed, which concerned streamwise vortex formation. Finally, the model has been used in Task 5.3 for the simulations of a 3-D wing with transition control.

The basic research in WP-1 and WP-2 is closely related to other three Work Packages concerning technical applications where the location of transition is a dominant issue in their further improvement of performance. These technical cases concerned flow in transonic compressor cascade, WP-3; high pressure turbine cascade with cooling system, WP-4; and a transonic laminar wing, WP-5. Each of WPs include only those transition control devices which are technically realisable in their particular case.

In all investigated cases, experiments are accompanied by numerical simulations. The realisation of each of the cases follows a similar structure:

- design of new research models (blade, wing profile) **by the industry partners**
- design of test sections for basic investigations using CFD, manufacturing and integration with wind tunnel and measurement systems
- measurements, data acquisition and data processing with special interest in transition location, interaction structure, separation scope and interaction unsteadiness
- numerical simulations with basic numerical approaches as RANS, URANS with a number of approaches concerning turbulence and transition
- simulation of chosen cases using advanced methods such as DNS and LES.

In all of the investigated cases, the final conclusion was expected to indicate the most downstream location of transition at which shock wave-boundary layer interaction still maintains a typical turbulent character. The results were however limited to several test cases and the optimal search for transition location was not possible. But obtained results showed the degree of general improvement in the finally chosen transition location in relation to fully laminar and fully turbulent interaction.

5 Work Package Descriptions

5.1 WP-1 Leader: IUSTI

Reference Cases of Laminar and Turbulent Interactions

The main objective of the project was to study the effect of transition location on the structure of interaction between a shock wave and a boundary layer. But the background knowledge needed is the actual interaction with laminar and with turbulent boundary layer. WP-1 task is to deliver the characterisation of these two limiting cases. By the way in this WP the development of appropriate test sections with shock generators is an indirect objective.

- Task 1.1 Test section design and construction

The same test section type was used allowing to, study normal and oblique shock waves, through the use of two different shock generator configurations.

- Task 1.2 Fully laminar interaction

One of the limiting cases of the research proposed by TFAST was fully laminar interaction. This does not mean that it was expected to maintain the laminarity throughout the interaction. This is rather impossible due to strong disturbance induced by the shock wave. The "fully laminar interaction" means that the boundary layer just upstream the interaction is still laminar. This is only possible if the natural transition does not take place earlier.

- Task 1.3 Fully turbulent interaction

Another limiting reference case was the interaction with turbulent boundary layer. Suggested was to use tripping wire or tripping strip just downstream of the leading edge. Thanks to this the distance between the leading edge and the shock wave would be unchanged in comparison to laminar case (Task-1.2). It was also important that the reference turbulent case has the most developed turbulence of all investigated cases.

5.2 WP-2 Leader: UCAM

Basic Investigation of Transition Effect

Main question was how close to the shock wave, the induced transition may be located, while still maintaining a typical turbulent character of the interaction. This question was investigated in WP-2 and also in the flow cases characterising different applications in WP-3, WP-4 and WP-5.

An important goal of the TFAST project was to study the effect of the transition location on the separation size, shock structure and unsteadiness of the interaction area. Boundary layer tripping (by wire or roughness) and flow control devices (Vortex Generators and cold plasma) had been used for boundary layer transition induction. Measurements were carried out in the test sections designed in Task-1.1.

- Task 2.1 Study of operation of transition control devices

Flow control devices investigated in TFAST were selected on the basis of minimum disturbance in laminar boundary layer. Therefore the fixed vain type vortex generators were not included. UFAST project has also shown that traditional Synthetic Jets have very small effect on supersonic boundary layer. Therefore TFAST was focusing on the following flow/transition control methods: tripping by wire, roughness patches, fluidic streamwise vortex generators and cold plasma.

- Task 2.2 Interaction sensitivity to transition location

The objective of this Task is the most downstream location of transition.

The experience and adjustment of CFD in Task-2.1 should allow in the present Task to find out at which location upstream of the shock, transition still makes the interaction turbulent, hence counteracts separation and the formation of large λ-foot.

- Task 2.3 Physical modelling of the control devices

The main objective of this Task was the introduction of some equivalent source terms (source terms which mimic the effect of pneumatic VG's on the incoming flow) for Navier-Stokes equations, in order to avoid specific (and costly) meshing of a great number of real VG's.

5.3 WP-3 Leader: RRD

Internal Flows—Compressors

In the case of a civil turbofan engine operating at particularly high altitudes the Reynolds number can drop by a factor of 4, when compared to the see level values. The laminar boundary layer on the transonic compressor rotor blades interacts with shock waves and as a result a strong boundary layer separation is formed. This can seriously affect the aero-engine performance and operation. One way to avoid strong separation is to ensure that the boundary layer upstream of the shock wave is turbulent. Forcing transition within the boundary layer can be achieved through the application of a surface roughness or a turbulator patch. Although such passive control methods are already in use, the mechanism of the shock wave-laminar boundary layer interaction, and in particular the source of the strong shock unsteadiness are still not well understood.

- Task 3.1 Single passage test section

A singe passage test section will be designed, built and installed in the wind tunnel of IMP PAN. Upstream of the blades a uniform supersonic stream had been formed. The supersonic Mach number of the incoming stream has been defined by RRD during the design stage.

Transition was induced by three methods:

- a step obtained by a tripping strip
- roughness patches
- streamwise vortex generators AJVG.

The aim of investigations carried out in Task 3.1 besides the improved understanding of the flow structure was to get knowledge how each of these methods should be implemented in the cascade experiments in the Task 3.3.

- Task 3.2 Basic compressor cascade flow

Low pressure compressor in the reference cascade was studied without any special inlet unsteadiness. The test were carried out for natural transition and for the fully turbulent case. The experimental test facility was located at DLR Köln. This transonic compressor cascade wind tunnel was equipped with blade profiles especially designed for TFAST project by RRD (Task 3.1) representing the state of the art in transonic compressor design.

- Task 3.3 Compressor cascade flow with transition control

The main objective was to reduce cascade losses by appropriate location of the transition. Having both limiting cases from Task 3.2 the effectiveness of the transition control methods was verified. There were two control methods considered: one is tripping/ roughness and the other one are AJVGs. Their location in relation to the shock wave was suggested on the basis of the investigations in the Task 3.1.

5.4 WP-4 Leader: DLR Göttingen

Internal Flows—Turbine

Modern HP turbine stages consist of highly loaded aerofoils, including transonic and even supersonic flow regions. In terms of stators (NGV) a normal shock wave in the passage throat is formed, stabilizing the flow conditions at the operating point. In addition to these flow phenomena, the strong acceleration along the upper part of suction side leads to a relaminarisation of the flow, which in turn has a strong impact on the size of the shock induced separation. Finally, the injection of film coolant via rows of cooling holes further influences the boundary layer state. As heat transfer and film cooling effectiveness are of crucial importance in high pressure turbines, an in-depth understanding of transition mechanisms is needed for a competitive design.

- Task 4.1 Turbine blade suction side model

The main objective is to study in details flow development on the blade suctions side. The simplified test section at IMP PAN will give access to the important flow details. The blade model will be an NGV with cooling. The objectives are manifold:

- basic flow investigations with smooth bade—location of natural transition
- effect of cooling on transition and shock interaction
- interaction with turbulent boundary layer obtained by tripping
- effect of AJVG on transition location with and without cooling
- choice of AJVG location for cascade tests in Task-4.3.

The gained knowledge may help to design new and improved blades.

- Task 4.2 Basic turbine cascade flow

A high pressure turbine blade has been studied without any special inlet unsteadiness.

The start-up point for WP-4 was the design of a turbine cascade in accordance with the design strategy developed at RRD for HP cascades with cooling system.

The main objective was to investigate the effect of boundary layer state and of coolant injection on the shock wave/boundary layer interaction. The tests were carried out for natural transition and for the fully turbulent case.

- Task 4.3 Turbine cascade flow with transition control

The control of the interaction was obtained with the help of AJVG, which have proven to be effective in the case of turbulent boundary layer in AITEB-2 project.

Using the film-cooled HP turbine of Task-4.2 as starting point, AJVGs were optimised by IMP and included in the DLR turbine cascade. The results are used to reveal the effect of the AJVGs on the boundary layer, shock structure and associated losses in comparison to the basic results of Task-4.2. Task-4.3 also continues the numerical investigations of Task-4.2, using the same software package and numerical methods.

5.5 WP-5 Leader: Dassault Aviation

External Flows—Wing

The aim was to study the transition location effect (from natural transition to fully turbulent) on separation size, shock structure and unsteadiness. Boundary layer tripping (by wire or roughness) and flow control devices (VG) were used for boundary layer transition induction.

Although this type of flow had been studied widely in the past, there remained considerable uncertainty on the effects of transition on transonic airfoil performance. In particular how close to the shock location transition has to take place to avoid detrimental effects associated with laminar shock-induced separation. Finally, CFD methods were particularly challenged by such transitional flows.

- Task 5.1 Transonic flow over laminar wing

In the UCAM wind tunnel two-dimensional transonic airfoils were tested at Reynolds numbers up to 3 million (based on chord length). Available measurement techniques included traditional pressure sensors, laser flow diagnostics (LDA, high-speed PIV), pressure sensitive paint and surface shear sensors (hot film). There were also a number of flow visualization techniques available (high-speed schlieren/shadowgraph, shear stress sensitive liquid crystals, surface oil flow).

- Task 5.2 Buffet limit for laminar profile

Experiments were conducted in the 0.6 $\times$ 0.6 m trisonic wind tunnel in Institute of Aviation (IoA) in Warsaw. Geometry of new laminar airfoil was provided by Dassault Aviation. The model of airfoil with chord equal 200 mm and span 600 mm was designed and manufactured. The model was fixed inside walls of wind tunnel and its span was equal to the width of the test section. It was equipped with both miniature pressure scanners mounted inside the model (for the static pressure distribution measurements on the upper and lower model surface) and Kulite pressure sensors mounted just under model upper surface (for unsteady pressure measurements in the vicinity of the shock location). Apart from steady/unsteady pressure measurements, flow visualisation was performed to explore an influence of laminar-turbulent transition on airfoil aerodynamic characteristics and transonic shock wave boundary layer interaction.

- Task 5.3 Numerical simulation of 3-D wing

Based on the experience gathered by CFD from simulations in Task-5.1 and 5.2 a purely CFD task was undertaken in simulating a 3-D laminar wing. The goal was to understand transition control effect on a real wing with a leading edge sweep and a tip vortex.

5.6 Graphical Presentation of the Project Components Showing Their Interdependencies

There are several quite obvious interdependencies which are expressed by the contents of the Work Packages and related tasks. Moreover, all experiments are carried out with close cooperation with CFD. This coupling is necessary to design the experiment and to better understand the results.

WP-1 Task-1.2 and Task-1.3 provides the basis of the investigation for laminar and turbulent interaction, producing the reference data for WP-2 Task 2.2, which is focused on the different transition control methods and transition location effect on the interaction.

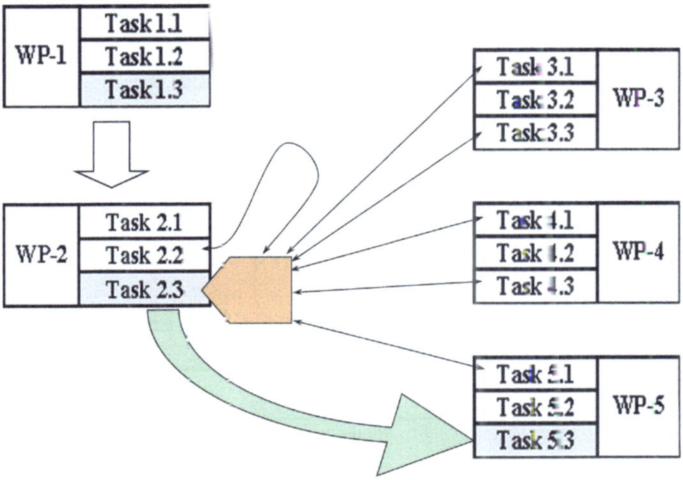

Fig. 1 Interdependencies between WPs and tasks

In WP-3 and WP-4 there are internal interdependence between Task-x.1 providing results from a basic test section and tasks x.2 and x.3 dealing with cascade application. Especially the results concerning flow control devices, carried out in the "basic Task" x.1 will provide knowledge to design flow control devices to be used in Task x.3.

The worked out models of control devices will be, first of all, checked by the partners who provided the knowledge base to Task 2.3.

It is of great importance that the worked out models of transition control devices were used in Task 5.3 (Fig. 1). This is a purely CFD Task concerning **utilisation of the TFAST results on the 3-D laminar wing** designed by Dassault-Aviation. The experimental study of such wing would be too expensive for a Level-1 project as TFAST. All the verification work carried out gives confidence that the Task 5.3 has delivered reliable results for the phenomena taking place at the laminar wing, when the transition control is applied.

6 Project Organisation

6.1 Management Structure

Management of the UFAST project was particularly taking care of the project's complexity. A special effort have been directed towards harmonic work progress across work packages in relation to the investigated flow phenomena. It is important to enable significant inter-change of results and ideas between WP's. Within each WP and between WPs it was necessary to establish links between the experiments and CFDs, inspiring the necessary circulation of scientific information.

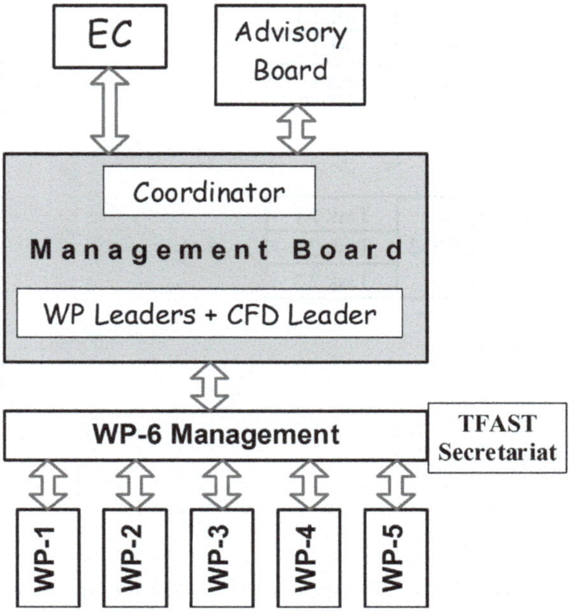

Fig. 2 Sketch of the management structure

These objectives had been realized by the appropriate management system and frequent meetings between partners (every six month). The meetings took place as plenary sessions, so that all partners could participate in discussions and therefore all partners were aware of the work progress across the project. As the experience of UFAST showed, this approach was very effective in cross-fertilisation between WPs. The management structure sketch is shown in Fig. 2.

6.2 Coordinator

On behalf of IMP PAN Piotr DOERFFER assumed responsibility for the technical, financial and administrative management of the TFAST. He was a nominated professor at IMP PAN and a head of the Department of Transonic Flows and Numerical Methods. He took part in many research projects and organised several international conferences. On a European level one could mention participation in EUROSHOCK I and II, organisation of the 5th ISAIF in Gdansk with the INCO programme support and participation in the AITEB project of the 5th FP. In the 6th FP he participated in: AITEB-2, TLC and FLIRET. Also In the 6th FP he coordinated UFAST project which was a basis of the present project. In the 7th FP he has realised a Regpot-2 project "IMP-Survey" within the Capacity Area. In the Cooperation Area of 7th FP he is a partner in ERICKA project, coordinated by RR.

He was the chairman of the Project Management Board, and the official contact point for the European Commission.

In particular, he was responsible to carry the overall responsibilities for:

- Efficient planning, monitoring and reporting of progress, deliverables and milestones.
- Initiating and leading meetings.
- Ensuring communication between teams.
- Acting as an intermediary between the EC and the consortium (transfer of information, reporting at the end of each scientific period, etc.).
- Taking care of any emerging risks and implementation of recovery plans.
- Overseeing the promotion of gender equity in the project and other societal issues.
- Managing the financial modalities between participants (distribution of funds, collection of financial reports) and informing the European Commission of all transfers.

For this important role, he had the full support of additional resources within IMP PAN, according to the needs. In the second part of the project realisation he had transferred his responsibilities to prof. Pawel Flaszynski, his close co-worker fully involved in TFAST research and realisation.

6.3 *Management Board*

The Management Board (MB) was formed by the coordinator, the WP leaders and the CFD Leader. This compact MB allowed for an effective decision making and fast reaction in case problems arose with the progress of individual tasks. Necessary "recovery" actions were realised effectively by this small body. During each six-month meeting a MB session was taking place. All partners were invited to these meetings, therefore everyone was aware of management actions and was contributing to discussions.

The management board personal responsibilities were changing during the project realisation. Due to the amount of work done and concern of all people involved it is fair to mention these people in the Table 2.

The task of project realisation was very difficult to manage due to the large scope of CFD and experimental issues. Therefore in the middle of the project duration additional people agreed to help in managing the project in specified topics. These persons and topics are presented Table 3.

6.4 *Advisory Board*

The Management Board was supported by the Advisory Board, consisting of a four recognised European scientists in the research field of the TFAST project who have

Table 2 List of coordinators

Role	Partner short name	Name
Coordinator	IMP PAN	Piotr Doerffer, Pawel Flaszynski
WP-1 leader	IUSTI	Jean-Paul Dussauge
WP-2 leader	UCAM	Holger Babinsky
WP-3 leader	RRD	MaciejOpoka Patrick Grothe
WP-4 leader	DLR	Friedrich Kost, Peter Giess, Ingo Röhle, Anna Petersen
WP-5 leader	Dassault Aviation	Jean-Claude Courty Vincent Levasseur, Flavien Billard
CFD Leader	NUMECA	Charles Hirsch

Table 3 List of additional coordinators

New flow case	Leader name
Normal shock wave Covering work done in WP-1 and WP-2	Sergio Pirozzoli
Oblique shock wave Covering work done in WP-1 and WP-2	Experiments—Pierre Dupont CFD—Lionel Larcheveque
Models for flow control devices Covering Task 2.3 and Task 5.3	George Barakos

agreed to participate. They were independent persons without any employment links to the TFAST consortium participants (see Table 4). First three advisors are pensioned and the last one is still active with large contribution to boundary layer transition problems, who is the chairmen of the ERCOFTAC SGI on Transition. They have attended the regular project meetings, contributing to the project realisation through discussions and advices.

Table 4 Members of the advisory board

	Name	Link to
1	Werner Haase	EADS-M
2	Jean Delery	ONERA
3	Claus Sieverding	VKI
4	Erik Dick	Ghent University

6.5 WP Leaders

Chosen WP leaders were scientists of the highest rank, who ensured that the output of TFAST was of very high value. These were the persons who contributed the most to the planning and realisation of the project objectives.

6.6 General Assembly

Another body of the TFAST project was the General Assembly (GA) which consisted of one representative of each Partner institution. This body had the main competence of strategic decision making.

Basic Flow Cases

WP-1 Reference Cases of Laminar and Turbulent Interactions

Jean-Paul Dussauge, Reynald Bur, Todd Davidson, Holger Babinsky, Matteo Bernardini, Sergio Pirozzoli, Pierre Dupont, Sébastien Piponniau, Lionel Larchevêque, Rogier Giepman, Ferry Schrijer, Bas van Oudheusden, Pavel Polivanov, Andrey Sidorenko, Damien Szubert, Marianna Braza, Ioannis Asproulias, Nikos Simiriotis, Jean-Baptiste Tô, Yannick Hoarau, Andrea Sansica, and Neil Sandham

Abstract In order to be able to judge the effectiveness of transition induction in WP-2, reference flow cases were planned in WP-1. There are two obvious reference cases—a fully laminar interaction and a fully turbulent interaction. Here it should be explained that the terms "laminar" and "turbulent" interaction refer to the boundary layer state at the beginning of interaction only. There are two basic configurations of shock wave boundary layer interaction and these are a part of the TFAST project. One is the normal shock wave, which typically appears at the transonic wing and on the turbine cascade. The characteristic incipient separation Mach number range is about M = 1.2 in the case of a laminar boundary layer and about M = 1.32 in the case of turbulent boundary layer. The second typical flow case is the oblique shock wave reflection. The most characteristic case in European research is connected to the 6th FP IP HISAC project concerning a supersonic business jet. The design speed of this airplane is M = 1.6. Therefore the TFAST consortium decided to use this Mach number as the basic case. Pressure disturbance at this Mach number is not very

J.-P. Dussauge (✉) · P. Dupont · S. Piponniau · L. Larchevêque
Aix-Marseille University, CNRS, IUSTI, Marseille, France
e-mail: jean-paul.dussauge@univ-amu.fr

R. Bur
Office National d'Etudes et de Recherches Aérospatiales, Meudon, France

T. Davidson · H. Babinsky
University of Cambridge, Cambridge, UK

M. Bernardini · S. Pirozzoli
University of Rome, La Sapienza, Rome, Italy

R. Giepman · F. Schrijer · B. van Oudheusden
Delft University of Technology, Aerodyn. Lab., Delft, The Netherlands

P. Polivanov · A. Sidorenko
Russian Academy of Sciences, Siberian Branch, Inst.of Theor. App. Mech. Novosibirsk, Russia

D. Szubert · M. Braza · I. Asproulias · N. Simiriotis · J.-B. Tô · Y. Hoarau
Institut de Mécanique des Fluides de Toulouse, UMR 5502 CNRS-INPT-UT3, Toulouse, France

A. Sansica · N. Sandham
University of Southampton, Southampton, UK

P. Doerffer et al. (eds.), *Transition Location Effect on Shock Wave Boundary Layer Interaction*, Notes on Numerical Fluid Mechanics and Multidisciplinary Design 144,
https://doi.org/10.1007/978-3-030-47461-4_2

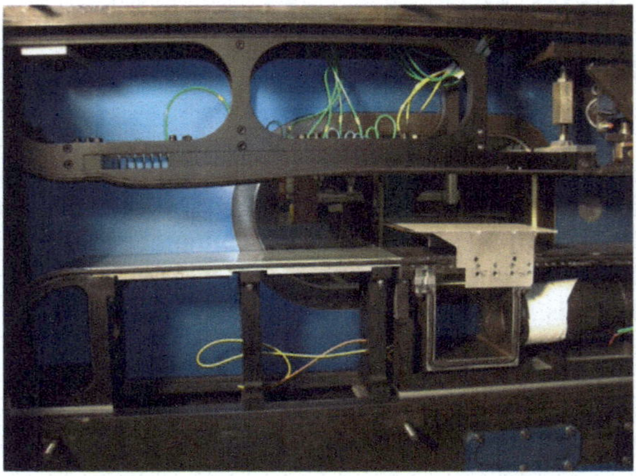

Fig. 1 Test set-up with the Mach 1.3 half nozzle in the S8Ch wind tunnel (w/o shock generator wedge)

high and can be compared to the disturbance of the normal shock at the incipient separation Mach number mentioned earlier. As mentioned earlier, shock reflection at M = 1.6 may be related to incipient separation. Therefore two additional test cases were planned with different Mach numbers. ITAM conducted an M = 1.5 test case, and TUD an M = 1.7 test case. These partners have also previously made very specialized and successful contributions to the UFAST project.

1 Normal Shock Wave

1.1 Test Section Design and Construction

1.1.1 ONERA M = 1.3

A test set-up has been manufactured to study the effect of a normal shock on the boundary layer transition process. Figure 1 shows the test set-up with the Mach 1.3 half nozzle configuration in the S8Ch wind tunnel. This half nozzle configuration (compared to the full nozzle one for the Mach 1.6 case) is imposed by the characteristics of the wind tunnel pumps: $Q = f(\Delta p)$.

1.1.2 UCAM M = 1.3

Supersonic Wind Tunnel

Experiments were conducted in the CUED No. 1 Supersonic Tunnel (pictured in Fig. 2), an intermittent blow-down facility with run times up to 30 s, a working section 179 mm high & 114 mm wide and capable of operating at Mach numbers between 0.5 and 3.5. Free-stream Reynolds numbers are between 13 and 55 million

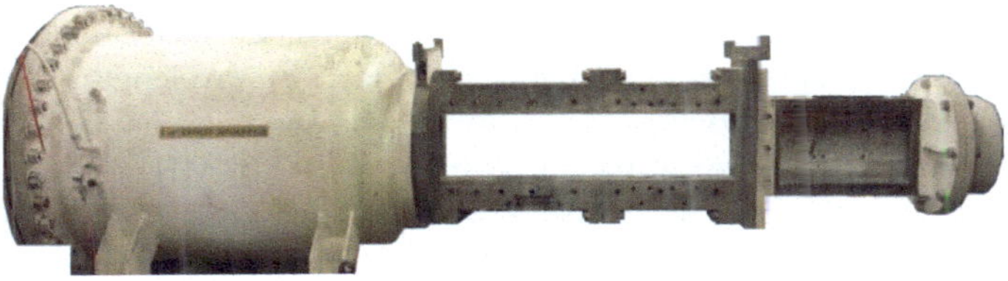

Fig. 2 CUED No. 1 SST, showing the settling chamber, working section and first diffuser

per metre, with the stagnation pressure measured in the settling chamber and manually kept constant by trained technicians, with <0.1% fluctuation. The stagnation temperature (measured as the average of four T-type thermocouples in the settling chamber) increases during a typical run from around 294 K to 300 K. The free-stream turbulence level of the tunnel in transonic mode was measured as 0.7% at Mach 0.7 (WP-5) and 0.5% at Mach 0.8 (WP-1&2) [1].

Experimental methods

A 'Z-type' schlieren system with a horizontal knife-edge is used to visualise the flow, with high frame rate images captured by a Photron FASTCAM ultima APX high speed camera. Imperfections in the optical path are removed by subtracting a 'wind off' background image from each 'wind on' picture. Variation in intensity along a row of pixels for each frame is used to extract shock position versus time, and the fluctuation throughout a run is analysed to measure any oscillation amplitudes and frequencies.

Pressures are recorded with small tappings in the walls and test models connected to a NetScanner 9116 pressure transducer system with a quoted accuracy of 0.05%. The total experimental error of measured pressures is then <0.2%, and as such the error in calculated Mach numbers is <0.5%. A greater contribution to inaccuracy in Mach number is slight fluctuations in free-stream conditions, which for the most unsteady flows can result in r.m.s. variations of ±0.03. Surface pressure data can be taken over a wide area by using pressure sensitive paint (PSP). When illuminated with UV, the paint luminesces with an intensity related to surface pressure. An ISSI LM2x-400 serves as the light source, with images taken by an Apogee Alta U2000 cooled CCD camera. The intensity also varies with the thickness of the paint layer, so 'wind on' images are referenced to 'wind off' background images to remove this dependency, and additional images are taken without UV to subtract any stray light. The relative intensity is then calibrated with pressure taps. The paint coat is rougher (Ra 0.8 μm) than the hand polished surface, and so causes the boundary layer to transition (evidenced by the lack of lambda structure seen in schlieren imaging).

A two-component laser Doppler velocimetry (LDV) system from TSI is used in forward-scatter mode to measure velocities. The current implementation uses a Coherent Innova 70C-5 Argon-Ion laser at 1.5 W, with beam wavelengths of 514.5 & 488 nm; an FBL-2 fiberlight box for beam manipulation, containing a 40 MHz Bragg cell; and a TR260 emitter with a TLN06-350 lens (focal length 363 mm), mounted so the measured vectors are at 45° to x & y and the lower beam pair is angled slightly below horizontal. This lens is chosen for its small measurement volume, producing beam overlap waists of 90 & 85 μm for the two wavelengths respectively. The corresponding fringe spacings are 3.7 & 3.6 μm, and the beam overlap lengths are 1.3 & 1.2 mm (span-wise). A band pass filter of 40–120 MHz allows velocities from 0 m/s up to 300 & 284 m/s respectively to be measured. A TLN06-500 lens (focal length 512 mm) is used for the aerofoil wake traverses, where instantaneous velocities can frequently be negative, as at the maximum bandwidth setting (20–175 MHz) this allows velocities between-106/-100 & 712/675 m/s to be measured. This lens gives beam waists of 127 & 120 μm, with fringe spacings of 5.3 & 5.0 μm, and overlap lengths of 2.6 & 2.5 mm.

The receiving head is an RV70 with a 500 mm lens, mounted to point 11.25° down and 10° downstream which reduces reflections and decreases the effective length of the measurement volume. The receiver is connected to a PDM 1000-2 photo-detector box and an FSA 4000-2 frequency/burst spectrum analyser, with data recorded by TSI's FlowSizer software.

Droplets of paraffin, 0.5 μm in diameter (as determined by a shock-lag method), are injected through a rake in the settling chamber and provide good seeding levels around the centre-span, following the flow well. [2] In the free stream, valid burst rates are in the range of 25–40 kHz, although this drops dramatically in the boundary layer. Averages are taken over 0.2 s at each point of the traverse, which typically ensures at least 5,000 samples in the bulk flow. However, very near the wall (within a few dozen microns) this can fall as low as 500 per data point. Samples outside of three standard deviations are rejected for the purposes of determining the mean and r.m.s. velocities.

The manufacturer's quoted accuracy for the optics and burst analysis is 0.1%, but larger errors are present due to the physical geometries of the emitting head (for example, aligning the emitter with the flow and determining in the angle between the beams), which result in an estimated uncertainty of 1.5% for measured velocity components away from the wall [2]. Of greater impact are the difficulties in measuring the flow within a millimetre or so of the wall. Here, errant reflections contribute to a drop in signal-to-noise ratio; low seeding levels reduce particle counts and the accuracy of the measured mean and fluctuating velocities; velocity variation across the measurement volume can be significant in smaller boundary layers, giving increased scatter; and at the extreme end of the measurement range the volume can actually intersect with the floor, exacerbating all of these problems. As such, when the boundary layer thickness is under 0.5 mm (as is typical upstream of the shock in the clean cases), confidence in the data is much reduced and the technique is only trusted to give an indication of δ, and not expected to accurately determine profiles and integral parameters.

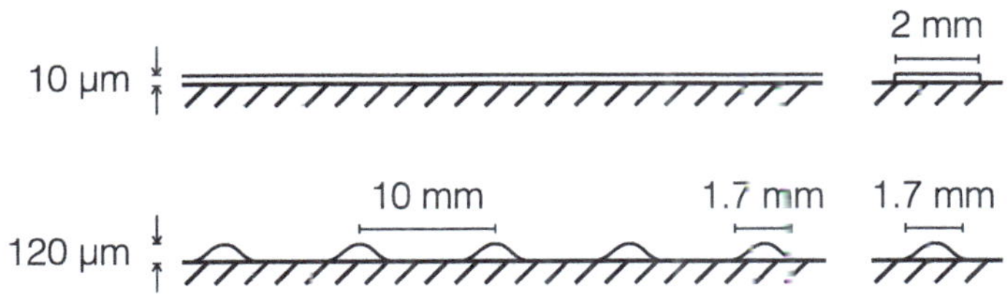

Fig. 3 Sketches of the two trip styles: (top) 2D trip—paint strip; (bottom) 3D trip—glue dots (mean values given); (left) stream-wise view; (right) span-wise view

Integral quantities (δ_i^*, θ_i & $H_i = \delta_i^*/\theta_i$) are important parameters in understanding the boundary layer health. Incompressible quantities are used here as they are less dependent on Mach number and so a better universal parameter, [3]. The resolution of LDV measurements is too low to integrate directly and so where the boundary layer is turbulent, profiles are fitted according to the theories of Sun & Childs [4] for the logarithmic law-of-the-wall & the Coles wake function; and of Musker [5] for the viscous sub-layer. For full details, see Davidson & Babinsky [6].

Two styles of trip were added to the models during the experimental campaigns, sketched in Fig. 3. The first is a two-dimensional step, created by a strip of gloss spray paint approximately 2 mm wide and 10 μm high. The second tripping method is three-dimensional and involves carefully applying drops of poly (vinyl acetate) glue at centre-span ($z = 0$) and then at 10 mm intervals along the span. Dot heights are between 90 and 180 μm with an average of 120 μm, and the diameters vary from 1.4 – 2.2 mm with a mean of 1.7 mm.

Experimental configuration

The transonic test section used for the flat plate study is sketched in Fig. 4. The plate has a 6:1 elliptical profile around the leading edge, shaped by wire-erosion; is 12 mm thick; and was hand-polished to a surface roughness of Ra < 0.5 μm. The subsonic free-stream is accelerated by the curvature of the ellipse and a supersonic region develops around the leading edge of plate, terminated by a shock. The profile ensures a favourable pressure gradient (FPG) throughout the initial boundary layer development to maintain laminarity. The rest of the plate has a flat surface, to remove any effects of curvature from the SBLI. The plate has a slight angle of attack (~+0.2°) to counteract the blockage effect of boundary layer growth along the tunnel side-walls and the floor below is cut away to avoid choking the flow.

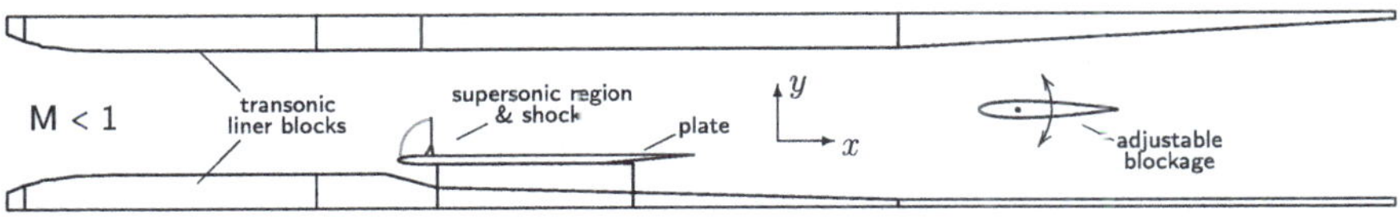

Fig. 4 Tunnel layout for the transonic flat plate investigation. The co-ordinate system is shown, with $x = 0$ at the leading edge, $y = 0$ at the plate's flat surface & $z = 0$ at centre-span

The experiment as first planned employed a sharp leading-edge plate mounted in a supersonic flow, with a shock-holding plate installed above, as sketched in Fig. 5. However, initial testing revealed that the boundary layer developing on the lower plate was turbulent, as detailed in Davidson & Babinsky (2014) [1], in spite of a very smooth surface finish. Natural transition is thought to have been promoted by instabilities arising as the flow negotiated the leading edge. Results from this configuration are discussed below as an example of a turbulent SBLI at similar Mach and Reynolds numbers to the transonic plate experiments.

A third plate was manufactured with a modified super-ellipse (MSE) profile around the leading edge. This shape, defined by Eq. (1) and sketched in Fig. 6a, avoids the discontinuity in curvature at the end of the ellipse (as shown in Fig. 6b) and so has had success in other experiments in further delaying natural transition and maintaining laminarity [7].

$$\left(\frac{y}{b}\right)^2 \tag{1}$$

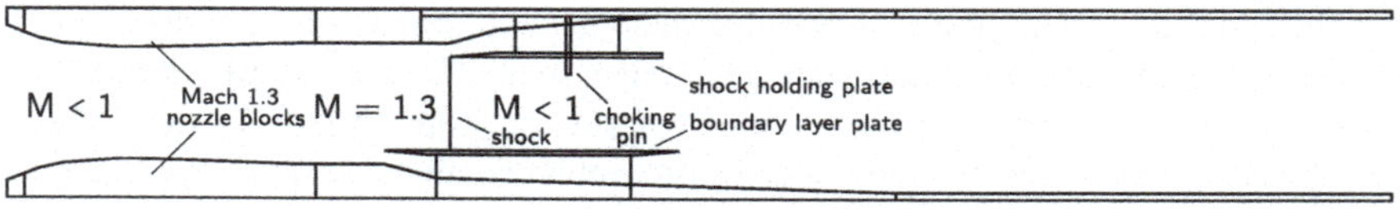

Fig. 5 The supersonic flat plate configuration that gave rise to the turbulent interaction

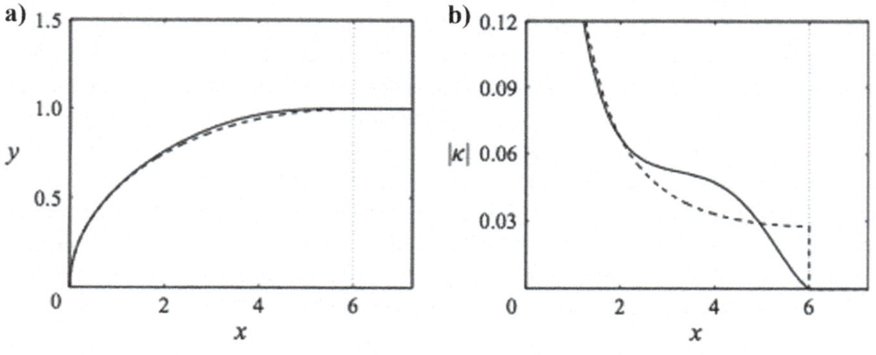

Fig. 6 Modified super-ellipse details: **a** Leading edge profiles for MSE (–) and ordinary ellipse (- -) shapes with an aspect ratio of 6. **b** Surface curvature. **c** Equation [7].

1.2 Experimental Investigation at Laminar and Turbulent Conditions

1.2.1 ONERA-DAFE (M = 1.3)

Tests of feasibility have been made on: (1) wind tunnel start; (2) control of the blockage effect only due to the flat plate, without the shock generator wedge into the test section. Figure 7 shows Schlieren visualizations for the natural boundary layer transition case. At the beginning of the tests, one has noticed the presence of humidity into the wind tunnel due to a failure of its drying device. A large quantity of particles of water leads to a strong modification of the flow pattern (see Fig. 7a): curvature of shock and waves, dissipation of the Mach wave at the leading edge of the flat plate. Moreover, one observes a blockage of the secondary flow, even if the suction device under the flat plate is active. So, one has to stop the tests in order to repair the drying device of the wind tunnel. The flow pattern has been improved when the atmospheric air is correctly desiccated but a blockage effect persists (see Fig. 7b). To suppress it, a suction device has been installed on the upper wall of the test section, above the flat plate: the shock moves downstream thanks to the suction, but stays into the test section far from the trailing edge of the flat plate, even for the maximum mass flow rate available.

At this nominal Mach number of 1.3, the theoretical ratio between the section of the exit nozzle plane and the throat section of the nozzle, A/A_{throat}, is equal to 1.07 (very close to 1). So, the presence of the flat plate (with a thickness of 3 mm) into the test section strongly contributes to the appearance of blockage strengthened by the viscous effects. Moreover, it will be impossible to install the shock holder to generate the normal shock-wave. To conclude, the lack of power of the set of pumps avoids to successfully start the wind tunnel.

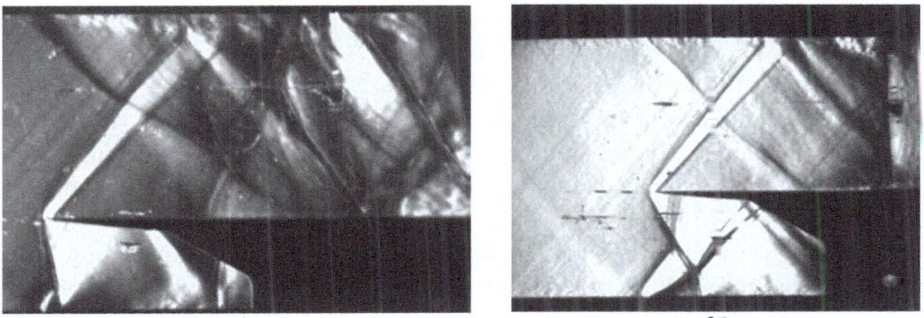

a) b)

Fig. 7 Schlieren visualizations for the natural boundary layer transition case **a**—before reparation of the drying device, **b**—after reparation of the drying device

1.2.2 UCAM

Results

A schlieren image of the interaction on the clean transonic plate is shown in Fig. 8 (a). White expansion waves can be seen around the elliptical profile up to the start of the flat section of the plate (x = 36 mm, marked by a vertical white line on the image) as the flow accelerates up to supersonic speeds. Four millimetres downstream of this there is a dark wave (from x = 40 mm at the plate) which forms the front leg of a lambda shock structure (which has its intersection around y = 18 mm), implying the presence of a separation bubble, just as in Ackeret [8]. The rear leg of the shock structure appears stronger than the front leg, and is rotated back to point slightly downstream from the intersection. There is no secondary shock present, contrary to the observations of Ackeret et al., which suggests that it was induced by the curvature of the plate in that experiment and was not inherent to the laminar SBLI.

Measuring the wave angle at the end of the end of the expansion fan in the schlieren image predicts a Mach number around 1.16. A lambda shock has never been observed at a Mach number below 1.29 for a turbulent boundary layer [2], indicating that the boundary layer here ahead of the shock is laminar, or possibly transitional. The schlieren image also suggests that the boundary layer ahead of the shock is less than 1 mm thick, whereas the interaction is more than 10 mm long, and so several times larger, providing further confirmation that the boundary layer is not turbulent, as even separated turbulent interactions are rarely more than O (5δ) and attached turbulent interactions are typically even smaller [9, 10].

LDV can be employed to gather more information about the strength of the interaction. Figure 9a presents a traverse at 1 mm above the plate surface through the flow of Fig. 8a, which gives the shock Mach number as 1.18. Further traverses are built into a pressure contour map in Fig. 9b. Much as in Ackeret et al.'s total pressure survey, the front wave is much more smeared in quantitative measurements than in the schlieren image. The results indicate that Mach number drops to 1.14 through the first wave and from 1.11 down to 0.93 through the rear wave, giving nominal pressure rises of 5 & 23% in p/p_0 respectively.

This provides further evidence that the boundary layer is laminar or transitional, as it has only required a small pressure increase to separate, characteristic of laminar behaviour. In contrast, separated turbulent interactions have a larger pressure jump across the leading shock leg due to turbulent boundary layers' greater resistance to APGs.

This low pressure rise through the front leg also provides an explanation as to why the separation of Fig. 8a is benign (without the extreme flow deviation and unsteadiness that was feared) and very shallow: a weak oblique separation shock corresponds to a small amount of flow turning, and so the bubble only grows gradually. With hindsight, this was also apparent in the experiments of Ackeret et al., but its significance remained unnoticed. As such, the separation can be many boundary layer thicknesses long, but the LDV results (Fig. 10) and schlieren image suggest the bubble height is still less than δ_1 by the rear leg location (as the boundary layer edge has only reached twice the original thickness).

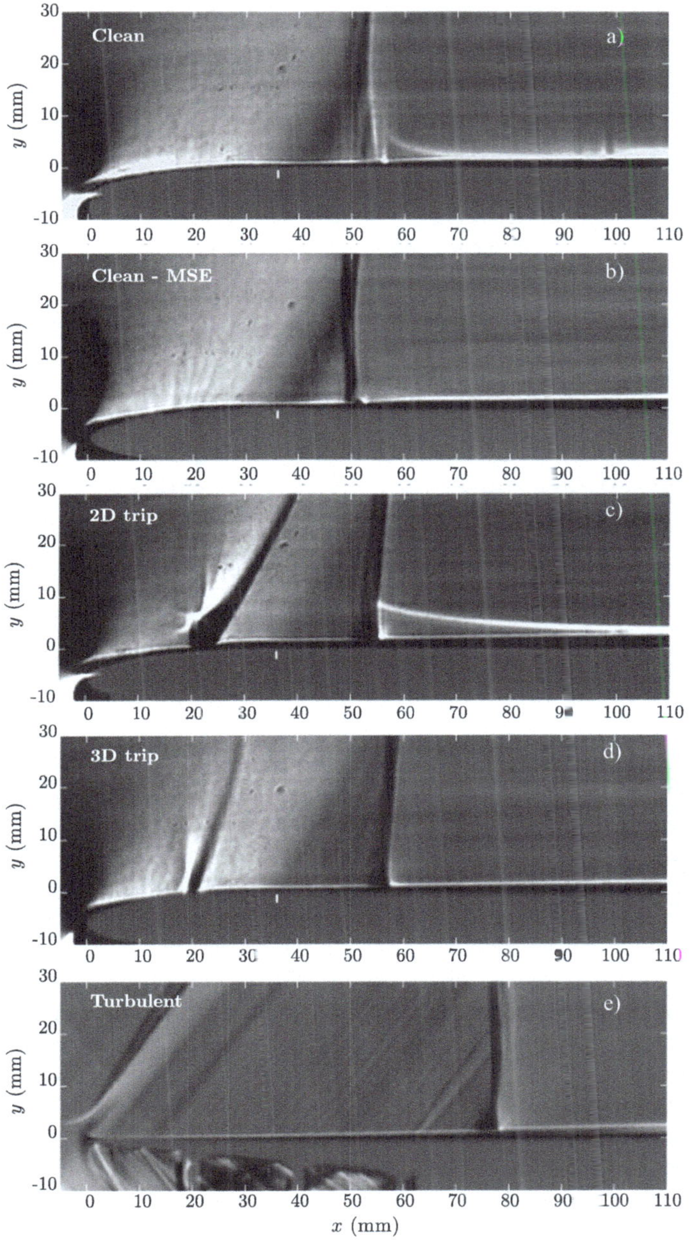

Fig. 8 Schlieren images of interactions on: **a** the clean ellipse plate; **b** the clean MSE plate; **c** the standard ellipse with paint strip at x = 20 mm; **d** the standard ellipse with trip dots at x = 20 mm, z = ..., −10, 0, 10,... mm; **e** the supersonic plate of Fig. 5. (The start of the flat surface on the transonic plates, at x = 36 mm, is marked with a white line in (**a**–**d**))

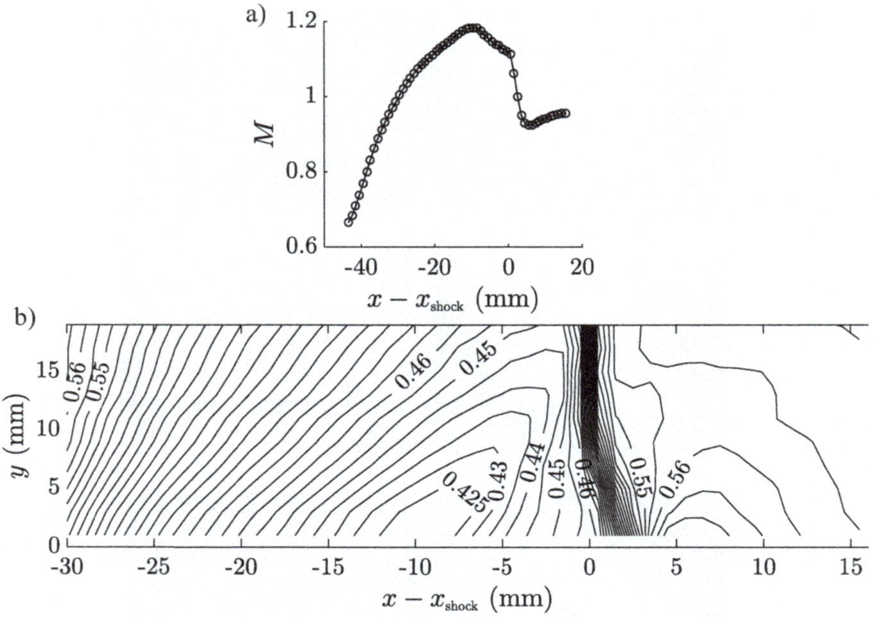

Fig. 9 LDV traverses through the clean interaction of Fig. 8a: **a** at y = 1 mm; **b** calculated isobars of p/p_0 (0.005 between contours)

It is difficult to determine where transition occurs in the flow, as upstream of and through the interaction the LDV system is only trusted to give an indication of the boundary layer height, due to the seeding problems associated with laminar flow and separation bubbles and compounded by the small size of the flow features. However, the presence of a lambda shock structure many times larger than δ, at a low Mach number, with little pressure rise through the front leg, effectively implies that the boundary layer is not turbulent ahead of the shock. By contrast, LDV shows a nominal 23% rise in p/p_0 through the rear leg, with no dramatic increase in thickness or separation size in the schlieren image, which suggests that transition has occurred by this location: if the boundary layer were not turbulent, a much greater impact would be expected at the pressure jump. It is suspected that transition is promoted by the front shock and the shear layer at the edge of the separation bubble, allowing the boundary layer to uneventfully navigate the rear shock.

Figure 8b shows a schlieren image of the flow with the MSE profile for comparison with the standard ellipse, with little variation seen between the two interactions. The expansion fan seems to end before the start of the flat surface of the plate, but this is because, as seen in Fig. 6b, the curvature significantly drops off towards the end of the profile.

Figure 8c shows the resulting flow when a stripe of paint is added at x = 20 mm to cause transition. Although the trip is only 10 μm high, a dramatic wave is seen in the schlieren, possibly indicating a small local separation, along with an increase

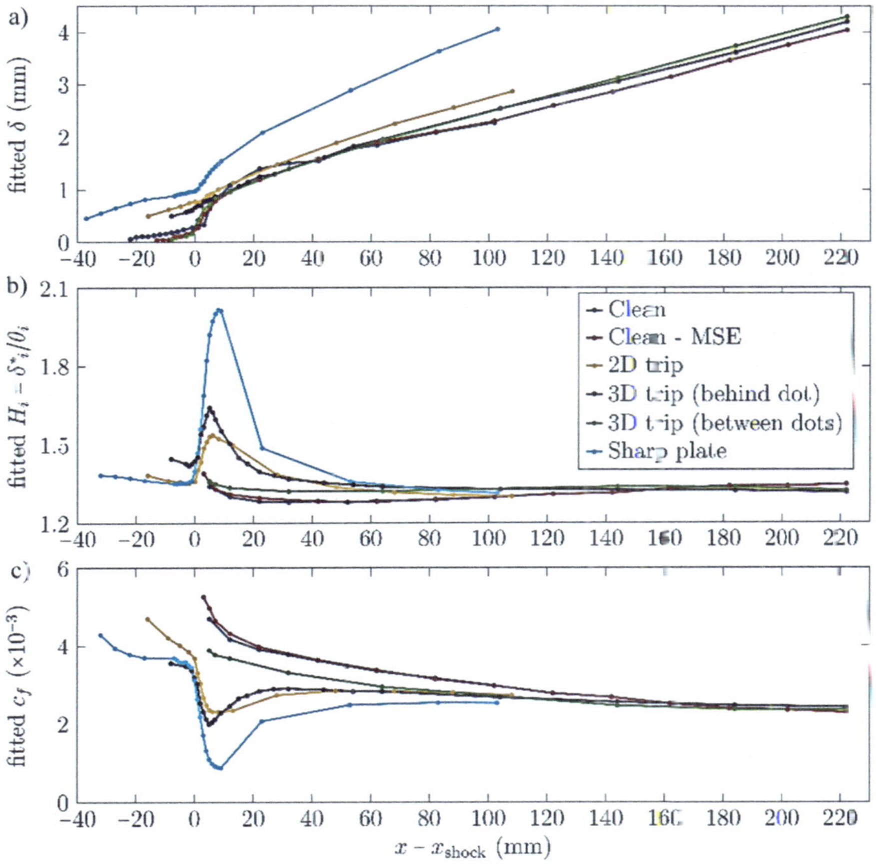

Fig. 10 Boundary layer properties for the flat plate interactions: **a** δ; **b** Hi; **c** c_f. (Other parameters shown only where δ < 0.5 mm.)

in boundary layer thickness (whilst not a direct measurement, the boundary layer increases from 6 to 9 pixels across the trip in the schlieren image). The interaction between the normal shock and the boundary layer is then much smaller than in the clean cases, with only a small amount of smearing visible at the shock foot. A very similar interaction is seen in both Fig. 8d, e on the standard ellipse plate with trip dots (Fig. 2) and the supersonic configuration of Fig. 5 (where the boundary layer is turbulent soon after the leading edge [1] respectively.

The LDV system has been used to investigate the boundary layer development for each flow case. Figure 10 shows the measured boundary layer thicknesses and calculated incompressible shape factors. In many of the traverse locations, δ is under 1 mm (compared to a measurement volume height of 0.09 mm) and so these data points have a high degree of uncertainty, particularly for integral properties, and are marked with dotted lines in the figure. This fact notwithstanding, in both the naturally turbulent and the tripped cases, the boundary layer becomes less full through the interaction and then fills out again in a long post-shock rehabilitation process, only reaching an equilibrium turbulence state a long way downstream. Conversely (although the measurements are less trustworthy due to the smaller δ) the boundary layer downstream of the clean interactions are most full just after the shock, and then relaxes back to a higher Hi profile. This is corroborated in Figs. 11 and 12, which show normalised profiles and fluctuation intensities at two locations downstream of the shock. Every boundary layer is turbulent downstream of the interactions, with transition in the clean cases presumed to occur as the flow is lofted over the separation bubble.

Crucially though, at the far downstream locations the boundary layer state is the same for the clean and the tripped interactions, as evidenced by the shape factors. This suggests that the presence of separation in the clean cases has had no long-term effect on the flow. This is attributed to the small size of the separation, which is due to the low pressure rise required for a laminar/transitional boundary layer to separate, whereas it was anticipated that the ease of separation would cause a large bubble,

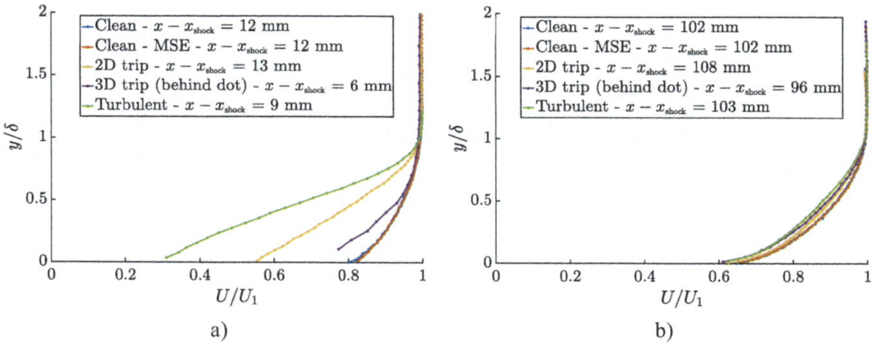

Fig. 11 Flat plate boundary layer profiles: **a** $x - x_{shock} \sim 10$ mm; **b** $x - x_{shock} \sim 100$ mm

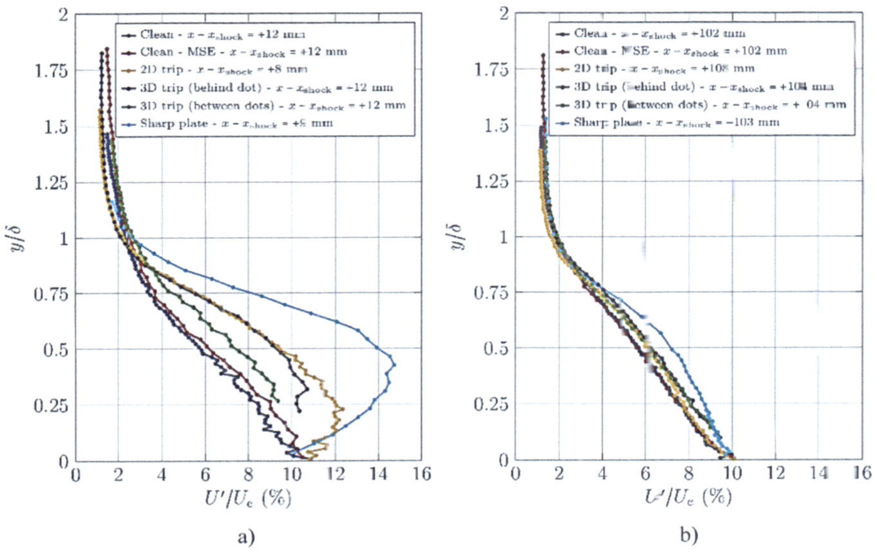

Fig. 12 Boundary layer fluctuation profiles:) $x - x_{shock} \sim 10$ mm; **b** $x - x_{shock} \sim 100$ mm

with associated losses and unsteadiness. In fact, due to the increase in δ across the paint strip, the 2D tripped case is actually less desirable than the clean cases, contrary to expectations (Table 1).

Table 1 Flat plate interaction details calculated from LDV & pressure data (δ_1 is the boundary layer thickness measured just upstream of the interaction), and so its value has a high degree of uncertainty in the 'Clean' and 'MSE' cases)

Interaction	M_∞	M_{shock}	P_0 (bar)	Unit Re (/m)	Re_{xshock}	$Re_{\delta 1}$	δ_1(mm)
Clean	0.85	1.18	2.05	28.9×10^6	1.59×10^6	2,800	0.1
MSE	0.81	1.22	2.05	28.1×10^6	1.35×10^6	<2,800	<0.1
2D trip	0.85	1.19	2.05	28.9×10^6	1.50×10^6	20,200	0.7
3D trip	0.82	1.21	2.05	28.3×10^6	1.53×10^6	8,500	0.3
Turbulent	1.3	1.27	1.7	26×10^6	2.00×10^6	23,400	0.9

1.3 CFD of the Laminar and Turbulent Interaction

1.3.1 URLMS

Physical model and numerical method
The numerical code of URMLS solves for the full three-dimensional Navier-Stokes equations for a compressible, viscous, heat-conducting gas

$$\frac{\partial\rho}{\partial t}+\frac{\partial(\rho u_j)}{\partial x_j}=0 \tag{2}$$

$$\frac{\partial(\rho u_i)}{\partial t}+\frac{\partial(\rho u_i u_j)}{\partial x_j}+\frac{\partial p}{\partial x_i}-\frac{\partial\sigma_{ij}}{\partial x_j}=0 \tag{3}$$

$$\frac{\partial(\rho E)}{\partial t}+\frac{\partial(\rho E u_j+p u_j)}{\partial x_j}-\frac{\partial(\sigma_{ij}u_i-q_i)}{\partial x_j}=0 \tag{4}$$

where ρ is the density, u_i is the velocity component in the i-th coordinate direction, E is the total energy per unit mass, p is the thermodynamic pressure, and

$$q_j = -\lambda\frac{\partial T}{\partial x_j} \tag{5}$$

$$\sigma_{ij}=2\mu S_{ij}-\frac{2}{3}\mu S_{kk}\delta_{ij} \tag{6}$$

are the heat flux vector and the viscous stress tensor respectively, and S_{ij} is the strain-rate tensor. The molecular viscosity is a function of the temperature T through the Sutherland's law, and the thermal conductivity λ is related to μ through the relation $\lambda = \frac{c_p\mu}{P_r}$ (the molecular Prandtl number P_r is assumed to be 0.72).

The Navier-Stokes equations are discretized on a Cartesian mesh and solved by means of a conservative finite-difference approach. The current version of the flow solver, extensively used in the past for the analysis of wall-bounded flows [11] and shock wave/boundary layer interactions [12] relies on sixth-order central discretization of the convective terms of the Navier-Stokes equations cast in split form [13], and it allows shock capturing through localized application of fifth-order WENO reconstructions controlled by the Ducros shock sensor. The diffusive terms are expanded to Laplacian form for improved stability, and also approximated with sixth-order central-difference formulas. The resulting semi-discrete ODE system is advanced in time through an explicit low-storage third-order Runge-Kutta algorithm [14]. We point out that the numerical methods particularly suitable for the simulation of transitional flows (which are expected in WP-2), because it is based on the concept of preservation of the total kinetic energy at the discrete level (in the limit of inviscid, incompressible flow and vanishing integration time step), thus yielding stable numerical solutions without needing extra numerical dissipation. In our opinion this is a mandatory pre-requisite for numerical simulations of transitional flows to be reliable, since spurious numerical dissipation can artificially impact the evolution of disturbances, thus preventing or delaying the (possible) breakdown to turbulence. As explained in the next Section, boundary layer transition is promoted in the computations through vortex generators. In our numerical approach the presence of such devices is handled through the immersed-boundary (IB) method, which allows to deal with embedded geometries with arbitrary shape on a Cartesian mesh. The IB method is here implemented according to the approach proposed by [15] and adapted to steady compressible flows by [16]. Specifically, at each Runge-Kutta sub-step the velocity at grid points lying inside the body is modified to make it consistent with the local body velocity, which is zero in the case of a stationary body. Furthermore, the velocity at the fluid points nearest to the body surface is extrapolated in such a way that the no-slip condition is satisfied at the boundary points, thus providing second-order accuracy in the boundary treatment.

Computational arrangement and simulation parameters
The numerical simulations foreseen in WP-1 and WP-2 include the presence of a normal shock wave that interacts with a boundary layer which can be fully laminar, fully turbulent or transitional.

Results for laminar interaction
A sequence of flow snapshots for the $M = 1.2$ interaction are shown in Fig. 13, where we show contours of the density fields. In the same figure results are also shown from a purely two-dimensional simulation, to more clearly highlight the effects of three-dimensionality within the interaction zone. Starting from the same (unbalanced) initial conditions, the two simulations exhibit in the first stages ($t = 113$) the formation of a lambda-shock structure, and a precursor shock traveling to the right.

No difference is observed between 2D and 3D at this stage. Further on ($t = 284$), the lambda shock is seen to be expanded significantly. Clear Tollmien-Schliching waves are observed at this time in the 2D simulation, with matching pattern of eddy radiating waves, whereas the 3D simulation already shows signs of break-up to a three-dimensional state, and TS waves are much fuzzier in this case.

At $t = 454$ a distinct turbulent pattern is observed past the shock foot, whose effects translate into reduced spreading of the lambda pattern. At later times the reduced size of the interaction zone in the 3D case is more evident. Further, it is interesting to note the apparent alternation between periods characterized by the formation of large rollers (as at $t = 625$), which periodically break up into developed turbulence (as at $t = 795$). From a qualitative standpoint it appears that the most important effect of flow three-dimensionality is a reduction of the region of flow reversal in the 3D case, which implies that the shock foot remains much flatter than in the strictly 2D case, and it reached very close to the wall. A closer view of the interaction zone is shown in Fig. 14, which more clearly highlights the transition of the boundary layer to a turbulent state taking place shortly upstream of the impinging shock foot, which, at this specific instant, is seen to be bent somewhat downstream. It is interesting to observe that transition occurs well within the region of flow reversal, and reattachment follows quite rapidly, followed by an extended recovery region in

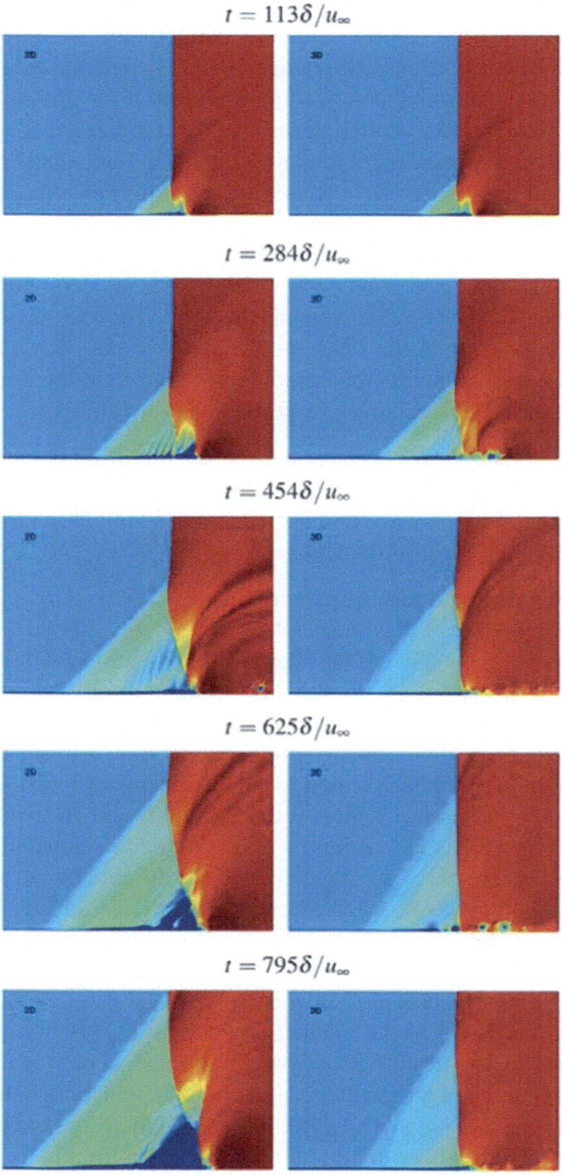

Fig. 13 Sequence of flow snapshots for numerical simulations of fully laminar interaction at $M_s = 1.2$. Two dimensional simulation result are shown in the left column, and three-dimensional results are shown in the right column. Density contours are shown in colour scale from blue (low density) to red (high density)

which large structures (see the organization of the pressure field) develop. The time-average skin friction and pressure coefficients, shown in Fig. 15, further highlight the complex, transitional nature of the interaction. Specifically, the pressure rise starts at about 150 δ upstream of the nominal impingement point, and flow separation occurs at x = 200 δ well in the laminar regime. A long shallow laminar separation

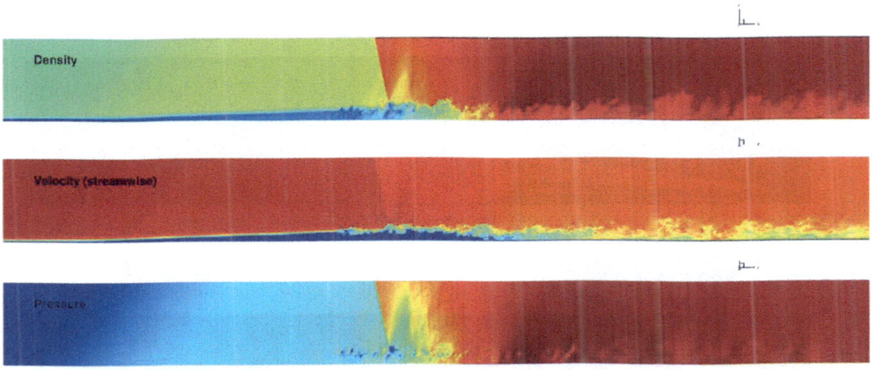

Fig. 14 Instantaneous x-y slices of density, streamwise velocity, and pressure in the vicinity of the shock foot, for fully laminar interaction at $M_s = 1.2$

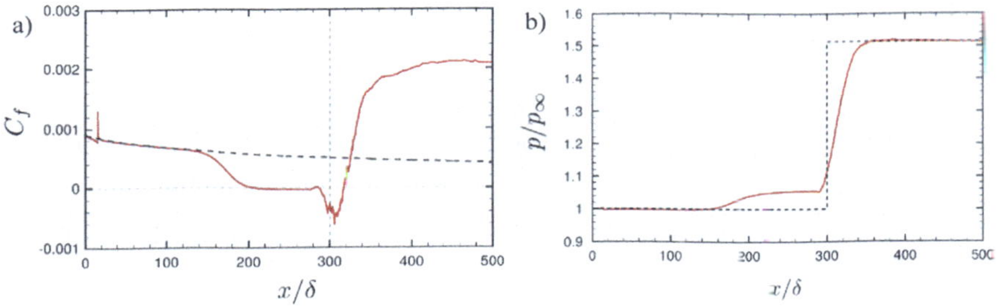

Fig. 15 Distribution of average skin friction (**a**) and pressure (**b**) in fully laminar interaction at $M_s = 1.2$

bubble is observed, in which pressure attains a plateau, and the friction coefficient has small negative values. Transition to turbulence takes place at $x - x_s = 12\delta$, where the friction coefficient is nearly zero, and then undergoes an abrupt increase in absolute value. Turbulent reattachment takes place at $x - x_s = 17\delta$, followed by a recovery region where the skin friction significantly overshoots its laminar value, indicating that the state of the downstream boundary layer is turbulent. A slight overshoot of pressure in the recovery region is also interesting, which is typical of shock/turbulence interactions. Overall, the observed pattern closely corresponds to the one observed in transitional separation bubbles in the incompressible regime [17, 18].

Turbulence statistics in the interaction zone are shown in Fig. 16. Notably, inception of transition to turbulence is associated with a distinct increase of the streamwise velocity fluctuations just above the separated shear layer. On the other hand, all other Reynolds stress components are only excited after the transition process is completed, and their amplification is clearly connected with the sudden reattachment of the boundary layer past the shock foot. An instantaneous three-dimensional view of

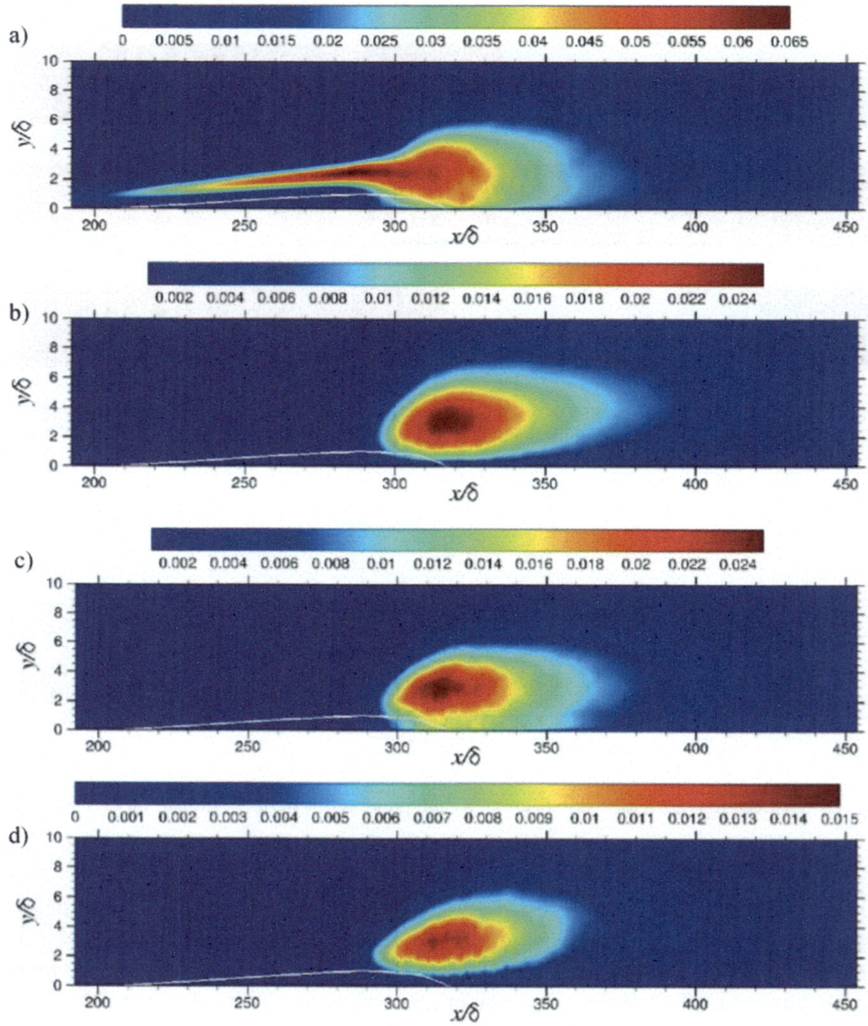

Fig. 16 Distribution of Reynolds stresses for laminar interaction at M_s= 1.2. **a** $\overline{u'^2}$; **b** $\overline{v'^2}$; **c** $\overline{w'^2}$; **d** $-\overline{u'v}$. All stresses are scaled with respect to u_∞^2. The white line mars the boundary of the region of flow reversal ($\overline{u} = 0$). The vertical line denotes the nominal location of the interacting shock

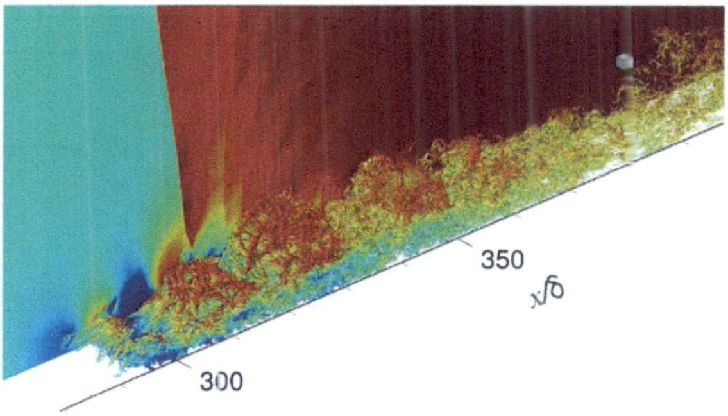

Fig. 17 Flow visualization for laminar interaction at $M_s = 1.2$. Vertical structures are educed through iso-surfaces of the swirling strength. The pressure field in also shown in a $x - y$ plane slice

the interaction zone at M = 1.2 is shown in Fig. 17, where we show iso-surfaces of the imaginary part of the complex conjugate eigenvalue pair of the velocity gradient tensor, or swirling strength [19]. Hairpin-shaped vortex tubes are observed to be generated near the transition point, at some distance from the wall, as in the typical by-pass scenario of all shear flows. The hairpin vortices are observed to form clusters downstream, which correspond to the rollers previously noticed in the instantaneous side views of the interaction. This 2D-like organization is lost further downstream, where no large-scale structures are observed, corresponding to the typical pattern observed in wall turbulence.

Results for turbulent interaction

Numerical parameters

The computation of the fully turbulent interaction has been carried out for a value of the normal shock strength equal to M = 1.3.

The computational domain employed for the simulation has a size 150 × 200 × 11 δ and it has been discretized with a grid consisting of 10240 × 640 × 1024 points. The mesh points are non-uniformly spaced in the streamwise direction to allow better resolution in the region occupied by the VGs. Points are also clustered in the wall-normal direction to resolve the boundary layer, and the compression fan of the shock wave up to y = 5 δ. Finally, the grid points are equally spaced in the spanwise direction. The size of the computational domain has been selected through a series of preliminary calculations. In particular, the length of the computational domain is dictated by the necessity to accommodate the entire interaction zone, including the upstream influence region and a large part of the recovery zone past the impinging shock. With regard to the extent of the domain in the wall-normal direction, we have found that it is critical to have it as large as possible to prevent choking of the computational duct, thus avoiding bulk motion of the impinging shock. Placing the upper boundary of the domain at y = 200 δ prevents the occurrence of computational

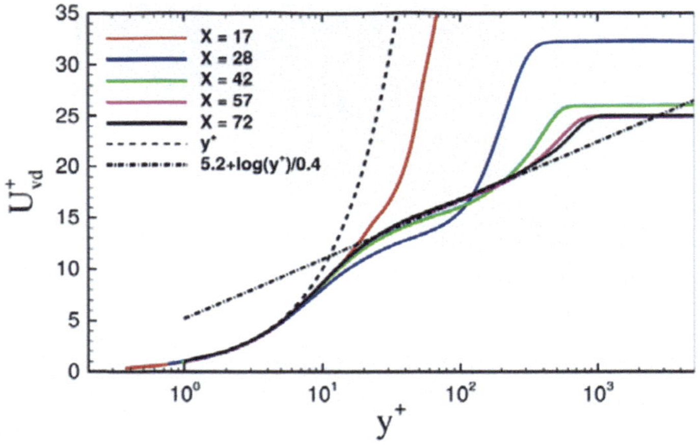

Fig. 18 Distribution of Van Driest transformed mean streamwise velocity in inner scaling at reference stations

Table 2 Properties of incoming boundary layer at the reference station $x_{ref} = 17.3\ \delta_{in}$. $Re_\theta = \rho_e u_e \theta/\mu_e$; $Re_{\delta 2} = \rho_e u_e \theta/\mu_w$; $Re_\tau = \rho_e u_\tau \theta/\mu_w$; $H = \delta^*/\theta$; $H_i = \delta_i^*/_i\theta$

M_e	Re_θ	$Re_{\delta 2}$	Re_τ	Cf	δ^*/δ	θ/δ	H	H_i
1.2	3770	3020	1003	2.73×10^{-3}	0.381	0.188	2.03	1.36

choking over much longer times than those necessary to achieve stationarity of the flow and statistical convergence.

Characterization of the incoming boundary layer

The statistics of the incoming flow are first analyzed to check that the structure of the boundary layer immediately upstream of the interaction is that of a canonical zero-pressure-gradient (ZPG) boundary layer. To that purpose, the distribution of the Van Driest-transformed mean streamwise velocity

$$U_{vd} = \int_0^u \sqrt{\rho/\rho_w} du \tag{7}$$

is reported in Fig. 18 in inner scaling at some streamwise stations taken past the vortex generators. The figure well describes the evolution of the (initially laminar) boundary layer in the streamiwse direction, which, owing to the presence of disturbances generated by the VGs, experience transition to a fully turbulent state, which is achieved approximately at x = 57 δ, where the shape of the velocity profile well conforms to that of a canonical TBL, with a nearly logarithmic region between $y^+ = 30$ and $y^+ = 200$, characterized by a von Karman constant k = 0.41 and additive constant C = 5.2. The global boundary layer properties at a location immediately upstream of the interaction (x = 80 δ), are listed in Table 2. The thickness of the boundary layer is determined as the point where u = 99% u_e, and the displacement

(δ^*) and momentum (θ) thicknesses are defined as

$$\delta^* = \int_0^{\delta_e} \left(1 - \frac{\overline{\rho}}{\rho_e}\frac{\overline{u}}{u_e}\right) dy \tag{8}$$

$$\theta = \int_0^{\delta_e} \frac{\overline{\rho}}{\rho_e}\frac{\overline{u}}{u_e}\left(1 - \frac{\overline{u}}{u_e}\right) dy \tag{9}$$

where δ_e is the edge of the rotational part of the flow field, and u_e and ρ_e are the corresponding external mean velocity and density. The 'incompressible' boundary layer thicknesses (pedex i) and the associated shape factor (H) are also determined from the above equations by setting the density ratio to unity.

Mean wall properties

The mean wall pressure across the interaction zone is here compared with the experimental data of Delery & Marvin. Note that the former were obtained in a transonic channel with a wall-mounted bump to accelerate the flow and form a supersonic region terminated by a quasi-normal shock wave. Hence, owing to the presence of the bump, the boundary layer upstream of the shock wave develops under favourable pressure gradient, and it exhibits a fuller profile than in a ZPG boundary layer at the same Reynolds number. As a consequence, the wall pressure distribution is not constant upstream of the interaction, and the incompressible shape factor has a relatively small value (see Table 2). Because of the differences in the Reynolds numbers, and the sensitivity of the flow details on the downstream conditions, comparison with experiments should only be interpreted in qualitative sense.

Let the interaction length-scale L be defined as the distance between the sonic point location x_1 (i.e. the streamwise station where the mean wall pressure equals the critical pressure $p^* = 1.46p_e$ at M = 1.3), and the origin of the interaction x_0 (i.e. the point where the wall pressure starts to rise) For weak-to-moderate interactions, Delery & Marvin showed that scales with the upstream boundary layer properties according to

$$L \approx 70\delta_0^*(H_{i0} - 1) \tag{10}$$

As observed in Table 3, the computed interaction length-scale agrees fairly well with the equation above; Delery & Marvin also showed collapse of the wall properties

Table 3 Interaction parameters for DNS and experiments (the subscript refers to the orgin of the interaction)

	M_0	$Re_{\theta 0}$	H_{i0}	L/δ_0^*	$L/\delta_0^*/(H_{i0}-1)$
DNS	1,3	3770	1,36	25,5	70
Delery and Marvin [1986]	1,3	7526	1,3	21	70

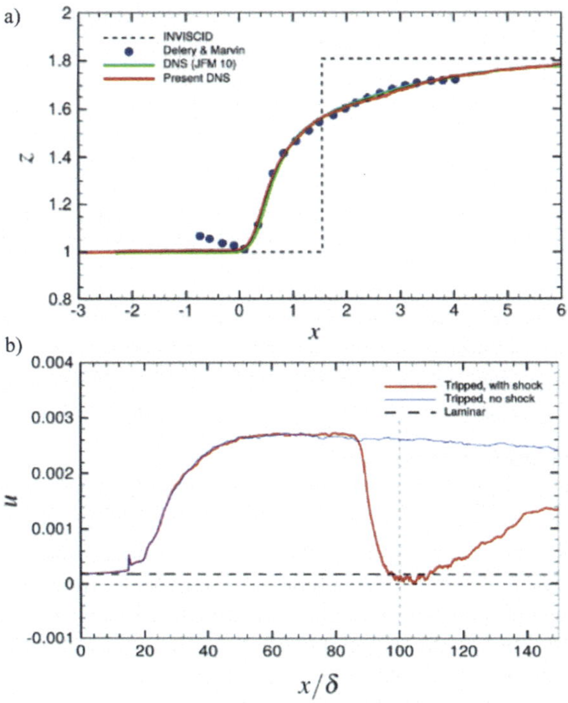

Fig. 19 Distribution of mean wall pressure (**a**) and skin friction coefficient (**b**). *R—H* denotes the inviscid distribution resulting from the Rankine-Hugoniot jump conditions. Refer to Table 3 for nomenclature of symbols

(at various Re and M) when reported in the scaled interaction coordinates $x^* = \frac{x-x_0}{L}$, $y^* = y/L$. In the following, for comparison purposes, the results are then reported in terms of x^* *and* y^* and we refer to three distinct zones: the upstream ZPG region ($x^* < 0$); the supersonic adverse-pressure-gradient (APG) region ($0 < x^* < 1$); and the subsonic APG region ($x^* > 1$).

The scaled mean wall pressure is reported in Fig. 19a together with the inviscid pressure jump predicted by the Ranking-Hugoniot relations. It exhibits a sharp rise in the supersonic APG region, in excellent agreement with experiments, and a milder increase in the subsonic APG region. Note that, remarkably, the distribution obtained in the present DNS is nearly identical to that of a previous computation [12], performed with the recycling/rescaling method, without taking into account the full spatial evolution of the incoming boundary layer, from a laminar to a fully turbulent state. This agreement confirms the validity of the approach here followed, which has the advantage to be more general and to allow the opportunity of simulating transitional interactions.

The distribution of the mean skin friction coefficient is reported in Fig. 19b together with that obtained in the absence of the normal shock. These curves are nearly identical up to (), which marks the location of the shock upstream influence and are characterized by a rapid rise of the friction coefficient at x, corresponding to the flow breakdown and the transition to a turbulent state. While in the absence of the normal shock the skin friction slightly decreases in the fully developed turbulent region, in the case of SBLI, starting from x = C_f exhibits a sudden drop, and attains a minim (close to zero), before showing a slow recovery past the interaction zone. According to Delery & Marvin the present case is to be classified as an incipient separation one.

Flow visualizations: instantaneous fields
The computed instantaneous density, streamwise velocity and pressure fields are reported in Fig. 14 in the streamwise, wall-normal plane at a given time frame.

As observed by [20] in the case of an impinging shock/boundary layer interaction at supersonic Mach number, and by [21] for a low-speed turbulent boundary layer under adverse pressure gradient, the instantaneous pressure field highlights the formation of pressure minima associated with the shedding of eddies that form in the proximity of the outermost inflection points of the mean velocity profiles. As better appreciated from inspection of the flow animations (avai able on the TFAST website), such vortices lift off from the wall approximately in the middle of the interaction zone, and propagate downstream giving rise to a turbulent mixing layer. Sharp density interfaces are also observed in the outermost part of the boundary layer, separating boundary layer turbulence from the outer, essentially inviscid flow, and that become more convoluted past the interaction zone.

2 Oblique Shock Wave Reflection

2.1 Test Section Design and Construction

2.1.1 IUSTI M = 1.6

Experiments are performed in the S7 test section of the IUSTI supersonic wind tunnel. It is a closed-loop continuous hypo-turbulent facility (intensity of the pressure fluctuations is 0.12% for a stagnation pressure of 0.4 *atm*); it can be operated continuously during several hours without pressure drift. The operating points correspond to stagnation pressures ranging from 0.15×10^5 to 0.9×10^5 *Pa*. The stagnation temperature depends on the ambient temperature, and it is maintained nearly constant by a cooling water system, with a typical drift less than 1 K/hour, The test section is 150 mm wide (spanwise) and 80 mm high and the nominal Mach number is 1.68.

The studies deal with a laminar boundary layer developed over a flat plate on which an oblique shock wave impinges. The model is composed of a flat plate of 175 mm long and 150 mm wide. Its leading edge is placed at 110 mm from the exit of the wind tunnel nozzle and it is placed at 20 mm above the wind tunnel bottom wall. A sharp-edged shock generator 30 mm long is installed at 72 mm downstream of the leading edge of the flat plate and at a distance of 30 mm above the flat plate. The shock generator and the flat plate span entirely the test section. For such Mach numbers, special attention must be paid to reduce the risks of blockage

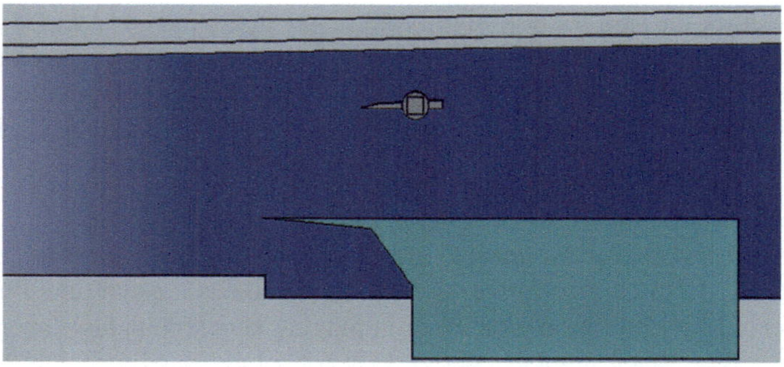

Fig. 20 Sketch of the test section with the model

effects. RANS simulations were used to obtain a design minimizing blockage effects. The final design is reported in Fig. 20. The leading edge of the flat plate makes an angle less than 5° and a cavity is placed underneath the flat plate to avoid chocking this secondary channel. A progressive recompression is generated downstream of the plate. The shock generator is directly housed in the Plexiglas windows: this suppresses the presence of masts to fix the generator and contributes to avoid blockage of the flow over the shock generator; of course, this design precludes to use Schlieren visualizations. The origin of the abscissa is taken at the leading edge of the flat plate.

Total pressure was measured with a flattened Pitot probe: the thickness of the probe is of 0.3 mm, and its opening is of 0.1 mm high. As the boundary layer thickness is very small (typically less than 0.7 mm) only external measurements, over the shear layer, have been performed. Hot wire measurements have been made at the same positions than the pressure measurements with a "StreamLine" Constant Temperature Anemometer (CTA) system from *Dantec Dynamics*. The wire length is 1 mm, its diameter is 5 μm. It is operated with an overheating ratio $a_w = 0.8$. These external measurements have been supplemented with in-flow Laser Doppler Anemometry measurements. One component system has been used in order to allow very near wall measurements. An Argon-ion laser from *Spectra-Physics (6 W, Stabilite 2017)* has been used. Only the most energetic wave length ($\lambda = 514.5\,nm$) is used, to measure the longitudinal component of the velocity. The emitted beam is of 2.25 mm in diameter and a FiberFlow system from *Dantec Dynamics* is used. A Bragg Cell imposes a shifting frequency of 40 MHz to one of the beam, allowing the system to resolve negative velocities. A beam-spacer is used to adjust the distance between the two beams to 19.2 mm. Two different emitting head have been used, with respectively one and two beam expanders with 1.98 ratio, both with a 50 *cm* converging lens, leading to a measurement volume of respectively 70 μm and 38 μm diameter. The fringe spacing is respectively 6.70 and 5.97 μm. The 70 μm was initially used for mean velocity measurements. Unsteady measurements (second to fourth order moments) have been obtained with the 38 μm system. The laminar state of the boundary layer has been validated from LDA measurements. The profiles obtained in various location and for stagnation pressures ranging from 0.4 to 0.8 atm are

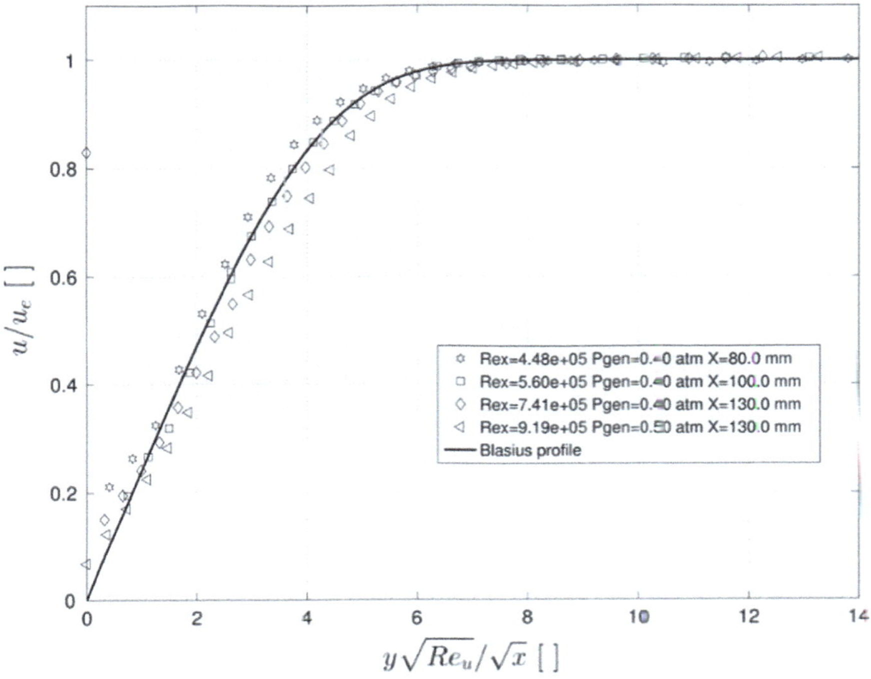

Fig. 21 Velocity profiles in normalized representation—LDA measurements

reported in Fig. 21 similarity variables. They compare very well with the theoretical Blasius profile. The dimensionless thickness of the boundary layer has been derived

$$\tilde{\delta} = \delta\sqrt{\frac{\mathrm{Re}_u}{x}} = 6.67 \tag{11}$$

The receptive head is positioned backward and off-axis. The off-axis angle is about, in order to receive the maximum light emanating from particles in the backward configuration and then to optimize the signal-to-noise ratio. We used particles of incense smoke whose average size is less than one micrometer to seed the flow. Particles are injected through a seeding cane placed at the end of the plenum chamber, on the wind tunnel axis. The seeding system is adjustable vertically in order to seed the interaction zone.

2.1.2 ONERA-DAFE M = 1.6

Introduction

Shock-wave/boundary layer interaction plays a major role in any circumstances where the flow becomes supersonic, either locally or in totality. This phenomenon

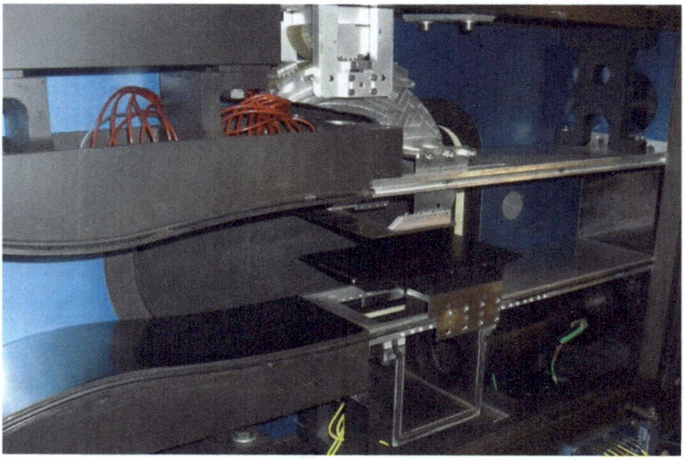

Fig. 22 Test set-up with the Mach 1.6 full nozzle in the S8Ch wind tunnel

is not clearly understood when the transitional regime (from laminar to turbulent) of the boundary layer appears during the interaction process, which is the case for compressor or turbine cascades configurations and for laminar transport/business aircraft wing. It is proposed to execute an experimental investigation of the shock-wave effect on the boundary layer transition occurring at a moderate supersonic Mach 1.6 number. Then, control devices—two types of 3-D turbulator—are tested to act on the boundary layer transition submitted to a shock-wave.

Firstly, the test set-up arrangement in the wind tunnel with means of investigation and flow parameters are presented. The boundary layer transition is detected by using several techniques: Schlieren visualizations, IR (Infra-red) thermography and TSP (Temperature Sensitive Paint) measurements. Results on the natural boundary layer transition are discussed. Then, the transitional interaction is described with respect to variation of flow parameters and to influence of control devices.

Test set-up arrangement and means of investigation

Wind tunnel and test set-up arrangement

The experimental investigation is executed in the S8Ch wind tunnel of the ONERA Meudon Centre. This facility is a continuous wind tunnel supplied with desiccated atmospheric air. The stagnation conditions are near ambient pressure and temperature: $p_{st} = 0.96 \times 10^5 \pm 300$ Pa and $T_{st} = 300 \pm 10$ K. It is constituted by a rectilinear channel having a test section with a height of 120 mm and a span of 120 mm (in full nozzle configuration). The test section side walls and upper wall are equipped with high quality glass windows to allow using optical techniques, mainly to detect the location of the boundary layer transition.

Figure 22 shows the test set-up with the Mach 1.6 full nozzle configuration, the unit Reynolds number being around 14×10^6 m^{-1} for this Mach number. The shock generator wedge (with its displacement device) has been installed on the upper wall of the test section. It is supported by lateral masts and has reduced dimensions to alleviate the aerodynamic loads when the angle of attack is increasing during the run. The flat plate under study is mounted above the lower wall and the re-generated boundary layer will interact with the shock-wave. The location of the flat plate has to be movable in both the longitudinal and vertical directions to allow the study of the boundary layer behaviour with respect to the shock impingement. The leading edge shape of the flat plate is designed to take into account of the bluntness effect on the boundary layer transition and to allow the shock to stay attached: the flat plate thickness is equal to 3 mm; its leading edge radius is around 20 µm and the bevel angle is equal to 6°. The flat plate length, equal to 200 mm, is dimensioned by both the location of the natural transition and the side wall boundary layers contamination, and its span is equal to the test section width (120 mm). In order to perform IR (Infra-red) thermography measurements, the material of the flat plate is chosen as Ino × 304L. The surface finish is achieved by hand work using emery cloth and the averaged rugosity (Ra) is estimated at 0.8 µm. Moreover, the expansion fan emanating from the shock generator trailing edge influences the flow on the study plate far downstream of the region of interest. Lastly, the test set-up is located in such a way that the leading edges of the two plates remain in the Mach number rhombus, in order to avoid any perturbation of the flow.

Suction through the upper and lower walls of the test section is necessary to remove the incoming wind tunnel floor boundary layer and to avoid blockage effect of the flow due to the presence of the flat plate and the shock generator wedge. Auxiliary pumps and high pressure air supply are available to perform this fluidic control of the flow (see the open areas, the cavity and the duct on Fig. 22). Despite the beneficial effect of suction, several modifications of the test set-up are carried out to postpone blockage effect of the secondary flow under the flat plate; i.e., a reduction of the lateral masts section supporting the flat plate and a 12°-inclination of the lower wall upstream of the cavity to accelerate the secondary flow. Finally, the suction mass flow rate necessary to maintain the secondary flow supersonic is equal to 120×10^{-3} kg/s under the flat plate and 50×10^{-3} kg/s above the shock generator wedge.

Means of investigation

The detection of the boundary layer transition region is performed by using several techniques whose results will be compared and cross-checked to accurately determine the transition process on the flat plate. So, these techniques are:

- Schlieren visualizations (continuous and spark) obtained by using an high-speed Phantom V710 camera located on the side of the wind tunnel.
- IR (Infra-red) thermography measurements. Figure 23 shows the IR thermography arrangement used. It is constituted of a camera FLIR SC7000 (size image: 640 × 512 pixels with 1 pixel equal to 14 µm; lens of 50 mm; integration time: 2 ms and frame rate: 100 Hz) and of a IR window mounted on a sidewall of the test set-up.

- TSP (Temperature Sensitive Paint) measurements. Figure 24 shows the TSP arrangement used. It is constituted of a UV-lights source located in one side of the test set-up (right picture) and of a camera placed in the other side (left picture). The flat plate is entirely painted excepted near the leading edge in order to not modify its shape, because the bluntness effect is sensitive on the transition location.

2.1.3 TUD M = 1.7

The experiments were performed in the blow-down transonic-supersonic wind tunnel (TST-27) of Delft University of Technology. The facility is able to generate flows in the Mach number range 0.5 to 4.2 in the test section. The Mach number is set by means of a continuous variation of the throat section in combination with flexible nozzle

Fig. 23 Infra-red thermography arrangement near the test set-up of the S8Ch wind tunnel

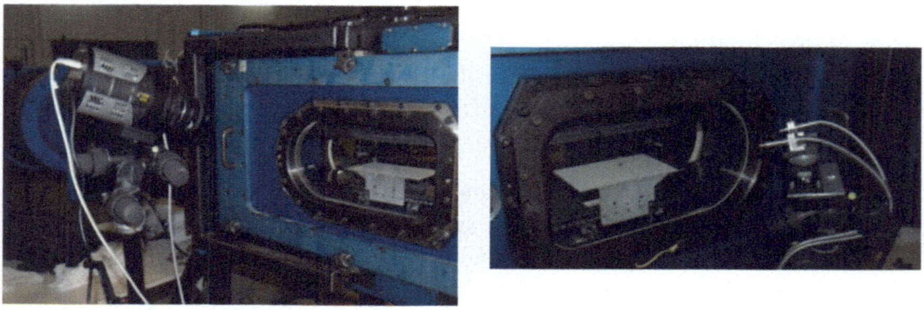

Fig. 24 Temperature Sensitive Paint arrangement near the test set-up of the S8Ch wind tunnel

walls. Air is supplied from a storage vessel of 300 m^3 charged to 42 bar and stored at ambient temperature (280–290 K typical), which allows a blow-down operating use of the tunnel of approximately 300 s maximum (depending on operating conditions), before recharging is required. In the present experiments, the tunnel is operated in the low-supersonic regime at a Mach number of 1.7. In order to reliably operate the wind tunnel at this Mach number a (minimum) stagnation pressure of 230 kPa is required which yields a unit Reynolds number in the test section of approximately $34 \times 10^6\ m^{-1}$.

The settling chamber of the wind tunnel has a square cross section of 800 × 800 mm. In the first section of the converging nozzle the flow channel contracts in lateral direction only, to the final (constant) channel width of 280 mm. Subsequently, the nozzle upper and lower walls provide the (continuously adjustable) contoured converging-diverging shape, symmetrical with respect to the tunnel centre line, to produce the required Mach number in the test section. The distance from the throat to the centre of the test section measures approximately 2 meters. The major components and dimensions of the nozzle and test section are indicated in the technical drawing of Fig. 25.

The SWBLI geometry consists of a full-span flat plate with a sharp leading edge ($R \sim 150\ \mu m$), which is used for generating a laminar boundary layer, and a symmetric partial-span shock generator (see Fig. 26). The relative longitudinal position of these elements can be adjusted to locate the incident shock wave in the laminar, transitional or turbulent region of the boundary layer. The angle of attack of the shock generator can be changed, allowing the investigation of shock strengths in the range of 1° – 6°. For the baseline TFAST experiments, the flow deflection angle is set to 3°, which corresponds to an inviscid pressure ratio of $p_3/p_\infty = 1.35$. After several iterations a flat plate—shock generator configuration was obtained which achieved a correct (non-chocked) operation of the wind tunnel, as can be verified from the Schlieren visualization in Fig. 27.

A combination of experimental techniques has been employed to characterize boundary layer transition and the shock wave-boundary layer interactions. Particle

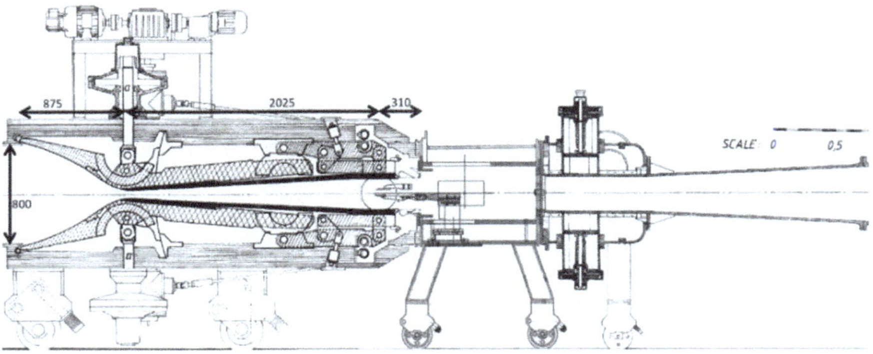

Fig. 25 Characterization of wind tunnel overall geometry

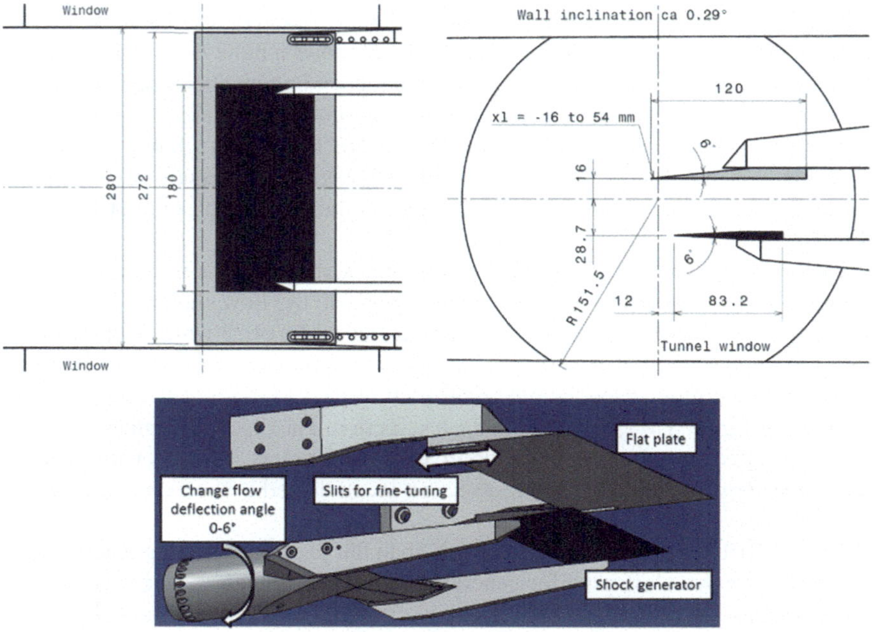

Fig. 26 Side view (top left) and bottom view (top right) of the wind tunnel configuration (in grey: flat plate model, in black: shock generator), all dimensions are in mm; the bottom picture shows an isometric view of the flat plate, shock generator and supports (Catia model)

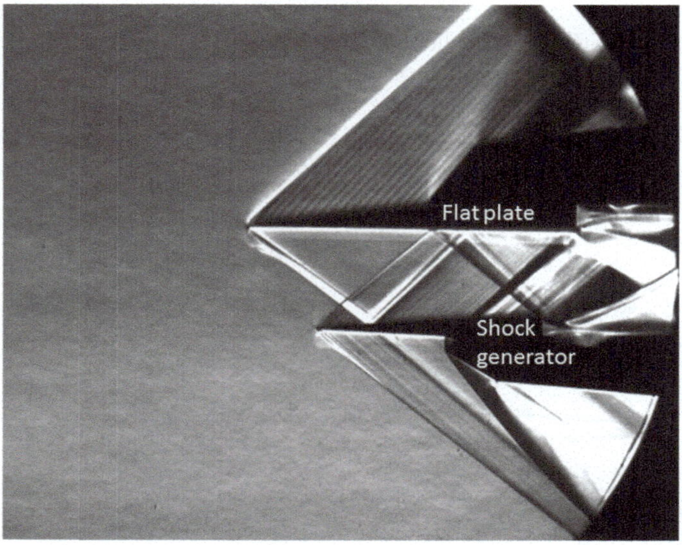

Fig. 27 Schlieren visualization of the full configuration

image velocimetry (PIV) was used as the main diagnostic too . Additional use was made of hot wire anemometry to determine the freestream turbulence level. Spark-light Schlieren visualizations and infrared thermography measurements were used to determine the transition location. Oil-flow visualizations were performed to assess the two-dimensionality of the interaction and to reveal the wall flow-topology.

2.1.4 ITAM M = 1.5

The experiments were performed in wind tunnel T-325 of ITAM for Mach number $M_\infty = 1.47$, total temperature $T_0 = 291$ K and total pressure $P_0 = 0.55 \div 1.0 \times 10^5$ Pa. The test section of the wind tunnel had a rectangular cross section of 160 × 200 mm. The nozzle was specially designed and manufactured for TFAST experiments. It consists of symmetric upper and lower parts providing uniform flow with designed Mach number. The experimental setup is presented in Fig. 28. Experimental setup consisted of a plate with sharp leading edge occupying full span of the test section and a wedge generating a shock wave. The flat plate and the wedge were attached to the lower and upper boxes extending the nozzle halves and used for the flow bypass. There were two bypass channels providing start of supersonic flow in the test section one beneath the model and another above the wedge. The flow bypass was designed so that does not influence the flow in the area of interest. The boundary layer formed on the lower wall of the nozzle went to the channel below the model and did not influence the flow in the interaction zone. The boundary layer developed on the upper nozzle wall went above the wedge. The channels were shaped to provide supersonic

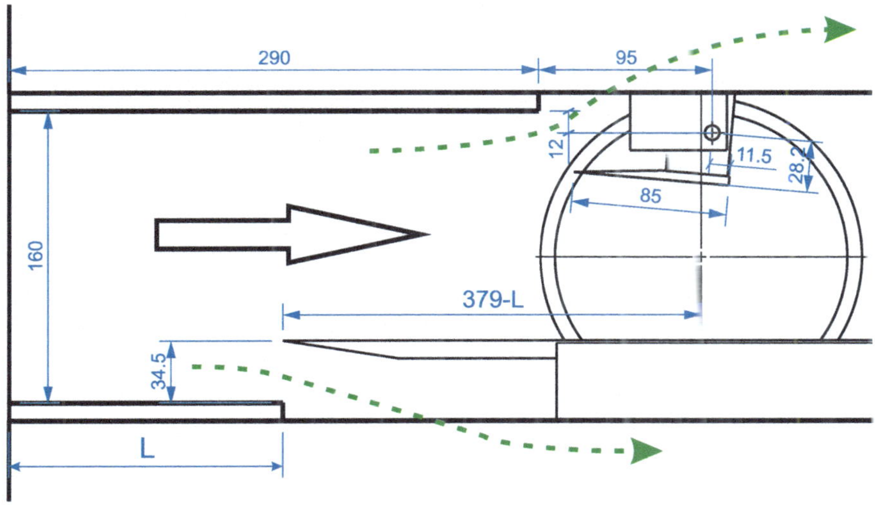

Fig. 28 Draft of the experimental setup

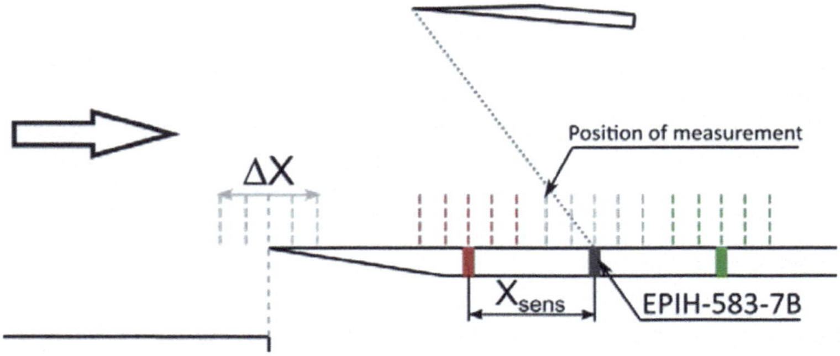

Fig. 29 Scheme of the experiment

flow at least in their intake regions so the shock at the wedge is attached to its leading edge. The flow near the model LE was supersonic too.

There were three flat plate models of the same shape prepared for the experimental campaign. The first model was made from steel and intended for PIV measurements, hot-wire and BL transition measurements. The second model (see Fig. 29) was equipped by pressure ports for static pressure measurements and EPIH sensors for unsteady surface pressure measurements. It had a line of pressure ports for static pressure measurements displaced from the centerline for 14 mm. The line of unsteady pressure sensors EPIH-583-7B was placed parallel to centerline at a distance 14 mm. The third model was made from plastic and was intended to install plasma control devices. The flat plate models were installed by means of two long pylons and could be fixed in any desired streamwise position.

The wedge assembly was installed in the upper box. To provide the space for the flow bypass in the box, the wedge was held by two pylons of streamlined cross-section. Position of the wedge rotation axis was chosen to provide minimum shift of the interaction point with variation of the wedge angle of incidence. The wedge frame was driven by pneumatic cylinder and could be fixed in two extreme positions defined by the stoppers. The range of wedge position was $-2 \div +5°$. The wedge was initially installed in the nose-down position providing start of the flow in the test section and then rotated to the test position (for example $+4°$) for the measurements.

The optical windows of diameter 220 mm were installed into the side walls. Position of the wedge was chosen to provide intersection of the shock wave with the flat plate model approximately in the center of the optical window. These windows were used for schlieren visualization and PIV measurements. The laser sheet for PIV entered through the window downstream and reflected by the prism attached to the traversing gear.

PIV was chosen as the main flow diagnostic technique. The flow seeding was performed in the wind tunnel settling chamber by DEHS particles of diameter 0.9 μm. PIV measurements were done in the region of the interaction at the centreline of the model. In the case of flow control experiments the PIV measurements were synchronized with plasma discharge.

2.2 *Experimental Investigation at Laminar and Turbulent Conditions*

2.2.1 IUSTI

The experiments performed at IUSTI are summarized Table 4. For each experiments, the flow deviation angle is fixed at an angle Θ_i by use of the shock generator. The real flow deviation has been defined Pitot measurements trough the incident shock wave, by analysis of the pressure jump, at each stagnation pressure. The real flow deviation angle is then reported for each experiment (see Table 4). A slight influence of the stagnation pressure on the effective flow deviation has been observed: about $\pm\, 0.5°$ from the intermediate pressure of 0.6 *atm*.

As previously mentioned, no Schlieren visualizations were available, therefore Pitot measurements have been used to derive the global organization of the interaction. The pressure measurements along the interaction, performed at 5 mm from the wall, are plotted on Fig. 30. In this figure, a first smooth pressure rise corresponding to reflected compression waves can be observed. This area is followed by a pressure plateau, characteristic of separated interactions (Délery and Marvin [22]), which is

Table 4 Matrix of experiments

Pgen Θ	Θ_1	Θ_2	Θ_3
0.4	3.67 ▼	4.40 ■	4.96 ●
0.6	4.30 ▼	4.92 ■	5.08 ●
0.8	4.62 ▼	5.28 ■	5.88 ●

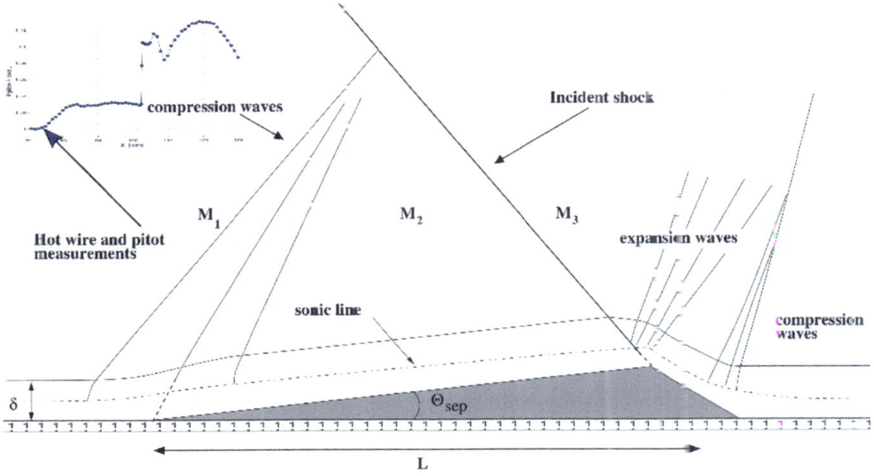

Fig. 30 Sketch of a separated laminar boundary layer interaction and longitudinal Pitot pressure through the interaction, measured at 5 mm from the wall. $P_0 = 0.8$ *atm*, Θ_3

ended by the incident oblique shock wave. The plateau pressure extends downstream from the incident shock up to the expansion wave. The last pressure rise corresponds to the reattachment compression waves.

The flow deviation at the separation point (Θ_{sep}) can be derived from the Pitot measurements, as well as from the Mach numbers downstream from the compression waves (M_2) and from the incident shock (M_3). The length of interaction L is defined as the distance between the extrapolation down to the wall of the center of the compression waves and the incident shock. The associated sketch of the interaction is presented in Fig. 30.

Length scales characterization

The locations of the inflection point of the compression waves and of the incident shock have been carefully evaluated. For the compression waves and the incident shock, the equivalent angles of deviation were derived from Pitot measurements. Then, the extrapolation down to the wall has been evaluated using the theoretical angles of the equivalent shock waves. Additionally, the length L deduced from Pitot measurements and from Hot Wire measurement are in a good agreement. The non-dimensional longitudinal coordinate X^* is used and is defined as:

$$X^* = \frac{X - X_{imp}}{L} \tag{12}$$

The position of the incident shock at the wall is then $X^* = 0$ and the compression waves are centred around $X^* = -1$. An important parameter to evaluate to compare the different experiments is the separation criterion, meaning the minimum pressure jump that separates the boundary layer. The pressure jump required for boundary layer separation is higher in turbulent case than in the laminar one: velocity profiles are fuller near the wall and the boundary layer can sustain larger positive pressure gradients.

Katzer et al. [23] proposed a formula based on free interaction theory to estimate the pressure jump for incipient separation in laminar flow:

$$\frac{p_{inc} - p_0}{p_0} = \frac{1}{2}\gamma M_0^2 K_{inc}\left[C_{f_{imp}} / (M_0^2 - 1)^{1/2}\right]^{1/2} \tag{13}$$

where $C_{f_{imp}}$ is the skin friction coefficient of an undisturbed flat-plate boundary layer at shock impingement x_{imp}, and the subscript 0 designates the upstream conditions. The constant K_{inc} is: $K_{inc} = 1.85\sqrt{2}$.

Others authors have attributed a different value to this constant: for Greber et al. [24], $K_{inc} = 2\sqrt{2}$ whereas Rizzeta et al. [25] proposes $K_{inc} = 1.57\sqrt{2}$. For our conditions, the relation (13) suggests an onset of separation of about 1°. It is worth notice that such small angles are not realized with the current experimental set-up: the flow over the shock generator is blocked. Nevertheless, TU-Delft achieved experiments which seem to validate the relation (13) for large unit Reynolds numbers.

This expression is based on similar assumptions than used to derive the classical Free Interaction Theory (Chapman et al. [26]). This theory suggests that separation

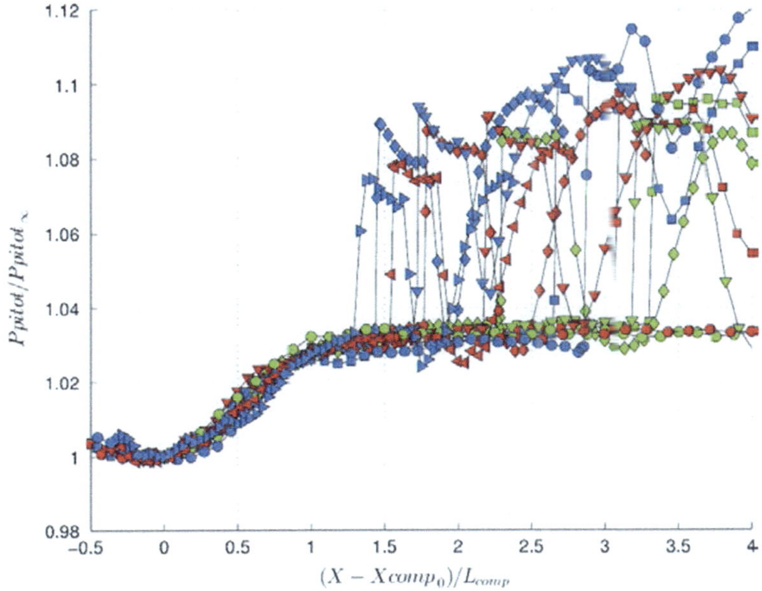

Fig. 31 Longitudinal Pitot profiles measured at $y = 5$ mm from the wall for different stagnation pressures and shock generator angles

conditions (pressure, flow deviation) are independent from the imposed flow deviation (Θ_{sg}): this is clearly verified by the present longitudinal Pitot measurements as shown Fig. 31. The section x is centered on the initial rise of pressure and normalized by the length of compression L_{comp}: the compression waves extend from 0 to 1 in this representation.

In Fig. 32, is plotted the non-dimensional interaction length L/δ^*_{0i} against the separation parameter. The dimensionless separation parameter is calculated as $Se = \Delta p/(p_{inc} - p_0)$. Here, Δp is defined as the equivalent total pressure jump of an inviscid interaction. δ^*_{0i} is the compressible displacement thickness of the unperturbed laminar boundary layer at the location X_0. Experiments with tripping devices (symbols labeled B2 and B3, see 32) are also reported. We notice a very large aspect ratio for all interactions configurations, varying between 120 and 170, and a nearly linear increase of the dimensionless interaction length with the separation criteria.

Mean velocity fields and standard deviation longitudinal velocity profiles are reported on Fig. 33a for several sections along the interaction for a Θ_3 flow deviation and total pressure 0.4 atm. Superimposed on the figure, the blue line represents the location of the 99% boundary layer thickness, and the black line give a first approximation of the recirculation zone: it corresponds to the location of the first positive velocity measured. Technically, in the first part of the recirculation zone, no—or very few- particles are present, and the black line is located at the last position where measurements were possible. At the end of the recirculation, the black line

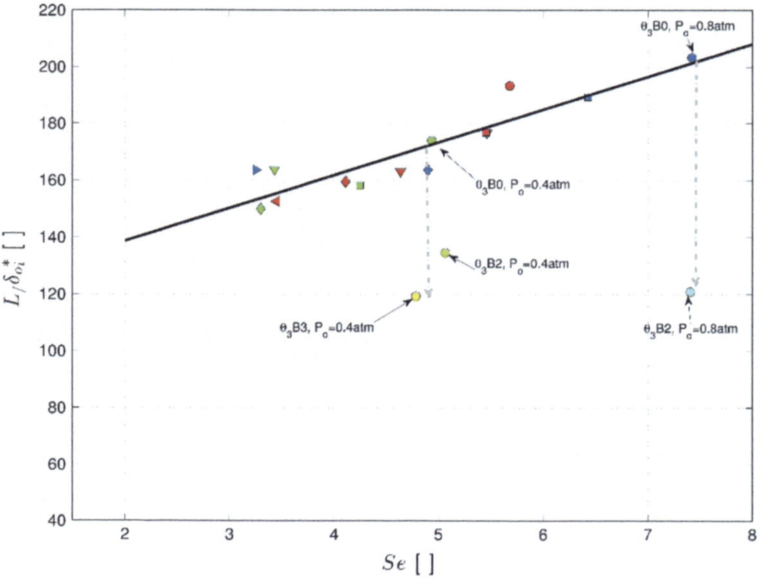

Fig. 32 Influence of interaction strength on separation length. Symbols as Table 4

match the zero-velocity line. The apex (maximum elevation of the separated bubble) is found at $X^* = -0.2$ at $y = 1.3$ mm.

Downstream from the separation point, typical profiles of mixing layer can be observed. The maxima of the longitudinal velocity RMS levels are reported in the figure (magenta curve): they follow an exponential shape. They vary from about 1% of the external upstream velocity (which is the noise level of these LDA measurements) to 17% near the reattachment region. This behavior is in agreement with an amplification of perturbations along the shear layer. The longitudinal evolution of the maxima of the RMS longitudinal velocity fluctuations are reported Fig. 33b. They are normalized with the saturation level (maximum of the RMS along the interaction). An exponential amplification is observed in the second part of the interaction ($X^* > 0.5$). Superimposed on the figure are the IUSTI's LES simulations results for similar interactions. Very similar observations can be derived from numerical results.

Unsteadiness characterization

The LDA measurements provided broadband velocity fluctuations energies. In order to describe in more details the longitudinal amplification of the shear layer fluctuations, hot wire measurements have been carried out. Measurements have been done at $y = 5$ mm above the wall along the interaction to evaluate the pressure fluctuations radiated in the supersonic part of the flow. A longitudinal space correction has been done, using the characteristic angles. The Power Spectral Density (PSD) of the external radiated fluctuations have been documented. We report on Fig. 34 the pre-multiplied PSD for positions located in the upstream boundary layer and

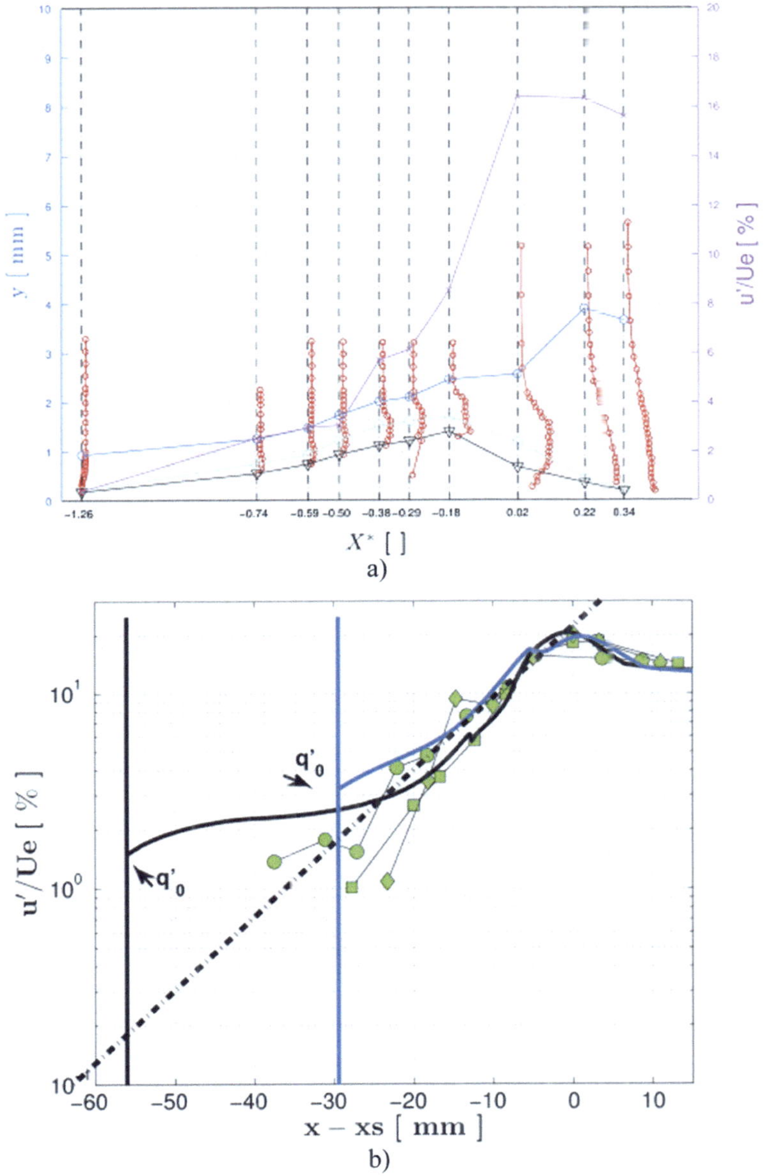

Fig. 33 Turbulent longitudinal velocity LDA measurements with the 38 μm probe volume, Θ_3 angle, $p_0 = 0.4\ atm$. **a** profiles along the interaction; **b** longitudinal evolution of the maxima of RMS along the interaction. Symbols: experiments; lines: Large Eddy Simulations

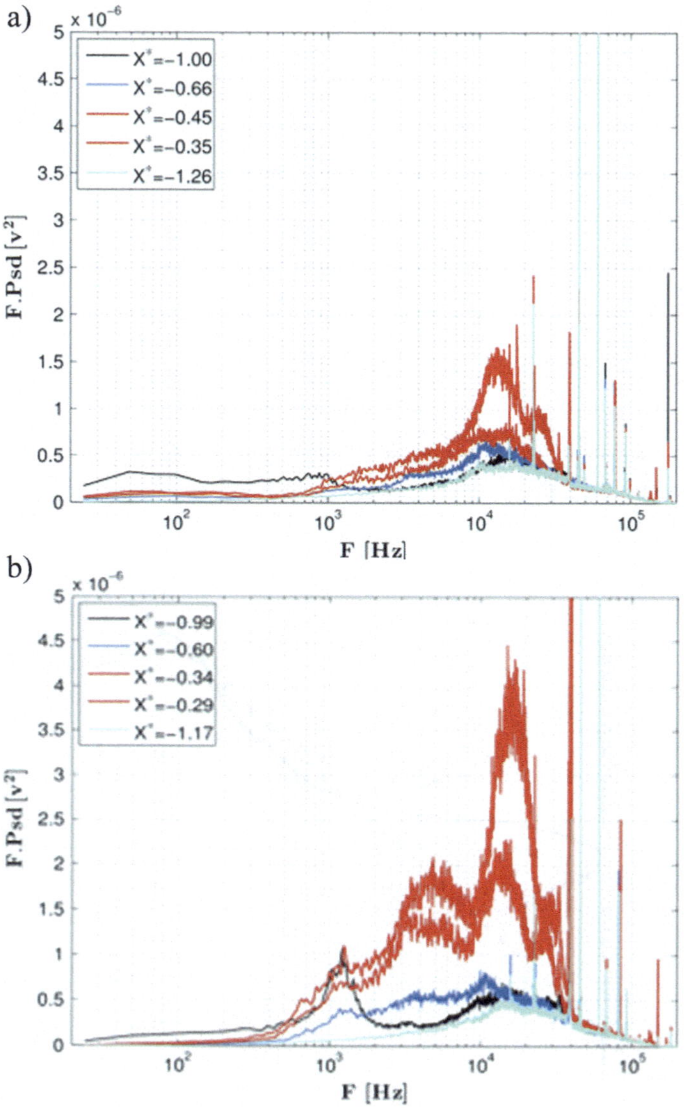

Fig. 34 Pre-multiplied PSD through the interaction for the for the reference flow deviation Θ_1 (**a** $p_0 = 0.4\ atm$, $p_0 = 0.8\ atm$) and for the largest flow deviation Θ_3 (**b** $p_0 = 0.4\ atm$, **d** $p_0 = 0.8\ atm$)

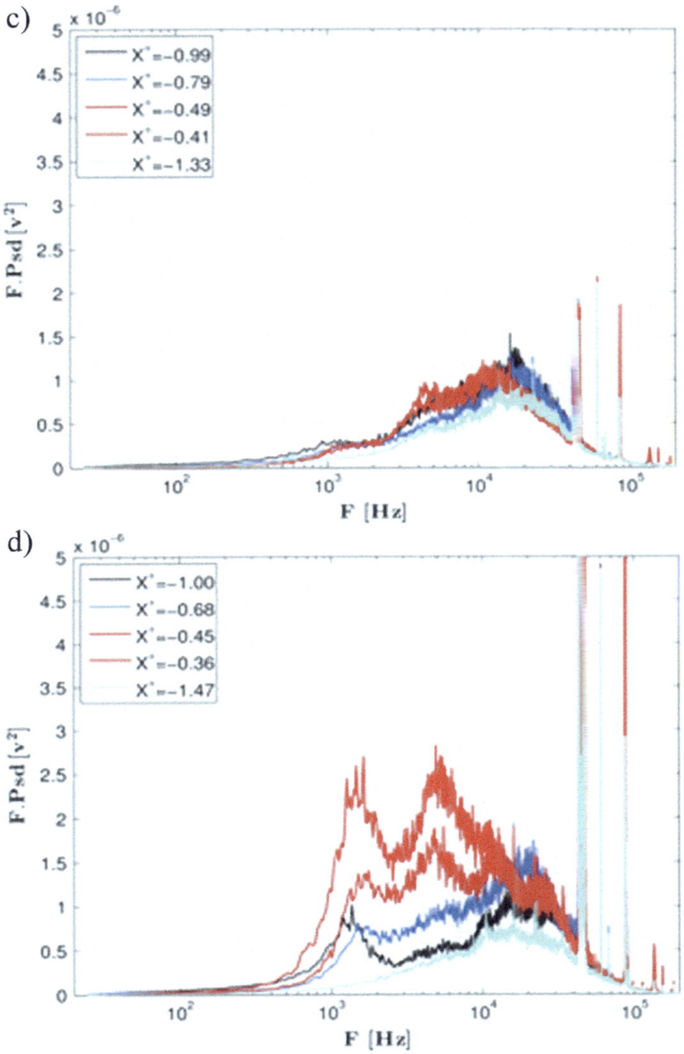

Fig. 34 (continued)

along the interaction, for several stagnation pressure, and for the angles Θ_1 and Θ_3 respectively. In this representation, the energy is proportional to the area under the curve. The upstream boundary layer frequencies ($X^* < -1$) involve energetic scales over 10 kHz, whatever the unit-Reynolds number. For the Θ_1 case, these scales are amplified along the shear regions of the interaction, but lower frequency scales (ranging from 1 to 10 kHz) are also developed. Same results are observable for the largest flow deviation Θ_3. It is clear that for such flow deviation, the amplification rate of the different frequency scales is larger: in case of the largest flow deviation and largest unit Reynolds number, the low frequency scales ($1 < f < 10$ kHz) become even more energetic than the high frequency scales ($f > 10$ kHz) which were already developed in the upstream boundary layer (see Fig. 34d).

To quantify these different amplification rates, the transfer function $H(f)$ has been evaluated for the same cases as in Fig. 35. The transfer function is defined as the ratio of the local PSD with the referent one, measured in the upstream boundary layer. Results are reported on Fig. 36.

These figures put in evidence the dependence of the amplification rate on the flow deviation. It is also clear that the lowest frequencies (around 1 kHz) are most amplified in the initial compression waves, then are only slightly amplified along the interaction. On the opposite, frequencies between these lowest frequencies and the upstream energetic scales of the boundary layers (frequencies ranging between 2 and 10 kHz) are amplified only in the interaction, downstream from the compression waves.

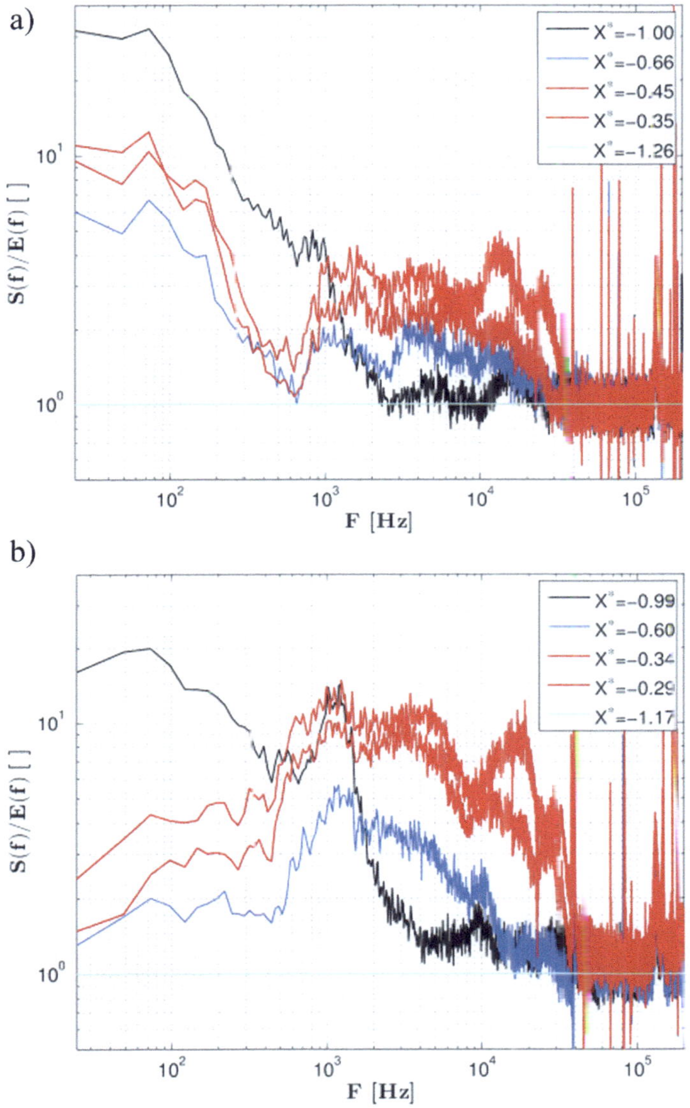

Fig. 35 Transfer function through the interaction for the reference flow deviation (**a** $p_0 = 0.4\ atm$, **c** $p_0 = 0.8\ atm$) and for the largest flow deviation Θ_3 (**b** $p_0 = 0.4\ atm$, **d** $p_0 = 0.8\ atm$)

c)

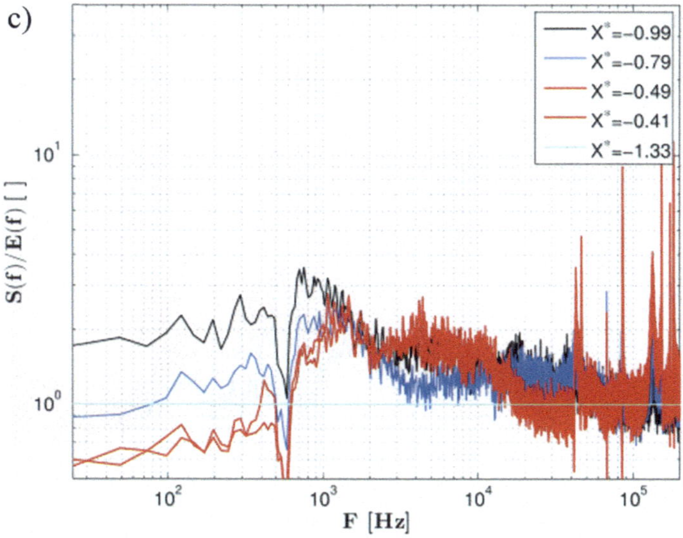

d)

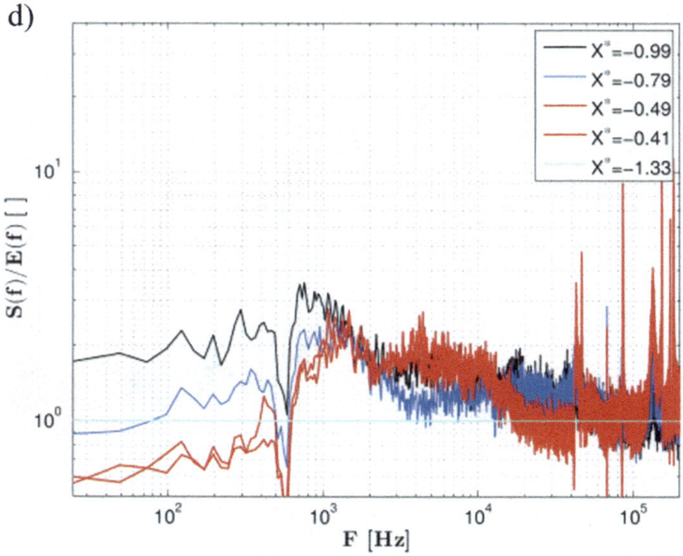

Fig. 35 (continued)

These results show that transitional interactions are clearly unsteady and involve characteristic scales that are about one decade lower than the upstream energetic scales. Moreover, these unsteadiness depend clearly on the flow deviation (or the interaction intensity): the amplification is about 10 time larger for the largest flow deviation.

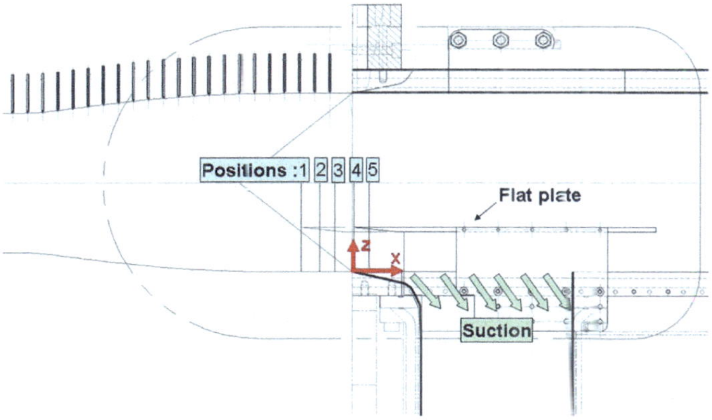

Fig. 36 Sketch of the test set-up with the five flat plate longitudinal positions under study

2.2.2 ONERA-DAFE

Tested configurations and flow parameters under study

Emphasis is placed on parameters influencing the viscous interaction, namely:

- The location of the theoretical (i.e., for a non-viscous flow) shock impingement point by translating the flat plate. This location is characterised by its abscissa X_I from the study plate leading edge, or in a non-dimensional form, by the Reynolds number computed with X_I. For the five longitudinal positions of the flat plate under study (see Table 5), the corresponding values of X_I lead to a related Reynolds number range of: $0.42 \leq Re_{XI}\ (\times 10^6) \leq 1.12$.
- The shock wave intensity, represented by the value of the flow deflection through the incident shock. For each position of the flat plate (and corresponding value of X_I), the angle of attack α of the shock generator wedge is increased from 0 to 5°.

Figure 36 shows a sketch of the test set-up with the flat plate longitudinal positions and the open cavity under the flat plate to control the secondary flow. Table 5 summarizes the longitudinal and vertical positions of the flat plate that have been tested in order to check if the secondary flow is staying supersonic thanks to flow control. For the flat plate at the altitude of $Z = 30$ mm, only three longitudinal positions are available. When the flat plate is at the more upstream location, the secondary flow is choked. On the contrary, when the flat plate is moving in the downstream

Table 5 Longitudinal and vertical positions of the flat plate under study

Z(mm)\X(mm)	−33.95	−21.6	−11.45	0.9	11.05
30		×	×	×	
40	×	×	×	×	×
50	×	×	×	×	×

direction, it is very difficult to start the wind tunnel because the static pressure in the cavity is of the same order as in the test section ahead of the flat plate, i.e., too weak with respect to the auxiliary pumps capacity. When the flat plate is elevated to $Z = 40$ and 50 mm, the needed minimum value of the suction mass flow rate is decreasing and, so, the range of longitudinal locations of the flat plate is increasing from $X = -33.95$ mm to $X = 11.05$ mm. Moreover, the reflected shocks could disturb the flow in the vicinity of the flat plate in a more upstream region if the flat plate is at the altitude of $Z = 50$ mm. Then, the retained altitude of the flat plate is at $Z = 40$ mm for the five longitudinal positions under study (see Table 5). The leading edge of the shock generator wedge is located near the test section entrance plane, at $X = 3$ mm and $Z = 84.5$ mm.

Detection of the natural boundary layer transition on the flat plate
Before testing the shock wave impact on the boundary layer properties, a detection of the natural transition location on the flate plate is carried out by removing the shock generator wedge from the test set-up.

Figure 37 shows a Schlieren visualization of the flowfield along the flat plate under study. Reflected oblique waves under the flat plate indicate that the secondary flow is staying supersonic. The reflection (barely visible) on the upper wall of the oblique shock emanating from the flat plate leading edge (i.e., a Mach wave) is far downstream of the flat plate. However, one has to take care to perturbations emanating from compression waves generated by the junction between the nozzle exit and the test section walls: they accelerate the triggering of the boundary layer transition. Moreover, the Schlieren apparatus is adjusted with the vertically-oriented

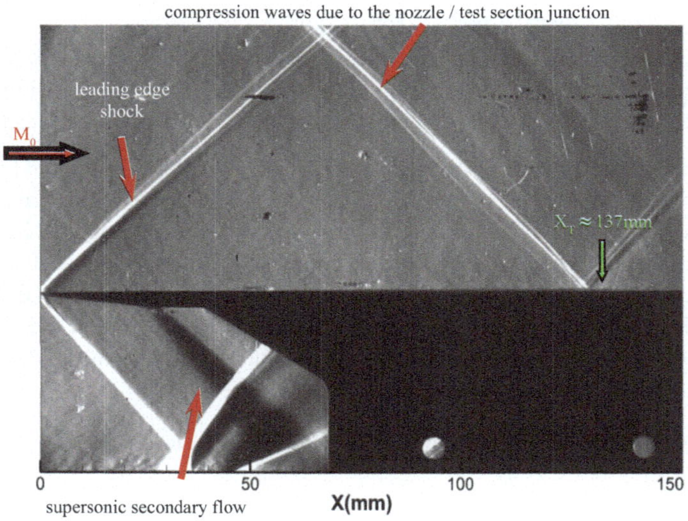

Fig. 37 Schlieren visualization used to detect the natural transition of the flat plate boundary layer

knives position in order to mainly visualize the X-gradient of the flow and, then, to point out the emergence of vortex structures in the boundary layer transition region.

Finally, the transition « point » of the boundary layer is estimated at $X_T \approx 137$ mm from the flat plate leading edge, which may in fact correspond to the end of the transition process (see the green arrow location on Fig. 37). So, the corresponding Reynolds number of transition is equal to: $Re_{XT} \approx 1.9 \times 10^6$.

This first result on the boundary layer transition location has to be strengthened by IR thermography and TSP measurements on the flat plate. Indeed, the transition process is developing along the plate according to a longitudinal range $[X_t, X_T]$. This range is obtained by extracting some temperature profiles along the plate near the mid-span test section. At the abscissa X_t, it corresponds the starting process of transition with a rapid increase of the wall temperature and at the abscissa X_T, the wall temperature does not change which means the end of the transition process and the emergence of vortex structures visible in the Schlieren visualization.

Main results obtained by both IR thermography and TSP measurements are shown in Figs. 38 and 39, respectively. Figures 38a and 39a show a IR thermography image and a TSP one on the flat plate (its leading edge being at $X = 0$) obtained after a registration process and scaled in temperature. For both measurements, the boundary layer seems to stay laminar at least for $X \leq 95$ mm (colour moving from blue to red), then the wall temperature is increasing due to the starting of the transition process. Moreover, a spanwise evolution of the temperature is observed due to the sidewall boundary layers contamination.

A more precise analysis is carried out by extracting the temperature profile along the mid-span of the flat plate (at $Y = 0$, see Figs. 38b and 39b) and by quantifying the slope variation in using its derivative versus the longitudinal direction, $\partial T/\partial X$. At the beginning of the plate (for $X \leq 20$ mm, not plotted here), temperature is rapidly decreasing from the stagnation temperature value to an equilibrium value behind the leading edge shock. Near $X = 30$ mm, temperature fluctuations are observed for the TSP case: at this abscissa starts the paint of the flat plate, which generates a compression wave clearly visible on the Schlieren visualization (see Fig. 39a). This low-intensity wave does not affect the flow properties; the boundary layer is in a laminar state and the temperature is staying at the same value since $X = 40$ mm. However, the longitudinal evolution of the temperature is very sensitive to the plate flatness: one notices a temperature variation from $X = 40$ to 85 mm due to a slight bump-shaped form of the Ino $\times$ 304 L plate used for IR thermography measurements (see Fig. 38b). Then, temperature is increased for $X > 95$ mm which means that the transition process has started, and the increasing is stopped around $X = 125$ mm, just downstream of the impingement of compression waves emanating from the nozzle/test section junction. Finally, the longitudinal range for the transition region $[X_t, X_T]$ is estimated by both measurement techniques around 30 mm. The temperature drop ΔT during the transition process is measured at 1.5° by IR thermography and at 2.5° by TSP. For comparison, the Crocco's law based on adiabatic assumption of the flow gives: $\Delta T \approx 5°$ at Mach number of 1.6. Finally, the Reynolds number Re_{Xt} related to the starting process of transition is equal to 1.3×10^6 and the one Re_{XT} related to the end of transition is estimated in the range of 1.75×10^6 to 1.9×10^6.

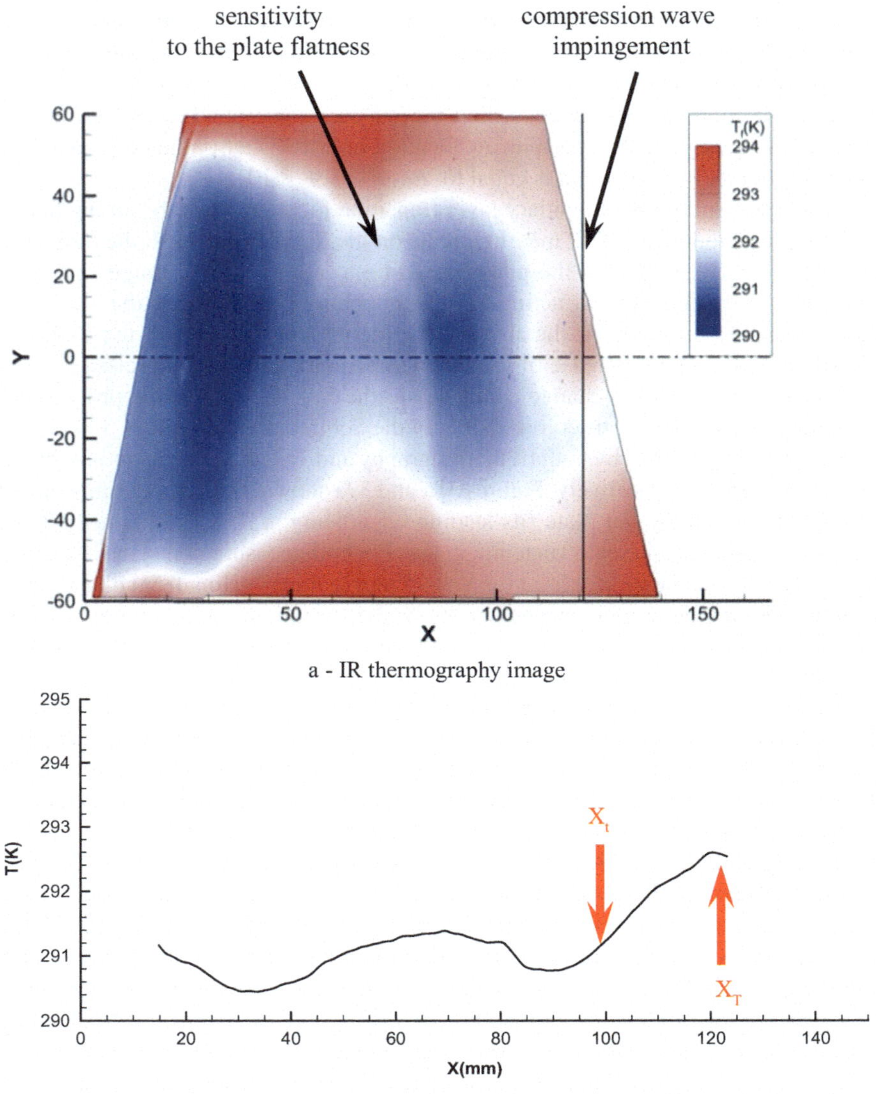

b - Temperature profiles along the mid-span of the flat plate (Y = 0)

Fig. 38 Infra-red thermography measurements. **a**—IR thermography image, **b**—Temperature profiles along the mid-span of the flat plate (Y = 0)

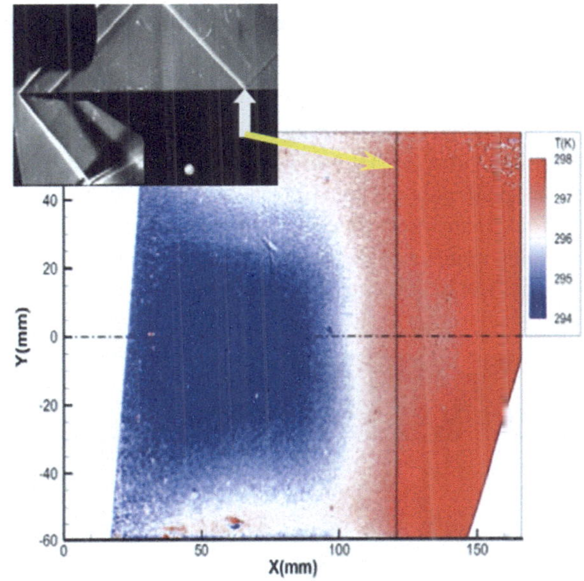

a - TSP image (with corresponding Schlieren visualization)

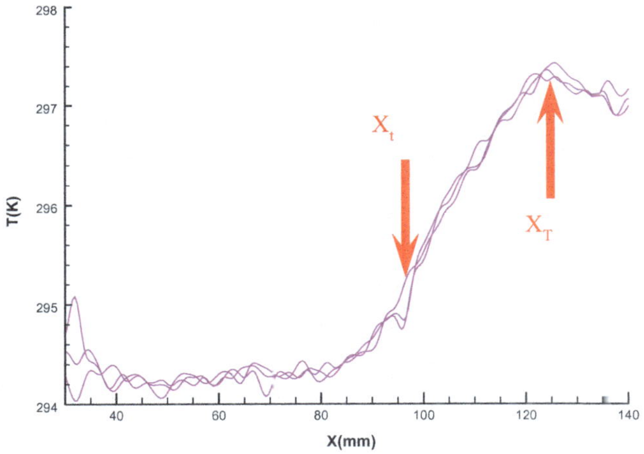

b - Temperature profiles along the mid-span of the flat plate (Y = 0)

Fig. 39 Temperature Sensitive Paint measurements. **a**—TSP image (with corresponding Schlieren visualization). **b**—Temperature profiles along the mid-span of the flat plate (Y = 0)

Shock intensity effect and flat plate location influence on the boundary layer transition

Effect of the impinging shock intensity

Figure 40 shows Schlieren visualizations for a fixed flat plate location near the test section entrance plane, at $X = 0.9$ mm (see on Table 6), and for several angles of attack of the shock generator wedge (from $\alpha = 1.5°$ to $4°$) to quantify the effect of the impinging shock intensity on the incoming boundary layer. For these configurations, the Reynolds number based on the impact location of the oblique shock on the flat plate, X_I, is nearly constant and equal to: Re_{XI} 0.61×10^6. In fact, a slight

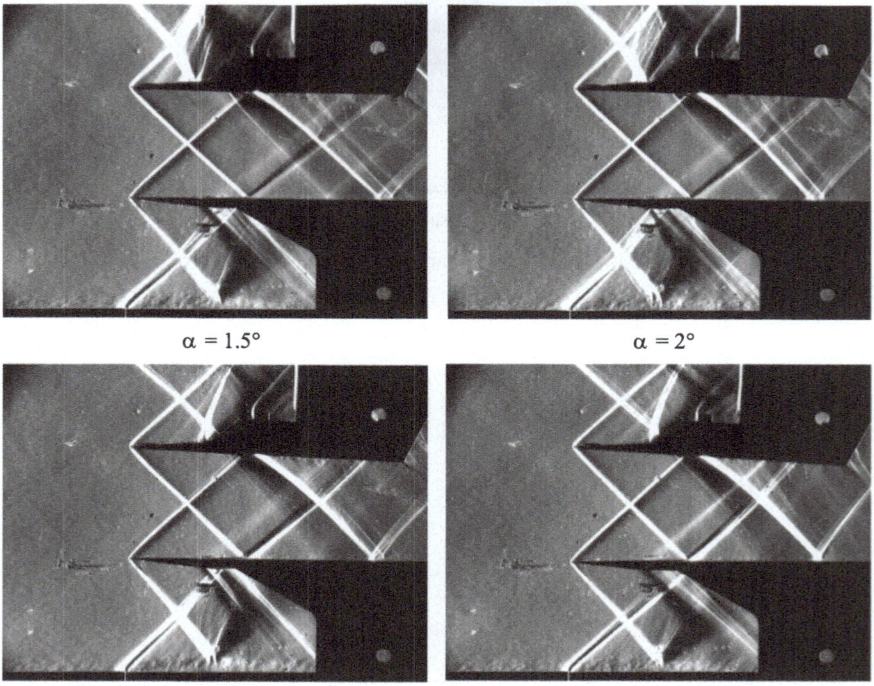

α = 1.5° α = 2°

α = 3° α = 4°

Fig. 40 Schlieren visualizations for a fixed flat plate location at $X = 0.9$ mm. Effect of the impinging shock intensity—$Re_{XI} \approx 0.61 \times 10^6$

Table 6 Evolutions of the boundary layer separation (X_S), the shock impingement (X_I) and the transition point (X_T) abscissas with respect to the shock intensity (α)

α(°)	1.5	2	3	4
X_S (mm)	34	33	32	31
X_I (mm)	46	45	44	43
X_T (mm)	55	53	47	43

displacement of X_I around its mean location when the angle of attack α varies is inherent to the experimental arrangement in which the shock intensity is adjusted by a simple rotation of the shock generator wedge. The apparent spreading of the shock waves observable in the pictures is due to the interaction with the boundary layer of the test section sidewalls. Experiments confirm that the destabilization of the laminar boundary layer in the interaction domain linked to the shock reflection has a determining effect on transition, which in turn can notably modify the flow structure. Moreover, the turbulence of the non-viscous part of the flow is modified by the crossing of the shocks.

As illustrated in Table 6, an in-depth analysis of the images indicates that the laminar boundary layer is separated due to the impact of the shock wave, even for the weaker shock strength tested ($\alpha = 1.5°$); the interaction length is growing with the shock intensity. For very small shock intensities ($\alpha = 1.5°$ and $2°$), a laminar course seems to remain downstream of the interaction domain which is typically laminar with incipient separation. An expansion fan is generated by the propagation of the shock in the boundary layer and induces, as the incident shock, a deviation of the flow towards the wall. The initial direction is finally recovered through a spread out compression. If the shock intensity is progressively increased, transition very rapidly moves upstream and reaches the interaction domain to which it remains attached for moderate shock intensity corresponding to $\alpha \geq 3°$. Then, its upstream displacement becomes very slow and is strongly linked to the increasing extent of the interaction region. For the strongest shock strength tested ($\alpha = 4°$), one observes a laminar separation well upstream of the shock impact with an associated compression wave propagating in the outer flow, and the emergence of large vortex structures behind the interaction domain. The destabilisation due to separation and the extent of the separated region is sufficient for the compression at reattachment to trigger transition. So, the perturbation imparted by the shock to the boundary layer contributes to the destabilisation process and to the creation of turbulence.

Influence of the flat plate location

Figure 41 shows Schlieren visualizations for a fixed shock intensity (angle of attack of the shock generator wedge equal to: $\alpha = 3°$), and for several flat plate locations corresponding to a range of Reynolds number at the shock impact point: $0.61 \times 10^6 \leq Re_{XI} \leq 1.09 \times 10^6$. This range of the Reynolds number allowed by the test set-up was roughly equal to half the values for the natural transition range, ignoring the small values of X_I which would introduce a coupling between shock interaction and leading edge phenomena.

As summarized in Table 7, transition is always attached to the viscous interaction during which it takes place for all the tested Reynolds number Re_{X_T}. It reveals that the transition abscissa X_T is close to the abscissa X_I of the theoretical reflection and that the Reynolds number effect is almost inexistent. Moreover, the distance between the separated point X_S and the impingement point of the shock X_I is growing when the incoming boundary layer is developed on a greater longitudinal range. This means that the size of the laminar separated bubble, and associated viscous domain, is increasing with the Reynolds number Re_{XI}.

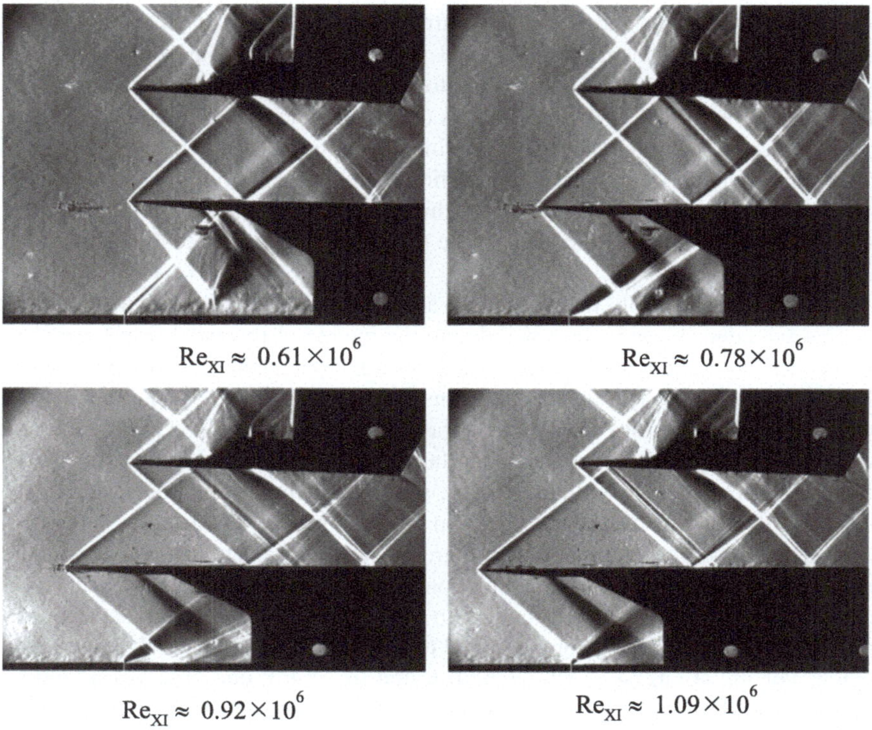

Fig. 41 Schlieren visualizations for a fixed shock intensity (alpha = 3°). Influence of the flat plate location and related Reynolds number at the shock impact point Re_{XI}

Table 7 Evolutions of the boundary layer separation (X_S), the shock impingement (X_I) and the transition point (X_T) abscissas with respect to the Reynolds number at the shock impact point (Re_{XI})

Re_{XI} × 10^6	0.61	0.78	0.92	1.09
X_S (mm)	32	42	51	60
X_I (mm)	44	56	66	78
X_T (mm)	47	59	69	79

Analysis by an inviscid flow approach

The shock intensity could be represented in a more quantitative approach by computing the pressure ratio p_2/p_0 (state *0* for the upstream condition, state *2* for the condition downstream of the shock system; see sketch of the flow on Fig. 42), using the Rankine-Hugoniot equations for the oblique (incident and reflected) shocks.

Then, analysis of results concludes that the transition « point » is located in the interaction region (the transition abscissa X_T being nearly coincident with the shock impingement abscissa X_I) for moderate shock intensities obtained with: $\alpha \geq 3°$, which corresponds to pressure ratio values: $p_2/p_0 \geq 1.35$.

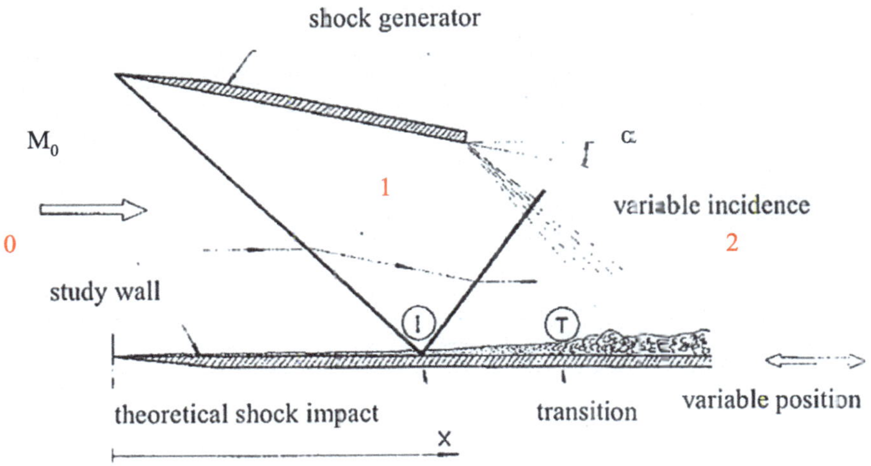

Fig. 42 Sketch of the investigated flow

Conclusions

An experimental study on a Mach 1.6 shock-wave/boundary layer interaction in a channel flow has been performed to analyze the transition process of the boundary layer growing on a flat plate. The detection of the boundary layer transition region is obtained by Schlieren visualizations, IR (Infra-red) thermography and TSP (Temperature Sensitive Paint) measurements, whose cross-checking of results allows to accurately determine the transition on the flat plate. Flow analysis shows that the transition process is not very sensitive to the flat plate boundary layer growth, but is strongly dependent to the oblique shock intensity. In fact, transition moves upstream and reaches the viscous interaction region for a moderate shock intensity given by the angle of attack of the shock generator wedge equal to 3°, which corresponds to a pressure ratio through the incident-reflected shock system of 1.35. Beyond this value, one concludes that the shock-wave triggers the transition.

2.2.3 TUD

Free stream turbulence level

Mass flux and velocity fluctuations in the free stream were measured by means of HWA and PIV, respectively. The perturbations in the freestream mass flux are mostly of an acoustic nature. The HWA measurements conducted at high overheat ratio (0.93) provide fluctuation levels of $(\rho u)' = 0.77\%$ of the freestream mass flux, yielding $p' = 1.9\%$ of the freestream pressure. Furthermore, from PIV measurements with a long pulse separation distance it was found that $u'/U_\infty = 0.57\%$ and $v'/U_\infty = 0.43\%$. For more information regarding the assessment of the free stream turbulence level the reader is referred to [27].

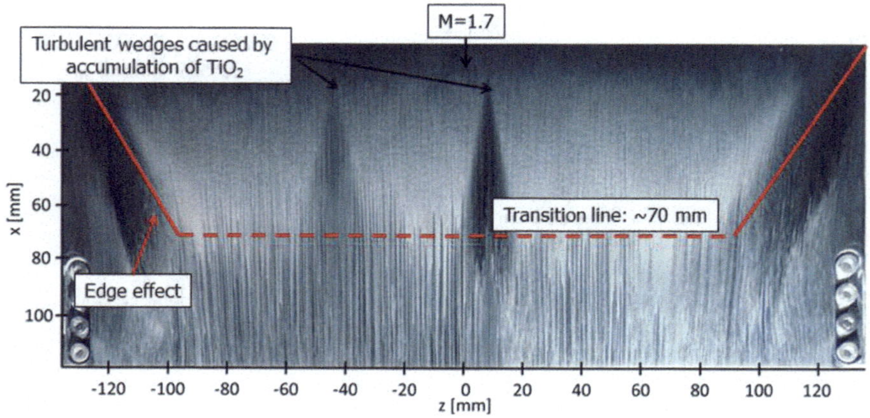

Fig. 43 Oil flow visualization for natural transition without shock interaction

Natural transition location

The transition location has been determined with four different methods: oil flow visualization, spark-light Schlieren visualization, infrared thermography, and particle image velocimetry. Results of the first two methods are given directly below; the infrared approach is not addressed here for brevity (but can be found in [28] while the PIV measurements will be described in the subsequent section in connection to the boundary layer characterization.

The oil flow visualization pattern in Fig. 43 indicates that boundary layer transition occurs at approximately 70 mm from the leading edge of the plate. A nominal 2D transition front is obtained over a spanwise length of ± 90 mm away from the centreline (~66% of the plate span). Clearly visible in the image are two turbulent wedges that originate close to the leading edge of the plate and which are triggered by small agglomerations of TiO_2 pigment particles in the oil film applied to the surface. These are not caused by plate imperfections, as they appear on different locations for different runs.

The spark-light Schlieren measurement visualizes instantaneous turbulent features in the flow and therefore allows to distinguish the laminar from the turbulent part of the boundary layer based on the granularity of the features detected therein. The transition location is found (see Fig. 44) to fluctuate between 55 and 65 mm during the course of a wind tunnel run. The Schlieren image presents a spanwise averaged image of the flow and any turbulent portion on the plate will overshadow the laminar portion of the plate. The Schlieren images therefore are indicative of the transition onset location x_T, rather than the average (50% intermittency) transition location $x_\gamma = 0.5$.

The results for the transition region characterization by the four different techniques are summarized in Table 8. The transition onset location x_T is found to be 55 mm ($Re_{x_T} = 1.9{\cdot}10^6$) from both the spark-light Schlieren visualizations and the PIV measurements. The average transition location $x_{\gamma=0.5}$ has been determined by

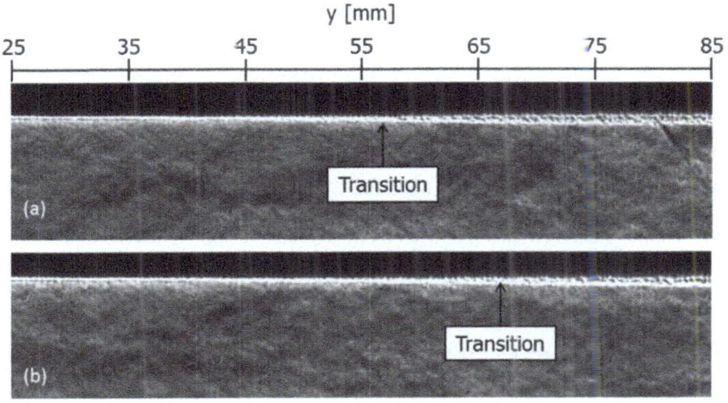

Fig. 44 Spark-light Schlieren visualization of boundary layer transition. Two snapshots are shown to illustrate the intermittent behaviour of the transition region

Table 8 Comparison of the transition locations found with the four different techniques

	x_T	$x_{\gamma=0.5}$	L	$dx_{\gamma=0.5}/dT_w$
Spark-light Schlieren	55 mm			
Oil-flow visualizations		70 mm		
Infrared thermography		68 mm		0.6 mm/°C
Particle image velocimetry	55 mm	71 mm	35 mm	

three techniques (oil-flow visualization, infrared thermography and PIV) and equals approximately 70 mm ($Re_{x_{\gamma=0.5}} = 2.5 \cdot 10^6$). From PIV also the transition length L has been obtained, which equals 35 mm. This implies that the process of transition approximately extends over the region of x = 55 to 90 mm on the flat plate, or in terms of Reynolds numbers from 1.9 to $3.2 \cdot 10^6$. Finally, infrared thermography also delivered the sensitivity of the average transition location to the average upstream wall temperature, as ~0.6 mm/°C.

Boundary layer characterization:

High-resolution PIV was applied to characterize the boundary layer. The measurements were performed with two Imperx Bobcat IGV-B1610 cameras. These cameras have a CCD chip of 1624 × 1236 pixels, which is cropped in the wall normal direction from 1236 to 651 pixels to speed up the data acquisition process (acquisition frequency of 10.2 Hz). The cameras are operated at a magnification of 0.57 and a spatial resolution of 130 pixels/mm. A major advantage of these cameras is the very short interframe time of 430 ns that can be achieved. At the given spatial resolution, this value results in a freestream particle displacement of ~24 pixels between frames. Illumination is provided by a double-pulse Nd:YAG Spectra Physics Quanta Ray PIV-400 laser, which is operated at a laser power of 140 mJ per pulse. The pulse duration is less than 7 ns, which translates into a particle displacement during illumination of less than 0.4 pixel and therefore incurs negligible particle blur. The flow is seeded with TiO_2 particles (30 nm crystal size), which have a response time of 2.5 μs.

In view of the very small boundary layer thickness only a limited number of pixels is available that can be used to reconstruct the velocity profile. At 40 mm from the leading edge, the laminar boundary layer is ~0.2 mm thick, which at the given spatial resolution of 130 pixels/mm translates into just 26 pixels. So, for that reason elongated interrogation windows of 8×256 pixels were used, which size corresponds to 0.06 mm in the wall normal direction and 1.97 mm in the streamwise direction. A window overlap of 75% was used in the streamwise direction and an overlap of 87.5% was used in the wall normal direction, resulting in vector pitches of 7.7 μm and 0.49 mm, respectively. Before applying cross-correlation to obtain the velocity fields, several pre-processing steps were applied to correct the images for camera read-out noise, plate vibration and large differences in particle intensities [27]. Because of the thin interrogation windows and the relatively low seeding density close to the wall, it is not possible to determine the instantaneous velocity field from a single image pair. Instead an ensemble correlation approach is used, which cumulatively builds up the correlation plane by calculating and summing up the correlation results for all the image pairs. All of the processing is performed with an in-house iterative multi-grid window-deformation PIV code (Fluere).

To track the development of the complete boundary layer 12 field of views are combined. For every run, two cameras were operated simultaneously, one on each side of the tunnel, with a typical overlap of 4 mm between the cameras FOVs to allow stitching of the velocity fields. A total of 300 image pairs was obtained per location.

The major challenge of the PIV measurements proved to be the adequate seeding of the boundary layer. Using a TiO_2 seeder operated at an overpressure of 1 bar with respect to the pressure in the settling chamber, a satisfactory seeding density of 60–70 particles per mm^3 was obtained for the freestream. Figure 45 shows the seeding distribution over the flat plate and is obtained by calculating the maximum number of counts minus the minimum number of counts per pixel location for the entire dataset. This visualization shows that for the upstream laminar part boundary layer, no seeding is present in the near-wall region, while further downstream the seeding distribution is homogenized by the mixing that takes place in the turbulent boundary layer.

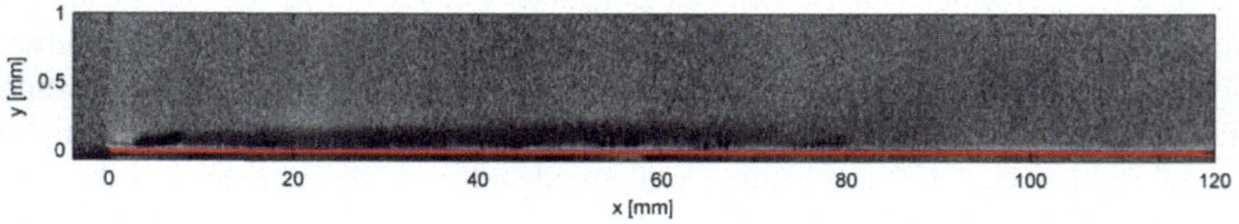

Fig. 45 Seeding distribution along the flat plate. The red line indicates the location of the wall

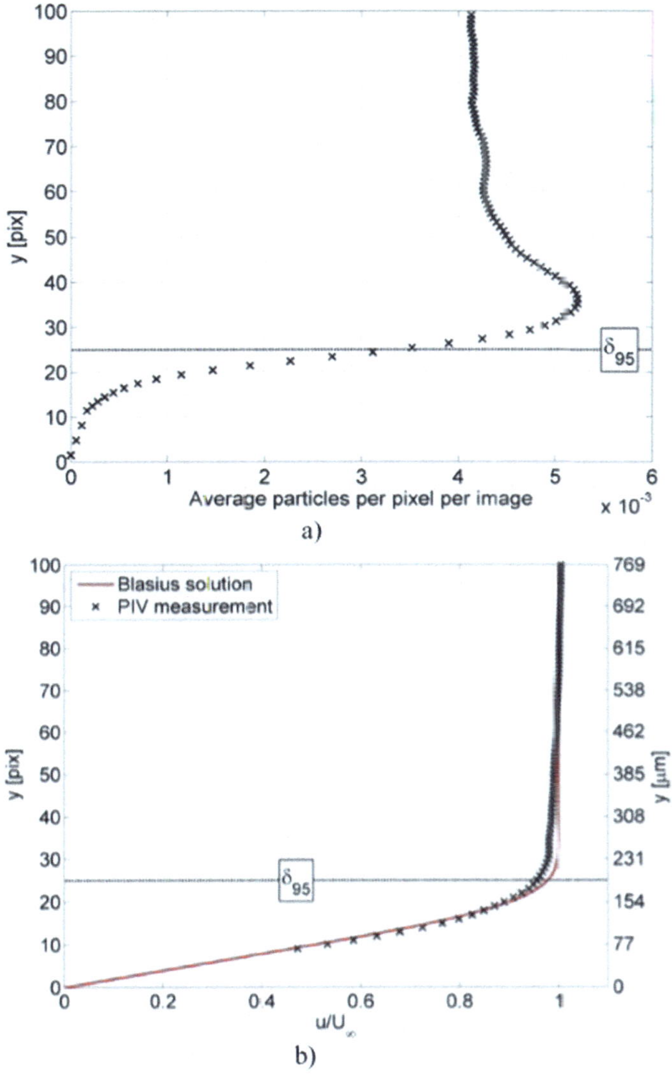

Fig. 46 Average particle distribution in the laminar boundary at x = 40 mm (**a**) and the corresponding velocity profile (**b**)

Figure 45 shows that the gap without seeding gradually disappears in the region of 60 to 80 mm from the leading edge, which agrees with the transition region determined by the other experimental techniques. This visualization provides an averaged view of the laminar-to-turbulent transition process. Inspection of the individual images reveals an intermittent behaviour, in that the gap is present in some images and absent in others. To quantify this intermittency, the ppp levels were calculated in windows covering the region from 5 to 15 pixels away from the wall for every individual image. If the ppp level in the window has a value higher than 30% of the freestream ppp level for that particular image, the boundary layer state is qualified as turbulent and otherwise as laminar. Figure 47 shows the results of this analysis and indicates a smooth transition taking place over approximately 35 mm. The distribution exhibits an inflection point around x = 71 mm, at an intermittency level of 50%. The experimental data is compared to the semi-experimental/semi-theoretical intermittency distribution of Dhawan and Narasimha [29].

A further illustration of the seeding distribution is given in Fig. 46a, which shows the average pixel density at 40 mm from the leading edge. Virtually no seeding is detected in the first 10 pixels from the wall (~40% of δ_{95}), whereas just outside of the boundary layer an accumulation of migrated particles is found. Notwithstanding that the lower 10 pixels provide virtually no information, the remaining 15 pixels still allow to extract the upper 60% of the boundary layer profile. The resulting boundary layer profile is presented in Fig. 46b in comparison to the compressible Blasius solution, with properties documented in Table 9.

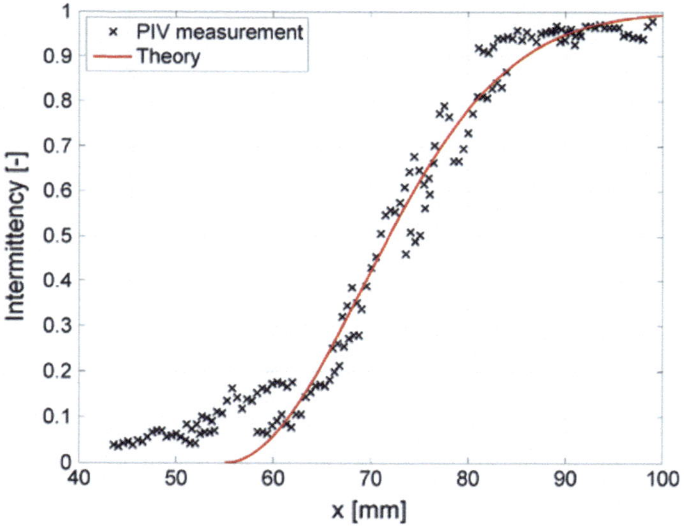

Fig. 47 Development of the intermittency throughout the transition region

Table 9 Boundary layer properties at $x = 40$ mm

	PIV	Blasius
δ_i^*	87 ± 5 μm	80 μm
θ_i^*	30 ± 2 μm	29 μm
H_i	2.9 ± 0.1	2.7
δ_{95}	200 ± 8 μm	170 μm

The development of the integral parameters of the undisturbed boundary layer, in terms of the incompressible displacement thickness δ_i^*, incompressible momentum thickness θ_i and the incompressible shape factor H_i, is displayed in Fig. 48. To obtain these parameters for the laminar part, the velocity profile close to the wall was extrapolated using the compressible Blasius solution. The shape factor has a value of approximately 2.9 at 40 mm from the leading edge and from 55–85 mm

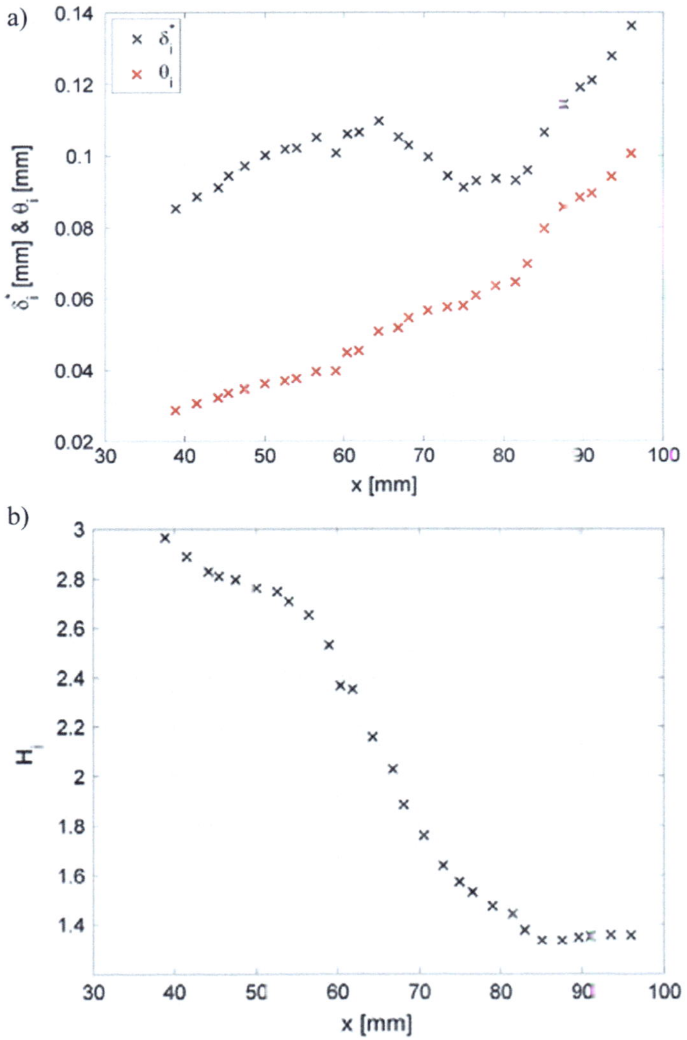

Fig. 48 Development of the integral parameters. Incompressible momentum ϑ_i and displacement thickness δ_i^* (**a**) and incompressible shape factor H_i (**b**)

displays a gradual decrease to a value of ~1.35, which are typical values for laminar and turbulent boundary layers, respectively.

Oblique shock wave reflection under natural transition conditions
To study the effects of boundary layer transition on an oblique shock wave reflection, three inflow conditions were considered: a laminar, transitional ($\gamma \sim 50\%$) and turbulent boundary layer, depending on the state of the undisturbed boundary layer at the shock impingement location.

For these cases, the incident shock wave (3° flow deflection angle) was positioned at, respectively, $x_{sh} = 51, 71$ and 101 mm from the leading edge. Schlieren visualizations of these cases are given in Fig. 49, while the mean velocity fields as measured with PIV are compared in Fig. 50.

For the **laminar case** (Fig. 51a) a large separation bubble is present which starts 8 mm upstream of the incident shock impingement location and ends 5 mm downstream. The total bubble length (10.5 mm) corresponds to approximately 110 times

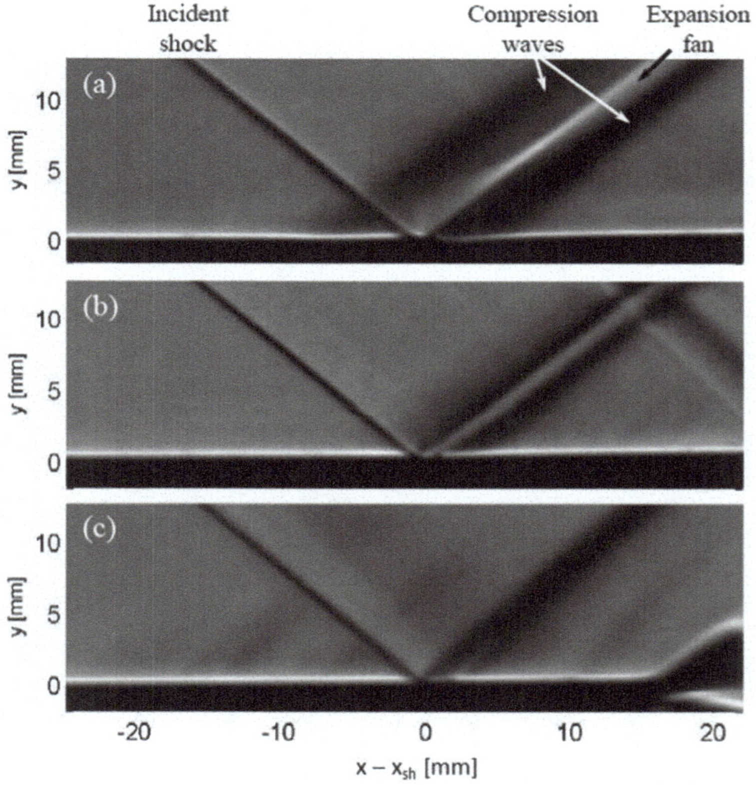

Fig. 49 Schlieren visualizations of an oblique shock wave reflection ($\theta = 3°$) with a laminar boundary layer, $x_{sh} = 51$ mm (**a**) a transitional boundary layer, $x_{sh} = 71$ mm (**b**) and a turbulent boundary layer, $x_{sh} = 101$ mm (**c**)

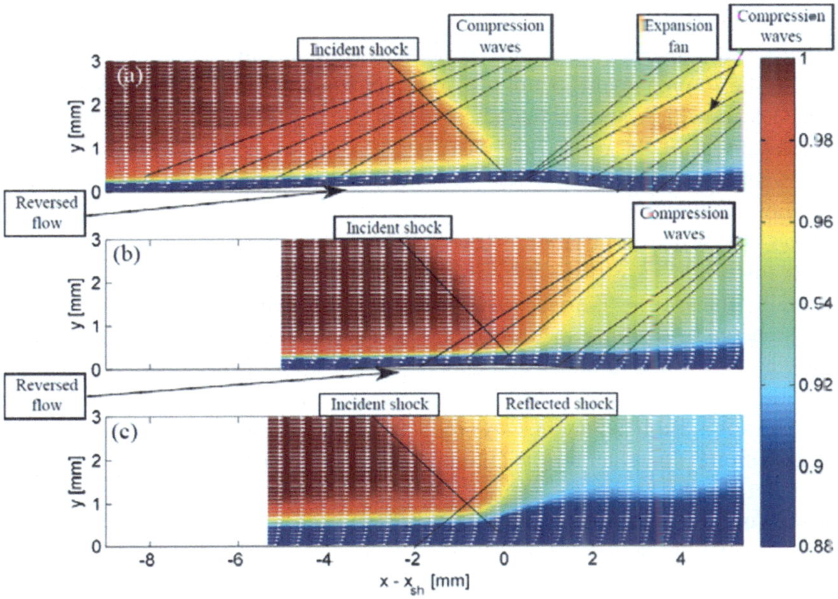

Fig. 50 Average velocity field of an oblique shock wave reflection with an incoming laminar (**a**), transitional (**b**) and turbulent (**c**) boundary layer. The corresponding shock locations are, respectively, $x_{sh} = 51$, 71 and 101 mm

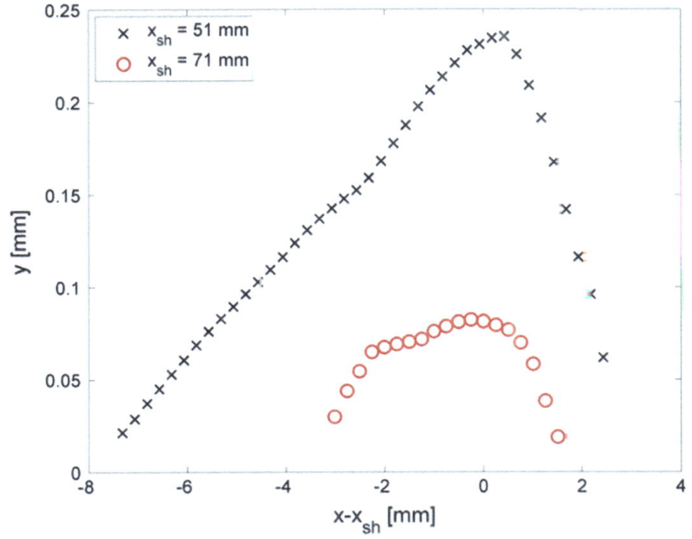

Fig. 51 Comparison between the size of the reversed flow region for an oblique shock wave reflection for a laminar and transitional incoming boundary layer

the displacement thickness δ^* recorded at $x = 40$ mm. The incoming laminar boundary layer is lifted over the separation bubble and remains in a laminar/transitional state up to the impingement location of the incident shock ($x \approx x_{sh}$), after which it undergoes transition and reattaches at the wall as a turbulent boundary layer. The displacement of the outer flow associated to the formation of the separation bubble results in the formation of subsequent series of compression and expansion waves. These features are typical for a separated laminar shock wave-boundary layer interaction and were also observed in the Schlieren visualizations.

The **transitional case** (Fig. 50b) also shows a region of reversed flow, though much smaller than for the laminar case. From the mean velocity field two regions of compression can be distinguished, one starting upstream of the incident shock wave and the other is formed around the reattachment location of the separation bubble. Similar observations were also made in the Schlieren and oil-flow visualizations for this configuration. Both techniques also indicate that the separation bubble is smaller for the transitional case than for the laminar case and that the compression waves are confined to a smaller region.

For the **fully turbulent interaction** (Fig. 50c) no separation is detected in the average velocity field, which is according to expectations based on theory. The inviscid pressure ratio that is imposed by the oblique shock wave reflection equals $p_3/p_\infty = 1.35$, whereas according to the free-interaction theory, a pressure ratio of at least $p_3/p_\infty = 1.77$ would be required to separate the turbulent boundary layer. Also the oil-flow visualizations (see section 4) that were performed for this configuration showed no indication of boundary layer separation.

The seeding difficulties that were encountered when investigating the undisturbed boundary layer were also present when investigating the laminar and transitional interactions. This factor, amongst others, prevents from obtaining an accurate estimate of the velocities inside the separation bubble under the present conditions. Therefore, a Falkner-Skan based extrapolation procedure was applied to the available velocity profile data to provide an estimate of the $u = 0$ isoline position. These results, indicative of the bubble shape are compared in Fig. 51 for the laminar and transitional case. The separation bubble for the laminar interaction is longer (10.5 vs. 5.3 mm) and higher (0.23 vs. 0.08 mm) than for the transitional case. For both cases there is a slow build-up in bubble height when approaching the incident shock wave. After passing the shock wave, the boundary layer undergoes transition and reattaches as a turbulent boundary layer. Notice also that the initial build-up in bubble height for the laminar case is observed to be nearly linear with streamwise distance.

Effect of shock strength

In addition to the nominal test condition case, a parametric investigation was conducted regarding the influence of shock strength, Mach number and Reynolds number. Here, the results on the shock strength are reported, obtained by varying the flow deflection angle between 1.0 and 5.0 degrees at Mach number of 1.7 (pressure ratio $p_3/p_1 = 1.11$–1.64). The corresponding reversed flow regions are presented in Fig. 52-left, where the length of the upstream portion, L_u, and the downstream portion, L_d, are defined by fitting straight lines to the experimental data points. For $\theta > 2°$

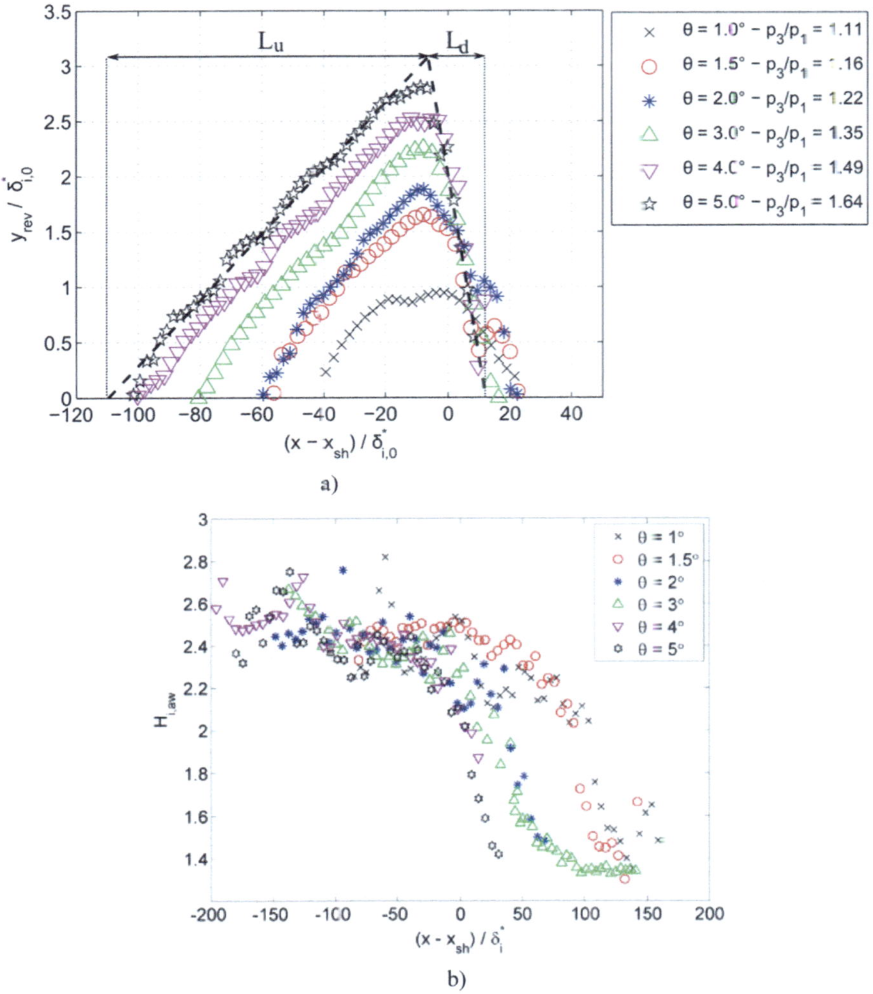

Fig. 52 The effect of shock generator angle on the reversed flow region size (left) and on the transition process (right); the latter is represented by the variation of the shape factor of the flow profile above the u = 0 isoline

($p_3/p_1 > 1.22$) the downstream portion of the separation bubble keeps a near-constant length of 20 $\delta^*_{i,0}$ (see also Fig. 53). The upstream portion of the separation bubble, L_u, on the other hand, shows a near-linear increases with the shock strength. For weak shock waves ($\theta < 2°$), however, these trends no longer apply and the upstream portion of the bubble is found to rapidly decrease in size with reducing shock strengths, while the downstream portion of the bubble increases in size. For shock strengths of $\theta = 1°$ – 1.3° (p_3/p_1 = 1.11–1.14) a near-symmetrical bubble is recorded, with comparable L_u and L_d.

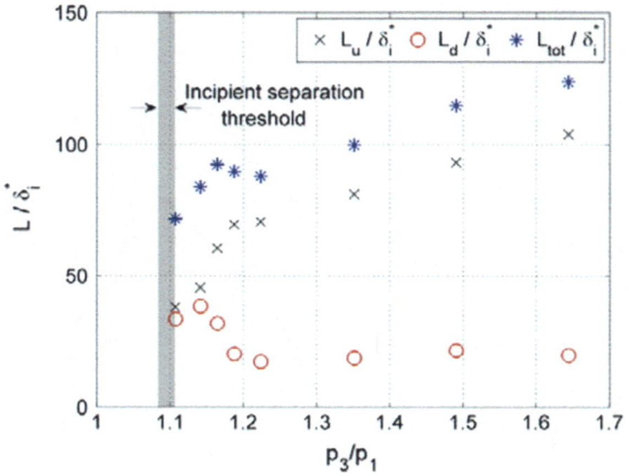

Fig. 53 The size of the reversed flow region as function of shock strength

The incipient separation threshold indicated in Fig. 53 is based on the free-interaction of Chapman [26].

The observed trends are strongly connected with the transition behaviour of the shear layer over the separated flow region, as supported by further investigation of the velocity profiles. This analysis (see Fig. 52-right) indicates that the transition location (where the incompressible shape factor decreases) is found to move upstream, yet even for the strongest shock wave ($\theta = 5°$) transition always occurs downstream of the shock impingement location. So, the boundary layer stays in a laminar state over the upstream part of the separation bubble. On the other hand, transition occurs significantly further downstream for weak shocks, corresponding to $\theta < 2°$. The data implies that for shock angles in the range of 1–1.5° the boundary layer remains in a close to laminar state throughout the entire interaction, which explains the longer downstream portion of the separation bubble.

Oblique shock wave reflection under turbulent conditions

Two approaches were followed to establish the presence of a fully turbulent boundary layer entering the interaction (referred to as case 1 and case 2, respectively):

- The boundary layer was tripped 5 mm from the leading edge with a 0.2 mm thick zig-zag strip. The incident shock wave is positioned 71 mm from the leading edge.
- The incident shock wave is positioned 30 mm downstream of the approximate natural transition location, so at x = 101 mm. At this location, the flow has left the intermittent transitional regime and a fully turbulent velocity profile is established.

As evidenced by the velocity profiles shown in Fig. 54, good agreement with the log law theory is found for both cases (with $H_i = 1.37$ and 1.32 for case 1 and 2, respectively).

In view of the similarity of the incoming boundary layer profiles for case 1 and 2, comparable flow fields are expected to occur throughout the interaction for both cases. This is confirmed by Fig. 55, from which it is clear that the impinging shock wave results in a thickening of the boundary layer.

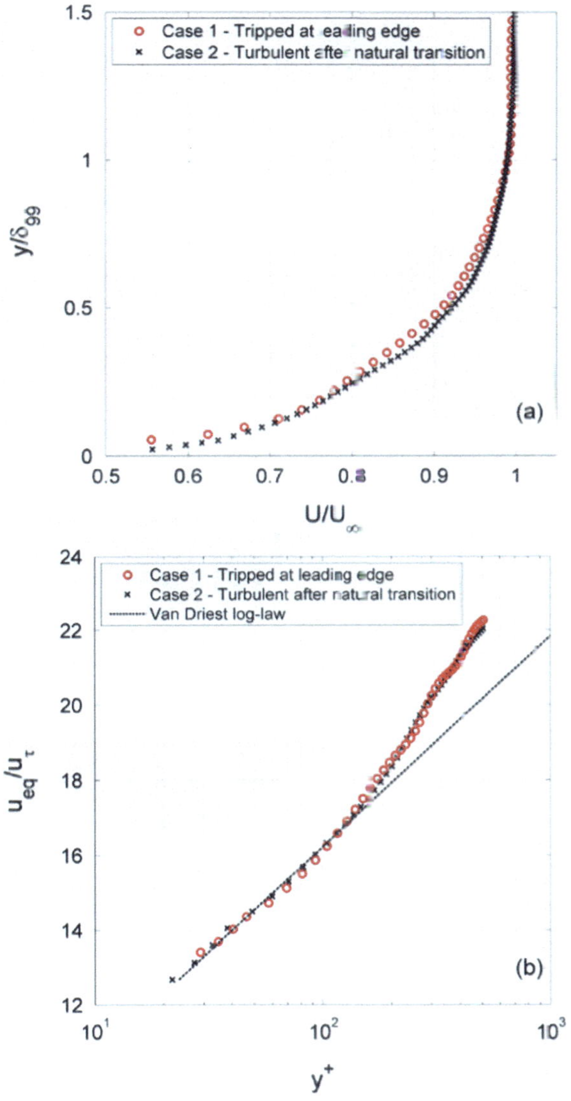

Fig. 54 Comparison between the velocity profiles measured 6 mm upstream of the interaction for case 1 and 2, in outer variables (**a**) inner variables (**b**)

This is even more evident when considering the development of the integral boundary layer parameters in Fig. 56. For case 2, the displacement thickness starts to increase approximately 2δ upstream of the incident shock and reaches its maximum value in close vicinity of the impingement point of the incident shock. Case 1 shows the same general trend, but the change occurs more gradual, with the displacement thickness already showing an increase 3δ upstream of the incident shock

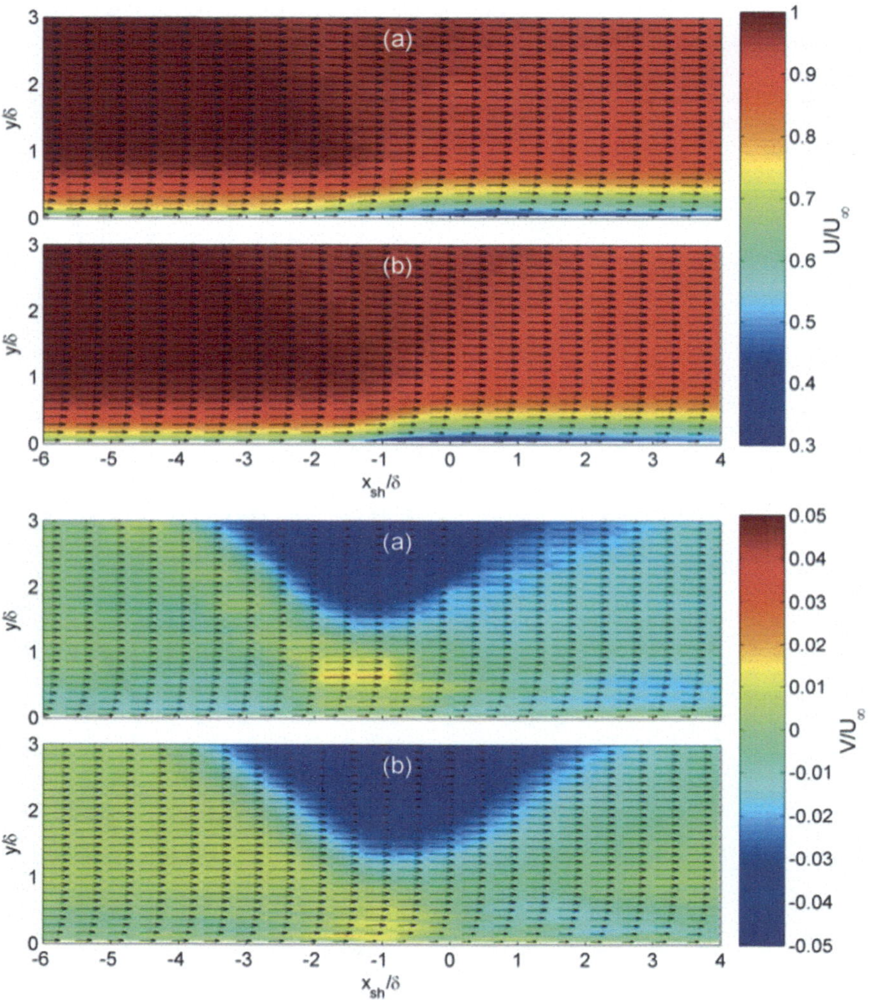

Fig. 55 The velocity components of the velocity field inside the interaction region. For case 1 (**a**) and case 2 (**b**)

wave. Downstream of the incident shock, the boundary layer recovers and the displacement thickness is again reduced, this in contrast to the momentum thickness, which remains relatively constant downstream of the incident shock.

The development of the shape factor shows a trend very similar to the one described for the displacement thickness, with the highest shape factors (1.6–1.65) being reached around the impingement point of the incident shock.

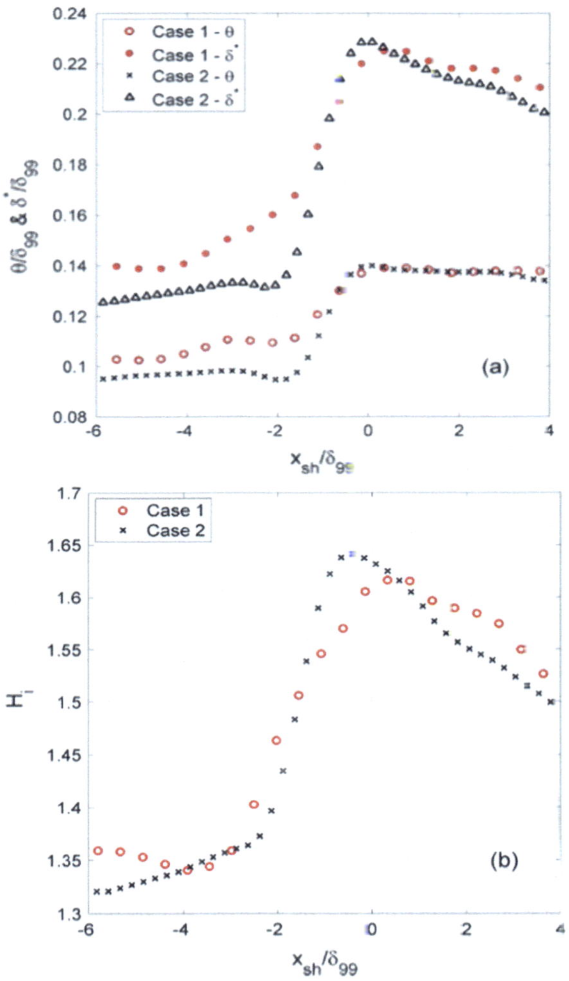

Fig. 56 Variation of the integral boundary layer properties through the interaction region: (**a**) incompressible momentum thickness θ_i and displacement thickness δ_i^* and (**b**) the incompressible shape factor Hi¬ (**b**)

2.2.4 ITAM

Laminar test case

Parameters of experiments carried out at natural laminar-turbulent transition are shown in Table 10. From the table it can be seen that the Mach number upstream of SWBLI (calculated on the basis of PIV data) is less than one measured upstream of the model (M = 1.47). This can be explained by a decrease of Mach number in a weak compression wave generated by the leading edge of the plate which crosses the flow several times due to its reflections from the walls. The value of X_{imp} corresponds to the point of intersection of the incident shock wave with plate for inviscid flow, measured from the leading edge of the plate.

Table 10 Parameters of experiments

M	P_0, bar	T_0, K	Re_1, 10^6 1/m	X_{imp}, mm	β,°	Re_{Ximp}, 10^3	δ*, mm	State of BL
1.42	0.694	286.4	10.9	288	3,4	3145	0.53	Turb.
1.42	0.847	280.8	13.7	290	3,4	3975	0.43	Turb.
1.42	0.984	284.2	15.6	290	3,4	4540	0.46	Turb.
1.43	0.551	291	8.5	132	3,4	1120	0.30	Lam.
1.43	0.694	290.4	10.7	133	1,2, 3,4	1425	0.27	Lam.
1.43	0.834	286.3	13.2	134	3,4	1755	0.25	Lam.
1.43	0.978	285.1	15.5	134	1,2, 3,4°	2070	0.22	Lam.

Fig. 57 Schlieren visualization

Example of the results illustrating the wedge angle effect on the interaction study are shown in Fig. 57 as instantaneous (exposure 1.5 μs) and the averaged schlieren images obtained for the laminar case. Origination of vortices in the zone of SWBLI can be clearly seen and indicates the laminar-turbulent transition. The adverse pressure gradient caused earlier turbulization of boundary layer. For a horizontal orientation of the knife in the averaged images the laminar boundary layer is observed as narrow bright line. Therefore we can qualitatively assume that the end of the bright line corresponds to the point of the boundary layer turbulization.

With the reduction of the wedge angle the size of the separation zone is reduced and the location of the laminar-turbulent transition is shifted downstream (Fig. 58). Apparently decrease of the pressure gradient in the shock wave results in weakening of disturbance growth in the interaction zone. Increasing of the Reynolds number from 10.7 up to $15.5 \cdot 10^6\ m^{-1}$ significantly reduces the size of the separation region, but has little effect on the position of the laminar-turbulence transition.

Figure 59 shows PIV velocity fields obtained for the same flow parameters. PIV measurements were performed in the region of the interaction at the centreline of the model. All the figures clearly show the formation of a weak shock wave arising at the beginning of the separation zone. The intensity of the separation shock is small

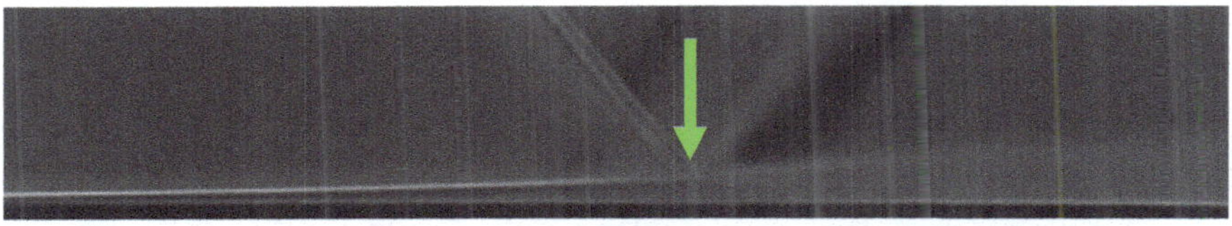

β = 4°

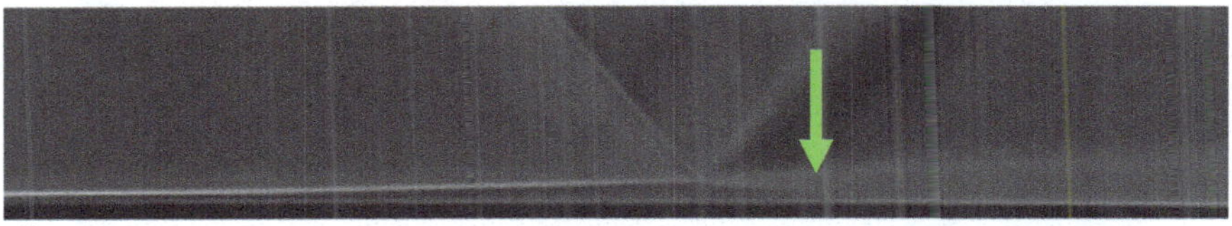

β = 3°

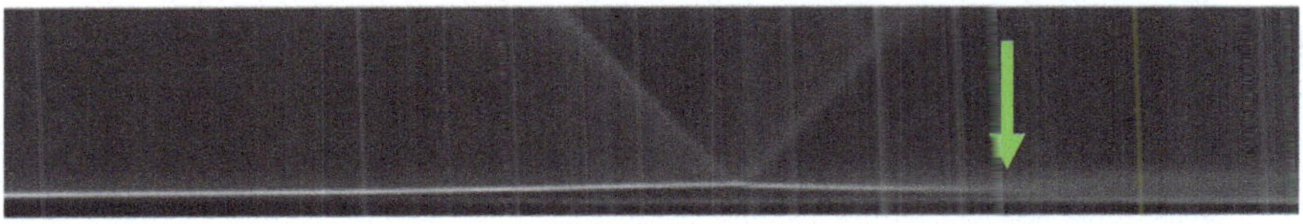

β = 2°

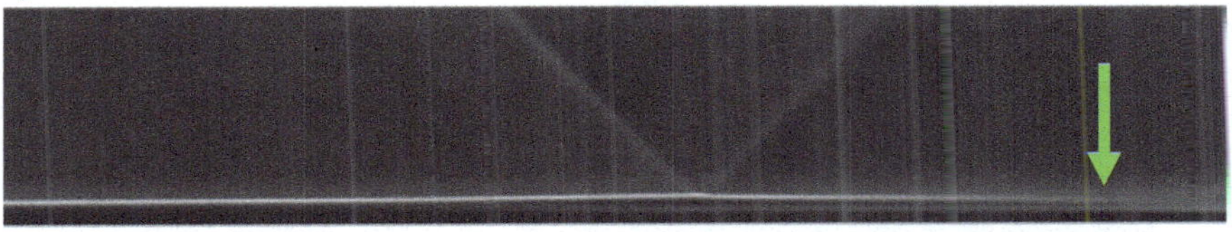

β = 1°

Fig. 58 Laminar case (L = 250 mm, Re_1 = 10.7e6 1/m)

due to a slight growth in the displacement thickness in the separation zone. There is only one separation shock wave on the velocity distribution, since the displacement thickness in the laminar separation increases approximately linearly. It can be seen that reduction of the incident shock wave angle just slightly changes the intensity (angle) of the separation shock. It means that the angle of the flow displacement by the separation weakly depends on the strength of the shock wave.

Let's consider a change of the separation zone length (length of interaction zone). If the wedge angle decreases from 4 to 3° the length of the separation zone remains almost constant. The figure clearly shows that decrease of the wedge angle is accompanied by a downstream shift of the separation beginning for about 10 mm, but the end of the separation is also shifted by the same distance. This is the result of the displacement of X_{imp} downstream due to changing of wedge angle β.

With further decrease of the incident shock wave strength, the length of the separation zone begins to decrease more significantly. This behavior is well explained by the turbulization of the boundary layer in the SWBLI zone. Laminar-turbulent transition in the interaction region leads to significant reduction of its length compared to the classical laminar case. Thus for the case of 4° the RMS velocity distribution

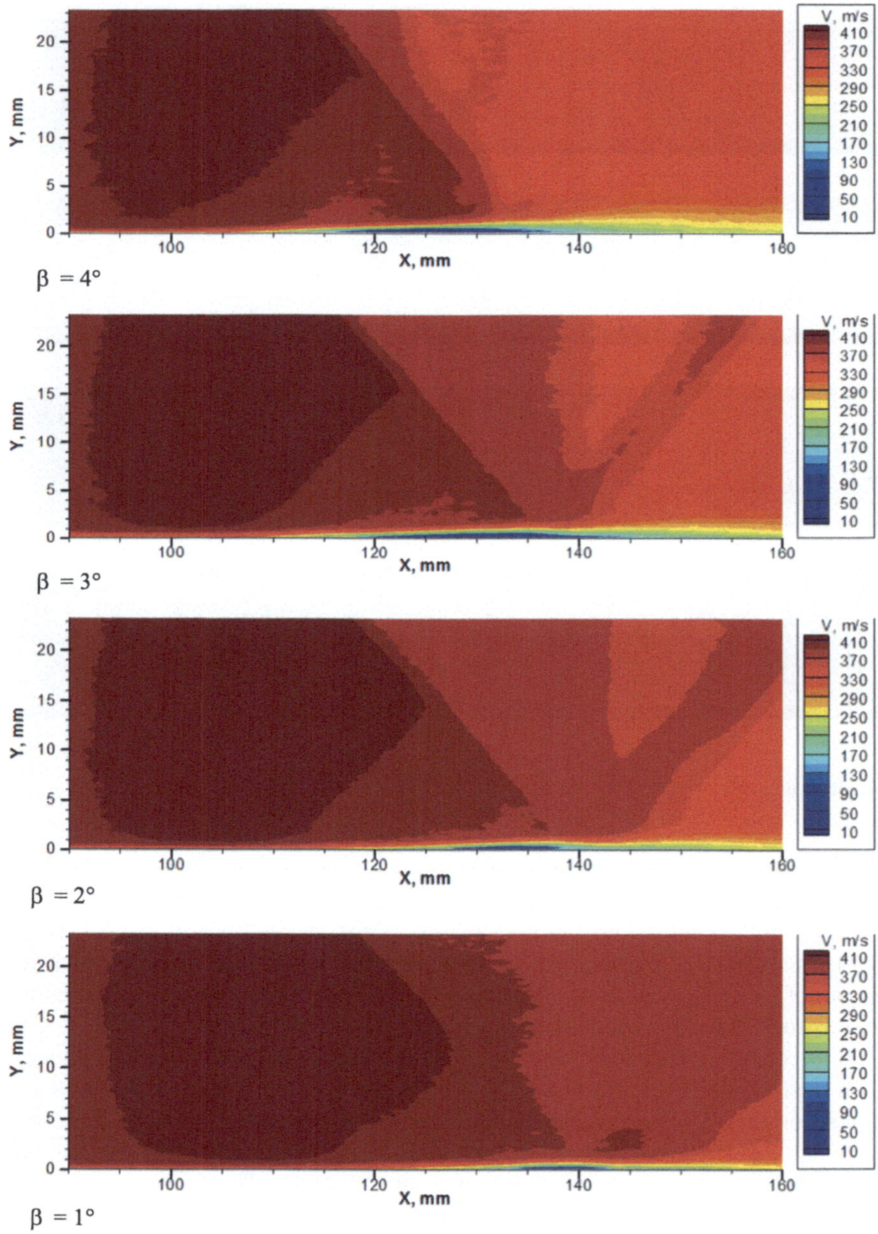

Fig. 59 Velocity field (Laminar case, L = 250 mm, Re_1 = 10.7e6 1/m)

shows that the transition occurs near the point of intersection of the incident shock wave with the model. The decrease in the strength of the incident shock wave is accompanied by a shift of the transition point downstream, which is clearly seen from the substantial decrease in the thickness of the wake boundary layer for $\beta = 1,2°$. As a result the laminar-turbulent transition process gradually diminishes its effect on the length of the SWBLI. Therefore the significant decrease of the wedge angle from 4 to 1° changes the size of the interaction zone only by a factor of ≈2. Figure 60 shows the distribution of the displacement thickness and the shape factor in the zone of SWBLI at $\beta = 4°$. The solid lines correspond to the data obtained in the zone of interaction, the dotted lines correspond to another experiment where the measurements in the wake were performed. It is clearly seen that the data obtained for the same flow conditions are perfectly matched. Integral parameters (momentum thickness, shape factor) vary only slightly. The sharp decrease of the shape factor at X_{imp} indicates the turbulization of the boundary layer, but its value indicates that the equilibrium state is not reached.

The spectra of wall pressure pulsations confirm this assumption. For laminar case at $\beta = 3, 4°$ peak of wall pressure pulsation was found near X_{imp} and it is most likely associated with the beginning of the turbulization of BL (see Fig. 61). The growth of pulsations in the wake mainly takes place only in some frequency band. This means that the turbulent boundary layer is not equilibrium. The characteristic peak of pulsations in the high-frequency range for the laminar case has a frequency 20–30 kHz which is substantially smaller than the characteristic frequency of the turbulent boundary layer. It can be attributed by the generation of a large-scale structures for the laminar case.

In the low-frequency region two peaks at 2 and 0.2 kHz were found. These pulsations are most probably the characteristic ones of the separation zone and the zone of interaction. From the correlation analysis it was discovered that these disturbances (up to 2 kHz) propagate upstream and originate near the point of the shock wave interaction with the boundary layer.

Figure 61b shows that the decrease of the total pressure was accompanied by a drop of the pulsations in the low-frequency region < 10 kHz. It can be concluded that there is an influence of the Reynolds number on the development of perturbations in the separation zone.

Turbulent Test Case

For Mach number $M = 1.43 - 1.47$ and turbulent state of the incoming boundary layer the adverse pressure gradient in the shock is insufficient for the flow separation. For the case of strong incident shock wave ($\beta = 4°$) and fully turbulent interaction it is possible to see the formation of small Mach stem (Fig. 62). Mach stem is generated here due to presence of strong reflected shock wave and leads to considerable increase of the local adverse pressure gradient. It is interesting to note that the cause of the formation of the strong reflected shock is higher resistance of the turbulent boundary layer to the separation. As a result the length of separation bubble is relatively small

Fig. 60 **a** The momentum thickness and **b** shape factor distribution along the SWBLI

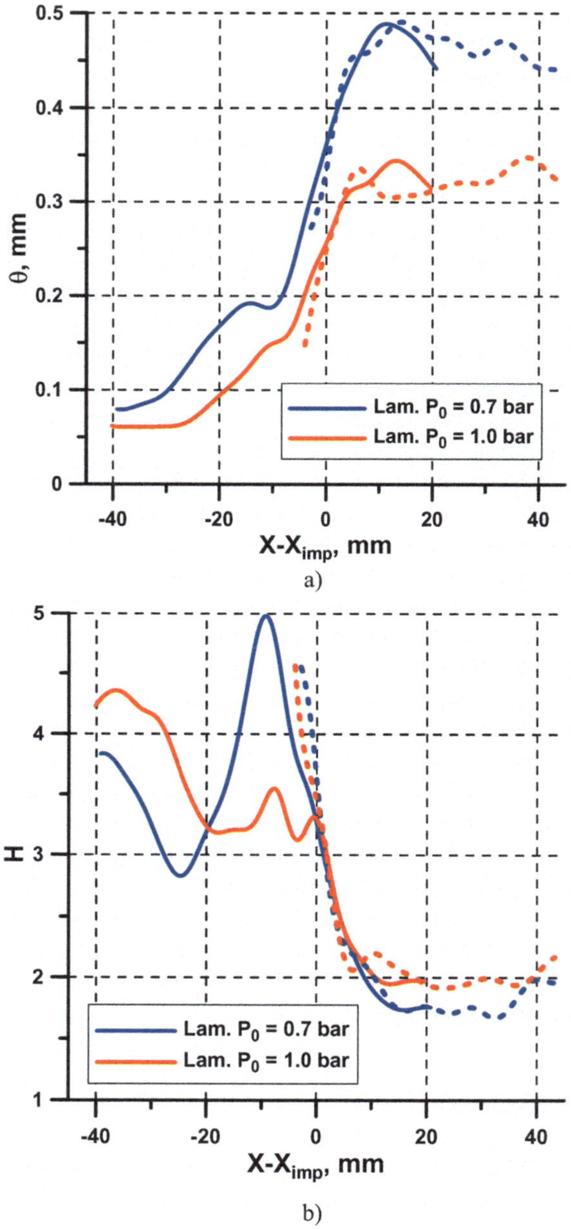

Fig. 61 *PSD·f* of wall pressure pulsations for laminar test case **a** along zone of SWBLI for P_0 = 0.7 bar and **b** at the point of maximum pulsation for various P_0

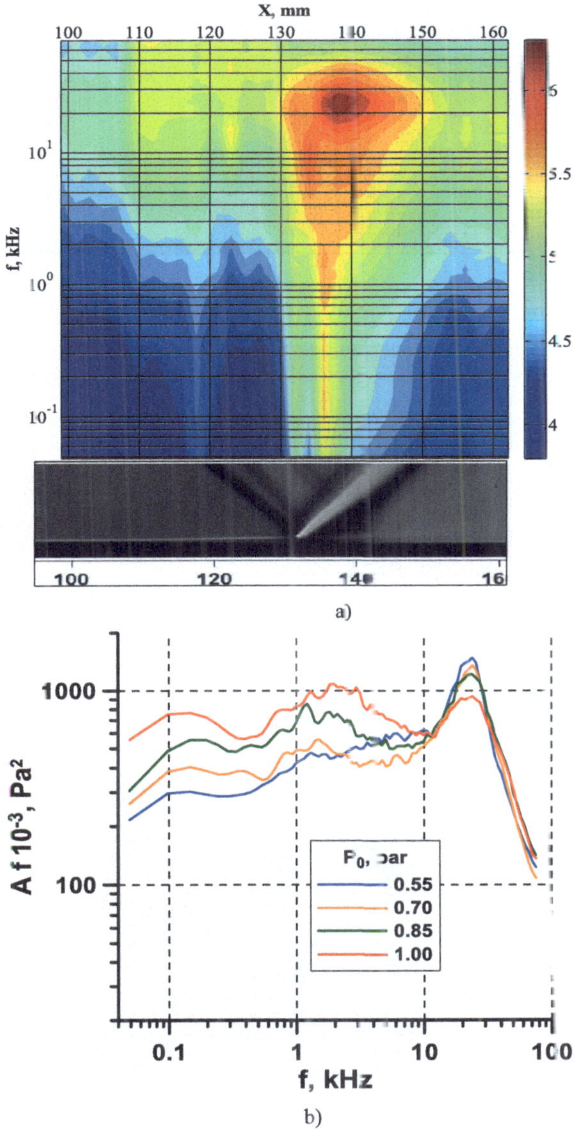

comparing to the laminar case and therefore there is very rapid growth of the boundary layer displacement thickness in the zone of SWBLI. This leads to the formation of a strong reflected shock wave. But if the wedge angle changes from 4 to 3° the Mach stem almost disappears. Note that the thickness of the boundary layer in the wake obtained for the turbulent case is close to the laminar one founded at $\beta = 4°$.

Example of RMS value of streamwise velocity pulsations is presented in Fig. 63. The level of pulsations upstream of the incident shock corresponds to a turbulent flow.

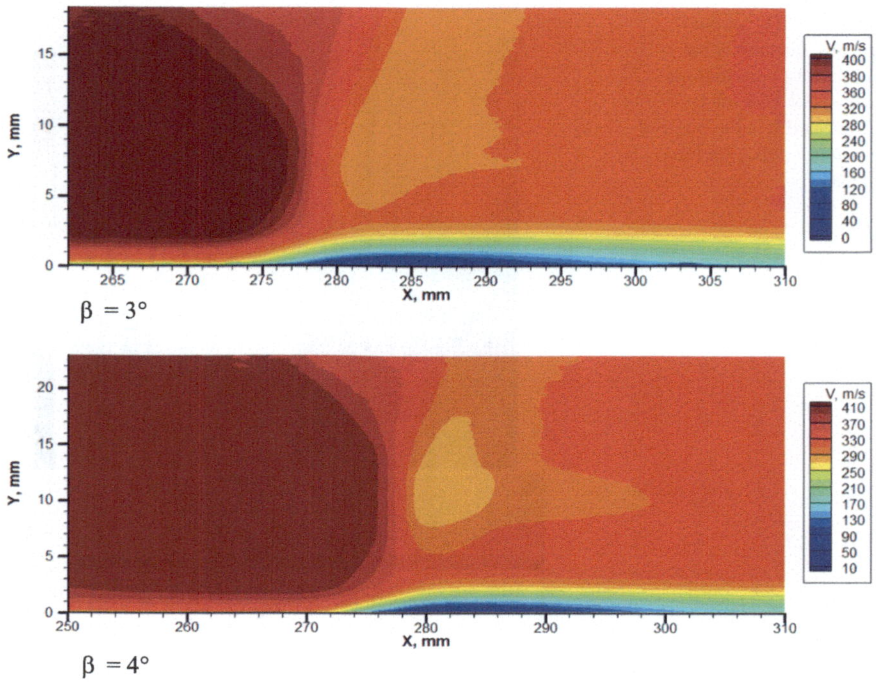

Fig. 62 Velocity magnitude at the centerline, L = 100 mm, natural turbulization $P_0 = 0.7$ bar

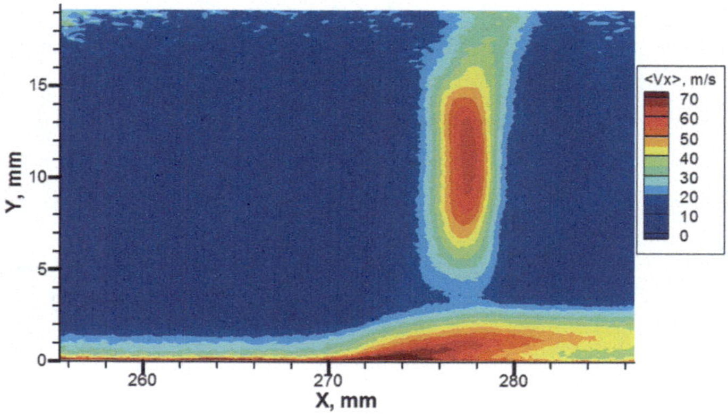

Fig. 63 RMS of streamwise velocity pulsations, L = 100 mm, natural turbulization

The streamwise distribution of pulsations in the zone of interaction corresponds to the published data. Pulsations in the zone of the Mach stem are result of low-frequency oscillations of the interaction zone.

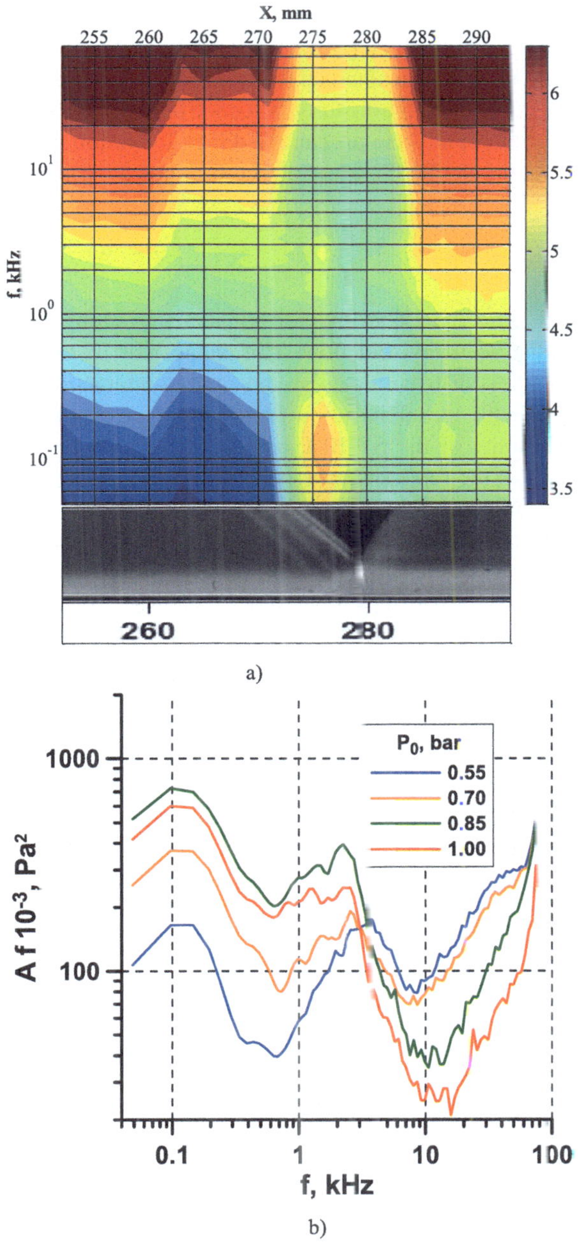

Fig. 64 *PSD·f* of wall pressure pulsations for turbulent test case **a** along zone of SWBLI (L = 100 mm, $\beta = 4°$, $P_0 = 0.7$ bar) and **b** at point of maximum pulsation for various P_0

An example of the wall pressure pulsations spectra distribution along SWBLI zone for the turbulent case is shown in Fig. 64. The data show that the level of pulsations upstream of the incident shock corresponds to a turbulent flow. The spectra in the wake are also turbulent. Near the reflected shock wave the low-frequency pulsations are generated which have been well studied in UFAST.

In view of data obtained for the laminar case it is possible to assume that the incoming pulsations significantly influence on the process of perturbations development in the zone of adverse pressure and their evolution in the wake.

Using the RMS values of U pulsations the following value can be calculated:

$$\int \rho u^2 \, dy \tag{14}$$

Figure 65b. The integral is calculated along the vertical coordinate up to the undisturbed inviscid flow. This value will be analyzed coupled with momentum thickness distribution along the zone of interaction (Fig. 65a).

For the turbulent inflow boundary layer the peak of pulsations was found at the beginning of the SWBLI zone near the reflected shock wave. Near this location there is the rapid increase of momentum thickness and corresponding loss of total pressure. It is obvious that the growth of pulsations is a result of convective processes and energy transfer in the shear layer which leads to increase of drag.

Growth of the pulsations for the transitional case is more gradual and accompanied by weaker growth of momentum thickness. For the laminar case the most dramatic growth of pulsations (more than in turbulent case) was obtained. It was accompanied by an increase of momentum thickness approximately up to the level of turbulent case. But for the laminar case the turbulent boundary layer in the wake is substantially more nonequilibrium as follows from the POD analysis and the spectra of pressure pulsations (Fig. 61).

It should be noted that due to low spatial resolution of the PIV method (for the selected scale), the momentum thickness for the laminar and transitional inflow boundary layer is overestimated. It can be concluded that for small supersonic Mach numbers and strong incident shock waves, the process of the laminar-turbulent transition has a very large effect on the mean and nonstationary parameters of the zone of SWBLI.

2.3 CFD of the Laminar and Turbulent Interaction

(IUSTI, ONERA, SOTON, IMFT)

2.3.1 IMFT

Introduction

The SWBLI of oblique shock is studied in comparison with the TUD experiments and in respect of the effect of fully turbulent boundary layer upstream and on the

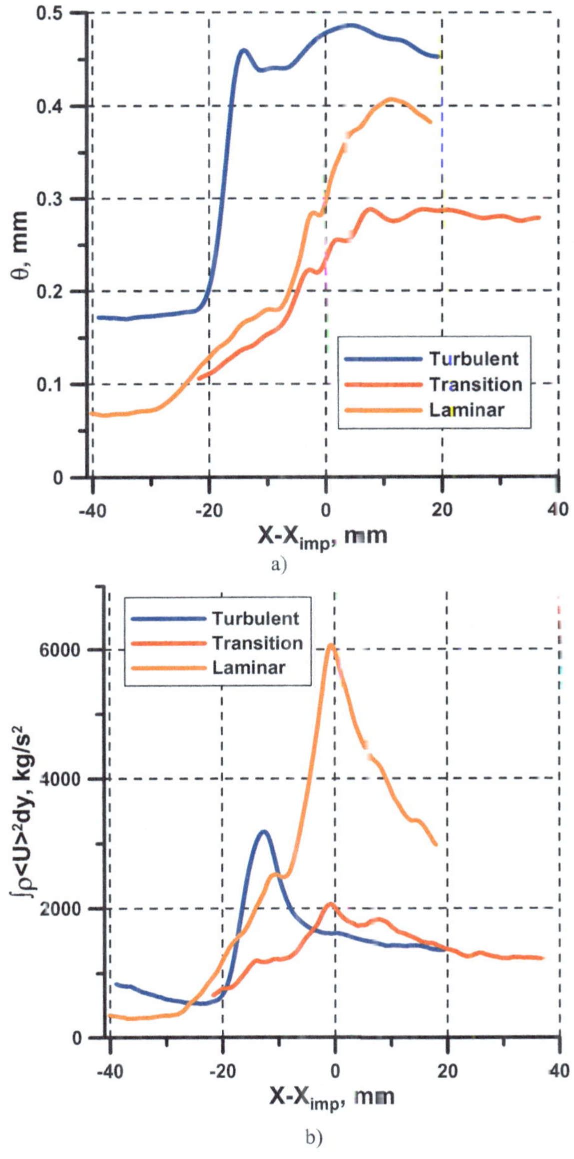

Fig. 65 **a** The momentum thickness distribution and **b** The integrated RMS of velocity pulsations along the SWBLI ($Re_1 = 13.2 \cdot 10^6$ m^{-1})

transition location by numerical tripping (turbulent viscosity value), by using the hybrid DDES—Delayed Detached Eddy simulation as well as the Organised Eddy Simulation approaches. These methods are also compared with the WM-LES (Wall-Model LES approach of Stanford, thanks to collaboration of IMFT with the group of Prof. P. Moin and our involvement in the CTR—Center of Turbulence Research programme of July-August 2014).

This study accounts on the efficiency of the DDES-OES in respect of other URANS approaches in capturing the SWBLI physics in comparison with the experimental data. To this end,

- integral quantities along the boundary layer and in the SWBLI are presented and compared with the TUD experiments
- grid sensitivity and influence of the upstream turbulence intensity
- skin-friction coefficient
- influence of fixed transition position at x/C = 33%.

The analysis of the shock wave-boundary layer interaction (SWBLI) is carried out in supersonic speeds regarding laminar wing technology of future aircraft design, in the context of the TFAST European project. Vision of H2020, whose objectives include the reduction of emissions and more effective transport systems, puts severe demands on aircraft velocity and weight. These facts require an increased load on wings and aero-engine components. The greening of air transport systems means a reduction of drag and losses, which can be obtained by keeping laminar boundary layers on external and internal airplane parts. Increased loads make supersonic flow velocities more prevalent and are inherently connected to the appearance of shock waves, which in turn may interact with a laminar boundary layer. Such an interaction can quickly cause flow separation, which is highly detrimental to aircraft performance and poses a threat to safety. In order to diminish the shock induced separation, the boundary layer transition upstream of the interaction should be optimized in respect of minimizing the skin-friction coefficient upstream and within the interaction.

Based on the natural flow developed, the laminar/turbulence transition can be imposed anywhere upstream of the SWBLI and the effects of various locations can be studied. In specific supersonic Mach number ranges, the boundary-layer structure within SWBLI can insipient separation accompanied by predominant unsteadiness and the influence of the transition location plays an important role for the design, concerning oblique shock interactions.

Simulations

Numerical method

The simulations have been performed with the Navier-Stokes Multi-Block (NSMB) solver. The NSMB solver is the fruit of a European consortium that included Airbus from the beginning of'90's, as well as main European aeronautics research Institutes, as KTH, EPFL, IMFT, ICUBE, CERFACS, Univ. of Karlsruhe, ETH- Zürich, among other. This consortium is coordinated by CFS Engineering in Lausanne, Switzerland. NSMB is a structured code including a variety of efficient high-order numerical schemes and turbulence modelling closures in the context of LES, URANS and RANS-LES hybrid turbulence modelling, especially DDES (Delayed Detached Eddy Simulation). NSMB highly evolved up to now and includes an ensemble of the most efficient CFD methods like URANS modelling for strongly detached flows, allowing for Detached Eddy Simulation for e.g. and the Delayed Detached Eddy Simulation

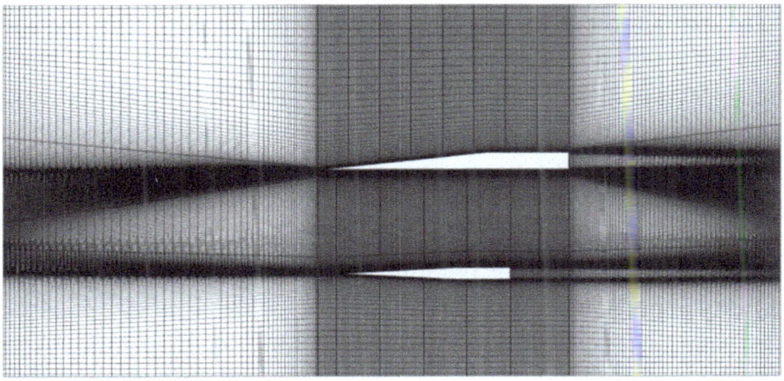

Fig. 66 Computational domain

(DDES), [30] as a hybrid method. NSMB solves the compressible Navier-Stokes equations using a finite volume formulation on Multi-Block structured grids.

In the two studies presented here, the time integration relies on a second-order backward Euler scheme based on the full matrix implicit LU-SGS (Lower-Upper Symmetric Gauss-Seidel) method and on the dual-time stepping, performing internal iterations, to reach convergence in each time step, which is 10^{-7} s.

Oblique Shock-Wave—Simulations Based on the TUD Configuration

The oblique shock wave test case has been studied in the conditions of the Delft experiment concerning the fully turbulent experimental case. The Reynolds number based on the flat plate length is ~4.08 million. The upstream isotropic turbulence intensity level is 0.56%.

For the 2D simulations, the structured mesh used has ~319000 cells. Due to convergence issues with two-equation models, this mesh has been refined around the trailing edge of the flat plate, giving a grid of ~25 M cells in 3D. The first grid has been extruded to a span length of 272 ~ mm, the span of the flat plate, giving a 3D mesh of ~31 million cells. The domain used for the computations is 1.5 height, 3.125 length (and 2.27 width in 3D) non-dimensioned by the length of the flat-plate (120 mm), and is also represented in Fig. 66.

The experimental conditions have been set up such that the effects of the upper and lower walls on the interaction region are limited. This helped the CFD calculations by placing the flat plate and the shock generator in free stream conditions, which define the external boundary conditions of domain. The geometrical elements are defined by adiabatic solid wall boundary conditions. Therefore, the inlet conditions are free-stream Dirichlet, where at the outlet boundary, characteristic velocity boundary conditions are considered. On the upper and lower boundaries, free-stream conditions are specified. In the spanwise direction, symmetry conditions have been used.

The AUSM spatial scheme has been first used but it displayed convergence insufficiencies. The Roe (1981) 3rd order scheme with van Leer (1978) MUSCL limiter has been used afterwards for all the computations. The diffusion terms have been discretized by means of central differencing. The time-step value was set after detailed tests to 0.5×10^{-7}.

Four turbulence models have been used for the 2D URANS computations: the one-equation Spalart-Allmaras [31], the *k*-ω—SST, Menter (1994) [32], the *k*-ε with Chien [33] low Reynolds number damping near the wall, as well as the *k*-ε-OES (Organised Eddy Simulation) model [34]. The use of the *k*-ω SST model displayed convergence issues on the coarse and finer grids. Therefore, the Spalart-Allmaras, the *k*-ε Chien and the *k*-ε-OES models have been finally used. The Spalart-Allmaras (SA) model has been used on the coarse grid. The *k*-ε-Chien (denoted hereafter as *k*- ε), as well as the *k*-ε-OES have been used on the finer grid. The results presented in this part will focus on these three models, by comparing the boundary-layer properties as well as the caption of the unsteadiness in the SWBLI.

Boundary-Layer Analysis

Figure 67 shows the integral parameters versus x in the boundary layer for the different turbulence models. The momentum thickness is over-predicted by the SA and *k*-ε models and underpredicted by the *k*-ε-OES, for which the displacement and momentum thicknesses are found closer to the experiments. The shape factor though is found in better agreement according to the two first models. The skin-friction coefficient displayed a decrease across the SWBLI region. Figure 68 shows the iso-pressure contours and probe—points for the spectral analysis.

Figure 69 shows the Power Spectral Density (PSD) at several selected positions. Point 1 is located in the beginning of the interaction. The models indicate formation of predominant frequency peaks around the frequency of order 5×10^4 Hz, being of the same order of magnitude as in experimental studies by Dupont, by means of TRPIV. In the following, the influence of a fixed transition position at 33 and 66% from the leading edge is studied by using the SA model (Fig. 70).

DDES and IDDES Simulations and Comparison with WM-LES

For a detailed view of this study, the reader can also refer to the edition of the Center for Turbulence Research—CTR, http://ctr.stanford.edu/publications.html.

WM-LES stands for Wall-Model-LES method of the group of Prof. Moin.

In Fig. 71, the mean stream-wise velocity profiles of the boundary-layer, around x_{sh}, are provided at eight different stream-wise locations and allow a more detailed comparison. The velocity profiles are normalized by the corresponding local free-stream velocities in the experiment at each location. In the DDES case, the mean stream-wise velocity is underestimated compared to the experiment, which can be understood as an overestimation of the development of the turbulence in the boundary

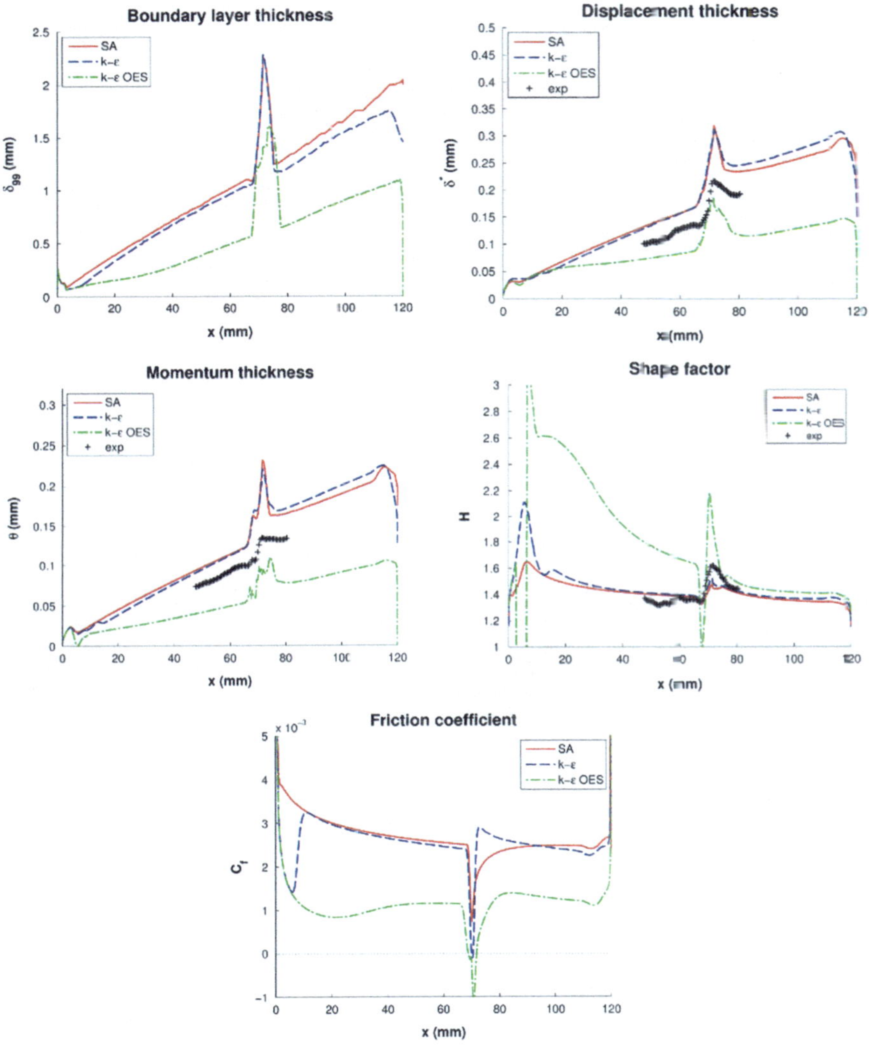

Fig. 67 Integral parameters in the boundary layer versus x in comparison with the TUD experiments

layer. The Spalart-Allmaras model induces a quasi-instantaneous laminar-turbulent transition from the leading edge in the RANS layer, while in the experiment, the transition is triggered in the zone of $x_{LE} = 5$—16 mm by the zig-zag tripping. The result of the transitional DDES matches better with the experiment by using the conditioning of the boundary layer, which delays its development to the turbulent state, until the flow approaches the interaction zone where the decrease in velocity observed in the experiment is underpredicted. WM-LES profiles matches well with the experiment, especially in the upstream and downstream directions of the SWBLI

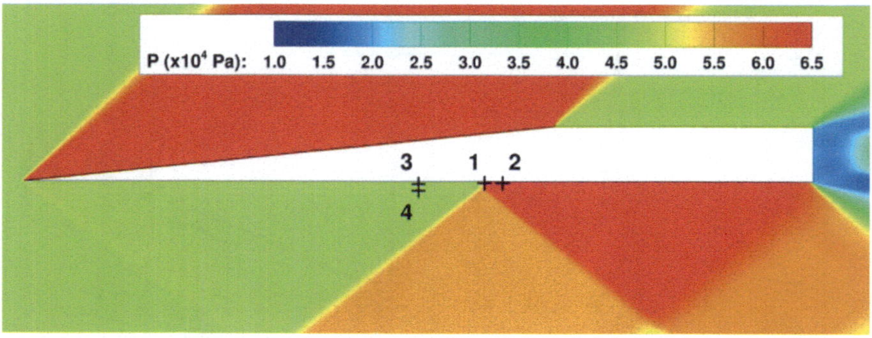

Fig. 68 Iso-pressure coefficients and probe (monitor) points locations

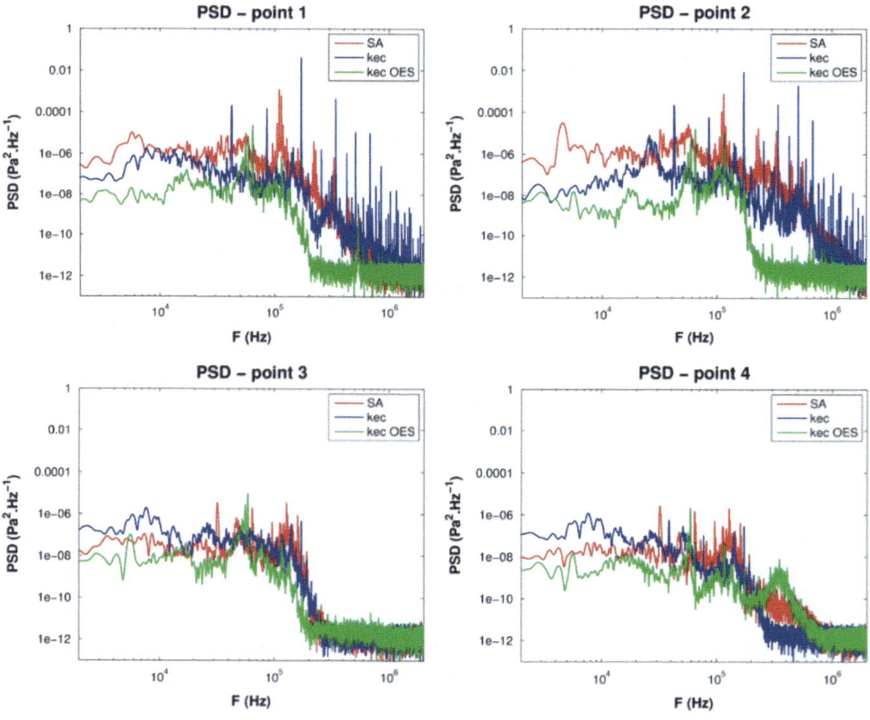

Fig. 69 Power Spectral Density PSD at selected monitor points as I, Fig. 68

zone. In the interaction zone $(x-x_{sh}) = 0.1$ and 4.8 mm), however, there are noticeable discrepançies from the experiment, similar to the transitional DDES. Since an equilibrium WM-LES formulation is used in this study, non-equilibrium effects such as strong pressure gradient and flow recirculation cannot be achieved in the wall model. Dawson [35] also observed poor predictions through interaction in their study of a

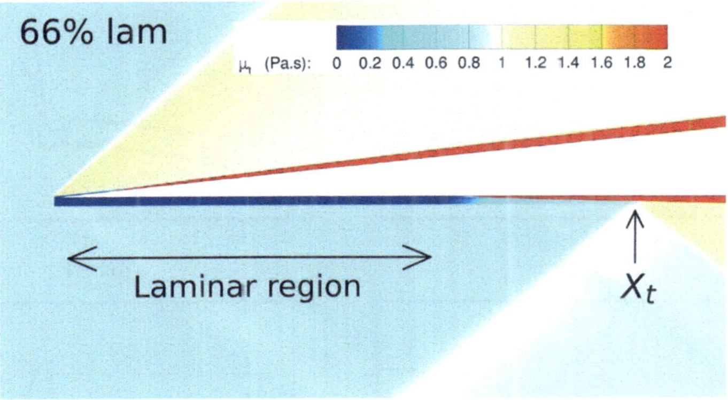

Fig. 70 Eddy-viscosity iso-contours illustrating the laminar-to-turbulent transition regions for a fixed transition position at 66%

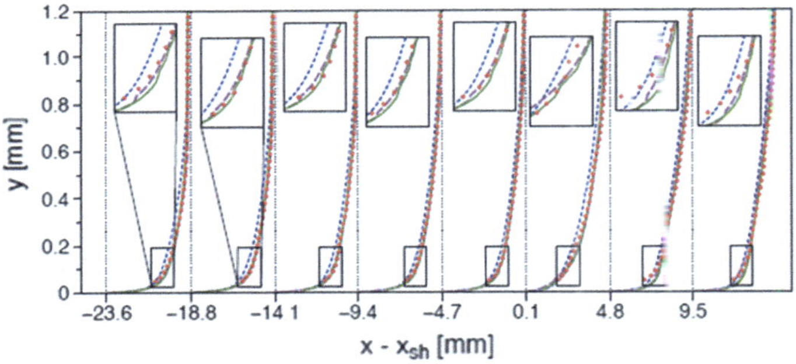

Fig. 71 Velocity profiles at 8 streamwise locations: normalized bu the experimental U∞ at each location. Blue line; DDES, violet line DDES with fixed transition at 33%. Green line: comparison with WM-LES commputations by the group of Prof. Moin, CTR 2014, Stanford. Dots: TUD experiment

supersonic compression ramp using a WM-LES. By investigating the magnitude of each term in a wall-resolved LES in the same configurations, they concluded that the convective and pressure gradient terms are dominant in near interaction zone. However, previous attempts to include dominant terms measured at the matching location (h_{wm}) in the equilibirum formulation such as that by Hickel [36] not only had difficulties in showing a satisfactory result but also suffered from numerical stability problems. As the flow goes downstream of the interaction and recovers equilibrium behavior, the WM-LES profiles is getting close to the experiment. Therefore, it may be necessary to solve the full non-equilibrium equations in the wall model. However, the accuracy of the PIV measurements in the SBLI region is reduced compared to that of the other regions of the boundary layer.

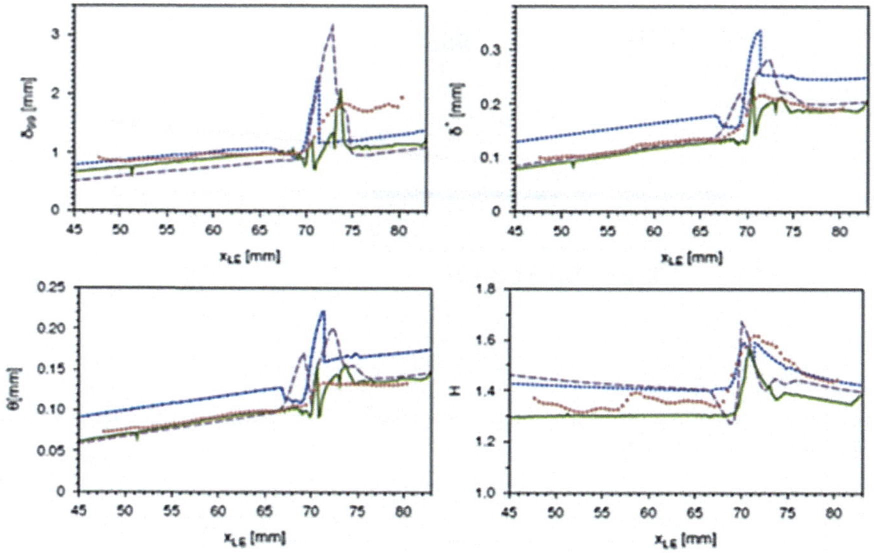

Fig. 72 Boundary layer thickness δ_{99} and displacement thickness δ^* (top). Momentum thickness, bottom. Line symbols as in previous figure

Figure 72 shows the distributions of boundary-layer thickness (δ_{99}), displacement thickness, momentum thickness and shape factor H, as a function of x_{LE}.

For $\delta\ \_{99}$, DDES and WM-LES match relatively well with the upstream of the SBLI, given the fact that in general 99 cannot be accurately defined for such complex flows. For and, however, the DDES slightly overestimates the integral values, which confirms the remarks of the previous paragraph: without any conditioning, the DDES generates an early development of the turbulent boundary layer compared to the experiment. This can be corrected by imposing the transition at $x_{LE} = 23$ mm, as explained above. In this case, the development of the boundary layer is delayed, as shown in all the graphs and the integral values downstream of the transition location get closer to the WM-LES and the experiment. In the interaction zone, none of the numerical methods can predict accurately the quantities.

In the downstream of the interaction, the WM-LES approaches the experimental values as well as the transitional DDES, as observed in Fig. 71. For the shape factor (H), both the transitional DDES and the WM-LES are reasonably close to the experimental value in x_{LE} < xsh. In the interaction zone, the transitional DDES and the WM-LES follows the general trend of the experiment but shows noticeable discrepancies from the experiment. In the downstream of the interaction zone, the transitional DDES shows a better agreement with the experiment. Interestingly, the DDES results are closer to the experiment for $x_{LE} \geq x_{sh}$ than for the other two calculations despite its poor predictions of the upstream flow for the other quantities

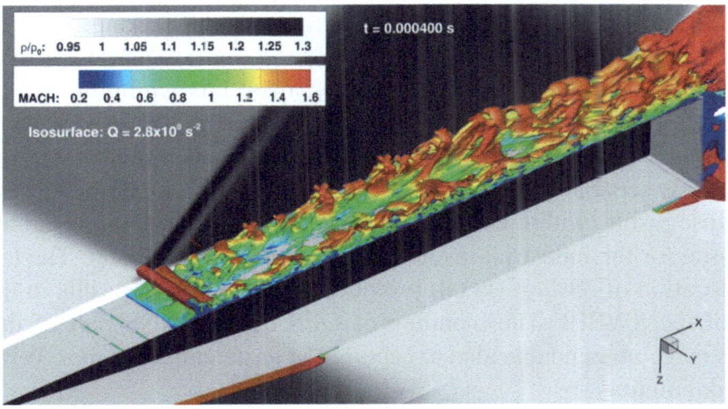

Fig. 73 Iso-Q criterion surfaces coloured by the Mach number showing the dynamics of the 3D vortex structures in the SWBLI and downstream, showing formation of 'hair-pin'vortices (grid of 13,5 M cells)

without conditioning. We briefly recall that the PIV measurements are less accurate in the SBLI region thanin the other regions of the boundary layer.

Figure 73 shows view of the 3D vortex structure dynamics in the SWBLI region and downstream of it, by illustrating formation of 'hair-pin' vortices, by means of Improved DDES (IDDES), Spalart [37] computations involving a specific wall model embedded LES in the near region. This model enhanced the turbulence intensity in all the flow field and produced higher amplitudes of *rms* than the experiments and the DDES results. These simulations allowed showing the 3D vortex structure within the interaction and past of it.

Conclusions

The present study analysed the SWBLI effect on the boundary layer subjected to supersonic (oblique shock) inlet Mach number conditions. A comparison of DDES and OES methods has been presented as well as in respect to experimental results is carried out concerning the supersonic interaction. A fair comparison with the TUD results is obtained. Comparison of DDES and of the WM-LES (Park, Moin, 2014), are used to predict the SWBLI in a Mach 1.7 flow. The flow is tripped very close to the leading edge in the experiment to insure a turbulent interaction, and both numerical approaches use different techniques to simulate the tripped fully turbulent boundary layer. While the results of the DDES modeling show an overestimation of the integral values of the boundary layer, the DDES with fixed transition at 33% and the WM-LES match well with the boundary-layer characteristics found in the experiment for the supersonic equilibrium flows. The results of standard DDES show an overestimation of the development of the boundary layer compared to the reference results. By using a preconditioning of the upstream boundary layer in an analogy with the WM-LES that used blowing and suction for the tripping in the experiment, a quite good behaviour is achieved

Within the SWBLI region, strong pressure gradient and complex flow features near the wall at the interaction cannot be represented in the numerical methods. The

WM-LES needs to incorporate non-equilibirum dynamics for strong non-equilibrium regions. A possible future approach is the non-equilibirum WM-LES formulation suggestedby Park & Moin (2014), which uses a full non-equilibrium formulation to calculate the transient wall shear stress and heat flux q_w. However, even the full non-equilibrium WM-LES formulation cannot guarantees a more exact prediction in some strongly separated flows. The DDES approach is quite promising to provide the most close results within the region of SWBLI by using more economic grids, a crucial issue for the industrial involvement in the TFAST project. The tripping at transition location of 33% of (x-xsl) provides also quite close results in the region upstream of the SWBLI. The blending of OES in the RANS part of the DDES approaches is recommendable also to capture the unsteadiness in the SWBLI.

Acknowledgements
The investigations presented in this section have been obtained within the European research project TFAST (Transition location effect on shock wave induced separation), coordinated by P. Doerffer, IMP-PAN, Gdansk Academy of Science, Poland. The computing CPU allocation of IMFT has been attributed by the French supercomputing Centres CINES in Montpellier), IDRIS (Paris) and CALMIP (Toulouse). The authors are grateful to Professor Moin for the invitation in the CTR—2014 of D. Szubert and M. Braza.

2.3.2 IUSTI

Numerical method

The shock-wave boundary layer interactions into consideration are both transitional and possibly prone to low-frequency unsteadiness. Direct Numerical Simulation is the modeling of choice when dealing with transitional flow. However the occurrence of low– to medium–frequency unsteadiness makes the computation of several dozen of periods of the lowest-frequency phenomenon mandatory in order to achieve statistical convergence. The computation cost of DNS appears consequently to be very high, preventing parametric study involving several computations to be carried out.

In that context, Large–Eddy Simulation was demonstrated to be an interesting modeling compromise between computation accuracy and statistical convergence when dealing with turbulent interactions (see [38, 39], among others). It is however well known that Large–Eddy Simulations are rather ill–suited to describe transition to turbulence in wall–bounded flow [40], with significant modification of the transition location when the grid is refined, unless to go up to quasi-DNS resolution. However it may be noted that mixing layer instabilities are less prone to such mispredictions than boundary layers instabilities because of their far higher amplification rates.

The strategy developed is to perform computations for various grid resolutions while adjusting the amplitude of some inflow perturbations in such a way that a similar separation length is achieved from one computation to another. The rationale behind this is that, for a given Reynolds number of the incoming boundary layer, the length of separation is mostly imposed by the location of the transition within the mixing layer developing over the bubble. The location of the transition is in turn governed by the non-linear saturation of the instable modes. Consequently, keeping constant the separation length should help ensuring that the transition process is not significantly altered when changing the grid resolution.

The various computations described hereafter are performed using ONERA's FLU3M solver that has been extensively used in the recent years to analyze successfully compressible flows either by LES and DNS [41, 42]. The numerical scheme is designed to be able to capture the shock while meeting the LES requirement of very low dissipation in the turbulent regions [43]. This is achieved by adding the dissipative part of the Roe scheme [44], modulated by Ducros' sensor [45], to a second order centered scheme. The subgrid filtering is implicitly provided by the mesh and the subgrid modeling relies on the selective mixed-scale subgrid model, well suited for compressible wall bounded flows [46].

Time integration is achieved by means of a second-order accurate implicit Gear scheme [47]. The timesteps of the various simulations have been selected in order to achieve maximum CFL numbers lower that 11, making the implicit time filtering negligible with respect to the implicit grid filtering. The resulting non-linear system is solved iteratively at every timestep with 7 sub-iterations.

Flow parameters selection and mesh design

Because of the difficulties encountered initially for the experiments, computations have been performed without knowledge of the definitive experimental setup. No measurements of the incoming boundary layer were therefore available to help setting the various parameters of the LES when the present numerical study has started. Consequently the incoming boundary layer has been assumed to match a Blasius profile for a Mach number equal to 1.63 and a stagnation temperature and a stagnation pressure equal respectively equal to $T_0 = 293.15\ K$ and $P_0 = 0.4\ atm = 50650\ Pa$. The boundary layer thickness at the inflow of the computational domain has then been evaluated by considering both the developing length and the location of the incident shock deriving from the initial experimental setup, resulting in Reynolds number equal to 1,400,000. A reference mesh with 26 M cells has been designed based on these information.

It has been used to perform parametric studies of the influence on the interaction region of either the shock angle or the inflow perturbation level. Inflow fluctuations are generated using a Synthetic Eddy Method with amplitudes one to two orders of magnitude lower than the ones retained for fully turbulent flows. Values of U_{rms} respectively equal to 0.125, 0.25, 0.5 and 1.0% of U_∞ have been tested for a shock generated by a 4.5° deviation of the flow, yielding interaction ranging from transitional separation bubbles to fully turbulent, attached interactions. Intermediate values of 0.25 and 0.5%, corresponding respectively to locations of the transition in the first half of the separation bubble and slightly upstream of attached interaction region, have been retained for the analysis of the influence of the shock strength. Three deviation angles ranging from 3.5° to 4.5° have been tested. Similar trends have been found for each angle, as seen in Fig. 74. Eventually, the 4.5° deviation angle has been

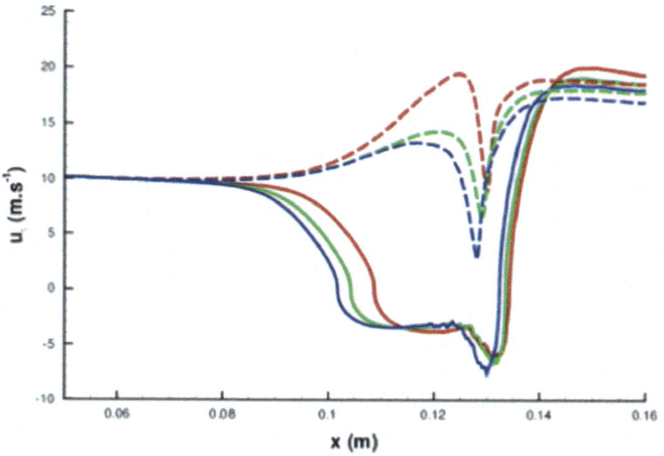

a) Friction velocity

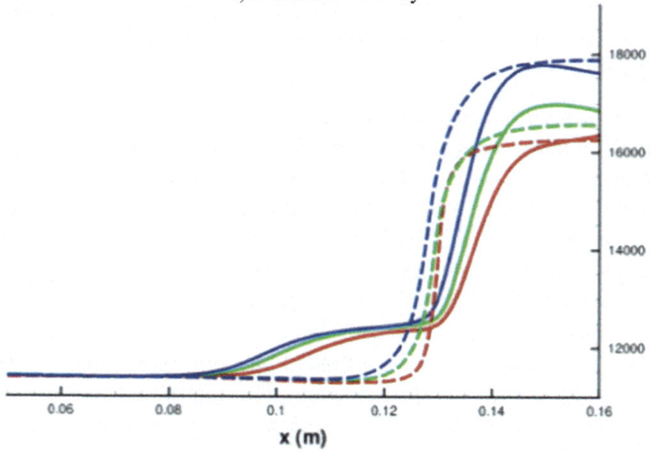

b) Wall pressure

Fig. 74 Influence of the shock deviation angle on the interaction region: 3.5° (red), 4.0° (green), 4.5° (blue). Solid and dashed lines correspond to inflow fluctuation levels of 0.25% and 0.5%, respectively

retained since it results in a separation length of about 30 mm, a dimension that was compatible with was could be inferred from the preliminary experimental set-up.

A refined mesh has then been designed in order to quantify the dependence of the results upon the grid. Cell counts were increased by 40% in the streamwise and spanwise direction and by 20% in the wall-normal direction. Because of the high sensitivity of the transition process to the grid resolution in LES, the inflow perturbation level has been adjusted in order to obtain the same separation length as for the computation from the reference mesh. It allows for comparisons between computations free from transition modelling considerations. However the fluctuation level had to be dropped by a factor of five to achieve a similar separation length when moving from the reference mesh to the refined one, possibly denoting a change in the numerically-driven physics of the transition process. This point has been addressed by locally increasing the streamwise resolution in the interaction region similarly for the reference and the refined case. It results in an increase of the cell count by about 10% whereas the inflow fluctuation level required for these new meshes to obtain the desired separation length differ by less than 35%, (0.082% of the external velocity for the reference mesh versus 0.061% for the refined mesh). The shape of the separation bubble and the resulting pressure rise obtained from the two meshes are in good concordance, as seen in Fig. 75.

An additional, third mesh has also been derived from the locally-refined reference one by doubling the span of the domain. These three meshes will be hereafter referred to as *reference*, *extended* and *refined* and includes 28 M, 56 M and 67 M cells, respectively.

CFD results

These experimental results are complemented by data coming from the LES performed using the flow parameters and mesh described in previous sectio for a simulated time of 20 *ms*. Such a duration makes it possible to encompass at least 10 periods of the low frequency oscillations, thus allowing spectral analyses of the low-frequency dynamics of the flow with acceptable statistical uncertainties

The streamwise evolution of the premultiplied wall pressure power spectra obtained from the three meshes is plotted in Fig. 76. Note that for each streamwise location the power spectra is normalized by the local value of the pressure variance in order to highlight the *relative* energy contribution of a given frequency range. The three computations results in a similar space–frequency distribution which bears similarities with the experimental power spectra plotted in Fig. 34. The region in the vicinity of the separation point is energetically dominated by the low-frequency content while the energy contained in higher frequency band close to 10 kHz prevails when moving up to the shock impingement location. One can nonetheless note that the typical low frequencies, when normalized using the interaction length and the external velocity, have two– to three–time lower values in the LES than in the experiments.

Influence of the inflow perturbations on the interaction region

Three additional computations have been carried out for each of the three meshes defined in previous section. They correspond to inflow fluctuation levels respectively multiplied by $\alpha_{u'}$ *inflow* $= 3, 5$ *and* 7 with respect to the LES previously described. The statistics have been gathered over 6.25 *ms*, a value large enough for convergence since all these cases yield attached interactions and do not exhibit low-frequency unsteadiness. For all computations the largest cell dimensions in wall unit are encountered in the fully turbulent region downstream of the interaction. Computations based the fine mesh fully fulfill the recommended criteria for LES of turbulent wall bounded flow while for the standard and enlarged meshes values in the wall–normal and streamwise direction are slightly above recommendations in that region ($\Delta y^+_{wall} \simeq 1.6$, $\Delta Z^+ \simeq 22$).

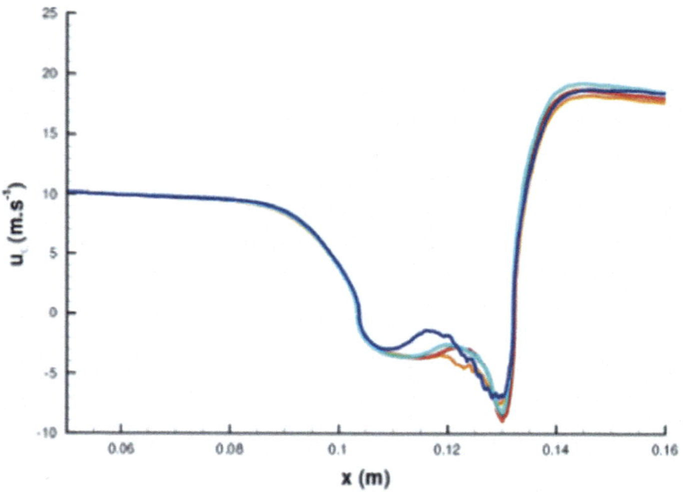

a) Friction

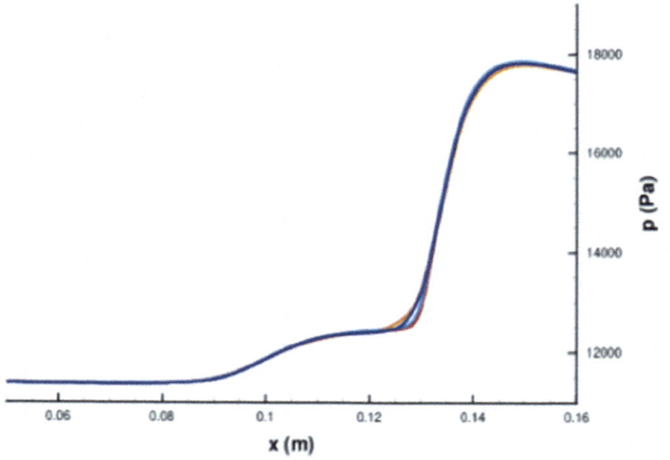

b) Wall pressure

Fig. 75 Influence of the computational grid on the interaction region: initial reference mesh (orange), reference mesh with local streamwise refinement (red), initial refined mesh (cyan) and refined mesh with local streamwise refinement (blue)

Since the scaling on the inflow velocity fluctuation was performed in order to achieve the same separation length for the three meshes, it does not necessarily result in similar locations of the transition for all the meshes when increasing the inflow perturbation level to trigger more upstream transitions. It can however be verified from the streamwise evolution of the compressible shape factor plotted in Fig. 77 that multiplying the reference perturbation level by a value $\alpha_{u'}$ *inflow* common to all three meshes indeed results in very similar transition locations. Moreover, the

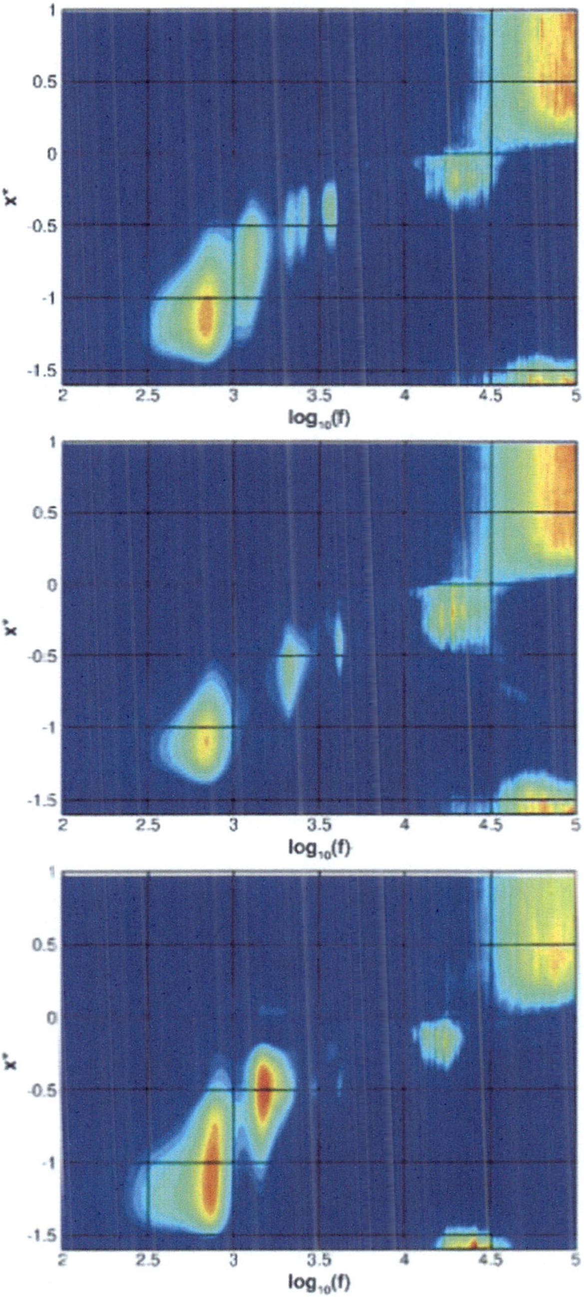

Fig. 76 Streamwise evolution of the pre-multiplied power spectrum of the wall pressure, normalized by the local variance for the reference mesh (left), the extended mesh (middle) and the refined mesh (right)

velocity profiles in wall unit sampled 1 *cm* upstream of the shock impingement, found in the remaining parts of Fig. 77, are in good concordance from one mesh to the other for the three values of $\alpha_{u'inflow}$ that have been tested. The most noticeable differences occur for the two fully turbulent cases above the buffer layer and are due to an underestimation by about 5% of the skin friction coefficient with respect to the fine mesh computations.

The nature of the incoming boundary layer can also be ensured by looking at the streamwise evolution of the skin friction velocity plotted in Fig. 78. These plots

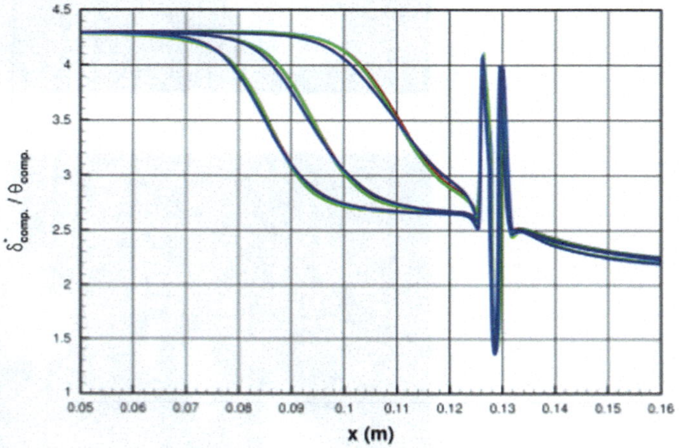

a) Compressible shape factor

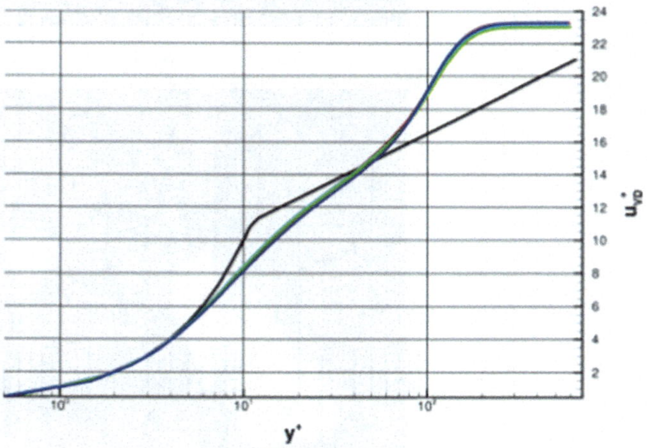

b) Profile for x=0.12 *m*.,

Fig. 77 Assessment of the numerical transition process: reference mesh (red), extended mesh (green) and refined mesh (blue). Results for $\alpha_{u'inflow} = 3$, 5 and 7 have been superimposed

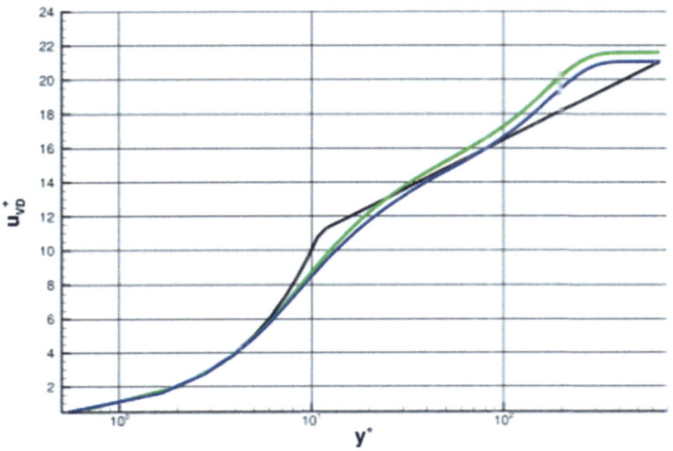

c) Profile for x=0.12

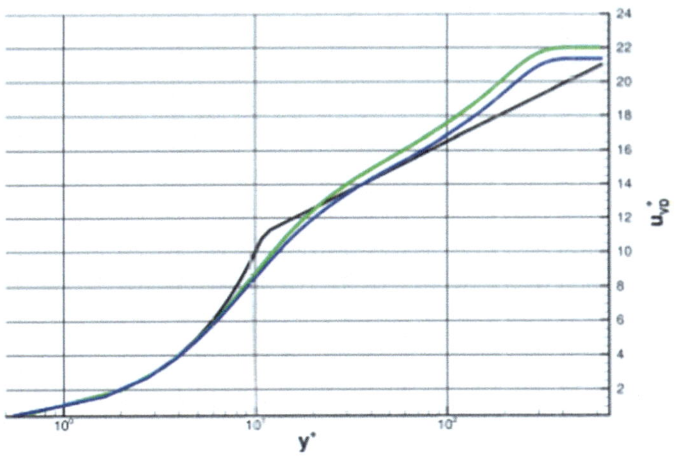

d) Profile for x=0.12 $m.$,

Fig. 77 (continued)

demonstrate that the transitional incoming boundary layer is as capable as the turbulent ones of inducing a reattachment of the flow in the interaction region. This point is confirmed by an additional computation performed using the reference mesh with $\alpha_{u'}$ *inflow* $= 2$, not shown here, which results in an early transitional stage for the incoming boundary layer (the skin friction is increased by 10% with respect to a laminar BL) preventing the flow separation.

It is however interesting to note that in the relaxation region downstream of the interaction, transitional cases result in value of the friction velocity higher by 5 to 10% to the ones obtained for the turbulent cases.

2.3.3 SOTON

Transition Location Effect on SWBLI

The effect of the transition location on the structure of interaction between an oblique shock wave and a boundary-layer at M = 1.5 is investigated. Three different types of interaction, denoted as laminar, transitional and turbulent based on the state of the boundary-layer just upstream of the interaction, have been considered. A naturally

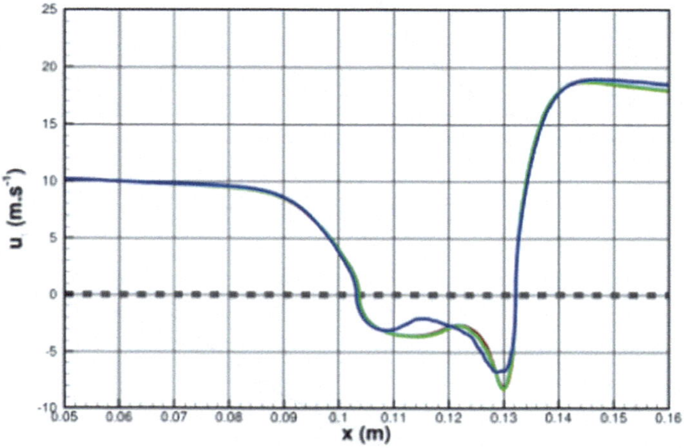

a) $\alpha_{u'inflow} = 1$

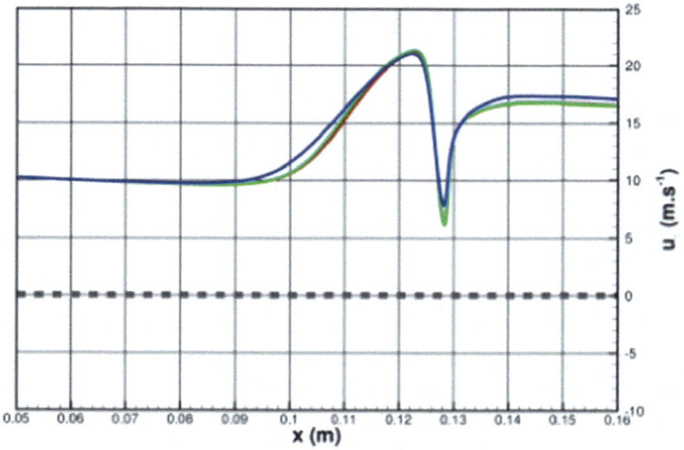

b) $\alpha_{u'inflow} = 3$

Fig. 78 Streamwise evolution of the friction velocity for various perturbation levels at the inflow: reference mesh (red), extended mesh (green) and refined mesh (blue)

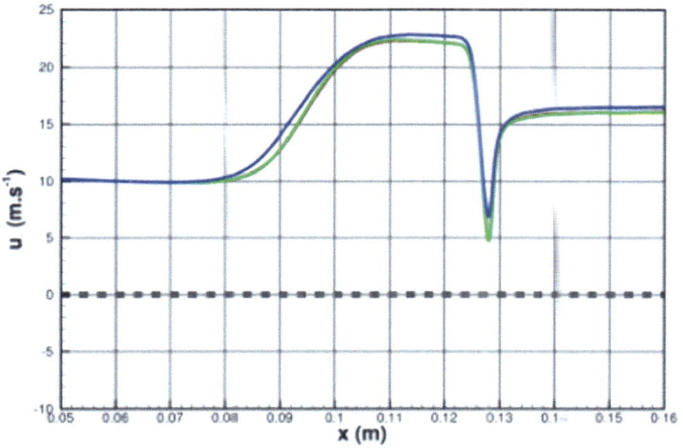

c) $\alpha_{u'\text{inflow}} = 5$

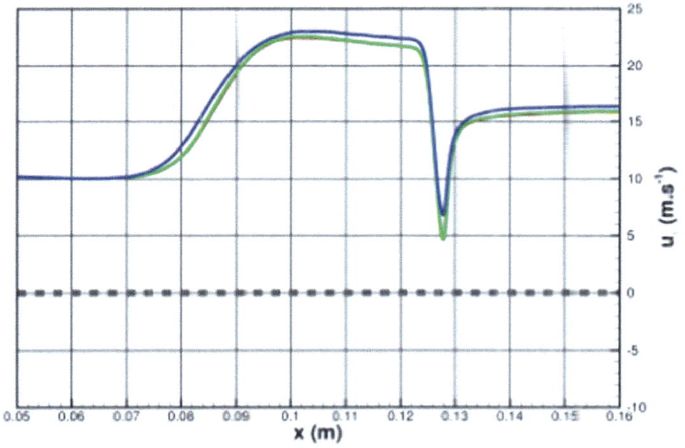

d) $\alpha_{u'\text{inflow}} = 7$

Fig. 78 (continued)

transitional boundary-layer can usefully be divided in three main regions (laminar, transitional and turbulent) and in this context a classification of the interaction can be made based on the state of the boundary-layer at the impingement location of the oblique shock wave. The selected laminar, transitional and turbulent interaction cases are based on the set of experiments run at the French aerospace laboratory (Office National d' Etudes et de Recherches A´erospatiales, ONERA), as part of the TFAST project, and modelled numerically by means of DNS. A modal forcing

Table 11 ONERA experimental setup

Exp. Facility	M	T_0 (K)	x_imp (mm)	Reximp × 10^6	P_0 (bar)	θ (deg)
ONERA	1.6	290–310	40–80	0.6–1.1	0.95–1.01	1.5–4

technique is used to promote transition in the boundary-layer and the effects of shock impingement location, shock strength and Reynolds number are investigated.

Experimental Setup

The flow conditions set by the TFAST project in the context of oblique shock reflection on a flat plate consider Mach number between 1.4 – 1.7, shock impingement Reynolds number Reximp = (0.4 – 3) × 10^6 and shock generator plate angles of 1 – 4 deg. The inflow conditions can slightly vary depending on the experimental facility considered. Attention is focused here only on the experiments of ONERA, whose Mach number M, stagnation temperature T_0, stagnation pressure P_0, impingement location x_imp, Reynolds number based on the impingement location Re_ximp and shock generator plate angle θ are reported in Table 11. The intention is to carry out stand-alone DNS calculations in a computationally affordable Reynolds number range, based on the ONERA experiments.

Numerical Setup

The most relevant aspects of the numerical setup are described here. Details on the nature of the forcing, domain size and grid resolution are presented in the following sections.

Inflow Conditions

The inflow conditions are set on the experiments conducted by ONERA. Although the measured Mach number in the experiments is either M = 1.6, for consistency with the previous set of simulations the Mach number is M = 1.5. For the considered shock strengths, the difference between the numerical and experimental Mach numbers in terms of pressure ratio p3/p1 (downstream of the reflected shock to upstream of the incident shock) is about 1%, therefore no significant effects on the interaction region are expected. The numerical inflow is placed at x_0 = 0.0518 m downstream of the flat plate leading edge where the displacement thickness is δ1,0 = 1.84 × 10^(−4) m. The unit Reynolds number is fixed to Re_1 = 10.7 × 10^6 m^ – 1 and the corresponding Reynolds number based on the displacement thickness at the inlet is Re_δ1,0 = 1971.07. Inflow profiles are given by a similarity solution using the Illingworth transformation. The angle of the shock generator is θ = 4 deg and the oblique shock is introduced by the Rankine-Hugoniot jump relations at the top boundary. For all the selected cases, the free-stream temperature is T∞ = 197.93 K and a Sutherland's law is used to describe the variation of viscosity μ with temperature (Sutherland's constant temperature TS = 110.4 K). The integration time step is Δt = 0.015. All the simulations are run until statistical convergence.

Domain Size, Grid Resolution and Boundary Conditions
While the domain size is kept fixed in the wall-normal (high enough to avoid the reflection of the wave system from the top boundary to impinge onto the boundary-layer) and spanwise (equal to one spanwise wavelength of the most unstable mode and is $Lz = 2\pi/\beta$) directions, the streamwise extent varies depending on the interaction type. The numerical inflow is kept fixed for all the cases studied, whereas the outflow is moved further downstream in order to let the boundary-layer become either transitional or turbulent. When no shock is introduced (ZPG case at $\theta = 0$), the numerical domain corresponds to the one for the turbulent interaction. For each case, the grid distribution in the wall-normal direction is stretched and clusters about 30% of the grid points within the boundary-layer at the inlet. The grid is stretched in the streamwise direction following a 10th-order polynomial distribution whose derivatives are continuous up to the 4th-order. The spatial step size Δx continuously decreases from the inflow up to either the shock impingement or the transition location (whichever comes first), after which uniform grids are used. The grid resolution in the streamwise direction changes depending on the interaction type. While for all the DNS cases, the gridresolution in the transitional/turbulent region is $\Delta x - = 4.8$, $\Delta z + = 4.8$ and $\Delta y + \text{wall} = 0.96$. The boundary conditions applied to the computational domain are no-slip and fixed temperature conditions (with temperature equal to the laminar adiabatic wall temperature) at the wall and time-dependent fixed inlet (where the modal forcing is applied). To minimise the reflection of waves into the domain, an integral characteristic method is applied to the top boundary and a standard characteristic boundary condition at the outflow.

Broadband Modal Forcing Technique
The modal forcing represents a very effective way to excite the unstable modes of the boundary-layer and eventually trigger transition. In order to mimic the broadband disturbances in the wind tunnel, the modal forcing is used by selecting a large number of stable and unstable modes. The transition location influences the size of the separation, character of the interaction and, consequently, the boundary-layer instabilities. Since the transition point is not known a priori, it is more convenient to select the boundary-layer unstable modes at the inlet of the numerical domain for the generation of the broadband modal disturbances. For all simulations, 42 eigenmodes are calculated for every combinations of $\omega = 0.02{:}\ 0.02{:}\ 0.12$ and $\beta = -0.6{:}\ 0.2{:}\ 0.6$. Since pairs of oblique modes are selected to construct the broadband modal disturbances, random phases are added to each mode in order to avoid any symmetry of the breakdown.

ZPG Boundary-Layer
The classification adopted for the definition of the type of interaction (laminar, transitional and turbulent) assumes a priori knowledge of the transition location. A ZPG boundary-layer is therefore forced with the previously described broadband modal forcing at an amplitude $Ao = 0.05$ (corresponding to a turbulence intensity at the inflow $\rho rms = 0.25\%$), for which transition is obtained approximately halfway down of the numerical domain. The time- and span-averaged skin friction distribution (a), contours of the time-averaged skin friction (b) and instantaneous streamwise velocity at the centreplane (c) reported in Fig. 79 show that transition starts at $Rex = 7 \times 10^5$

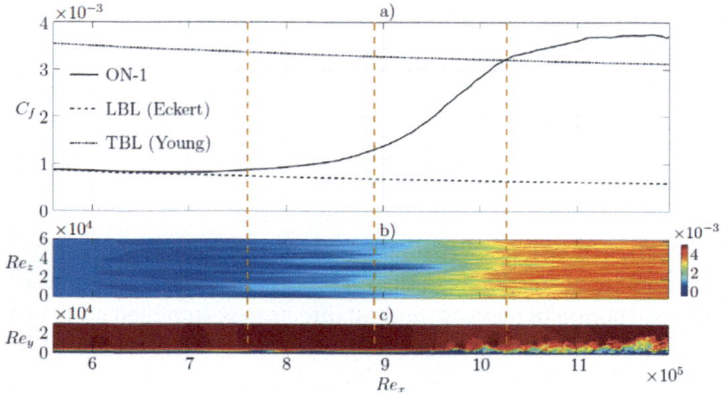

Fig. 79 **a** Time- and span-averaged skin friction distribution for the ZPG boundary-layer (black solid line) along with laminar (dashed-dotted black) and turbulent (dashed black) boundary-layer distributions by Eckert (1955) and Young (1989), respectively; contours of time-averaged skin friction (**b**) and instantaneous streamwise velocity (**c**). The vertical orange dashed lines indicate the shock impingement locations

and the turbulent state is reached at Rex $= 11 \times 10^5$. As seen for bypass transition in incompressible applications (Coupland, 1990, Bhushan and Walters, 2014), an overshoot of the skin friction with respect to theoretical distribution by Young (1989) can be seen at Rex $= 1.02 \times 10^6$ due to the high intensity structures formed during the breakdown to turbulence. The vertical orange dashed lines indicate the shock impingement locations for laminar, transitional and turbulent interactions, respectively. The laminar and turbulent interactions are, in a mean sense, at the boundaries of the transitional region. However, the main transition scenario is a bypass breakdown where intermittent turbulent spots are generated and move the transition point in the streamwise direction. Figure 80 shows the contours of wall-normal vorticity in the vicinity of the wall for three different time levels: t = 7,450 (a), t = 7,900 (b) and

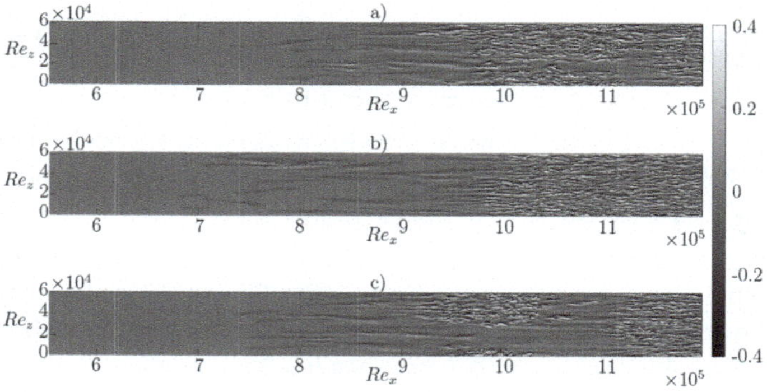

Fig. 80 Contours of wall-normal vorticity for the ZPG boundary-layer at time levels t = 7; 450 (**a**), t = 7; 900 (**b**) and t = 8; 350 (**c**)

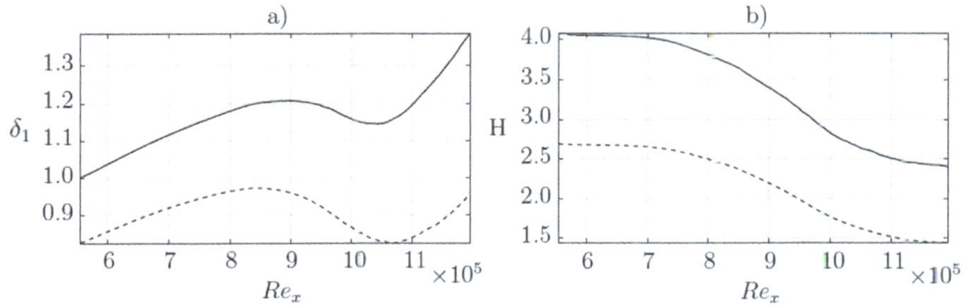

Fig. 81 Time- and span-averaged displacement thickness (**a**) and shape factor (**b**) distributions calculated with the incompressible (dashed lines) and compressible (solid lines) formulations

t = 8,350 (c). The formation of a turbulent spot is visible in Fig. 80c between Rex = 9 × 10^5 and Rex = 10.5 × 10^5. The appearance of the turbulent spots is periodic and localised at around Rez = 0.1 × 10^5 and Rez = 0.5 × 10^5. These preferred locations are due to the choice of the random phases applied to the forcing to avoid a symmetric breakdown. Although random, these phases are fixed throughout the simulation.

Displacement Thickness and Shape Factor Calculations
The state of the boundary-layer can also be examined by calculating the time- and span-averaged displacement thickness and shape factor H (displacement thickness over momentum thickness ratio), as reported in Figs. 81a, b respectively. The kinematic formulation to calculate the displacement thickness and shape factor (dashed lines) is compared with the compressible one (solid lines). When the variable density is not taken into account the shape factor follows the theoretical distributions for incompressible boundary-layers and is about 2.7 near the inlet and 1.5 towards the outlet, in accordance with the laminar and turbulent state. The contribution of compressibility is to shift δ1 and H distributions towards higher values. With reference to the summary of Gatski and Bonnet (2013) on the effects of compressibility on TBLs H-factors, the value obtained in the turbulent region is reasonable for M = 1.5.

Intermittency Calculation
For this bypass-like transition scenario, it can be interesting to measure the intermittency, Γ, to define the state of the boundary-layer. The intermittency is defined as the fraction of time during which the boundary-layer is locally turbulent and typically involves setting arbitrary thresholds on particular measured data. An alternative measure of the intermittency is proposed to avoid problems related to the mean skin friction distribution. Based on the criteria proposed by Volino [48], first time derivatives of the skin friction time series are selected to define the state of the boundary-layer. Low and high fluctuation levels of the time derivatives are related to laminar and turbulent states, respectively. False laminar states, when the first time derivatives of the skin friction cross zero, can be avoided by calculating the second time derivatives of the skin friction. The intermittency is then calculated as the fraction of time during which the first or second time derivatives of the skin friction are

higher than specified thresholds for both ∂Cf/∂t and ∂2Cf/∂t2, respectively. To avoid the need to fix two thresholds, one can iteratively adjust on threshold in order to have the same time average for Γ f and Γ s [48]. The obtained intermittency could then be low-pass filtered in order to smooth the distribution and to prevent false turbulent points in laminar regions or false laminar points in turbulent regions to be captured. However, in the present study no significant differences were found and the filtering is not performed. By visual inspection of the data, the threshold on the first time derivatives is set to be $\Gamma = 6 \times 10^{-5}$. The sensitivity to the selected threshold is studied for $5 \times 10^{-5} < \Gamma < 7 \times 10^{-5}$ and the resulting span averaged intermittency distribution is reported in Fig. 82. The intermittency distribution resulting from the more commonly used criterion proposed by Schneider [49], for which a p.d.f. of the skin friction distribution needs to be calculated (definition A, black solid line with error bars) is compared to the one calculated with the current method (definition B, red solid line with error bars), providing a good agreement and confirming definition B as suitable for both ZPG and SWBLI cases.

SWBLI: Laminar, Transitional and Turbulent Interactions
Laminar, transitional and turbulent interactions are here examined. Time- and span-averaged skin friction distributions are reported in Fig. 83 for the laminar (red solid line), transitional (blue solid line) and turbulent (green solid line) cases. The ZPG boundary-layer solution is also plotted (black solid line) along with the shock impingement locations (vertical orange dashed lines). A first main observation is that when a laminar interaction occurs the boundary-layer separates, whileit remains attached for the turbulent interaction. For the transition case, a marginal separation occurs. As part of the TFAST project, Giepman et al. (2015) report a similar situation for laminar, transitional and turbulent SWBLIs at M = 1.7. While theseparated region is large for the laminar interaction, for the transitional case the zone of reversed flow is strongly reduced and no mean-flow separation is present for the turbulent interaction. The size of the interaction, for example starting where the skin friction deviates from the ZPG boundary-layer distribution and ending with the steep increase downstream

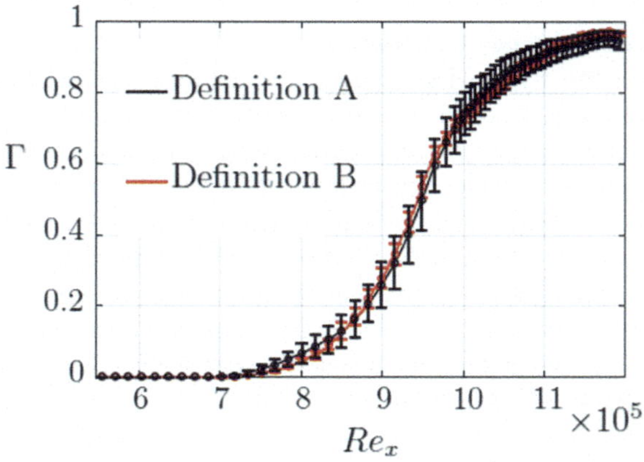

Fig. 82 Sensitivity to the denition for the intermittency calculation (denition A—black solid line, denition B—red solid line)

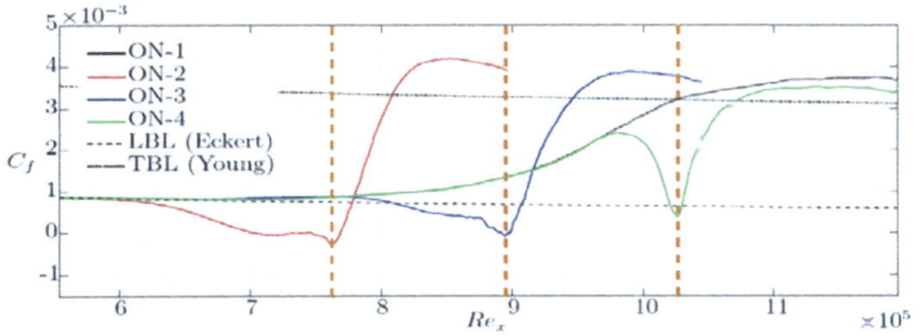

Fig. 83 Time- and span-averaged skin friction distributions for the laminar (red solid line—case ON-2), transitional (blue solid line—case ON-3) and turbulent (green solid line—case ON-4) interactions. The ZPG boundary-layer (black solid line—case ON-1) is also plotted along with laminar (dashed-dotted black) and turbulent (dashed black) boundary-layer distributions by Eckert (1955) and Young (1989), respectively. The vertical orange dashed lines indicate the shock impingement locations

of the impingement, decreases significantly with increasing Reximp. Although only marginally separated, the transitional interaction still presents a relatively large interaction size, which is very narrow in the turbulent case. Another important point, in qualitative agreement with the experiments of Giepman et al. (2015), is that a TBL is detected downstream of the impingement location for the laminar interaction and that transition is accelerated for the transitional and turbulent ones. The shock considered ($\theta = 4°$) is very strong and the boundary-layer becomes turbulent at the shock impingement location. Similarly to what happens to pressure and peak heat transfer distributions in hypersonic applications [50], the skin friction downstream of the impingement location overshoots the fully turbulent skin friction levels of the ZPG case. This is due to the enhanced energy transfer mechanism that precedes the turbulent state downstream of the impingement location. It is also interesting to notice that this overshoot decreases for increasing impingement location Reynolds number.

Displacement Thickness and Shape Factor Calculations

Compressible formulations of the displacement thickness (a) and shape factor (b) are reported in Fig. 84 for the laminar (red solid line), transitional (blue solid line) and turbulent (green solid line) interactions, along with the ZPG boundary-layer solution (black solid line). The APG introduced by the shock causes the appearance of a peak in the boundary-layer displacement thickness and shape factor at the impingement location. Similarly to the interaction size, the peaks get narrower for increasing impingement Reynolds number, allowing these quantities to be used to quantify the interaction size. Differently from the width, the peak value is very sensitive to the way the edge of the boundary-layer is chosen for the calculations of both displacement and momentum thickness. Here, the edge of the boundary-layer is identified by specifying a vorticity threshold, which represents a consistent choice everywhere but at the apex of the separation bubble. The shape factor confirms that the boundary-layer becomes turbulent downstream of the impingement but at a lower level with respect to the ZPG

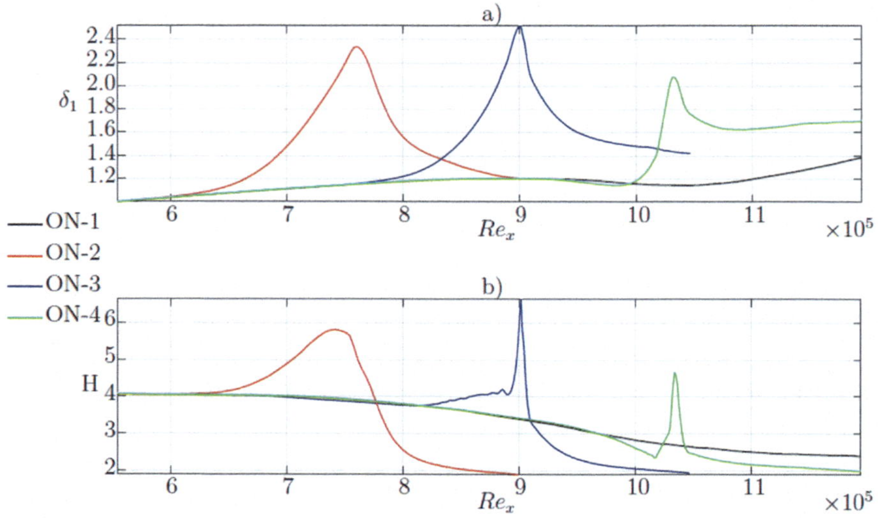

Fig. 84 Compressible displacement thickness (**a**) and shape factor (**b**) distributions for the laminar (red solid line—case ON-2), transitional (blue solid line—case ON-3) and turbulent (green solid line—case ON-4) interactions. The ZPG boundary-layer (black solid line—case ON-1) is also plotted

case, due to the TBL downstream of the impingement that is always significantly thicker than the case without shock.

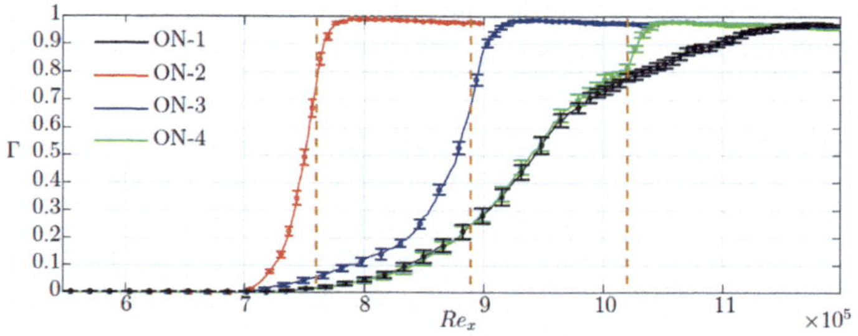

Fig. 85 DNS span averaged intermittency distributions (with error bars due to the threshold sensitivity) for the laminar (red solid line—case ON-2), transitional (blue solid line—case ON-3) and turbulent (green solid line—case ON-4) interactions. The ZPG boundary-layer (black solid line—case ON-1) is also plotted. The vertical orange dashed lines indicate the shock impinge-ment locations

Intermittency Calculation

Definition B is used to calculate the intermittency distributions for the laminar (red solid line), transitional (blue solid line) and turbulent (green solid line) in Fig. 85, where the sensitivity of the distributions to the threshold ($5 \times 10^{-5} < f < 7 \times 10^{-5}$) is indicated by the error bars. Similarly to what was observed from the skin friction and shape factor distributions, the intermittency distributions show a sharp increase towards unity downstream of the impingement location due to the turbulent character of the boundary-layer.

Upstream of the impingement, the effect of the shock-wave differs significantly depending on the interaction type. For the laminar interaction, the boundary-layer quickly switches from being laminar to turbulent when passing across the shock and this causes a very localised change of the intermittency. The transitional interaction affects the boundary-layer for a longer upstream extent and the increase of intermittency is more gradual. Although the boundary-layer is marginally separated, the presence of the interaction changes the instability of the boundary-layer and the Γ distribution deviates from the ZPG distribution. This also happens for the turbulent case but with less significant consequences since the size of the interaction is much smaller. The upstream response of the boundary-layer to the turbulent interaction is almost negligible and the major effects are only visible downstream of the impingement.

References

1. T.S.C. Davidson, H. Babinsky, An investigation of interactions between normal shocks and transitional boundary layers—control ID 1889354. *44th AIAA Fluid Dyn. Conf.*, no. June, pp. 1–16 (2014)
2. S.P. Colliss, Vortical structures on three-dimensional shock control bumps (2014)
3. N. Titchener, S. Colliss, H. Babinsky, On the calculation of boundary-layer parameters from discrete data. Exp. Fluids **56**(8), 1–18 (2015)
4. C. Sun, M.E. Childsf, A Modified Wall Wake Velocity Profile for Turbulent Compressible Boundary Layers, no. June, pp. 7–9 (1973)
5. T.B. Layer, Explicit Expression for the Smooth Wall Velocity Distribution in a Turbulent Boundary Layer, no. June, pp. 655–657 (1979)
6. T.S.C. Davidson, H. Babinsky, Influence of Boundary-Layer State on Development Downstream of Normal Shock Interactions, pp. 10–12
7. L.U. Schrader, L. Brandt, C. Mavriplis, D.S. Henningson, Receptivity to free-stream vorticity of flow past a flat plate with elliptic leading edge. J. Fluid Mech. **653**, 245–271 (2010)
8. J. Ackeret, F. Feldmann, N. Rott, Investigations of compression shocks and boundary layers in gases moving at high speed, *NACA Rep.*, no. 1113 (1947)
9. J.W. Kooi, Experiment on transonic shock wave boundary layer interaction, *AGARD Flow Sep.* (1975)
10. W.G. Sawyer, C.J. Long, A study of normal shock-wave turbulent boundary-layer interactions at mach numbers of 1.3, 1.4 and 1.5, *Tech. Rep. 82099, NASA* (1982)
11. S. Pirozzoli, P. Orlandi, M. Bernardini, The fluid dynamics of rolling wheels at low Reynolds number. J. Fluid Mech. **706**(July), 496–533 (2012)
12. S. Pirozzoli, M. Bernardini, F. Grasso, Direct numerical simulation of transonic shock/boundary layer interaction under conditions of incipient separation. J. Fluid Mech. **657**, 361–393 (2010)

13. S. Pirozzoli, Generalized conservative approximations of split convective derivative operators. J. Comput. Phys. **229**(19), 7180–7190 (2010)
14. M. Bernardini, S. Pirozzoli, A general strategy for the optimization of Runge-Kutta schemes for wave propagation phenomena. J. Comput. Phys. **228**(11), 4182–4199 (2009)
15. E.A. Fadlun, R. Verzicco, P. Orlandi, J. Mohd-Yusof, Combined immersed-boundary finite-difference methods for three-dimensional complex flow simulations. J. Comput. Phys. **161**(1), 35–60 (2000)
16. M.D. de Tullio, P. De Palma, G. Iaccarino, G. Pascazio, M. Napolitano, An immersed boundary method for compressible flows using local grid refinement. J. Comput. Phys. **225**(2), 2098–2117 (2007)
17. M. Alam, N.D. Sandham, Direct numerical simulation of 'short' laminar separation bubbles with turbulent reattachment. J. Fluid Mech. **410**, 1–28 (2000)
18. P.R. Spalart, M.K. Strelets, Mechanisms of transition and heat transfer in a separation bubble. J. Fluid Mech. **403**, 329–349 (2000)
19. S. Pirozzoli, M. Bernardini, F. Grasso, Characterization of coherent vortical structures in a supersonic turbulent boundary layer. J. Fluid Mech. **613**, pp. 205–231 (Oct. 2008)
20. S. Pirozzoli, F. Grasso, Direct numerical simulation of impinging shock wave/turbulent boundary layer interaction at M = 2.25, *Phys. Fluids* **18**(6) (2006)
21. P. Moin, K. Mahesh, Direct numerical simulation: a tool in turbulence research, Annu. Rev. Fluid Mech. **30**(1), pp. 539–578 (Jan. 1998)
22. J. Delery, J.G. Marvin, Shock wave—boundary layer interactions (1986)
23. E. Katzer, On the lengthscales of laminar shock/boundary-layer interaction, **206** (1989)
24. B.I. Greber, R.J. Hakkinen, L. Trilling, laminar boundary layer oblique shock wave interaction on flat and curved plates 1, **33**, pp. 312–331
25. R. Dp, B. Or, R. Jenson, Triple-deck solutions for viscous supersonic and l o w past corners hypersonic f, **89** (1978)
26. D.R Chapman, D.M Kuehn, H.K Larson, Investigation of separated flows in supersonic and subsonic streams with emphasis on the effect of transition (1957)
27. R.H.M. Giepman, F.F.J. Schrijer, B.W. van Oudheusden, High-resolution PIV measurements of a transitional shock wave–boundary layer interaction. Exp. Fluids **56**(6), 1–20 (2015)
28. R.H.M. Giepman, F.F.J. Schrijer, B.W. van Oudheusden, Infrared thermography measurements on a moving boundary-layer transition front in supersonic flow. AIAA J. **53**(7), 2056–2061 (2015)
29. S. Dhawan, R. Narasimha, Some properties of boundary-layer flow during the transition from laminar to turbulent motion. J. Fluid Mech. **3**(1951), 418–453 (1958)
30. P.R. Spalart, S. Deck, M.L. Shur, K.D. Squires, M.K. Strelets, A. Travin, A new version of detached-eddy simulation, resistant to ambiguous grid densities. Theor. Comput. Fluid Dyn. **20**(3), 181 (2006)
31. P. Spalart, S. Allmaras, A one-equation turbulence model for aerodynamic flows. *AIAA*, **439** (1992)
32. F.R. Menter, Two-equation eddy-viscosity turbulence models for engineering applications, **32**(8) (1994)
33. K.-Y. Chien, Predictions of channel and boundary-layer flows with a low-reynolds-number turbulence model. AIAA J. **20**(1), pp. 33–38 (Jan. 1982)
34. R. Bourguet, M. Braza, G. Harran, R. El Akoury, Anisotropic organised eddy simulation for the prediction of non-equilibrium turbulent flows around bodies. J. Fluids Struct. **24**(8), 1240–1251 (2008)
35. D.M. Dawson, S.K. Lele, J. Bodart, Assessment of wall-modeled large eddy simulation for supersonic compression ramp flows. In *49th AIAA/ASME/SAE/ASEE Joint Propulsion Conference*, American Institute of Aeronautics and Astronautics (2013)
36. S. Hickel, E. Touber, J. Bodart, J. Larsson, A parametrized non-equilibrium wall-model for large-eddy simulations (2015)
37. M. Shur, P. Spalart, M. Strelets, A. Travin, A hybrid RANS-LES approach with delayed-DES and wall-modelled LES capabilities. Int. J. Heat Fluid Flow **29**, 1638–1649 (2008)

38. E. Touber, N.D. Sandham, Low-order stochastic modelling of low-frequency motions in reflected shock-wave/boundary-layer interactions. J. Fluid Mech. **671**, 417–465 (2011)
39. L. Agostini, L. Larchevêque, P. Dupont, J. Debiève, J.-P. Dussauge, Zones of influence and shock motion in a shock/boundary-layer interaction. AIAA J. **50**(6), 1377–1387 (2012)
40. P. Durbin, X. Wu, Transition Beneath Vortical Disturbances (2007)
41. L. Agostini, L. Larchevêque, P. Dupont, L. Agostini, L. Larchevêque, P. Dupont, Mechanism of shock unsteadiness in separated shock/boundary-layer interactions Mechanism of shock unsteadiness in separated shock/boundary-layer interactions, **126103** (2015)
42. J. Riou, E. Garnier, C. Basdevant, Compressibility effects on the vortical flow over a 65° sweep delta wing. Phys. Fluids **22**(3), 2–13 (2010)
43. E. Garnier, Stimulated Detached Eddy Simulation of three-dimensional shock/boundary layer interaction, pp. 479–486 (2009)
44. P. Roe, Approximate riemann solvers, parameter vectors, and difference schemes. J. Comput. Phys. **43**(2), pp. 357–372 (Oct. 1981)
45. F. Ducros et al., Large-eddy simulation of the shock/turbulence interaction, **549**, pp. 517–549 (1999)
46. E. Lenormand, P. Sagaut, Subgrid-scale models for large-eddy simulations of compressible wall bounded flows, **38**(8) (2000)
47. C.W. Gear, *Numerical Initial Value Problems in Ordinary Differential Equations* (Prentice Hall PTR, Upper Saddle River, NJ, USA, 1971)
48. R.J. Volino, M.P. Schultz, C.M. Pratt, Conditional sampling in a transitional boundary layer under high freestream turbulence conditions. J. Fluids Eng. **125**(1), pp. 28–37 (Jan. 2003)
49. S.P. Schneider, Improved methods for measuring laminar-turbulent intermittency in boundary layers. Exp. Fluids **18**(5), 370–375 (1995)
50. H. Babinsky, J.K. Harvey, Shock Wave-Boundary-Layer Interactions (2011)

WP-2 Basic Investigation of Transition Effect

Holger Babinsky, Pierre Dupont, Pavel Polivanov, Andrey Sidorenko, Reynald Bur, Rogier Giepman, Ferry Schrijer, Bas van Oudheusden, Andrea Sansica, Neil Sandham, Matteo Bernardini, Sergio Pirozzoli, Tomasz Kwiatkowski, and Janusz Sznajder

Abstract An important goal of the TFAST project was to study the effect of the location of transition in relation to the shock wave on the separation size, shock structure and unsteadiness of the interaction area. Boundary layer tripping (by wire or roughness) and flow control devices (Vortex Generators and cold plasma) were used for boundary layer transition induction. As flow control devices were used here in the laminar boundary layer for the first time, their effectiveness in transition induction was an important outcome. It was intended to determine in what way the application of these techniques induces transition. These methods should have a significantly different effect on boundary layer receptivity, i.e. the transition location. Apart from an improved understanding of operation control methods, the main objective was to localize the transition as far downstream as possible while ensuring a turbulent character of interaction. The final objective, involving all the partners, was to build a physical model of transition control devices. Establishing of such model would

H. Babinsky (✉)
Department of Engineering, University of Cambridge, Cambridge, UK
e-mail: hb@eng.cam.ac.uk

P. Dupont
Aix-Marseille University, Marseille, France

P. Polivanov · A. Sidorenko
Khristianovich Institute of Theoretical and Applied Mechanics of the Siberian Branch of the RAS, Novosibirsk, Russia

R. Bur
Aeroelasticity and Aeroacoustics Department, ONERA - Aerodynamics, Meudon, France

R. Giepman · F. Schrijer · B. van Oudheusden
Technische Universiteit Delft, Delft, Netherlands

A. Sansica · N. Sandham
University of Southampton, Southampton, UK

M. Bernardini · S. Pirozzoli
Universita di Roma “La Sapienza”, Rome, Italy

T. Kwiatkowski · J. Sznajder
Łukasiewicz Research Network – Institute of Aviation, Warsaw, Poland

P. Doerffer et al. (eds.), *Transition Location Effect on Shock Wave Boundary Layer Interaction*, Notes on Numerical Fluid Mechanics and Multidisciplinary Design 144,
https://doi.org/10.1007/978-3-030-47461-4_3

simplify the numerical approach to flow cases using such devices. This undertaking has strong support from the industry, which wants to include these control devices in the design process. Unfortunately only one method of streamwise vortices was developed and investigated in the presented study.

1 Study of Transition Control Devices Operation

1.1 IUSTI

1.1.1 Tripping Devices Characteristics

Experiments achieved for the natural transitional SWBLI have shown that quite large aspect ratio were obtained $\left(L/\delta_0^* > 150\right)$ with evidences of unsteadiness depending on the shock strength and/or on the unit Reynold number. Moreover, the transition to turbulent flow is not observed along the interaction for this natural case and the reattachment point is found downstream of the impinging shock location. Large eddy Simulations achieved at the IUSTI for similar flows have shown that the size of the interaction as well as the location of the reattachment depends on the level of the upstream perturbations in the laminar boundary layer. The IUSTI's wind tunnel have significantly lower external perturbations than classical high pressure wind tunnel: this could explain the large differences observed on the length scales of the interaction. The upstream level of the perturbations could influence the length of transition along the interaction: higher levels will impose quicker transition, imposing the reattachment of the layer when the flow is turbulent and, as a consequence, a smaller interaction length. In order to validate these results, an attempt was made to modulate the upstream level of perturbations in the incoming boundary layer. Only one type of tripping device, a step, has been investigated for this study. The step spans the entire plate and several heights have been used: from $h = 0.05$ mm to $h = 0.36$ mm which corresponds to a height ratio of respectively $h\delta_h = 0.08$ and 0.56 at $P_0 = 0.4$ atm (respectively 0.11 and 0.8 at $P_0 = 0.8$ atm) where δ_h is the boundary layer thickness at the position of the step. The width of the step is of 2 mm. Main parameters are reported in Table 1.

1.1.2 Mean and Turbulent Velocity Fields

For limited heights, the upstream boundary layer remains laminar. Downstream from the separation point, typical profiles for mixing layer can be observed, as well as for the natural case, but RMS levels becomes slightly higher, with maxima near the reattachment of about 20% (to compare with 17% in the natural case). For all cases,

Table 1 Interaction parameters overview with tripped boundary layer

P_0 (atm)	Step	h (mm)	Rehi	x_{trip} (mm)	h/5 h	x_o (mm)
0.4	*Natural interaction*					
						68.4
	Tripped interaction					
	B2	0.100	142	52	0.156	
	B3	0.215	657	52	0.336	
	B4	0.300	1279	52	0.469	
	B5	0.360	1842	52	0.563	
	B6	0.250	1013	40	0.446	75.4
	B7	0.215	617	59	0.316	
0.8	*Natural interaction*					
						77.5
	Tripped interaction					
	B2	0.100	402	52	0.222	
	B3	0.215	1858	52	0.478	
	B4	0.300	3617	52	0.667	
	B5	0.360	5209	52	0.800	
	B6	0.250	2864	40	0.625	96.0
	B7	0.215	1744	59	0.448	

the flow remains clearly separated: backward flow is well observed with maximum reverse velocity up to 7% of the external velocity.

The length of the tripped interactions have been reported on Fig. 32 (chapter "WP-1 Reference Cases of Laminar and Turbulent Interactions"). The effect off the tripping device on the interaction length is clearly visible: the length is decreasing of ≈9% for the 0.4 atm case, but more than 25% in the 0.8 atm case. When normalized with the upstream incompressible boundary layer displacement thickness, the decreases is about 60%. This may be the result of a earlier turbulence transition inside the interaction, promoted by the upstream perturbations, as suggested from the Large Eddy Simulations performed at the IUSTI.

The Van Driest representation of the mean velocity profiles downstream of the tripped interaction (case B3 $P_0 = 0.4$ atm) is compared to the natural case on Fig. 1. We see that a well-defined log-law can be observed for the tripped case. In both cases, downstream from the interaction, the profiles exhibit large velocity fluctuations, higher than expected for a classical turbulent profile: these large fluctuations are the result of the relaxation of the flow, downstream of the interaction. Similar results were obtained in separated turbulent OSWBLI [1]. The near wall downstream boundary layer was recovering very quickly with the development of a new log-law region, based on the downstream wall conditions, while the external part of the layer was out of equilibrium with large energetic eddies shed in the downstream flow.

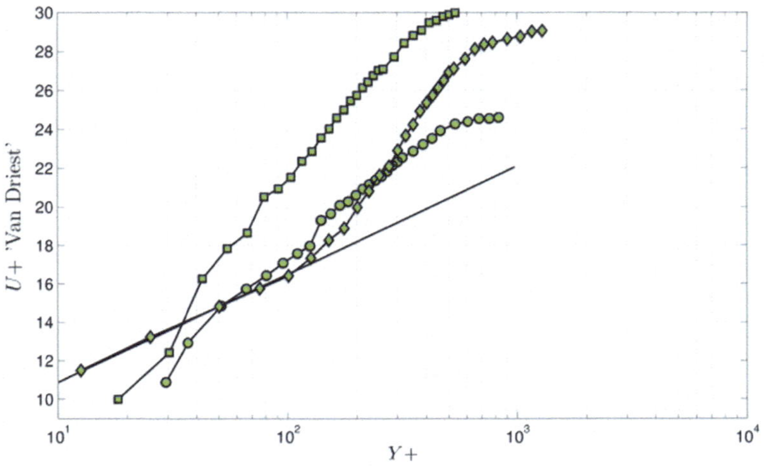

Fig. 1 Van Driest representation of the velocity profile downstream of the interaction ($X^* = 0.35$), Θ_3 angle: :B0, $P_0 = 0.4$ atm, □: B2, $P_0 = 0.4$ atm, ◇: B2, $P_0 = 0.8$ atm

The longitudinal evolution of the maximum value of u'/U_e along the interaction are reported Fig. 33 (chapter "WP-1 Reference Cases of Laminar and Turbulent Interactions") for the tripped case B2 at $P_0 = 0.4$ atm and $P_0 = 0.4$ atm. Very similar amplifications rate than for the natural case are obtained. In cases of tripped interaction, the maximum values are about 20% to compare to the 17% observed in the natural case. Over-imposed on the figure are also reported LES results for increased upstream perturbations (blue line). As for tripped interactions, similar amplification rates and saturation levels are obtained.

1.1.3 Unsteadiness Characterization

The Power Spectral Density (PSD) of the external radiated fluctuations along the interaction, for the B2 0.8 atm case, are presented on Fig. 2a, for the Θ_3 case. The cyan curve represents the upstream tripped laminar boundary layer. Compared to the natural case, we see that the frequencies amplified through the shear layer are similar. Nevertheless, the amplification is rather higher, especially in the 1–10 kHz domain and a typical frequency at 2 kHz is strongly amplified.

Transfer functions are presented on Fig. 15b. The representation is similar to the one used for the natural case. Again, a trend similar to that of natural case is observed. Nevertheless, the maximum amplification rate for frequencies of ≈1–10 kHz is one decade larger for the tripped boundary layer. In the compression waves zone ($X^* = -1$), only the 1–10 kHz domain is strongly amplified. Higher frequency scales are amplified downstream of the compression waves in a more moderate way, with an amplification rate from 2 to 3 times higher than the natural one. These results confirm the unsteady aspect of the transitional interactions, and the high sensitivity

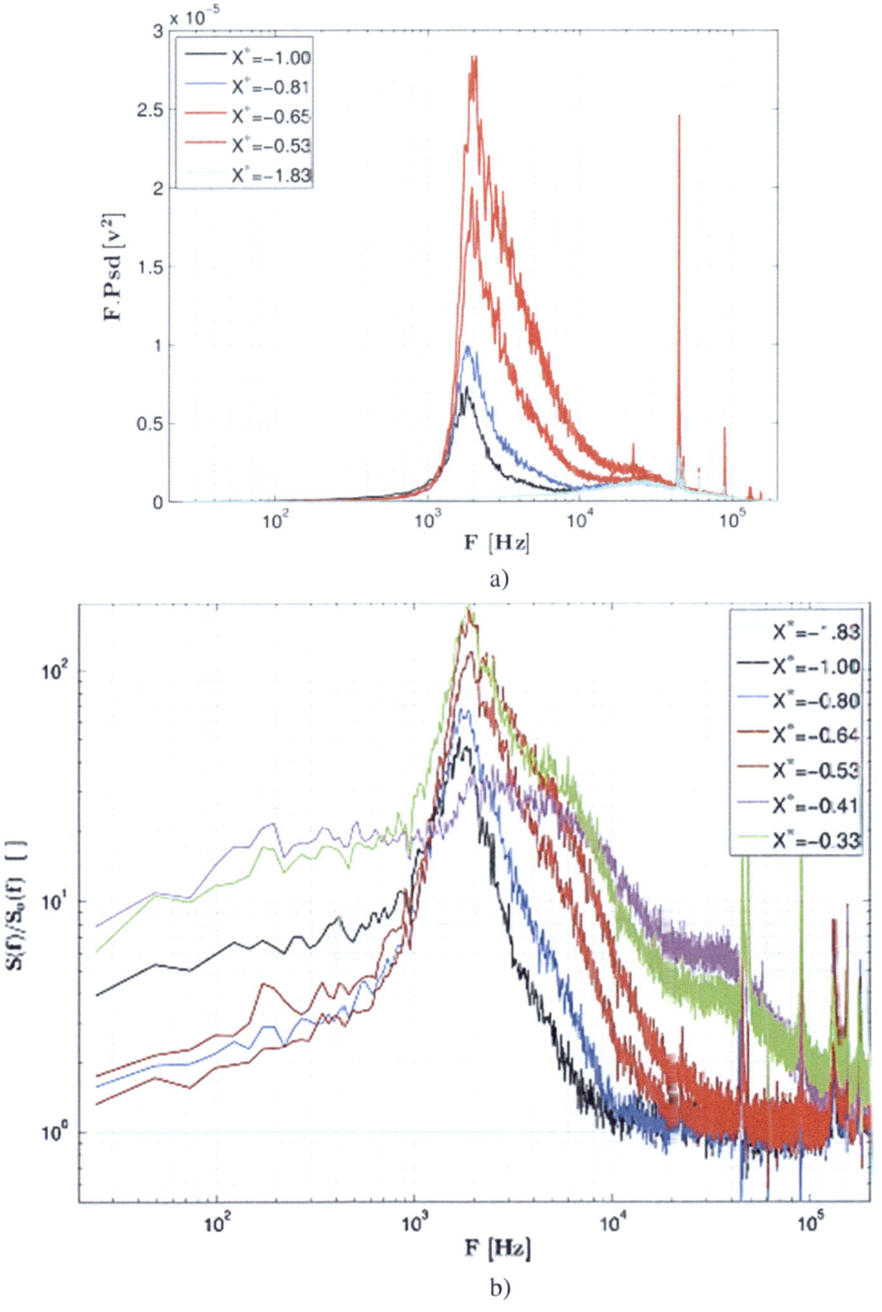

Fig. 2 Θ_3 flow deviation $P_{gen} = 0.8$ atm, tripped boundary layer; **a**: Pre multiplied PSD, **b**: Transfer function

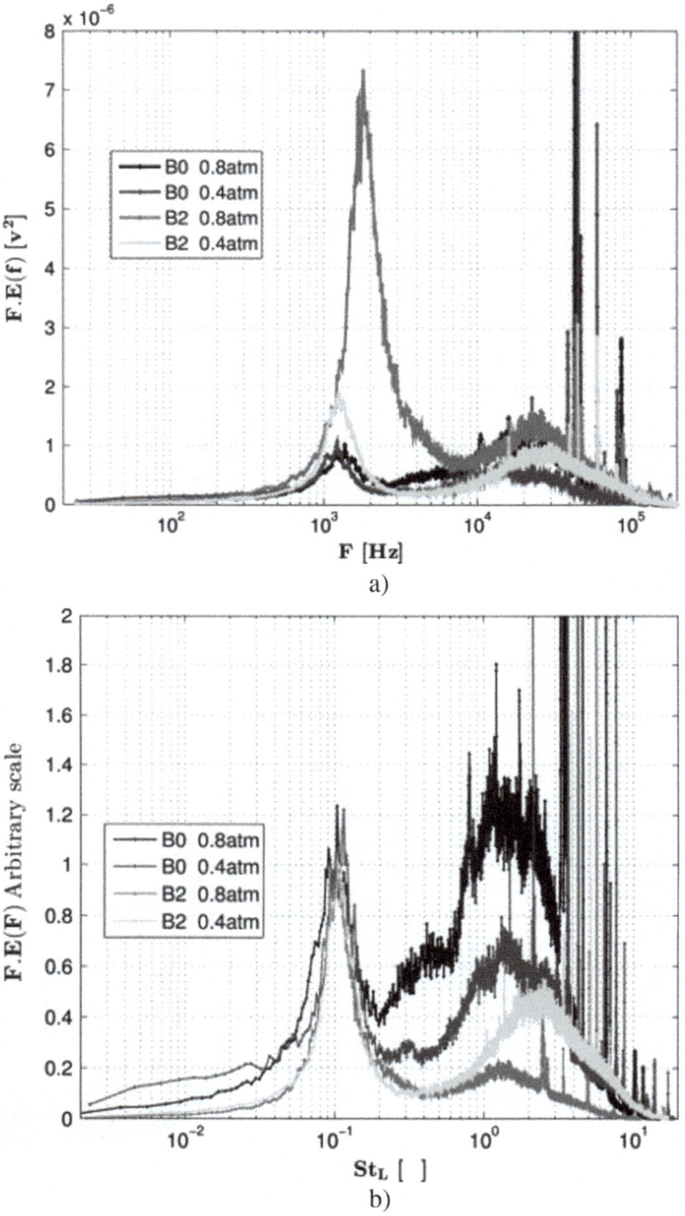

Fig. 3 Pre multiplied PSD, natural ($B0$) and tripped ($B2$) boundary layer, Θ_3 flow deviation, at $P_{gen}=$ 0.4 atm and $P_{gen}=$ 0.8 atm total pressure. **a**: pre-multiplied spectra vs frequency F, **b**: normalized pre-multiplied spectra vs Strouhal number $St_L = FL/U_1$

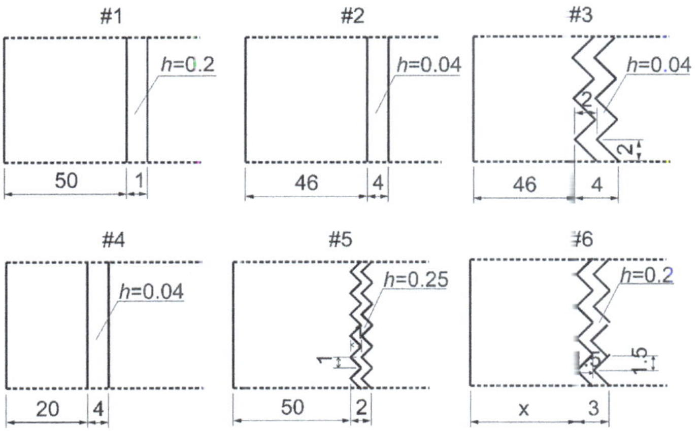

Fig. 4 Roughness elements

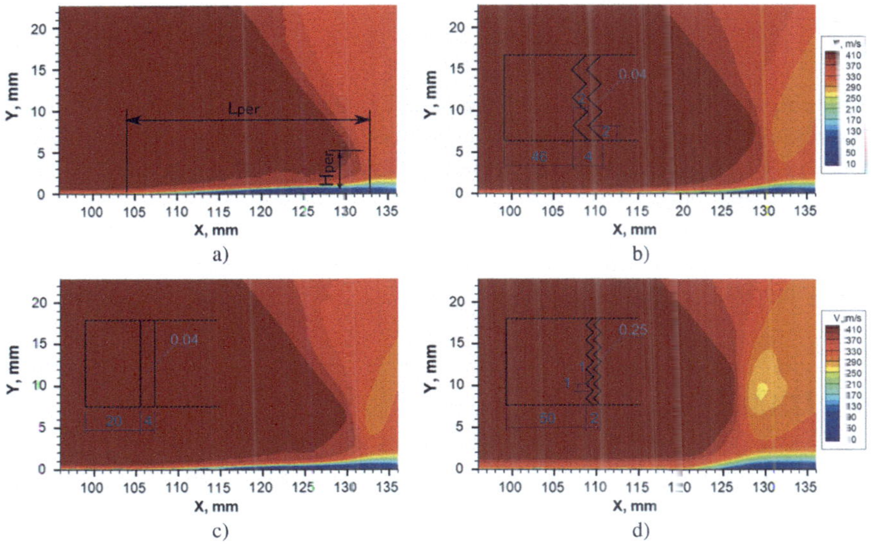

Fig. 5 Velocity fields for different roughness elements: **a** laminar case (reference), **b** #3, **c** #4, **d** #5 $Re_1 = 10.9 \cdot 10^6\ m^{-1}$

of such flow to the incoming unsteady conditions: the maximum amplification rate observed is multiply by a factor 10 (20–200), at a frequency of ≈ 2 kHz.

Figure 3 presents the pre-multiplied PSD measured at the middle of the compression zone ($X^* = -1$) for natural interaction and for *B2 case* tripped interaction, at stagnation pressure of 0.4 atm and 0.8 atm. This figure allows to notice the influence of the tripping on the unsteadiness of the compression zone. We first notice that the amplitude of the oscillation of the compression zone for the tripped interaction (*B2*

Fig. 6 Dimensionless parameters of SWBLI

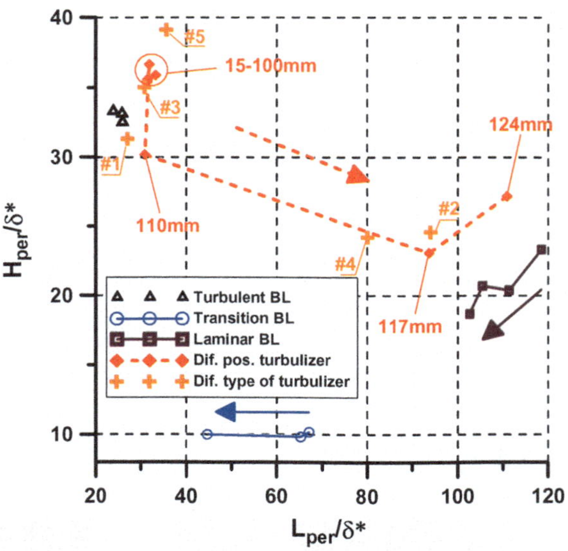

case) is higher than the one of the natural interaction (*B0 case*), both for 0.4 atm and 0.8 *atm* cases. Figure 3 is the dimensionless frequency representation of Fig. 3. The energy of the PSD are normalized to unity. This compilation gives a well-defined Strouhal number of $St_L = 0.10\ (St_L = FL/U_1)$ for the low frequency unsteadiness of the compression zone. However, this compilation suffers from the limited extent of the length of interaction variation: it varies from 42.7 to 25.3 mm from natural to tripped interaction.

1.2 ITAM

Passive and active methods were considered for flow control in SWBLI for the case of laminar incoming boundary layer. The main idea was excitation of additional disturbances by a roughness or an electric discharge which should lead to decrease of the separation zone length.

1.2.1 Passive Flow Control Devices

Study of the effect of the surface roughness shape and roughness location was carried out for the case of L = 250 mm and the stagnation pressure $P_0 = 0.7 \times 10^5$ Pa. In this study two types of roughness were considered (Fig. 4). The first type was a straight strip with rectangular cross-section (#1, #2, #4) and the second was a zig-zag strip (#3, #5, #6). As can be seen from the figure, several parameters of the roughness

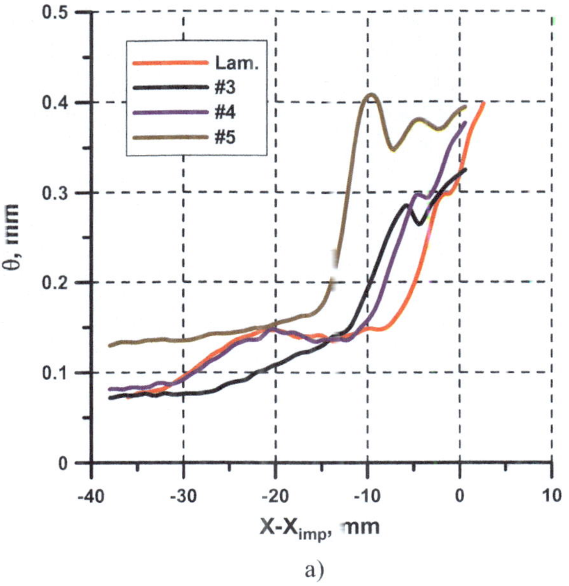

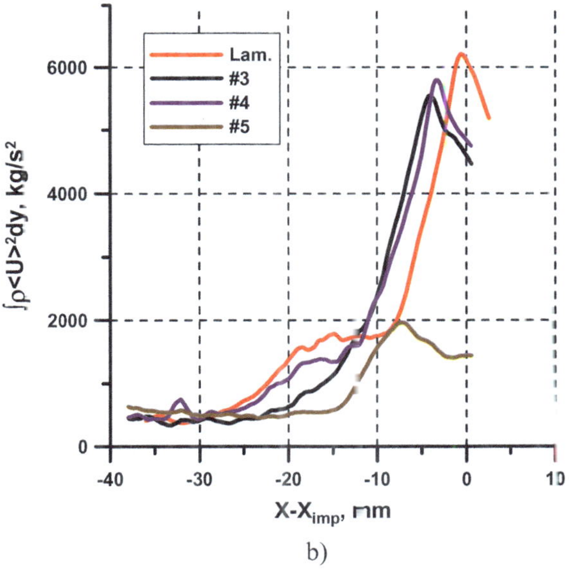

Fig. 7 **a** The momentum thickness distribution and **b** the integrated RMS of velocity fluctuations along the SWBLI for different types of turbulators ($Re_1 = 10.7 \cdot 10^6\ m^{-1}$)

were varied namely thickness, width, plane shape and location relative to the leading edge.

It was decided to use such parameters as the length (L_{per}) and height (H_{per}) of the interaction zone normalized by the displacement thickness of the inflow boundary layer ($\delta *$) to make comparative analysis (Fig. 5a). To find L_{per} from PIV data

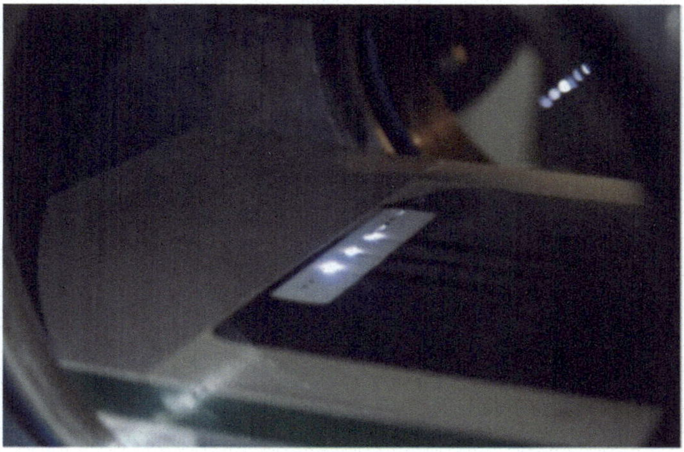

a)

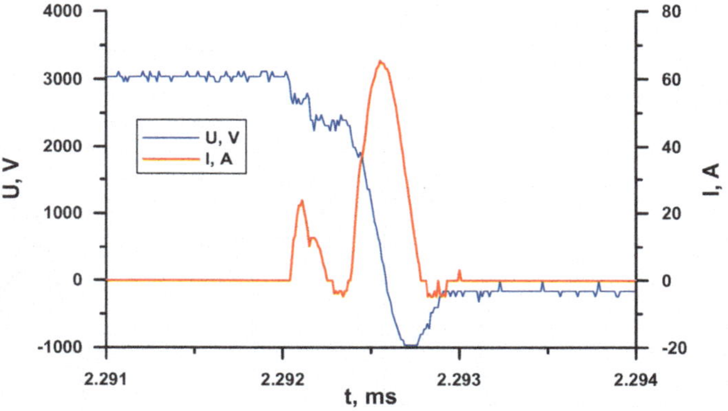

b)

Fig. 8 **a** Model with spark discharge actuator and **b** voltage and current oscillograms

Fig. 9 Spark discharge high voltage generator

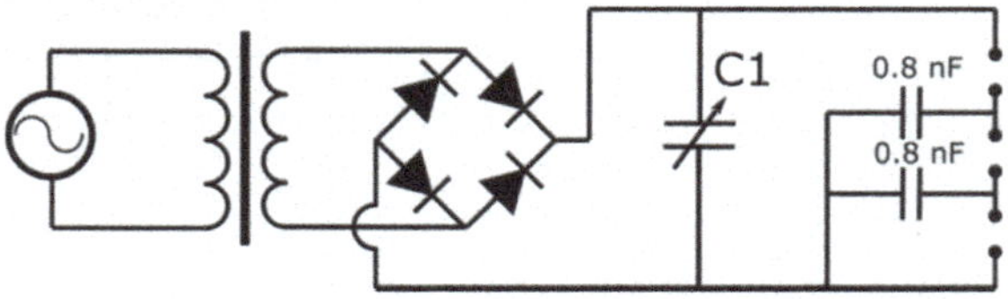

the minimal distance from the wall *Y* was determined where the reliable velocity measurements had been obtained. For this *Y* the longitudinal distribution of U was considered. Position where drop of velocity exceeds 5% of the corresponding inflow value was defined as the beginning of interaction. Significant growth of the velocity indicates the end of interaction. Upper border of the interaction H_{per} was defined as

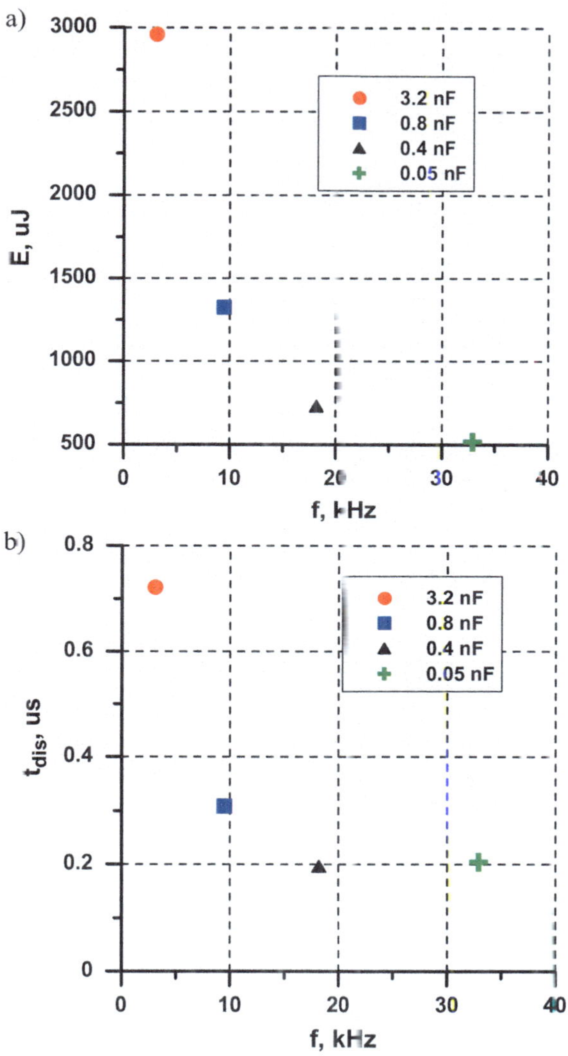

Fig. 10 Discharge characteristics: **a** pulse power and **b** pulse duration vs repetition frequency (average power was approximately 11 W)

the most downstream point where the flow speed decreased by 5% in comparison with inviscid upstream flow.

The effect the roughness shape on the SWBLI (denoted by plus in Fig. 6) was considered. Flat roughness of low thickness (#2, 4) does not lead to rapid turbulization of the flow (the inflow boundary layer remained laminar) and zone of interaction is close to the laminar case (Fig. 5c). But nevertheless some perturbations were excited leading to decrease of SWBLI length. Flat or zig-zag (#1, 5) roughness of large thickness quickly produced turbulent boundary layer (Fig. 5d). This results in a flow pattern typical for turbulent interaction.

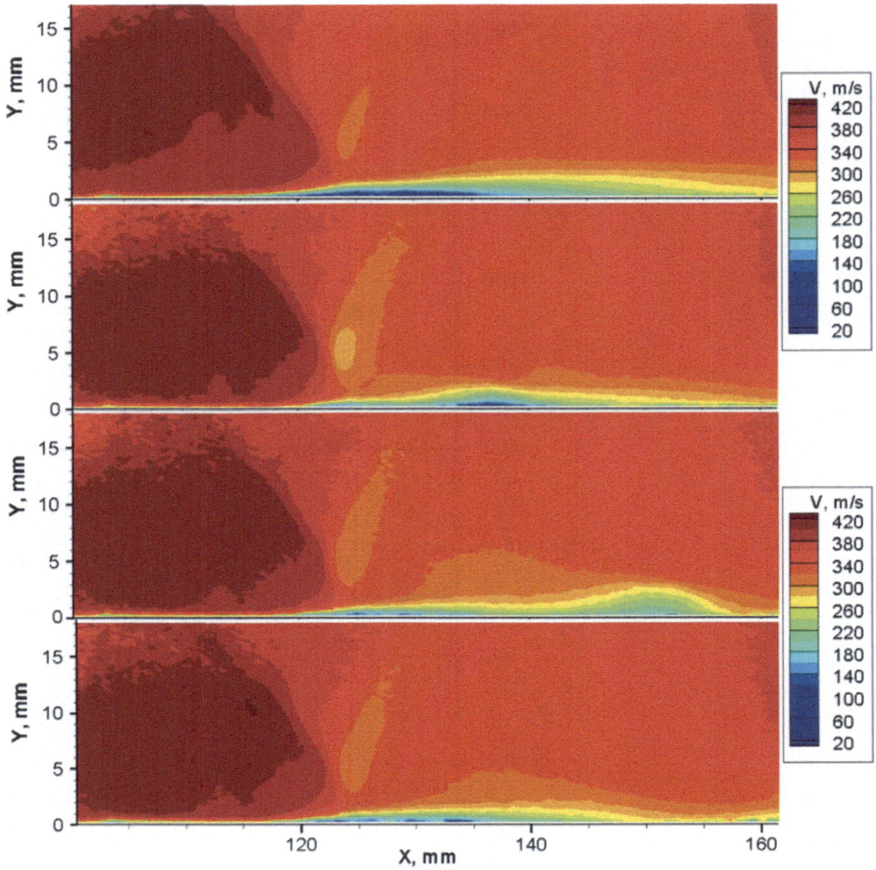

Fig. 11 Mean velocity fields at various discharge phase ($E_{dis} = 0.7$ mJ, $\beta = 4°$). Top to bottom: reference, $\Delta t = 100, 160, 220$ μs

The most interesting data were obtained for the zig-zag roughness of small thickness #3. This roughness does not lead to rapid turbulization of the flow and the boundary layer remained laminar up to the zone of interaction (Fig. 5b). But the length and height of the interaction is much closer to the turbulent case. Most probable that the reason of such behavior is the presence of additional perturbations given by the roughness. The power of perturbations is not sufficient for the flow turbulization flow without adverse pressure gradient but enough to trigger earlier laminar-turbulent transition in the shear layer with presence of adverse pressure gradient in the SWBLI.

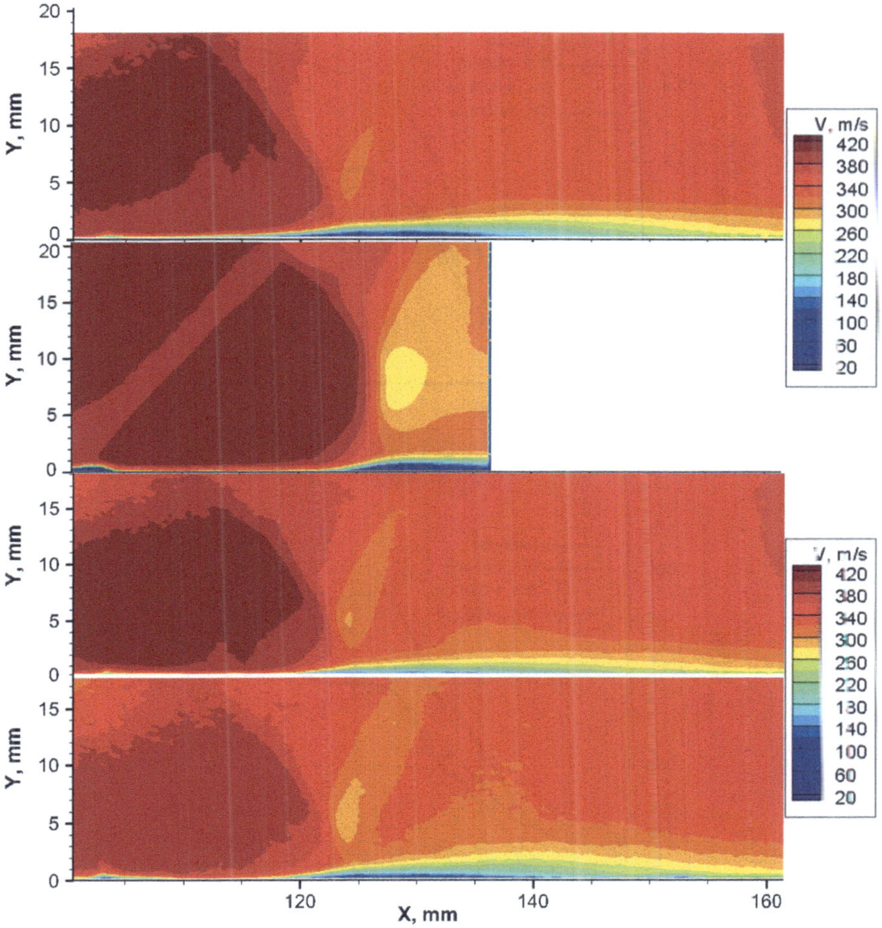

Fig. 12 Mean velocity fields at $\alpha = 4°$. Top to bottom: laminar case, turbulent case, average discharge case $E_{dis} = 0.7$ mJ, $E_{dis} = 1.3$ mJ

The next step was to study the effect of the turbulizer position on the SWBLI. The roughness sample # 6 providing rapid turbulization flow was chosen (denoted by diamond in the Fig. 6). Altogether 7 locations were investigated $X = 15, 50, 85, 100, 110, 117, 124$ mm (arrow in Fig. 6 indicates increase of distances). For all positions of the turbulizer upstream of separation bubble the SWBLI was typical for the turbulent case and dimensionless parameters were approximately the same. Since the thickness of the inflow turbulent boundary layer decreases with shift of the turbulizer downstream, the dimensional parameters also decreased. When turbulizer was placed at the beginning of the interaction zone H_{per}/δ^* decreased to 30. And the flow pattern was very similar to the case obtained with the roughness #3. Perhaps in this case we observed the similar processes. When the turbulizer #6 was in location

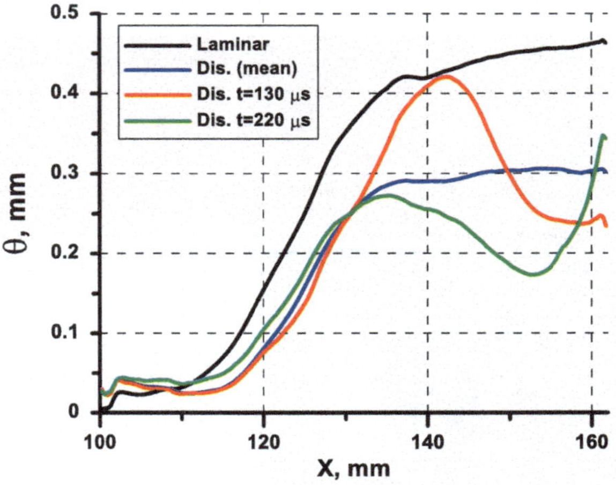

a)

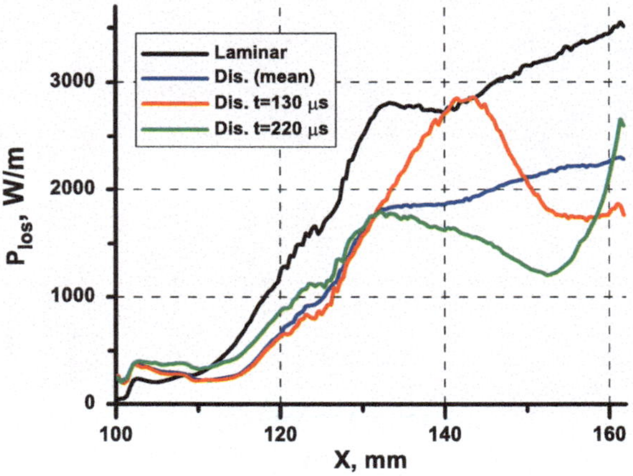

b)

Fig. 13 BL characteristics **a** momentum thickness and **b** energy losses along SWBLI ($E_{dis} = 0.7$ mJ, $\beta = 4°$)

near the SWBLI it was not able to provide the complete turbulization of the boundary layer. But it introduced significant perturbations which quickly grew in the interaction zone and resulting to turbulization of the flow in SWBLI.

In Fig. 7 one can see the effect of the some type of turbulators on momentum thickness. The momentum thickness distributions obtained for the turbulators, which do not cause immediate laminar-turbulent transition upstream of the interaction zone, are more or less similar to the laminar case. The momentum thickness for the natural laminar case at the end of the measurement area is close to the turbulent regime #5 but with higher level of pulsations. Taking into account the results obtained in

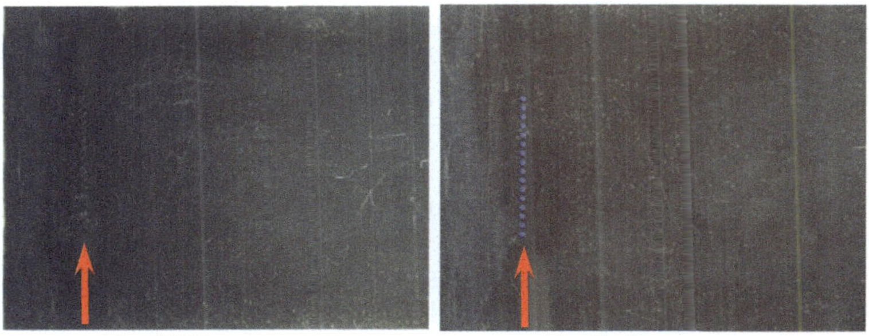

Fig. 14 Turbulator devices mounted on the flat plate

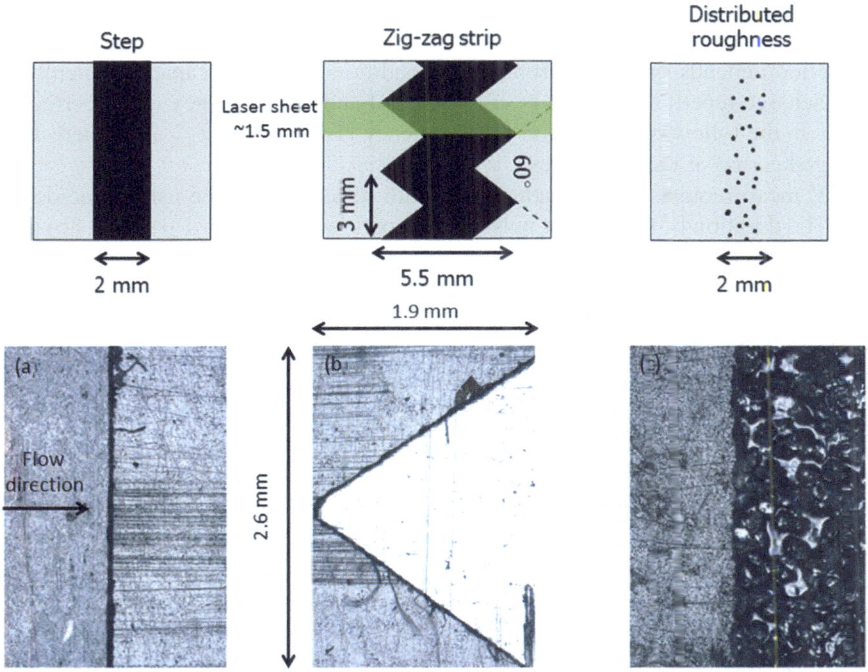

Fig. 15 Schematic representation of the tripping devices (top) and tripping devices imaged with a confocal microscope (bottom); **a** step **b** zig-zag tooth **c** distributed roughness

Chap. 2, it can be concluded that the turbulators did not allow to improve the flow in the SWBLI zone.

1.2.2 Active Flow Control Device

The plasma control devices were studied in the experiments. The spark discharge (SD) was chosen among others to achieve high concentrations of the energy in the plasma region. The model with installed SD actuator in shown in Fig. 8. It has ceramic insert made from MACOR holding a line of flush mounted electrodes. The electrodes are placed at distance 93 mm from the leading edge. There are three pairs of the electrodes with discharge gaps of 4.5 mm. Distance between the neighbor electrode pairs is 14.5 mm. All tree discharge gaps are connected in-series. The capacitors connecting the interim electrodes to the ground were used to assist the breakdown (Fig. 9).

Spark discharge actuator was fed by high voltage source using two transformers DAEHAN 15000 V/30 mA. The self-adapting scheme was used including a battery of capacitors C1 connected in parallel to the actuator (Fig. 9). This capacity is charged up to discharge level and consequently discharges. This process is periodic and the period depends on the environmental conditions and total capacity. Therefore the discharge repetition process is self-regulated and the frequency is not perfectly stable. In the following discussion this frequency is deduced as $f = 1/T$ where T is averaged period of the discharges.

PIV measurements were synchronized with plasma discharge using time delay unit. The duration of the current pulse in the spark was less than 1 μs (Fig. 8b) and average power for one discharge gap was estimated as $P_{dis} = 11$ W. Figure 7 shows pulse energy and frequency for several values of capacity C1.

The PIV data obtained for pulse energy of $E_{dis} = 0.7$ mJ ($f = 18.2$ kHz), $\beta = 4°$ are shown in Fig. 11 as the velocity fields corresponding to various discharge phases. High average frequency of the discharge means that the flow disturbances produced by the sparks travel downstream with small distance between them (≈15–20 mm). Since the high voltage system is self-regulated the breakdowns are not perfectly periodical. In the experiments PIV system was triggered by a discharge and the traces of the preceding discharges present in each single velocity distribution. However in the averaged data shown in Fig. 11 one can see only trace of the triggering breakdown since the delay between the sparks is not perfectly constant.

At the moment $\Delta t = 100$ μs the disturbance generated by the discharge passes the interaction. It can be seen that the compression waves upstream of the interaction are concentrated close to the interaction and they are more intense in comparison with the reference laminar case. This is evidence of diminishing or disappearance of the laminar separation zone upstream of the interaction. This is similar to the turbulent test case but the reflected shock wave is weaker. Fullness of the velocity profiles increases in the interaction and downstream.

The disturbance is represented by the area of low velocity in the boundary layer due to the hot spot with low density and high temperature. Decrease of velocity in this spot is amplified when it goes through the shock wave because of changes of the shocks configuration induced by the spot.

Figure 12 presents the comparison of laminar and turbulent (artificial turbulization, zig-zag trigger in position of $X = 100$ mm) test cases and cases with discharge excitation with various power. It can be seen that discharge actuator allows to achieve less intensive shock wave and slow growth of the wake. The best result was obtained for the lower value of spark energy.

Variation of the momentum thickness for the case of $E_{dis} = 0.7$ mJ, $\beta = 4^{\circ}$ is presented in Fig. 13. It can be seen that the spark discharge is able to reduce the average momentum thickness in the wake by 30% in comparison with laminar case.

Comparison of the averaged θ distribution with instantaneous ones presented in the same figure shows that in the region of the hot spot there is increase of losses. Therefore the positive effect of the disturbances provided by the discharge may be eliminated by the hot spot.

The flow control efficiency may be estimated basing on θ value at the end of measurement region as $\eta_{dis} = 0.5\rho U^3(\theta_{lam} - \theta_{dis})/P_{dis}$. For $\beta = 4^{\circ}$ the maximum value of efficiency $\eta_{dis}= 225\%$ was obtained for minimum spark energy $E_{dis} = 0.7$ mJ. If the spark energy was increased up to $E_{dis} = 1.3$ mJ the efficiency dropped to 167%. Increase of spark energy **up to 3 mJ resulted in negative efficiency – 55%. This means that the disturbance** generated by the spark is sufficient but there in negative effect provided by the hot spot increasing with the power. This conclusion agrees with results of computational study [1] where some optimum of pulse energy was found for the flow turbulization by a discharge. In the case of exceed of energy the effect diminishes due to heat spot formation. Since the flow parameters of study [1] and the presented experiments are close it can be assumed that decrease of energy by factor of 10 will allow to keep positive effect and increase the control efficiency.

1.2.3 Conclusions

Spark discharge actuators were tested and found to be effective to excite powerful periodic disturbances and control the interaction region. Basing on the quantitative analysis it may be concluded that spark discharge actuator improves the average flow in the interaction region. It was found that for all studied parameters of the discharge the pulse energy was sufficient for the generation of a turbulent spot and the suppression of the separation zone. However increasing of the discharge power leads to formation of the powerful hot spot which results in higher losses in the boundary layer. The analysis shows that active flow control may be more effective in comparison with passive control by the roughness.

1.3 ONERA-DAFE

1.3.1 Selected Configuration

A configuration has been retained for testing several control devices on the transition process, but also for related CFD simulations (RANS and LES) and flow stability analysis. In this configuration, the flat plate is located at $Z = 40$ mm from the test section lower wall to avoid blockage effect under the plate and at $X = -11.45$ mm into the Mach number rhombus ($X = 0$ is corresponding to the nozzle exit plane, see Fig. 36 and Table 5, chapter "WP-1 Reference Cases of Laminar and Turbulent Interactions") to avoid perturbation from the reflection of the Mach wave emanating from the plate leading edge (P3 position). The angle of attack α of the shock generator wedge is chosen equal to 2.5° to produce a moderate shock intensity. For this selected configuration, the laminar boundary layer separates and the transition process appears into the viscous interaction domain.

1.3.2 Control Devices Under Study

Several parameters are tested in order to optimize the effectiveness of control devices on the boundary layer transition, namely:

- Two types of 3-D turbulator device: a Cadcut device and a "ZZ" tape with a width of 6 mm (see Fig. 14).
- The height of device: the "ZZ" tape has two different heights h, 100 and 200 μm (called ZZ100 and ZZ200), leading to the Reynolds number values Re_h based on these heights respectively equal to 1400 and 2800; for comparison purpose, the height of the Cadcut device is chosen equal to 102 μm. These values are scaled to the compressible displacement thickness of the incoming boundary layer obtained by calculation (equal to 130 μm at 30 mm from the flat plate leading edge).
- The location of device with respect to the flat plate leading edge: the "ZZ" tape is located far upstream of the "averaged" natural (without the shock-wave) boundary layer transition position, but also upstream of the boundary layer separation region; then, two locations are selected, respectively at: $17 \leq X_{d1}$ (mm) ≤ 23 and at: $24 \leq X_{d2}$ (mm) ≤ 30 (corresponding to $Re_{xd} \approx 280{,}000$ and 380,000).

1.4 TUD

The basic test setup used for the study of the transition control devices is identical to that described in Sect. 2.3a and consists of a full-span flat plate and a symmetric partial-span shock generator, see [2] for more information. Earlier studies showed that the flat plate boundary layer remains laminar up to 55 mm from the leading edge

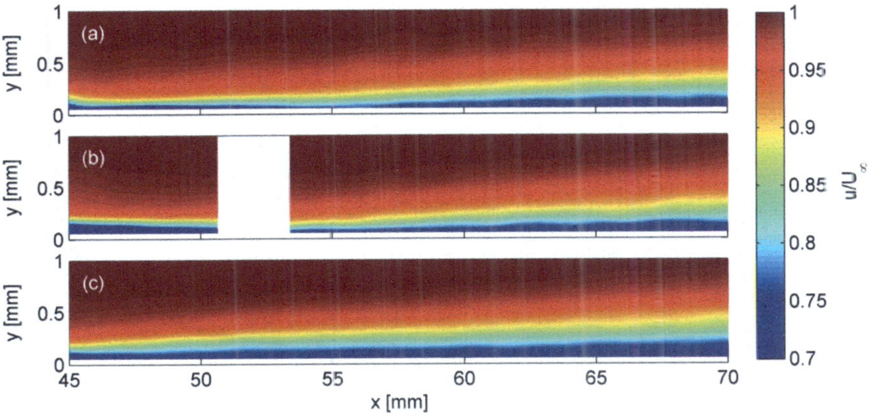

Fig. 16 Streamwise component of the velocity field downstream of the transition control devices. **a** zig-zag strip **b** step **c** distributed roughness patch

($Re_{x,tr} = 1.9 \cdot 10^6$). Natural transition takes place over a length of 35 mm and a fully turbulent boundary layer is established at approximately $x_{turb} = 90$ mm ($Re_{x,turb} = 3.2 \cdot 10^6$). In the SWBLI experiments the shock generator was set to a flow deflection angle of 3°, resulting in a theoretical (inviscid) pressure rise over the interaction of $p_3/p_1 = 1.35$.

The tripping devices were placed in the laminar regime of the boundary layer, at $x_{trip} = 40$ mm from the leading edge ($Re_{x,trip} = 1.4 \cdot 10^6$). The boundary layer at this location has been documented in Sect. 2.3b. Three types of tripping devices were investigated: a step-wise strip, a patch of distributed roughness (carborundum) and a zig-zag strip. Figure 15 shows a schematic representation of the trips and Fig. 16 shows the details of the trips as imaged with a confocal microscope. The devices spanned the entire plate width and had a roughness height of $k = 0.1$ mm, which corresponds to $Re_k = U_\infty k/\nu_\infty = 3.5 \cdot 10^3$ or $Re_{kk} = U_k k/\nu_k = 1.4 \cdot 10^3$. These values may be compared to the experimental study on tripping wires in supersonic flow [3], which delivers a critical $Re_k = 2.1 \cdot 10^3$ (equivalent $k = 0.06$ mm) for the present measurement conditions. It is therefore to be expected that the $k = 0.1$ mm trips introduce boundary layer transition in close proximity to the trip.

The velocity field downstream of the tripping devices is presented in Fig. 16 for the zig-zag strip (a), step (b) and distributed roughness patch (c). Velocity field data is missing for the step in the region from 50.6 to 53.4 mm due to a failed measurement (the incoming boundary layer was already transitional), but for all other datasets it has been confirmed that the incoming boundary layer is laminar. To provide more insight in the flow field downstream of the tripping devices, velocity profiles were extracted at $x = 45$, 50 and 55 mm (Fig. 17). The trips were centered around $x = 40$ mm and the trip to measurement station distance for Figs. 17a–c therefore equals 5, 10 and 15 mm, respectively. Or equivalently: 25, 50 and 75 undisturbed boundary layer thicknesses δ_{95} downstream of the centre of the trip.

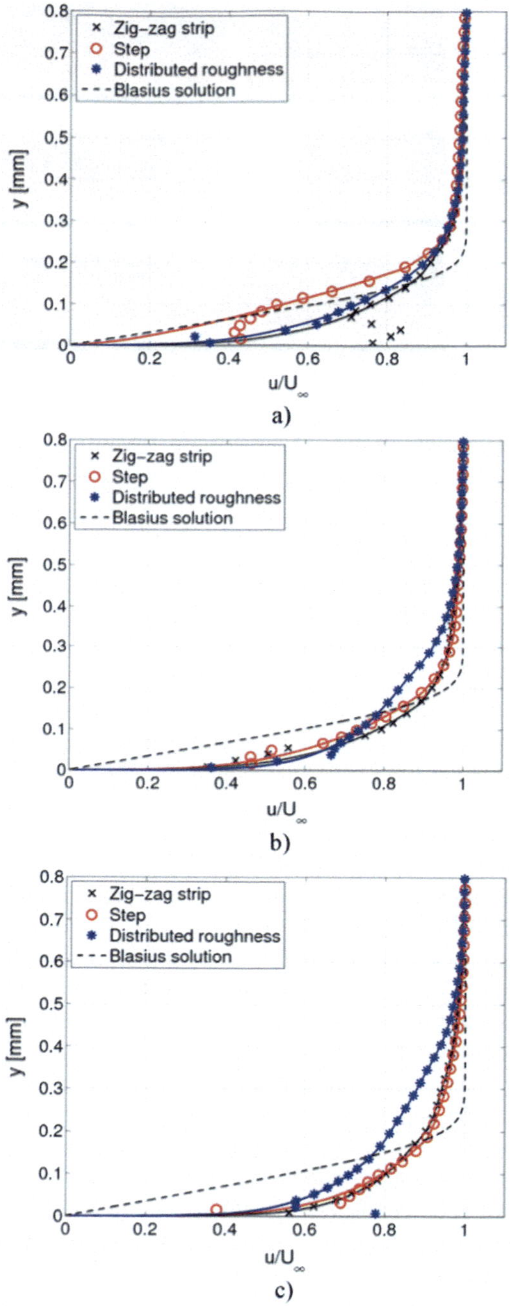

Fig. 17 Velocity profiles at x = 45 mm (**a**); 50 mm (**b**) and 55 mm (**c**)

The laminar boundary layer upstream of the tripping devices is lacking seeding in the near-wall region (y < 0.1 mm) of the flow. The boundary layer undergoes transition when crossing the trips, and due to the increased turbulent mixing the seeding conditions are found to improve gradually when moving downstream from the trip. For the first measurement station (x = 45 mm) it is however still difficult to accurately determine the velocity field in the near-wall region of the flow and some outliers may be observed in the data close to the wall. In order to calculate the integral boundary layer parameters from the experimental data it is therefore necessary to extrapolate the data towards the wall. The power law fit used for this purpose is indicated in the graphs.

At x = 45 mm (Fig. 38, chapter "WP-1 Reference Cases of Laminar and Turbulent Interactions") one may observe that the boundary layer displays a turbulent character for both the zig-zag strip and the distributed roughness, whereas for the case of the step the boundary layer is closer to a laminar profile. This difference disappears further downstream (x = 50 and 55 mm) and at these locations the boundary layer is turbulent for all trips, where it may be interesting to notice that the distributed roughness results in a substantially thicker boundary layer than the zig-zag strip and the step.

These observations are reflected in Fig. 18, which displays the development of the integral boundary layer parameters. For the computation of these properties, a power-law fit is used for the velocity data points close to the wall. The initial stage of the zig-zag strip and the step is different, but eventually both result in a turbulent boundary layer with approximately the same displacement and momentum thickness. The distributed roughness on the other hand results in a substantially thicker turbulent boundary layer compared to the other two tripping devices (displacement thickness is about 10 μm larger and momentum thickness about 8 μm larger). The scale on the right-hand side of the graphs indicates the development of δ_i^* and θ_i as a ratio of the undisturbed (laminar) values ($\delta_{i,0}^*$ and $\theta_{i,0}$) measured at the location of the trip. The displacement thickness temporarily becomes smaller than its undisturbed value due to the process of transition, which fills up the boundary layer profile. On the other hand, the momentum thickness downstream of the trip is always larger than its undisturbed value.

Finally, the shape factor development clearly shows that the zig-zag strip and the distributed roughness patch deliver a turbulent boundary layer much closer to the trip than the step. A shape factor of 1.4 is reached approximately 6 mm downstream of the centreline of the zig-zag strip and the distributed roughness patch. The same value is reached only after approximately 11 mm for the step. This implies that for the same degree of effectiveness the step should be placed further upstream of an impinging shock wave than the other two tripping devices.

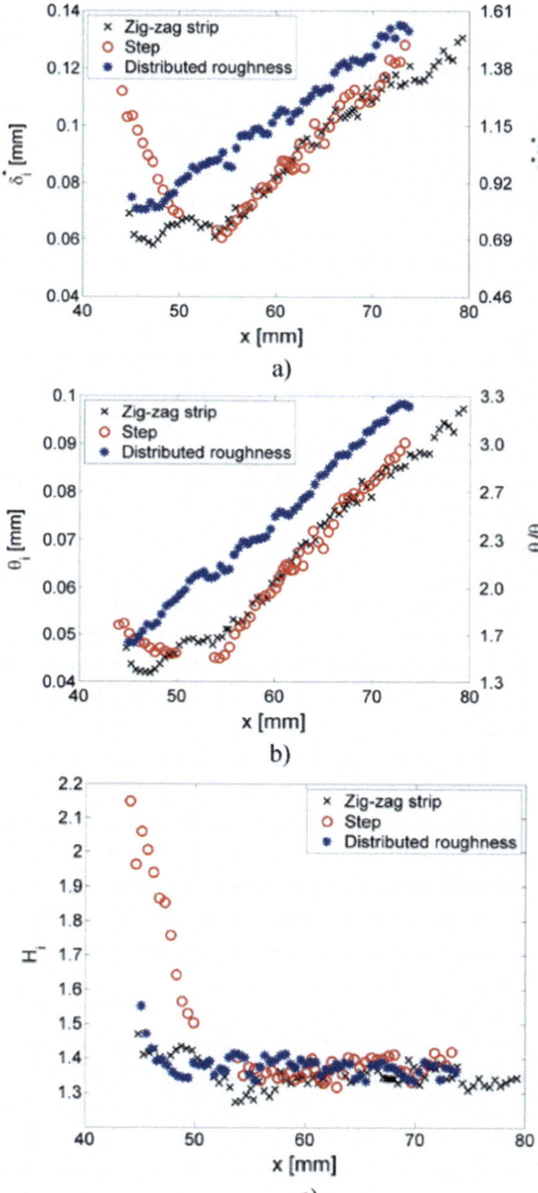

Fig. 18 Development of the boundary layer properties downstream of the tripping devices: **a** incompressible displacement thickness δ_i^*, **b** incompressible momentum thickness θ_i; **c** incompressible shape factor H_i

1.5 SOTON

1.5.1 Non-thermal Plasma Modelling

A linearised electric body force is used to model the effect of a non-thermal plasma flow control device [4]. The objective is to obtain a simplified representation of the body forces induced by the plasma actuator on the fluid without directly computing the electric field distribution. The Navier-Stokes equations, inclusive of the electrical body forces, become:

$$\frac{\partial \rho}{\partial t} + \frac{\partial (\rho u_j)}{\partial x_j} = 0 \tag{1}$$

$$\frac{\partial \rho u_i}{\partial t} + \frac{\partial \rho u_i u_j}{\partial x_j} = -\frac{\partial p}{\partial x_i} + \frac{1}{\mathrm{Re}}\frac{\partial \tau_{ij}}{\partial x_j} + D_c \bar{F}_i \tag{2}$$

$$\frac{\partial E_t}{\partial t} + \frac{\partial (E_t + p) u_j}{\partial x_j} = \frac{1}{\mathrm{Re}}\frac{\partial u_i \tau_{ij}}{\partial x_j} - \frac{1}{(\gamma - 1)\mathrm{Re}\,\mathrm{Pr}\, M^2 \partial x_j}\left[\kappa \frac{\partial t}{\partial x_j}\right] + D_c u_i \bar{F}_i \tag{3}$$

Apart from the usual notation, ρ_c is the dimensionless charge density and E_i ($i = 1, 2, 3$) are the three electric field components in the streamwise, wall-normal and spanwise directions, respectively. An alternative notation for the electric field distribution is $E = (E_x, E_y, E_z)$. The dimensionless parameter D_c is the ratio between electrical and inertial forces and can be expressed as

$$D_c = \frac{\rho^*_{c,ref} e^*_c E^*_{ref} \partial^*_{1,0}}{\rho^*_\infty {U^*_\infty}^2} \tag{4}$$

where $\rho^*_{c,ref}$ is the reference charge density, e^*_c is the electronic charge and E_{ref} is the reference electric field. The first modelling step is to linearise the distribution of the electric field that is mainly concentrated in the vicinity of the cathode and decreases in intensity moving downstream over the anode and away from the wall.

The electric field distribution can therefore be confined within a triangular region (in light blue) as schematically represented in Fig. 19.

The variation in space of the intensity of the electric field can be simplified as

$$|\boldsymbol{E}(x, y)| = E_0 - k_1(x - x_E) - k_2 y \tag{5}$$

with V the applied voltage and d the distance between cathode and anode in the streamwise direction. The electric field equals the breakdown electric field strength E_b at the boundary of the triangular region, allowing the constants k_1 and k_2 to be calculated as

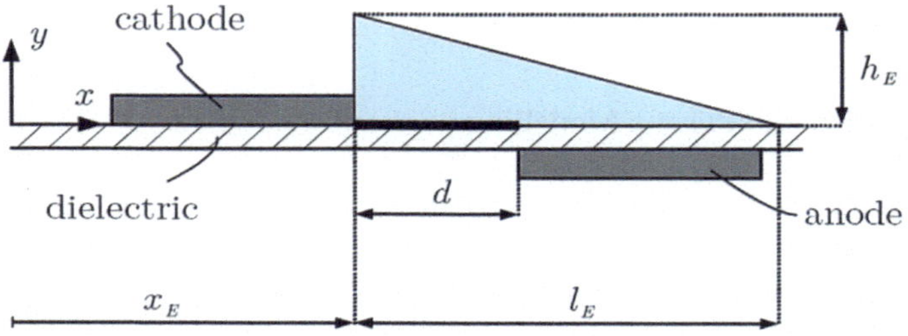

Fig. 19 Linearisation of the electric field

$$k_1 = \frac{E_0 - E_b}{l_E} \text{ and } k_2 = \frac{E_0 - E_b}{h_E} \tag{6}$$

where l_E and h_E are the streamwise length and wall-normal height of the linearised electric field distribution, respectively. Thus, the components of the electric field distribution (E_x, E_y) can be described as

$$\boldsymbol{E}(x, y) = |\boldsymbol{E}(x, y)| \left(\frac{k_2}{\sqrt{k_1^2 + k_2^2}}, \frac{k_1}{\sqrt{k_1^2 + k_2^2}}, 0 \right) = \left(E_x, E_y, 0\right) \tag{7}$$

Since the DBD is uniform in the spanwise direction, $E_z = 0$. In this way, it is possible to obtain the body forces (f_x, f_y) that are applied by the charged particles of the plasma to the neutral particles of the fluid as the parameter α is the collision efficiency and for simplicity can be set to unity [4], while the charge density ρ_c is assumed to be constant in the plasma region. The function δ ensures that the body forces are active only in the linearised-triangular region where the plasma is present and can be described as

$$f_x = \alpha E_x \rho_c e_c \delta \text{ and } f_y = \alpha E_y \rho_c e_c \delta \tag{8}$$

Although in the AC cycle during which the voltage is applied the plasma discharge happens only in a small fraction of time Δt, the frequencies of discharge are high under high applied voltage frequency and it is therefore possible to apply the averaged forces continuously over the whole cycle. The time-averaged body forces (F_x, F_y) can finally be written as

$$\delta = \begin{cases} 1, & |\boldsymbol{E}| \geq E_b \\ 0, & |\boldsymbol{E}| < E_b \end{cases} \tag{9}$$

where F_{ac} is the frequency of applied voltage. Although the real applied voltage has an unsteady nature, it is important to specify that this approach yields a steady representation of the actuator and its effects on the fluid. In addition, the cathode is assumed to have an infinitely small thickness in order to avoid the introduction of a surface step. The drawback to this simplified approach is the requirement of several input values that need to be specified from available experimental data. On the other hand, it avoids solving the Boltzmann and transport equations (per number of species, depending on the plasma chemistry) for electric field and species energy distributions that are typically computationally very expensive. For all simulations with plasma actuation the input parameters used here are taken from the experiments cited in [4].

$$\overline{F}_x = F_{ac} f_x \Delta t \text{ and } \overline{F}_y = F_{ac} f_y \Delta t \tag{10}$$

1.5.2 Thrust Direction, Unsteady Actuation and Spanwise Treatment

The general idea is to use the plasma actuator to force specific unstable modes of the boundary-layer, working as a transition tripping device. This can be done by changing the preferred direction of the electrical forces and applying a time and/or span variation to the body forces as described below:

- Thrust Direction: The effect of the DBD can be either flow-wise or flow-opposing, resulting in a "co-flow" or "counter-flow" actuation, respectively. Thus, the preferred thrust direction changes and momentum is either added to or subtracted from the boundary-layer.
  Unsteady Actuation and Spanwise Treatment: Linear stability theory provides frequencies and spanwise wavenumbers (ω, β) of the most unstable modes that can be used to apply a time and span modulation to the electrical body forces F_i as

$$F_i^{unsteady} = F_i^{steady} \cdot \left[1 + \sin\left(\beta_{pl} z\right)\right] \cdot \left[1 + \sin\left(-\omega_{pl} t\right)\right] \tag{11}$$

  where a factor of unity is added to the sine variation to avoid changes of sign in the body forces and provide the same amount of momentum input. The span variation of the body forces should not be confused with the spanwise shape of the electrodes. This model assumes that the electrodes are straight-edged elements (Fig. 20a) and the spanwise component of the electric field is equal to zero ($E_z = 0$). However, spanwise components of the electric field can be added in "serpentine" and "horseshoe" configurations (Fig. 20b, c), as suggested by [5].

In the current contribution, the effect of spanwise electric field components is investigated by modifying Eq. 2.8 to

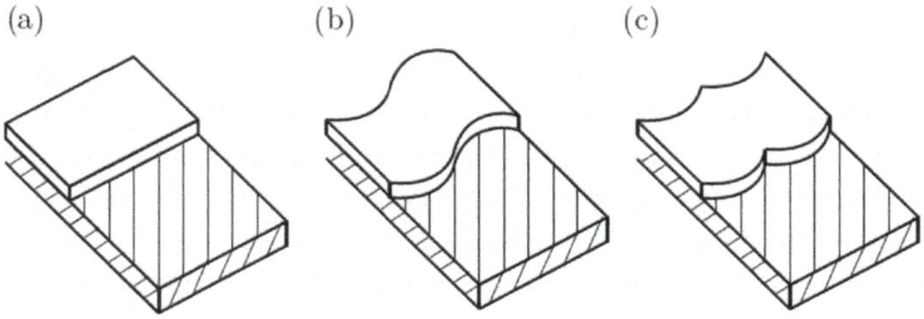

Fig. 20 DBD spanwise shapes for straight-edged (**a**), serpentine (**b**) and horseshoe (**c**) configurations

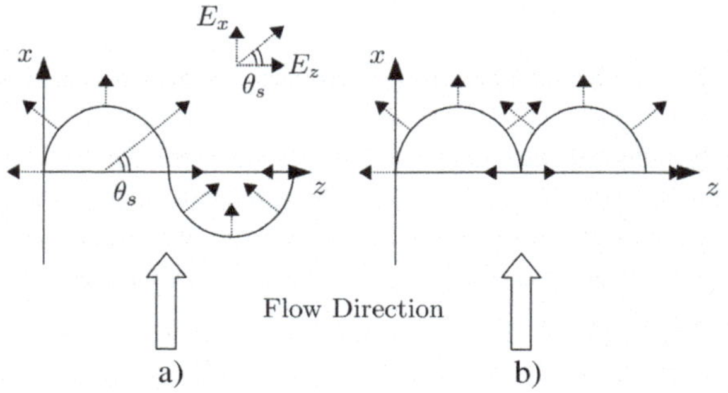

Fig. 21 Electric field spanwise component for co-flow serpentine (**a**) and horseshoe (**b**) configurations

$$\boldsymbol{E}(x, y, z) = |\boldsymbol{E}(x, y, z)| \left(\frac{k_2}{\sqrt{k_1^2 + k_2^2}} \sin\theta_S, \frac{k_1}{\sqrt{k_1^2 + k_2^2}}, \frac{k_2}{\sqrt{k_1^2 + k_2^2}} \sin\theta_S \right) \quad (12)$$

where θ's is the angle between the spanwise direction and the electric field direction on a generic x-z plane (as shown in Fig. 21a, b for the case of co-flow serpentine and horseshoe actuations, respectively).

1.5.3 Attached Boundary-Layer

Several DNS simulations are carried out to verify the effectiveness of a non-thermal plasma DBD as a flow control device. Different actuation configurations are tested with the intention of accelerating the transition to turbulence first in an attached zero-pressure gradient boundary-layer.

Inflow conditions, domain size and grid resolution are presented in Table 2. The domain height could not be reduced any further since a Mach wave is generated at the plasma location and its weak reflection from the domain top boundary needs to fall outside the numerical domain. The grid resolution is $(N_x, N_y, N_z) = (1680, 144, 240)$, corresponding to a grid size in wall units (when transition occurs) equal to $\left(\Delta x^+, \Delta y_w^+, \Delta z^+\right) = (5, 0.98, 5)$. The input parameters of the DBD plasma device

Table 2 Numerical setup for the zero-pressure gradient case

M	1.5
$\delta^*_{1,0}$	0.00075
$\mathrm{Re}_{\delta^*_{1,0}}$	75.00
T^*_∞	202.17
$L_x \times L_y \times L_z$	$310 \times 115 \times 27.32$
$N_x \times N_y \times N_z$	$168 \times 144 \times 240$
$\Delta x^+ \times \Delta y^+_w \times \Delta z^+$	$4 \times 0.85 \times 4$

are taken from the experiments used in Shyy et al. [4] (as described in Sect. 2) and give a dimensionless parameter $D_c = 1.75 \times 10^{-5}$ (applied voltage of 5 kV). The plasma actuator has an upstream effect on the boundary-layer, therefore the DBD is placed at $Re_x = 1.25 \times 10^5$ in order to allow the boundary-layer to follow the laminar skin-friction solution for about 30 $\delta_{i,0}$ upstream of the actuation location. Another important detail of the numerical setup is how the electric field distribution changes depending on the direction of the actuation. For all the cases with counter-flow actuation the electric field distribution is mirrored with respect to a vertical axis. All simulations presented here use unsteady plasma actuation with frequency equal to the most unstable modes predicted by LST (i.e. $\omega_{pl} = 0.1011$). The effects of the electric field spanwise treatment are also investigated. First, a spanwise variation of the electric field is applied according to the spanwise wavenumber of the most unstable mode predicted by LST ($\beta_{pl} = 0.23$) and compared to an actuation with no spanwise variation (i.e. $\beta_{pl} = 0$). For these cases, the electric field spanwise component (and related body-force) is zero. Secondly, different spanwise shapes of the dielectrics, such as straight-edged, serpentine and horseshoe, are used to study the effects of a spanwise component E_z of the electric field on the tripping. When the serpentine or horseshoe configurations are used, no additional spanwise modulation of the electric field distribution is introduced ($\beta_{pl} = 0$).

1.5.4 Plasma Regions Size

Regardless of the flow speed, higher plasma induced velocities can be directly translated into a more effective actuation. The highest induced velocities are typically no larger than 8–10 m/s [6]. For low subsonic flows, these values have been demonstrated to be high enough to obtain drag and separation reduction as well as turbulence tripping, but for supersonic applications this might no longer be true. As shown by [7], depending on the value of the applied voltage the size of the ionised region and induced velocities change accordingly. It is therefore important to reproduce the size of the plasma and the induced velocity field accurately. However, due to the lack of experimental data at M = 1.5, the size of the ionised flow is currently unknown. For this reason, the effect of the plasma region size on the induced velocity flow field is

Table 3 Induced flow velocity summary for plasma size investigation. Counter-flow actuation with straight-edged electrodes and applied voltage of V = 5 kV

(l_E, h_E)(mm)	U^*_{ind}(m/ s)	Transition
(3.00, 1.50)	14.1	No
(1.40, 0.30)	0.8	No
(0.07, 0.15)	0.2	No

here studied for counter-flow actuation with a straight-edged electrode and applied voltage of V = 5 kV.

Table 3 summarises the results of this investigation. The first selected plasma size corresponds to the original one from Shyy et al. [4] (l_E = 3.00 mm, h_E = 1.50 mm) and the results show a value of the maximum induced streamwise velocity of 14.1 m/s. In agreement with the general findings of other researchers who have of U_{ind} found that the modelled induced velocities and body-forces are usually overestimated [8], the value obtained is higher but not significantly different from what modern experimental plasma actuators can achieve. It is important to notice that the height of the plasma region is about 10 times larger than the boundary-layer thickness (δ 99 = 0.15 mm). This means that the actuation is more efficient inside the boundary-layer, but also that part of the energy is lost by doing work on the free-stream. Two other electric field distribution sizes are tested: l_E = 1.40 mm, h_E = 0.30 mm and l_E = 0.70 mm, h_E = 0.15 mm, where the latter corresponds to a plasma region whose height is rescaled to fit into the boundary-layer. It can be seen that the induced velocity field decreases when the size of the electric field distribution is decreased. This occurs despite the applied voltage being fixed, which is explained by the fact that the gradient of the linear distribution of the electric field is higher for decreasing plasma sizes. In this way, the region with high body-force intensity becomes concentrated very close to the wall. This study shows that the induced flow is very dependent on the modelled electric field distribution and on the definition of the experimental inputs. An optimisation of the electric field distribution with respect to the boundary-layer thickness could be carried out but it is not in the scope of this work. However, despite the increased induced velocity field, transition is not triggered for any of these configurations. With the intent to efficiently force only the boundary-layer, the size of the plasma region selected for the following investigations is l_E = 0.70 mm, h_E = 0.15 mm, where the electric field distribution does not extend beyond the boundary-layer thickness.

1.5.5 Induced Velocity Field

A forcing amplitude study is carried out for $D_c = 1.75 \times 10^{-5}$, 4.37×10^{-4} and 4.37×10^{-3}, corresponding to applied voltage rms-values V_{rms} = 5, 125 and 1250 kV, respectively. Although technically not feasible, boundary-layer transition only occurs when $U_{ind} \approx 60$ m/s is applied (this will be discussed in more detail in the next section where skin friction distributions are reported). Table 4 summarises for each actuation

Table 4 Induced flow velocity summary for each actuation configuration

Case no.	Direction	V_{rms} (kV)	Span treatment	u	Transition
1	Co-flow	5	$E_z = 0, \beta_{pl} = 0.23$	0.2	No
2	Co-flow	1250	$E_z = 0, \beta_{pl} = 0.23$	48.1	No
3	Counter-flow	5	$E_z = 0, \beta_{pl} = 0.23$	0.2	No
4	Counter-flow	1250	$E_z = 0, \beta_{pl} = 0.23$	53.1	No
5	Counter-low	1250	$E_z = 0, \beta_{pl} = 0.00$	35.5	No
6	Counter-flow	1250	$E_z \neq 0$-Serpentine	58.8	Yes
7	Counter-flow	5	$E_z \neq 0$-Horseshoe	0.2	No
8	Counter-flow	125	$E_z \neq 0$-Horseshoe	8.6	No
9	Counter-flow	1250	$E_z \neq 0$-Horseshoe	59.7	Yes

conguration the maximum induced streamwise velocities (Uin 1*) and whether transition occurred. Since the cases with low applied voltage did not yield to transition, only the results for $V_{rms} = 1250$ kV will be discussed for this attached boundary-layer (no-shock) case (i.e. cases 2; 4; 5; 6; 9).

Attention is focused first on the effect of the actuation direction (cases 2 and 4). The streamwise velocity profiles at different x-locations around the dielectric region (indicated with an arrow) at $z = L_z/4$ are reported in Fig. 22 for a case without actuation (black circles) and cases with co-flow (blue solid line) and counter-flow (red solid line) actuation. While the profiles upstream of the actuator are undisturbed, the downstream profiles are strongly distorted forwards or backwards for co-flow and counter-flow actuation, respectively. It is also possible to notice at $Re_x \approx 1.75 \times 10^5$

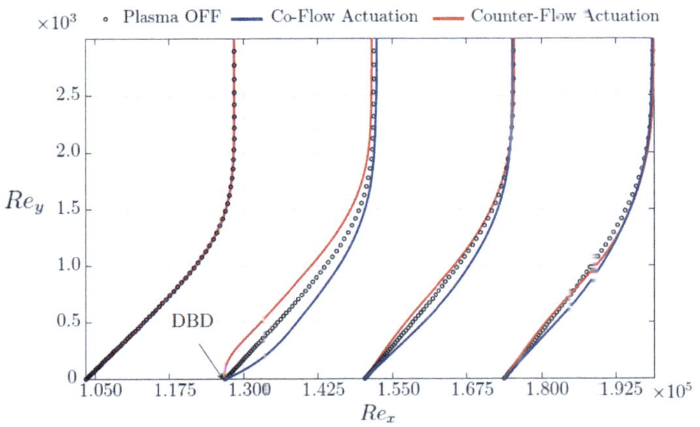

Fig. 22 Streamwise velocity profiles for no-actuation (black circles), co-flow (blue solid line, case 2) and counter-flow (red solid line, case 4) actuations with straight-edged electrodes. The DBD actuator location is indicated by an arrow

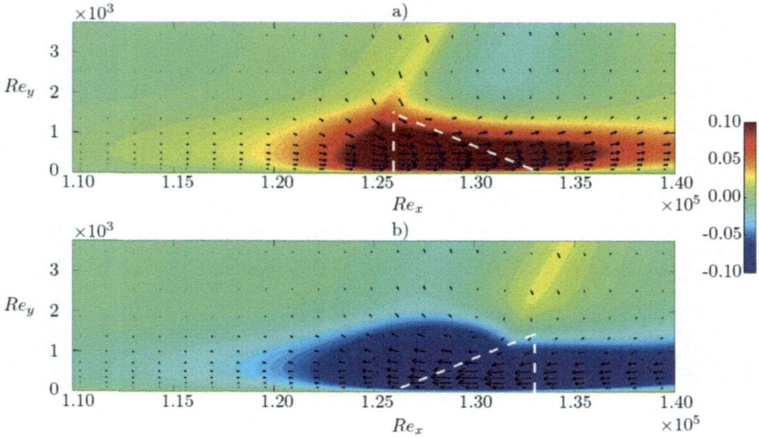

Fig. 23 Induced streamwise velocity vectors superimposed onto the induced streamwise velocity contours on an x-y plane at $z = L_z/4$ for **a** co-flow (case 2) and **b** counter-flow (case 4) actuations with straight-edged electrodes. The DBD position and shape is indicated by the white dashed lines

that the counter-flow actuation is very effective in the creation of a spanwise vortex that will develop downstream.

Induced streamwise velocity vectors are superimposed onto the induced streamwise velocity con tours on an x-y plane at $z = L_z/4$ in Fig. 23, for cases 2 (a) and 4 (b). The white dashed lines indicate the position and shape of the electric field distribution for the co-flow and counter-flow actuations. The preferred direction of the actuation is clearly visible in both cases, along with the generation of the spanwise vortex and a Mach wave above the plasma region.

The same analysis is repeated on an x-z plane at a distance 0.1 $\delta_{1,0}$ off the wall and reported in Fig. 24. Due to the electric field variation in the spanwise direction, a region of high induced velocity in the first half of the domain is created, producing strong tangential forces (the maximum induced streamwise velocities are $U^*_{ind} = 48.1$ m/s and $U^*_{ind} = 53.1$ m/s for cases 2 and 4, respectively), compared with the other half that has only a low induced velocity. The resulting wall-normal vortical structure (with opposite sign depending on the direction of the actuation) develops downstream and contributes to destabilising the boundary-layer. When the spanwise variation of the electrical field is removed from the counter-flow actuation (case 5), transition does not occur.

Induced flow fields for an x-y plane at $z = L_z/4$ and an x-z plane at a distance 0.1 $\delta_{1,0}$ off the wall are reported in Fig. 25a, b, respectively. Although the spanwise vortical structure still exists and the induced flow speed is 35.5 m/s, the wall-normal vortex associated with the spanwise variation of the electric field is absent. It is clear that this vortex plays a fundamental role in the destabilisation of the boundary-layer.

The same analysis is done for counter-flow actuation with serpentine (case 6) and horseshoe (case 9) shaped electrodes, with results shown in Figs. 26 and 27. With respect to the flow direction, while the shape of the serpentine electrode is represented

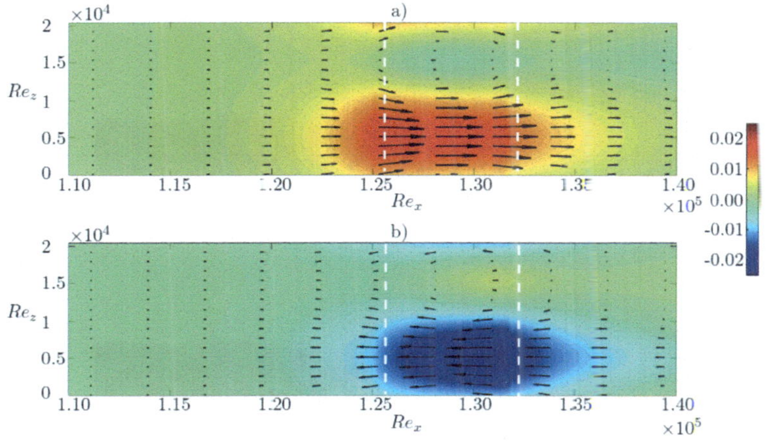

Fig. 24 Induced streamwise velocity vectors superimposed onto the induced stream for **a** co-flow (case 2) and **b** wise velocity contours on an x-z plane at y = 0.1 $\delta_{1,0}$ counter-flow (case 4) actuations with straight-edged electrodes. The DBD position and shape is indicated by the white dashed lines

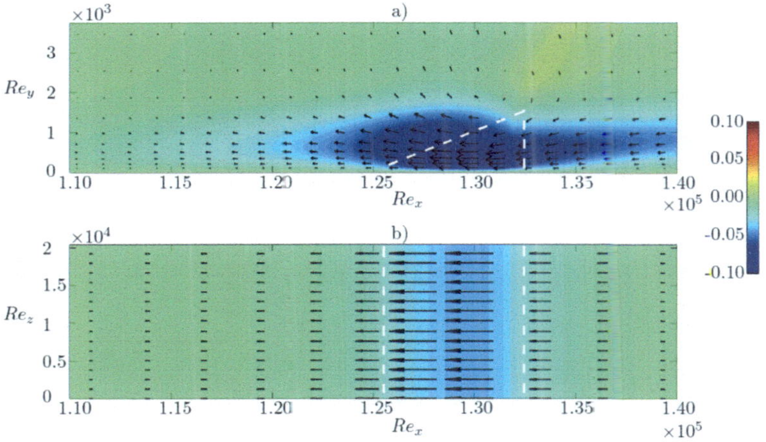

Fig. 25 Induced streamwise velocity vectors superimposed onto the induced streamwise velocity contours on **a** a x-y plane at z = $L_z/4$ and **b** a x-z plane at y = 0.1 $\delta_{1,0}$ for counter-flow actuation with straight-edged electrodes and no electric field spanwise variation (case 5). The DBD position and shape is indicated by the white dashed lines

by a concave part in $Re_z = (0 - 1) \times 10^4$ and a convex one in $Re_z = (1 - 2) \times 10^4$, the horseshoe configuration consists of two concave regions (see the white dashed lines in Fig. 27).

The introduction of spanwise components of the electric field distribution is beneficial for tripping purposes since both configurations lead to transition. While in the concave portion of the electrode the body forces are all acting towards the centre of

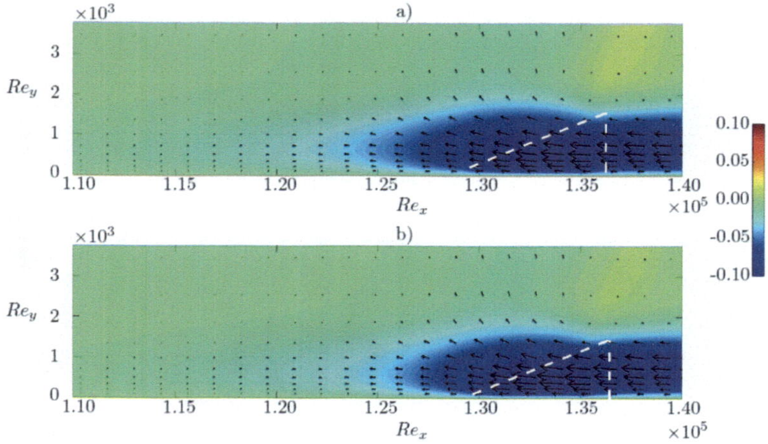

Fig. 26 Induced streamwise velocity vectors superimposed onto the induced stream-wise velocity contours on an x-y plane at z = $L_z/4$ for counter-flow actuation with **a** serpentine (case 6) and **b** horseshoe (case 9) electrodes. The DBD position and shape is indicated by the white dashed lines

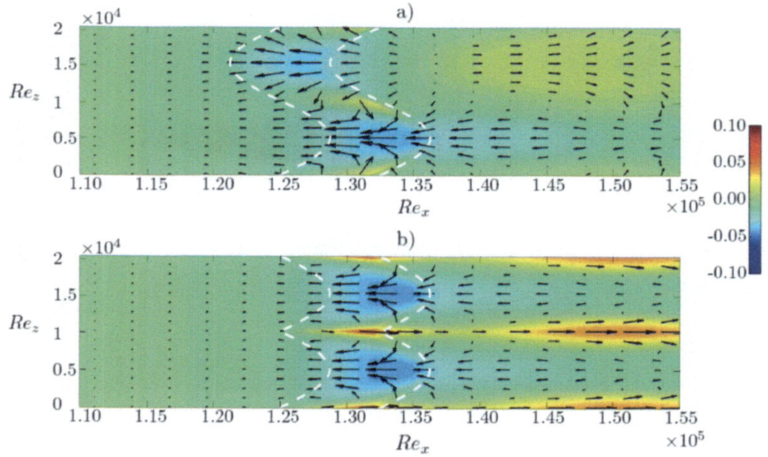

Fig. 27 Induced streamwise velocity vectors superimposed onto the induced streamwise velocity contours on an x-z plane at y = $0.1\ \delta_{1,0}$ for counter-flow actuation with **a** serpentine (case 6) and **b** horseshoe (case 9) electrodes. The DBD position and shape is indicated by the white dashed lines

the portion itself and generate an upward wall-normal jet, in the convex portion the opposite happens and a downward wall-normal jet is produced (Fig. 27a).

The fluid is strongly pulled upstream (or downstream) by the upward (or downward) wall-normal jet, also far downstream of the actuation region. For the horseshoe configuration, high speed streaks are produced at the conjunction of the two concave portions of the electrode and strongly contribute to the transition breakdown process (Fig. 27b).

1.5.6 Skin Friction Distributions

Time and span-averaged skin friction distributions are reported in Fig. 28 for cases 2 (orange solid line), 4 (red solid line), 5 (green solid line), 6 (black solid line) and 9 (blue solid line) along with the laminar (black dashed line) [9] and turbulent (black chain-dotted line) [10] boundary-layer distributions.

The presence of the actuators is shown at $\mathrm{Re}_x = 1.25 \times 10^5$, where spikes in the skin-friction distributions can be seen. Co-flow and counter-flow actuations with straight-edged electrodes (cases 2, 4) do not trigger transition, but it is clear that they both destabilise the boundary-layer and the skin friction distributions deviate from the laminar profile. When the spanwise variation of the electric field distribution is removed (case 5) the skin friction distribution follows the laminar solution, confirming that the generation of wall-normal vortical structure is necessary for a quicker destabilisation of the flow. It is also clear that the spanwise components of the electric field distribution (cases 6 and 9) accelerate the breakdown to turbulence. The generation of wall-normal jets destabilises the boundary-layer very quickly and transition is obtained. The transition scenario is shown in Fig. 29, where skin friction distributions are plotted in the whole wall-plane along with dashed white lines that indicate the position and shape of the electrodes. For straight-edged actuation

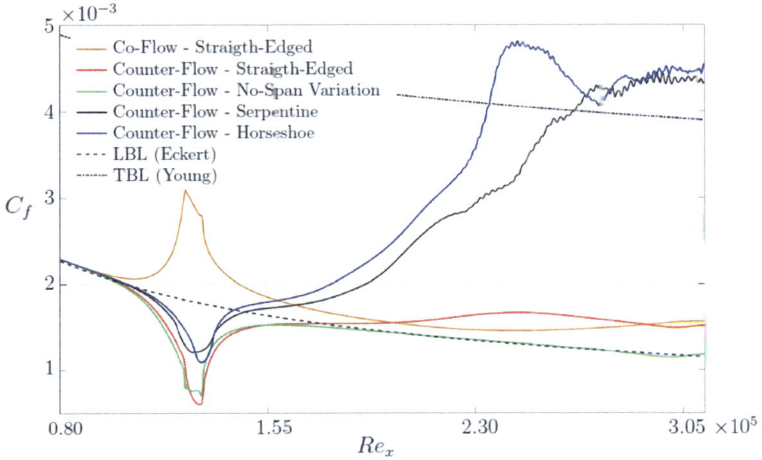

Fig. 28 Time and span-averaged skin friction distributions for different plasma actuation configurations. Cases: 2 (orange solid line), 4 (red solid line), 5 (green solid line), 6 (black solid line) and 9 (blue solid line). Laminar (black dashed line) [9] and turbulent (black chain-dotted line) [10] boundary-layer distributions are also plotted

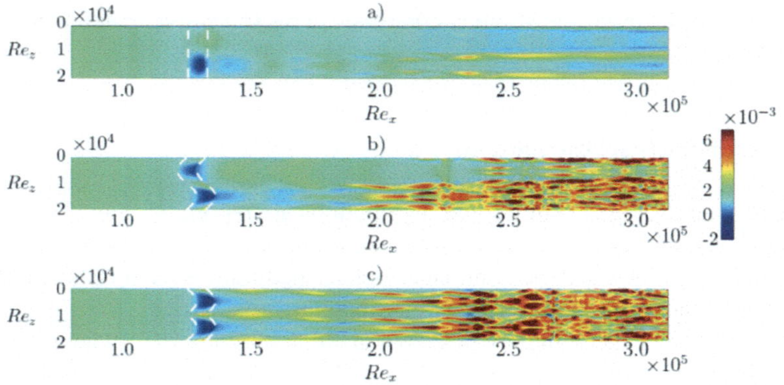

Fig. 29 Skin friction distributions for cases 4 (**a**), 6 (**b**) and 9 (**c**). The DBD position and shape are indicated by the white dashed lines

(Fig. 29a), the destabilisation of the boundary-layer is mainly localised in the portion of the domain where the spanwise variation of the electric field is higher. This suggests that if the actuation amplitude was further increased transition would occur relatively quickly. For the serpentine actuation (Fig. 29b), the breakdown is localised in the portion of domain corresponding to the concave part of the electrode. The body forces that act towards the centre of the electrode portion bring a strong contribution to the breakdown when an upward wall-normal jet is created and the actuator works like an unsteady blowing from the wall. With respect to the horseshoe configuration (Fig. 29c), transition starts at the same streamwise location but the turbulent state is reached further downstream due to this asymmetry. When the symmetry is restored with the horseshoe configuration, the turbulent state is reached earlier.

1.5.7 Transition Visualisation

A streamwise velocity x–y slice at $z = L_z/4$ is reported in Fig. 30 for the no-actuation (a), straight-edged (b), serpentine (c) and horseshoe (d) cases actuated with applied voltage of 1250 kV . For the actuated cases, it is possible to see the presence of vortex roll-ups that precede the breakdown. While for the straight-edged case the boundary-layer becomes unstable but stays laminar, for the serpentine and horseshoe configurations transition occurs at around $Re_x = 2 \times 10^5$.

The iso-surfaces of the Q-criterion for these three cases (Figs. 31, 32 and 33) show that aligned hairpin structures appear, suggesting that secondary instabilities set in and the harmonic or fundamental K-type behaviour precedes the final breakdown to turbulence.

Although the DBD tripping is effective, it is necessary to note once again that this occurs only for very high actuation amplitudes. When the applied voltage from Shyy

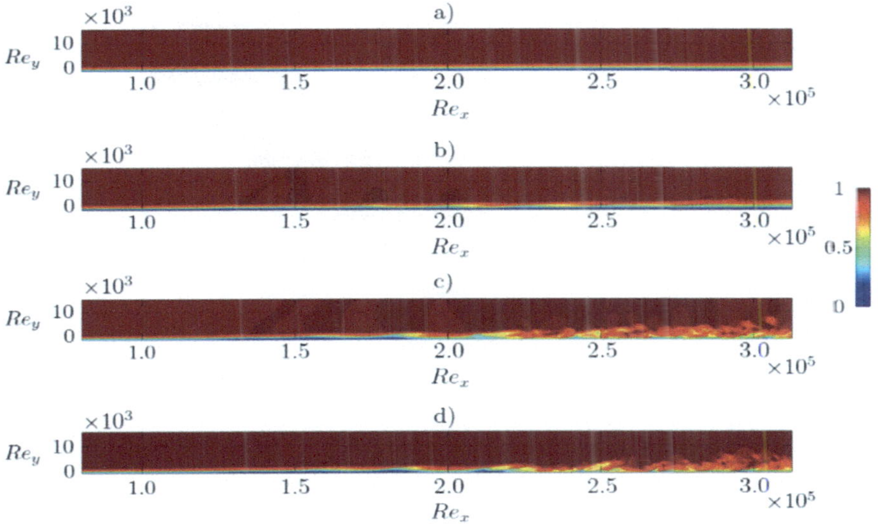

Fig. 30 Streamwise velocity x-y slice at z = $L_z/4$ for no-actuation (**a**) and straight-edged (**b**), serpentine (**c**) and horseshoe (**d**) counter-flow actuations

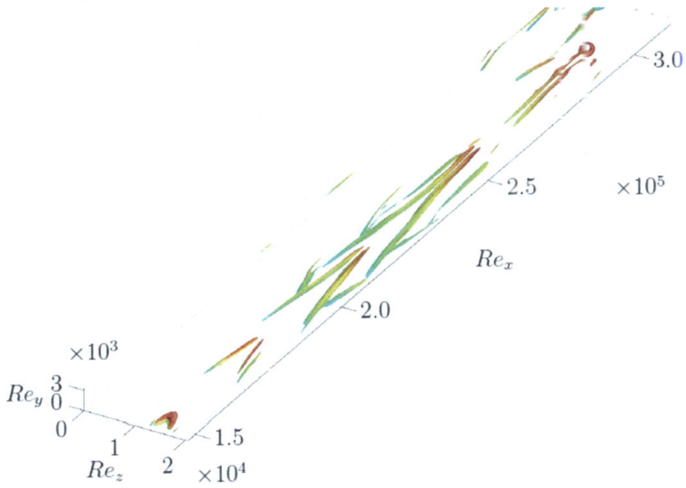

Fig. 31 Iso-surfaces of Q-criterion for straight-edged counter-flow actuation

et al. [4] is used, the plasma-induced velocities are too low to trigger transition and the boundary-layer maintains its laminar state.

This piece of work therefore demonstrates the theoretical potential of plasma actuation for tripping purposes in supersonic flows, but also shows that for real

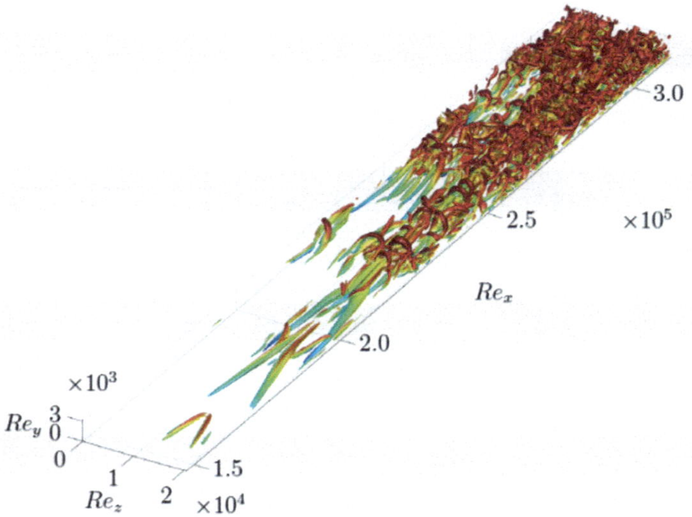

Fig. 32 Iso-surfaces of Q-criterion for serpentine counter-flow actuation

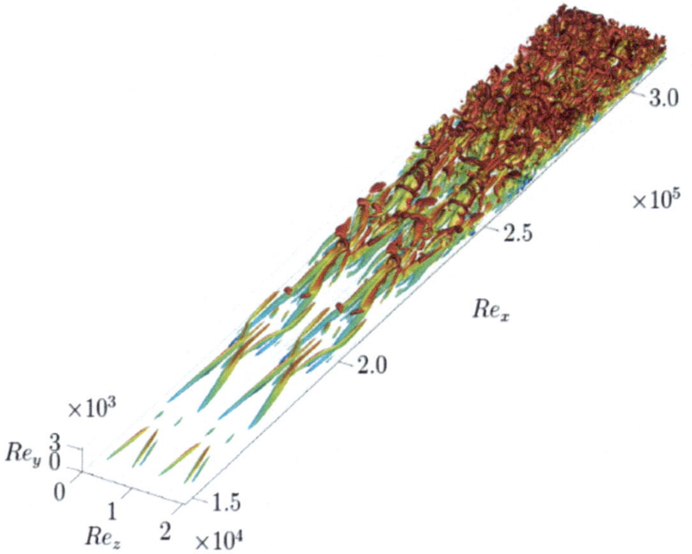

Fig. 33 Iso-surfaces of Q-criterion for horseshoe counter-flow actuation

applications this is not currently feasible since the induced velocities required are 5–10 times larger than those typically achieved in the experiments.

1.6 URMLS

1.6.1 Overview of the Physical Problem and State of the Art

The typical effect of a three-dimensional roughness element on a laminar boundary layer (see the sketch in Fig. 34) is to abruptly shift the transition location upstream with respect to the case of natural transition (i.e. for a smooth surface), with the amount of movement increasing with the roughness height (k). Early experiments [11] suggested that transition is determined by the flow properties of the undisturbed boundary layer evaluated at the edge of the roughness element (hereinafter denoted with the subscript k) through a roughness Reynolds number

$$\mathrm{Re}_k = \rho_k u_k k / \mu_k. \quad (13)$$

While Re_k well identifies the onset of transition in the incompressible regime (i.e. transition is observed when Re_k is greater than a critical value), this parameter cannot account for several effects, including compressibility, roughness shape and

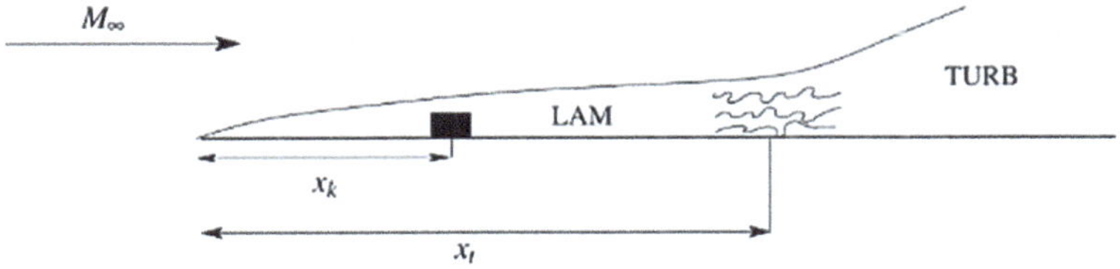

Fig. 34 Sketch of boundary layer transition induced by an isolated roughness element. M_∞ is the Mach number of the (compressible) boundary layer, x_k is the distance of the roughness element by the plate leading edge and x_t indicates the transition location

Fig. 35 Sketch of the computational arrangement used for DNS of transition past the 3D roughness element. The streamwise distance of the obstacle from the inflow station is $x_c = 15\,\delta$

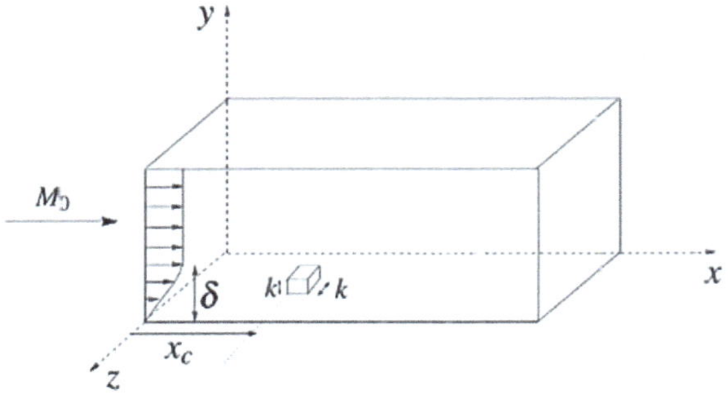

wall temperature. As a consequence, the current prediction of roughness-induced transition at high-speed heavily relies on empirical correlations, a popular one being the Re_θ/M criterion [12].

Past efforts have identified the typical paths to transition, especially in the low-speed regime. Experiments have shown that the flow around an isolated three-dimensional roughness element is characterized by the presence of a steady horseshoe vortex that wraps around the obstacle, with two steady counter-rotating vortices trailing downstream. The streamwise vortices and the associated low-momentum streak lead to a convective shear-layer instability in the wake of the roughness element, characterized by periodic shedding of hairpin-like vortical structures [13, 14]. According to [15], transition occurs when the growth of fluctuations is sufficient to trigger transition, and to penetrate the wall layer.

Recent direct numerical simulations (DNS) [16–17] performed in the supersonic and hypersonic regime indicate that the same scenario observed at low speed also holds at higher Mach numbers, with minor changes due to compressibility. The numerical studies have highlighted the importance of the unstable detached shear layer forming on the top of the roughness element, and have identified the wake behind the roughness element as the primary source for transition.

The activities of URMLS in the project focused on the generation of a DNS database to study the laminar-to-turbulent transition of compressible boundary layers over a flat plate induced by the presence of isolated three-dimensional roughness elements. The initial goal is to widen the range of flow conditions analyzed in literature and understand the effects of the relevant parameters affecting the transition process. The final goal for the WP will be to highlight the effect of different boundary layers states (laminar, transitional and fully turbulent) on the interaction with various shock waves configurations (oblique and normal shock).

1.6.2 Computational Setup and Code Validation

For the purpose of validating the flow solver, we have reproduced the numerical results of [18], who carried out DNS of a laminar boundary layer over a flat plate perturbed by an array of cylindrical roughness elements. The physical parameters for the simulation were selected by those authors to reproduce the experiments of Ergin and White [15], characterized by a roughness Reynolds number $\mathrm{Re}_k = 334$, roughness height $k = 0.332\delta$, cylinder aspect ratio $k/D = 0.15$ and spanwise spacing between the elements $\lambda_k = 3D$. Since the experiment was conducted at low speed, the Mach number for the simulation, performed with a compressible solver, was set at $M = 0.1$, which is the same value used here.

A sketch representing our computational arrangement is shown in Fig. 34. The computational domain is a Cartesian box, extending for $L_x = 70\delta$, $L_y = 20\delta$, $L_z = 6\delta$ in the streamwise (x), wall-normal (y) and spanwise (z) directions, where δ, the boundary layer thickness at the inflow station, is chosen as reference length. The grid points have been clustered toward the wall according to a hyperbolic sine mapping

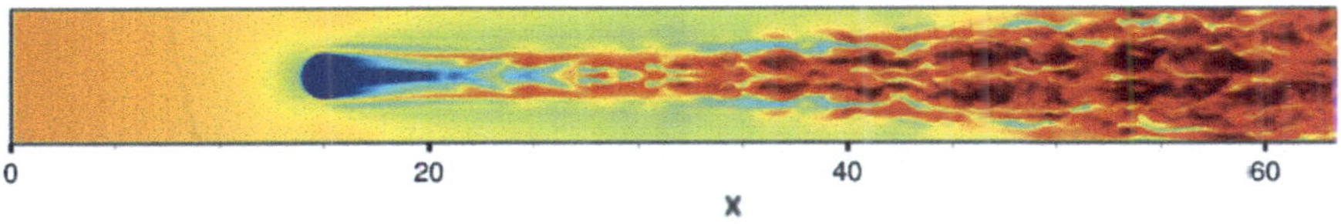

Fig. 36 Contours of the instantaneous streamwise velocity in a wall-parallel plane $y = k$

function, and non-uniformly distributed in the streamwise and spanwise directions to allow accurate resolution of the region close to the obstacle.

The roughness element, centered at $x_c = 15\delta$, $z_c = L_z/2$, perturbs the laminar boundary layer which develops over the flat plate. The initial condition is determined from a compressible similarity solution, and is also used to prescribe the inflow state. Radiative boundary conditions are assigned at the top and outflow boundaries, and periodicity is enforced in the spanwise direction, implying that the simulated flow mimics that around a periodic array of identical roughness elements.

According to the available experimental and numerical data, at the conditions selected for the validation test, the boundary layer past the roughness element is expected to exhibit bypass transition to a fully turbulent state. This is confirmed in our simulation, as visible in Fig. 36, where contours of the streamwise velocity field are shown in the wall-parallel plane at the edge of the roughness element. The velocity field reveals the presence of a low-momentum streak behind the cylinder, which is perturbed by the passage of an horseshoe vortex system, whose signature is visible in the range $x/\delta = 20$–35. As the flow evolves in the streamwise direction, the flow undergoes transition to a turbulent state, highlighted by the presence of multiple streaks for $x/\delta > 40$.

A quantitative comparison of our data with those of Rizzetta and Visbal [17, 18] is shown in Fig. 37a, where time-averaged velocity profiles are reported at various streamwise stations for the central plane of the computational domain ($z = L_z/2$). Good agreement is found at the various stations, which are representative of the boundary layer streamwise evolution. Additional comparison with the reference DNS data is also shown in Fig. 37b, showing the streamwise evolution of the planar integrated fluctuation energy, defined as

$$e(x) = \int\int u_{rms}^2 dydz \tag{14}$$

Again, our data are in good agreement with the reference data, predicting the rapid rise of the perturbation energy past the roughness zone, followed by a saturation zone starting at approximately $(x - x_c)/\delta \approx 30$.

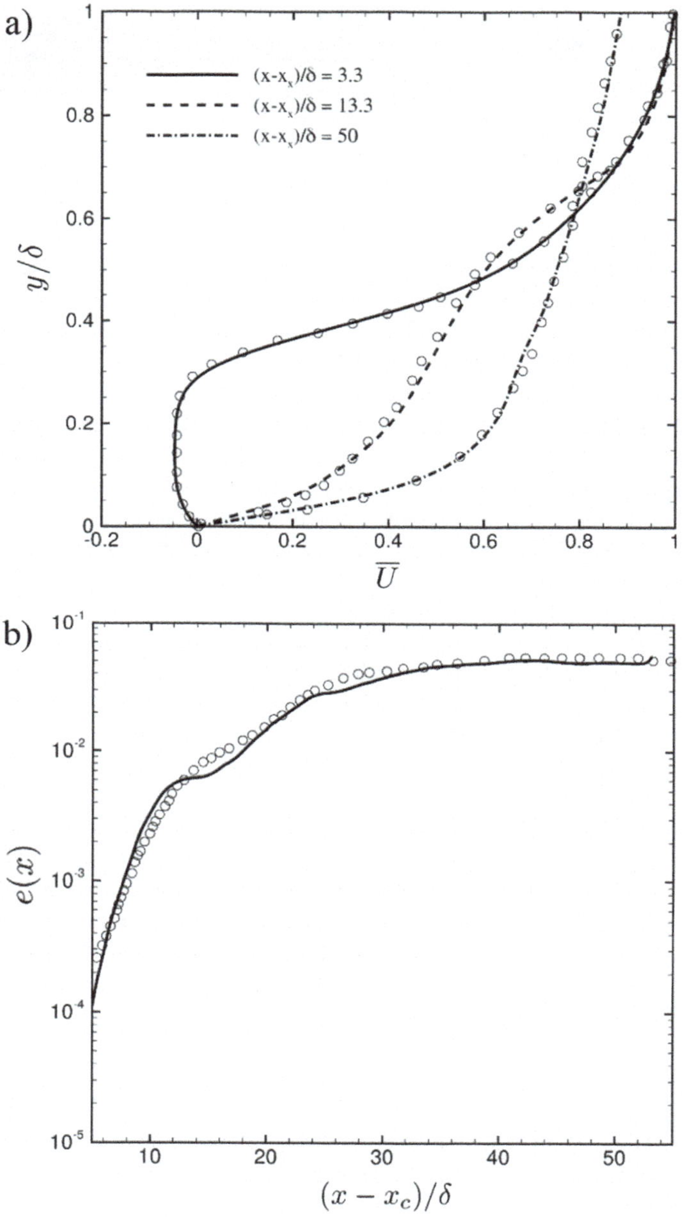

Fig. 37 Comparison with reference numerical data. Time-averaged streamwise velocity at various streamwise stations (**a**) and evolution of the integrated perturbation energy (**b**)

1.6.3 Database Description

Roughness-induced transition in the compressible regime is controlled by many parameters, which makes difficult the identification of general and effective criteria to predict the onset of transition.

In this work we made an effort to produce an extensive DNS database, attempting to cover a wide part of the available parameter space. In particular we have performed a series of simulations varying the following parameters: (i) Mach number (from M = 1.1 to M = 6); (ii) Reynolds number of the incoming boundary layer (the range in terms of roughness Reynolds number is $Re_k = 400–1300$); (iii) obstacle height as a fraction of the boundary layer thickness ($k/\delta = 0.15, 0.2, 0.25, 0.3, 0.4, 0.7$). (iv) roughness element shapes (we have used hemispheres cubes and cylinders). Moreover, for the cylindrical elements we also performed simulations changing the obstacle height/diameter aspect ratio (k/D = 0.5, 1, 2). A list of the simulations, all carried out with the same grid and computational setup of the validation test, is reported in Table 5.

It is important to remind that a critical issue in transitional flows is the characterization of the external disturbance environment [19], especially in the supersonic regime, where the boundary layer is quite receptive to free-stream disturbances [20]. In this case the type and the amplitude of the external disturbances can impact the bypass transition process [21], and they can be considered as an additional independent parameter for the problem under investigation. When dealing with numerical experiments, one has the advantage of working with controlled disturbances, and several options are available, as acoustic disturbances, random or well-organized perturbations.

In our simulations, disturbances are triggered within the incoming boundary layer in the form of random fluctuations of all three velocity components, with maximum amplitude of 0.5% of the free-stream velocity, a choice which clearly excites both the acoustic and the vortical modes of motion. Preliminary analysis performed for representative high-Mach number cases has shown that this level of perturbations does not promote the flow transition in the absence of the roughness element. However, the effect of different choices for the external perturbations is not included in this study, this being the main limitation of the present database.

1.6.4 Results—Compressibility Effects

To highlight the qualitative features of the transition process at the various flow conditions, we focus here on the hemispherical roughness element, under adiabatic wall conditions. The time-averaged skin friction coefficient in the symmetry plane is shown in Fig. 38. At all Mach numbers, two flow states occur, depending on the Reynolds number. In the sub-critical state, the boundary layer remains laminar and the skin friction decreases steadily past the obstacle all the way to the end of the computational domain. In the super-critical state, the amplitude of the disturbance caused by the roughness element is sufficient to promote flow breakdown, as indicated by

Table 5 Summary of parameters for DNS study

Shape	k/D	M_∞	k/δ	Re_δ	Shape	k/D	M_∞	k/δ	Re_δ
Hem	0.5	1.1	0.40	2500	Cyl	0.5	1.1	0.40	4300
Hem	0.5	1.1	0.15	6500	Cyl	0.5	2.0	0.40	6500
Hem	0.5	1.1	0.30	6500	Cyl	0.5	4.0	0.40	28000
Hem	0.5	2.0	0.40	4000	Cyl	1.0	1.1	0.40	3655
Hem	0.5	2.0	0.40	9000	Cyl	1.0	1.1	0.40	6500
Hem	0.5	4.0	0.40	15000	Cyl	1.0	2.0	0.40	7500
Hem	0.5	4.0	0.40	28000	Cyl	1.0	4.0	0.40	28000
Hem	0.5	6.0	0.40	40000	Cyl	2.0	1.1	0.40	2500
Hem	0.5	6.0	0.40	85000	Cyl	2.0	2.0	0.40	4000
Cub	1.0	1.1	0.40	3655	Cyl	2.0	2.0	0.40	9000
Cub	1.0	2.0	0.40	6500	Cyl	2.0	4.0	0.40	20000
Cub	1.0	2.0	0.40	14000	Hem	0.5	1.1	0.40	3655
Cub	1.0	4.0	0.40	38000	Hem	0.5	1.1	0.25	6500
Cyl	0.5	1.1	0.40	6500	Hem	0.5	2.0	0.40	3000
Cyl	0.5	2.0	0.40	9000	Hem	0.5	2.0	0.40	6500
Cyl	0.5	4.0	0.40	40000	Hem	0.5	4.0	0.70	5000
Cyl	1.0	1.1	0.40	5000	Hem	0.5	4.0	0.40	25000
Cyl	1.0	2.0	0.40	6500	Hem	0.5	6.0	0.70	14000
Cyl	1.0	4.0	0.40	20000	Hem	0.5	6.0	0.40	70000
Cyl	1.0	4.0	0.40	40000	Cub	1.0	1.1	0.40	2500
Cyl	2.0	1.1	0.40	6500	Cub	1.0	2.0	0.40	5000

(continued)

Table 5 (continued)

Shape	k/D	M_∞	k/δ	Re_δ	Shape	k/D	M_∞	k/δ	Re_δ
Cyl	2.0	2.0	0.40	6500	Cub	1.0	2.0	0.40	11500
Cyl	2.0	4.0	0.40	15000	Cub	1.0	4.0	0.40	28000
Hem	0.5	1.1	0.40	2900	Cyl	0.5	1.1	0.40	5000
Hem	0.5	1.1	0.20	6500	Cyl	0.5	2.0	0.40	7500
Hem	0.5	1.1	0.40	6500	Cyl	0.5	4.0	0.40	34000
Hem	0.5	2.0	0.40	5000	Cyl	1.0	1.1	0.40	4300
Hem	0.5	2.0	0.40	11500	Cyl	1.0	2.0	0.40	5000
Hem	0.5	4.0	0.40	20000	Cyl	1.0	2.0	0.40	9000
Hem	0.5	4.0	0.40	40000	Cyl	1.0	4.0	0.40	34000
Hem	0.5	6.0	0.40	50000	Cyl	2.0	1.1	0.40	3655
Hem	0.5	6.0	0.40	100000	Cyl	2.0	2.0	0.40	5000
Cub	1.0	1.1	0.40	6500	Cyl	2.0	2.0	0.40	11500
Cub	1.0	2.0	0.40	9000	Cyl	2.0	4.0	0.40	40000
Cub	1.0	4.0	0.40	20000					

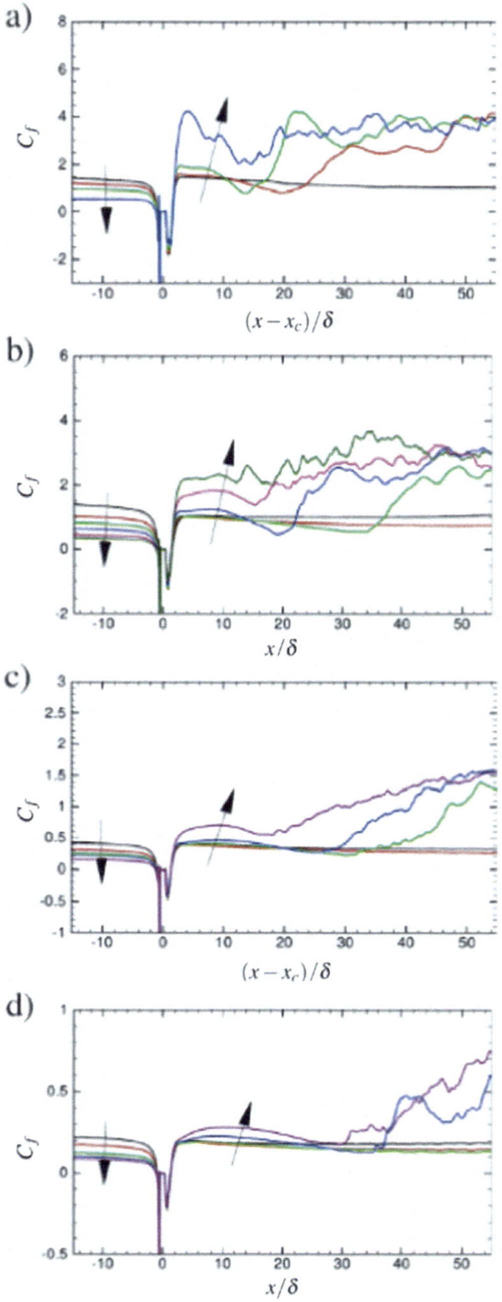

Fig. 38 Mean skin friction coefficient along the symmetry line for hemispherical roughness element with size $k/\delta = 0.4$ **a** $M_\infty = 1.1$, $\mathrm{Re}_\delta = 2500$, 2900, 3655, 6500; **b** $M_\infty = 2$, $\mathrm{Re}_\delta = 3000$, 4000, 5000, 6500, 9000, 11500; **c**, $M_\infty = 4$, $\mathrm{Re}_\delta =$ 15000, 20000, 25000, 28000, 40000; **d**, $M_\infty = 6$, $\mathrm{Re}_\delta =$ 40000, 50000, 70000, 85000, 100000. The arrows indicate the direction of increase Re_δ

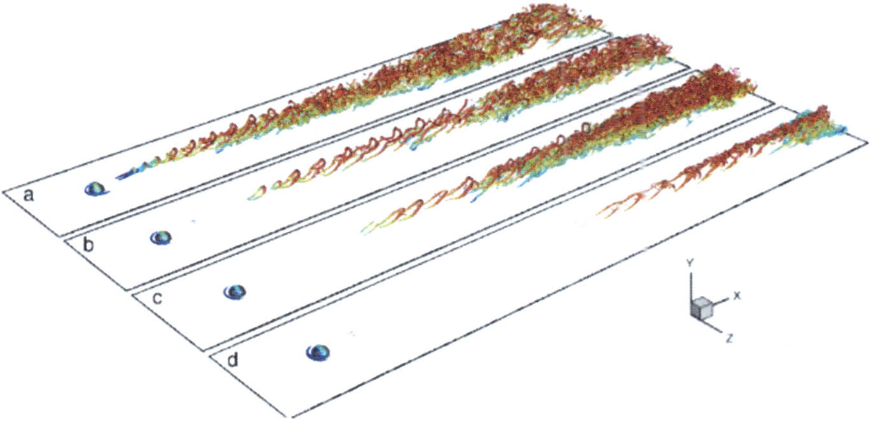

Fig. 39 Coherent vertical structures for flow around hemispherical roughness element. **a** $M_\infty = 1.1$, $\mathrm{Re}_\delta = 6500$; **b** $M_\infty = 2$, $\mathrm{Re}_\delta = 9000$; **c** $M_\infty = 4$, $\mathrm{Re}_\delta = 40000$; **d** $M_\infty = 6$, $\mathrm{Re}_\delta = 85000$. Vortices are educed as isosurfaces of swirling strength, and coloured with the streamwise velocity

the rapid rise of the skin friction coefficient. For a given Mach number, the transition point moves upstream as the Reynolds number is increased, attaining a limiting value for which the roughness is said to be 'effective', and further increase of the Reynolds number does not lead to movement of the transition point. A clear interesting effect of compressibility is observed on such limit position, which increases with the Mach number.

Similar considerations can also be drawn looking at the typical structures characterizing the roughness-induced transition process, shown in Fig. 39 through isosurfaces of the swirling strength (i.e. the imaginary part of the intermediate eigenvalue of the velocity gradient tensor) for representative super-critical flow cases. Regardless of the Mach number, the most prominent feature is the shedding of hairpin vortices past the roughness element, which propagate downstream leading to flow breakdown. The generation and the evolution mechanisms of these structures have been widely discussed in the low-speed regime by Acarlar and Smith [13] and Klebanoff et al. [14], and they are mainly related to the instability characteristics of the shear layer behind the obstacle. As for the skin friction, the main effect of the compressibility is to delay the flow breakdown, and the hairpin structures look more elongated as Mach number increases.

To further characterize the influence of compressibility on the transition process, we look at the frequency of hairpin shedding past the roughness element. For this purpose, the time-resolved streamwise velocity signal was stored along the centerline of the computational domain at an off-wall location corresponding to the edge of the roughness element (y = k). Pre-multiplied frequency spectra taken at different streamwise locations are shown in Fig. 40a for a super-critical flow case. As expected, the spectral densities immediately past the obstacle are characterized by a distinct peak, which is the signature of quasi-periodic shedding. The peak slowly vanishes at

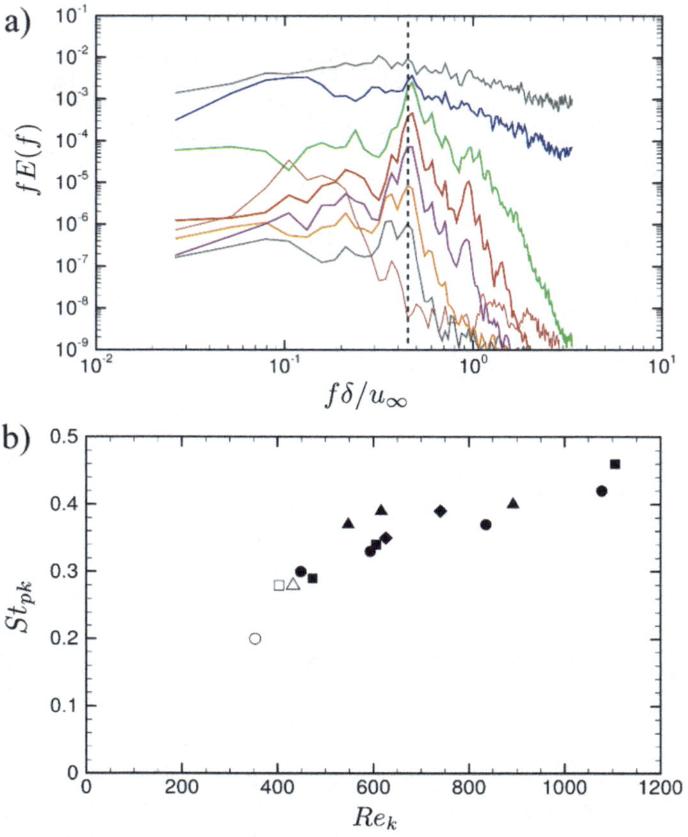

Fig. 40 Pre-multiplied frequency spectra of the streamwise velocity are shown in frame **a** at several streamwise stations ($x/\delta = 5, 17, 18, 19, 20, 23, 30, 40$, from bottom to top) in the symmetry plane at $x = k$, for hemisphere at $M_\infty = 4$, $\mathrm{Re}_\delta = 40000$. The vertical dashed line marks the peak frequency (f_{pk}). Frame **b** shows the peak Strouhal number ($St_{pk} = f_{pk}k/u_k$) derived from maps as in frame (**a**), for flow behind hemispherical obstacles, as a function of Re_k. Symbols: squares, $M_\infty = 1.1$; circles; $M_\infty = 2$; triangles, $M_\infty = 4$; diamonds, $M_\infty = 6$. Solid and hollow symbols correspond to super-critical and sub-critical cases, respectively

downstream stations, where the flow breakdowns, approaching a fully turbulent state, as indicated by the more broad-banded spectra. The typical Strouhal number ($St_{pk} = f_{pk}k/u_k$) associated with the vortex shedding is shown in Fig. 40b as a function of the roughness Reynolds number, for the hemispherical roughness element. Consistent with previous low-speed studies [13, 14], the Strouhal is found to increase with Re_k, in the present case ranging from $St_{pk} = 0.2$ to $St_{pk} = 0.4$. Apparently, the effect of Mach number is not significant when the roughness height and edge velocity u_k are used to form the Strouhal number.

1.6.5 Parametrization of the by-Pass Transition Process

Our goal here is to identify a (possibly) general threshold parameter for the onset of by-pass transition induced by the roughness element. Many generalizations of the roughness Reynolds number criterion to the compressible case have been proposed [22, 17, 19]. Based on the analysis of the present database, we find that a convenient generalization of Re_k is

$$e(x) = \int \int u_{rms}^2 dydz \tag{15}$$

where the density is evaluated at the obstacle edge, and the dynamic viscosity is taken at the wall. Note that Re_{kw} quantifies the ratio between the inertia forces at the edge of the obstacle and the viscous stress at the wall. The performance of Re_{kw} as a controlling by-pass transition parameter can be judged looking at the maps shown in Fig. 41 (left column), in which hollow symbols correspond to sub-critical flow cases, and solid symbols correspond to super-critical cases. To detect the transition onset we monitor the evolution of the time-averaged skin friction coefficient past the obstacle. While no doubts exist for super-critical cases, the stipulation of a flow as sub-critical is necessarily subjected to some uncertainty, owing to the finite streamwise extent of the computational domain.

The figure must then be interpreted with some caution in the sub-critical region. As evident in Fig. 41, the sub-critical and super-critical conditions are well segregated by Re_{kw}, which is promising in taking into account the effect of variations in the Mach number, Reynolds number and roughness height. Such segregation would not be obtained using the conventional definition of roughness Reynolds number. However, the critical value of Re_{kw} is not the same for all the roughness elements, and it ranges between 400 and 700.

limitations of the Re_{kw} criterion we propose an alternative controlling parameter, based on the observation that the main effect of the presence of the obstacle is a deficit in the mass flux (i.e. in the streamwise momentum) downstream of it, which translates into the formation of a low-momentum streak and of a pair of counter-rotating vortices. We then consider a Reynolds number formed with the momentum defect (Q, to be estimated), the maximum cross-stream section of the obstacle, and the wall viscosity,

$$\mathrm{Re}_Q = \frac{Q S_{yz}^{-1/2}}{\mu_w} \tag{16}$$

where $S_{yz} \sim k \cdot D$. It is possible to provide a rough estimate for Q by assuming that the upstream boundary layer is not affected by the obstacle. Assuming that the incoming velocity profile is linear up to the edge of the obstacle, it is easy to obtain the following estimate

$$Q \approx \rho_k k D u_k F(shape). \tag{17}$$

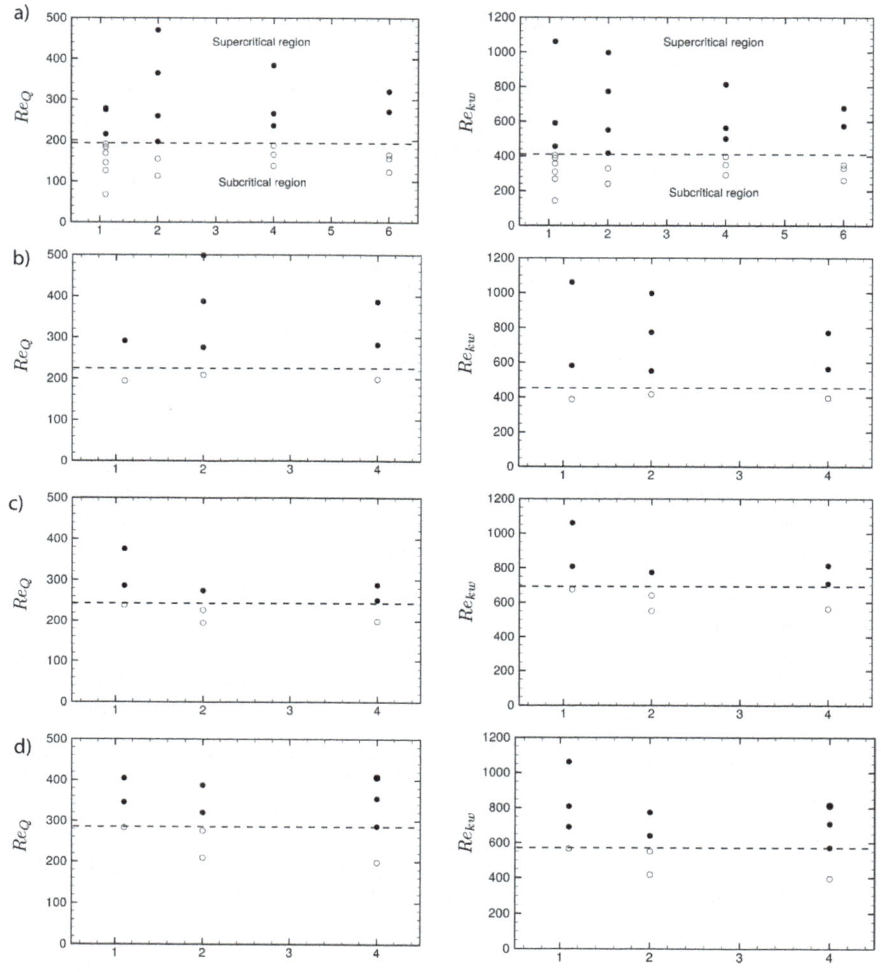

Fig. 41 Transition map as a function of Re_{kw} (left column). **a** hemispheres; **b** cubes; **c** cylinders ($D/k = 2$). Solid/open circles denote super/subcritical cases

Fig. 42 Sketch of the cross-sectional shape of the roughness element

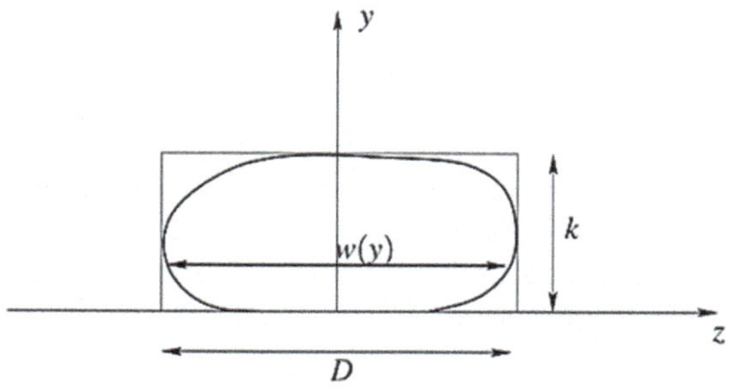

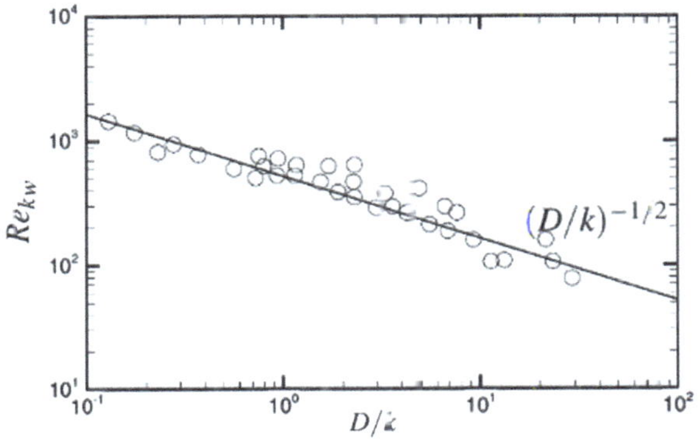

Fig. 43 Dependence of critical roughness Reynolds number on obstacle aspect ratio

where

$$F(shape) = \int_0^1 \eta w^*(\eta) d\eta \tag{18}$$

with $\eta = y/k$ and $w^*(\eta) = w(y)/D$ defines the cross-sectional shape of the obstacle (see Fig. 9). It is then easy to relate Re_Q to Re_{kw} through

$$\mathrm{Re_Q} = \mathrm{Re_{kw}} \cdot (\mathrm{D/k})^{1/2} \cdot \mathrm{F(shape)} \tag{19}$$

showing potential of Re_Q to incorporate the effects of the obstacle shape and aspect ratio. The effectiveness of Re_Q as a by-pass transition parameter can be appreciated from Fig. 41 (right column). It appears that, regardless of the object shape and/or aspect ratio, by-pass transition occurs if $Re_Q > 200 \div 280$, and the scatter is much less than with Re_{kw}.

Apparently, the by-pass transition criterion based on Re_Q correctly captures the experimentally observed trend with the obstacle aspect ratio. In Fig. 43 we show with symbols the critical roughness Reynolds number resulting from a series of experiments (van Driest and Blumer [19] with cylindrical obstacles with different aspect ratio. On the same graph, we show the trend of the critical roughness Reynolds number obtained from Re_Q, which predicts $Re_{kw} \sim (D/k)^{-1/2}$. Apparently, the trend is correct, and it is at least equally plausible as the frequently quoted empirical scaling $Re_{kw} \sim (D/k)^{-2/5}$ [23].

1.6.6 Computational Arrangement and Simulation Parameters

The numerical simulations foreseen in WP-1 and WP-2 include the presence of a normal shock wave that interacts with a boundary layer which can be fully laminar, fully turbulent or transitional. To arrange all the requested flow cases we have devised a simple and unified computational set-up, illustrated in Fig. 44.

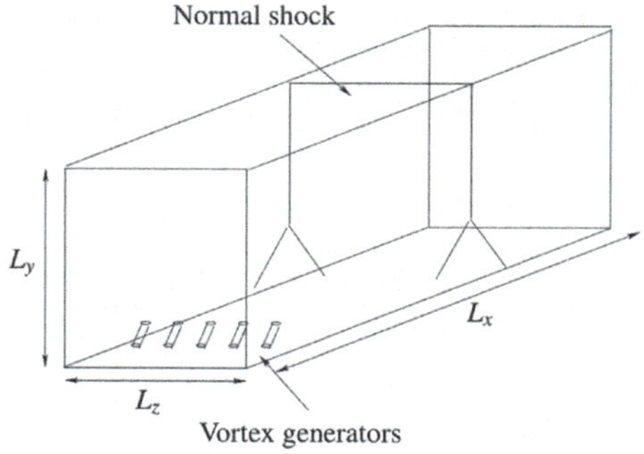

Fig. 44 Sketch of the computational arrangement used for the simulation of the interaction between a boundary layer and a normal shock wave

The domain is a Cartesian box with size $L_x \times L_y \times L_z$ in the streamwise (x), wall-normal (y) and spanwise (z) direction. The normal shock wave is located at some distance (x_s) from the inflow section, and it is made to interact with a boundary layer entering the domain from the inflow station. The leading edge of the plate is not directly simulated, and rather a compressible Blasius similarity solution is used as inflow velocity profile, and also to initiate the simulations. The inflow boundary layer thickness (based on 99% of freestream velocity) is hereafter indicated with, and it is used as the reference length. To control the state of the boundary layer at the beginning of the interaction (laminar, transitional or turbulent) an array of rod vortex generators is used, placed at distance x_r from the inflow.

The geometry of the VGs has been provided by IMP-PAN, and it is described in Fig. 45. Depending on the selected flow conditions (Mach and Reynolds number) and the relative size of the VGs with respect to the boundary layer thickness, the boundary layer may experience transition or not. The rod vortex generators are also used in the case of the laminar interaction to break the initial symmetry of the flow, and to allow the establishment of three-dimensionality across the interaction zone. In this case, care must be taken that the VGs are not effective in causing early transition to turbulence. Preliminary computations for the fully laminar case have shown that the effect of VGs is the same as with random inflow disturbances. Boundary conditions are prescribed by enforcing periodicity in the spanwise direction, and assuming

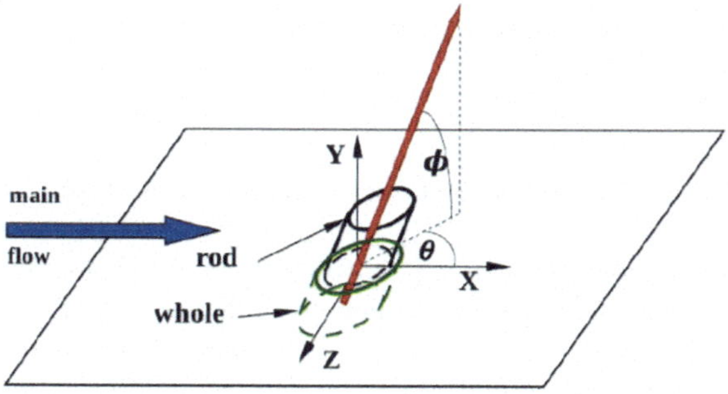

Fig. 45 Sketch of the vortex generator used in the computation

adiabatic no-slip wall conditions at the bottom of the computational domain. Non-reflecting boundary conditions are specified at the outlet and at the upper boundary to minimize spurious reflections of disturbances back into the computational domain [24]. The simulations are initialized by superposing a normal shock field onto the Blasius boundary layer, thus yielding a fictitious transient, after which flow statistics have been collected.

1.6.7 Numerical Parameters

Computations of fully laminar interaction have been carried out for a value of the normal shock strength equal to M = 1.2. The Reynolds number of the boundary layer at the inflow station is 4000, and the Reynolds number at the nominal shock impingement point is 1730000. An array of five VGs with diameter D = 0.11δ and height H = 0.22δ are located at $x_r = 15\delta$, equally spaced in the spanwise direction. The corresponding roughness Reynolds number is $e_k = 250$, at which the obstacles are ineffective in promoting transition.

The computational domain employed for the simulation has size 1000 × 600 × 11 δ and it has been discretized with a grid consisting of points 8192 × 512 × 256. The mesh points are non-uniformly spaced in the streamwise direction to allow better resolution in the region occupied by the VGs and in the interaction zone around the normal shock wave, whereas the spacing is increased towards the computational outlet, starting at x = 500 δ, to filter disturbances prior to hitting the boundary. Points are also clustered in the wall-normal direction to resolve the boundary layer, up to y = 15δ. Finally, the grid points are equally spaced in the spanwise direction. The size of the computational domain has been selected through a series of preliminary calculations. In particular, the length of the computational domain is dictated by the necessity to accommodate the entire interaction zone, including the large region of upstream influence (especially in the case of a laminar interaction) and a large part of the recovery region past the impinging shock. With regard to the extent of the domain in the wall-normal direction, we have found that it is critical to have it as large as possible to prevent choking of the computational duct, thus avoiding bulk motion of the impinging shock. Placing the upper boundary of the domain at 600 δ prevents the occurrence of computational choking over much longer times than those necessary to achieve stationarity of the flow and statistical convergence.

1.6.8 Results for Turbulent Interaction

Numerical parameters
The computation of the fully turbulent interaction has been carried out for a value of the normal shock strength equal to M = 1.3. The Reynolds number of the boundary layer at the inflow station is 20000. An array of ten VGs with diameter D = 0.11 δ and height H = 0.22δ is located at $x_r = 15\delta$, equally spaced in the spanwise direction.

The corresponding roughness Reynolds number is 950, at which the obstacles are expected to promote flow transition.

The computational domain employed for the simulation has a size $150 \times 200 \times 11\delta$ and it has been discretized with a grid consisting of $10240 \times 640 \times 1024$ points. The mesh points are non-uniformly spaced in the streamwise direction to allow better resolution in the region occupied by the VGs. Points are also clustered in the wall-normal direction to resolve the boundary layer, and the compression fan of the shock wave up to $y = 5\delta$. Finally, the grid points are equally spaced in the spanwise direction. The size of the computational domain has been selected through a series of preliminary calculations. In particular, the length of the computational domain is dictated by the necessity to accommodate the entire interaction zone, including the upstream influence region and a large part of the recovery zone past the impinging shock. With regard to the extent of the domain in the wall-normal direction, we have found that it is critical to have it as large as possible to prevent choking of the computational duct, thus avoiding bulk motion of the impinging shock. Placing the upper boundary of the domain at $y = 200\delta$ prevents the occurrence of computational choking over much longer times than those necessary to achieve stationarity of the flow and statistical convergence.

1.6.9 Characterization of the Incoming Boundary Layer

The statistics of the incoming flow are first analyzed to check that the structure of the boundary layer immediately upstream of the interaction is that of a canonical zero-pressure-gradient (ZPG) boundary layer. To that purpose, the distribution of the Van Driest-transformed mean streamwise velocity

$$U_{vd} = \int_0^u \sqrt{\rho/\rho_w}\, du \tag{20}$$

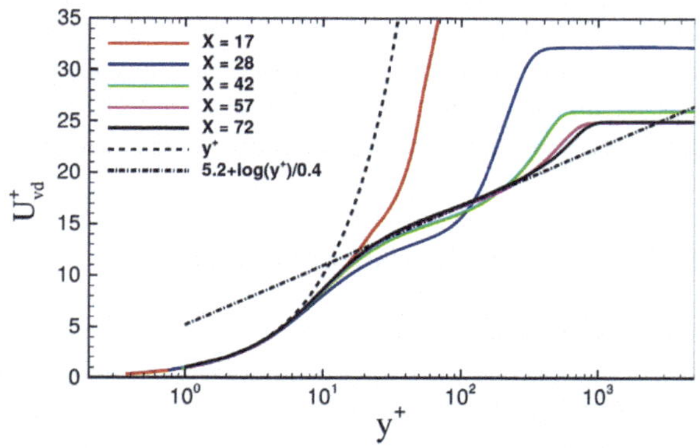

Fig. 46 Distribution of Van Driest transformed mean streamwise velocity in inner scaling at reference stations

Table 6 Properties of incoming boundary layer at the reference station $x_{ref} = 17.3\ \delta_{in}$. $\mathrm{Re}_\theta = \rho_e u_e \theta\ /\mu_e$; $\mathrm{Re}_{\delta 2} = \rho_e u_e \theta\ /\mu_w$; $\mathrm{Re}_\tau = \rho_w u_\tau \delta\ /\mu_w$; $H = \delta^*/\theta$; $H_i = \delta_i^*/\theta_i$

M_e	Re_θ	$\mathrm{Re}_{\delta 2}$	Re_τ	C_f	δ^*/δ	θ/δ	H	H_i
1.3	3770	3020	1003	$2.73 \cdot 10^{-3}$	0.381	0.188	2.03	1.36

is reported in Fig. 46 in inner scaling at some streamwise stations taken past the vortex generators. The figure well describes the evolution of the (initially laminar) boundary layer in the streamwise direction, which, owing to the presence of disturbances generated by the VGs, experience transition to a fully turbulent state, which is achieved approximately at x = 57 δ, where the shape of the velocity profile well conforms to that of a canonical TBL, with a nearly logarithmic region between $y^+ = 30$ and $y^+ = 200$, characterized by a von Karman constant k = 0.41 and additive constant C = 5.2. The global boundary layer properties at a location immediately upstream of the interaction (x = 80δ), are listed in Table 6. The thickness of the boundary layer is determined as the point where u = 99% u_e, and the displacement (δ^*) and momentum (θ) thicknesses are defined as

$$\delta^* = \int_0^{\delta_e} \left(1 - \frac{\overline{\rho}}{\rho_e}\frac{\overline{u}}{u_e}\right) \mathrm{dy},$$
$$\theta = \int_0^{\delta_e} \frac{\overline{\rho}}{\rho_e}\frac{\overline{u}}{u_e}\left(1 - \frac{\overline{u}}{u_e}\right) \mathrm{dy}, \qquad (21)$$

where δ_e is the edge of the rotational part of the flow field, and u_e and ρ_e are the corresponding external mean velocity and density. The 'incompressible' boundary layer thicknesses (pedex i) and the associated shape factor (H_i) are also determined from the above equations by setting the density ratio to unity.

1.6.10 Mean Wall Properties

The mean wall pressure across the interaction zone is here compared with the experimental data of Delery and Marvin. Note that the former were obtained in a transonic channel with a wall-mounted bump to accelerate the flow and form a supersonic region terminated by a quasi-normal shock wave. Hence, owing to the presence of the bump, the boundary layer upstream of the shock wave develops under favourable pressure gradient, and it exhibits a fuller profile than in a ZPG boundary layer at the same Reynolds number. As a consequence, the wall pressure distribution is not constant upstream of the interaction, and the incompressible shape factor has a relatively small value (see Table 7). Because of the differences in the Reynolds numbers, and the sensitivity of the flow details on the downstream conditions comparison with experiments should only be interpreted in qualitative sense.

Let the interaction length-scale L be defined as the distance between the sonic point location x_1 (i.e. the streamwise station where the mean wall pressure equals the

Table 7 Interaction parameters for DNS and experiments (the subscript 0 refers to the origin of the interaction)

	M_0	$Re_{\theta 0}$	H_{i0}	L/δ_0^*	$L/\delta_0^*/(H_{i0}-1)$
DNS	1.3	3770	1.36	25.5	70.
Delery and Marvin [25]	1.3	7526	1.30	21.	70.

critical pressure $p^* = 1.46 p_e$ at M = 1.3), and the origin of the interaction x_0 (i.e. the point where the wall pressure starts to rise) For weak-to-moderate interactions, Delery and Marvin showed that scales with the upstream boundary layer properties according to

$$\text{L} \approx 70\delta_0^*(H_{i0} - 1) \tag{22}$$

As observed in Table 7, the computed interaction length-scale agrees fairly well with the equation above; Delery and Marvin also showed collapse of the wall properties (at various Re and M) when reported in the scaled interaction coordinates $x^* = \frac{x-x_0}{L}$, $y^* = y/L$. In the following, for comparison purposes, the results are then reported in terms of x^* and y^* and we refer to three distinct zones: the upstream ZPG region ($x^* < 0$); the supersonic adverse-pressure-gradient (APG) region ($0 < x^* < 1$); and the subsonic APG region ($x^* > 1$).

The scaled mean wall pressure is reported in Fig. 47a together with the inviscid pressure jump predicted by the Ranking-Hugoniot relations. It exhibits a sharp rise in the supersonic APG region, in excellent agreement with experiments, and a milder increase in the subsonic APG region.

Note that, remarkably, the distribution obtained in the present DNS is nearly identical to that of a previous computation Pirozzoli et al. [26], performed with the recycling/rescaling method, without taking into account the full spatial evolution of the incoming boundary layer, from a laminar to a fully turbulent state. This agreement confirms the validity of the approach here followed, which has the advantage to be more general and to allow the opportunity of simulating transitional interactions.

The distribution of the mean skin friction coefficient is reported in Fig. 47b together with that obtained in the absence of the normal shock. These curves are nearly identical up to (x/δ ~ 90), which marks the location of the shock upstream influence and are characterized by a rapid rise of the friction coefficient at x, corresponding to the flow breakdown and the transition to a turbulent state. While in the absence of the normal shock the skin friction slightly decreases in the fully developed turbulent region, in the case of SBLI, starting from x = C_f exhibits a sudden drop, and attains a minim (close to zero), before showing a slow recovery past the interaction zone. According to Delery and Marvin the present case is to be classified as an incipient separation one.

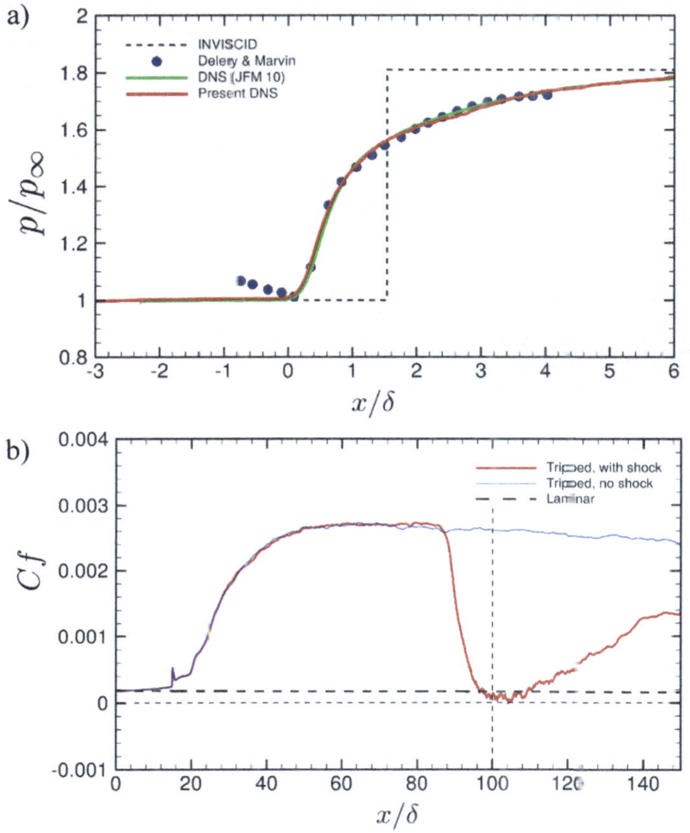

Fig. 47 Distribution of mean wall pressure (**a**) and skin friction coefficient (**b**). *R–H* denotes the inviscid distribution resulting from the Rankine-Hugo10niot jump conditions. Refer to Table 7 for nomenclature of symbols

1.6.11 Flow Visualizations: Instantaneous Fields

The computed instantaneous density, streamwise velocity and pressure fields are reported in Fig. 48 in the streamwise, wall-normal plane at a given time frame. As observed by Pirozzoli and Grasso [27]in the case of an impinging shock/boundary layer interaction at supersonic Mach number, and by Na and Moin [28] for a low-speed turbulent boundary layer under adverse pressure gradient, the instantaneous pressure field highlights the formation of pressure minima associated with the shedding of eddies that form in the proximity of the outermost inflection points of the mean velocity profiles. As better appreciated from inspection of the flow animations (available on the TFAST website), such vortices lift off from the wall approximately in the middle of the interaction zone, and propagate downstream giving rise to a turbulent mixing layer. Sharp density interfaces are also observed in the outermost

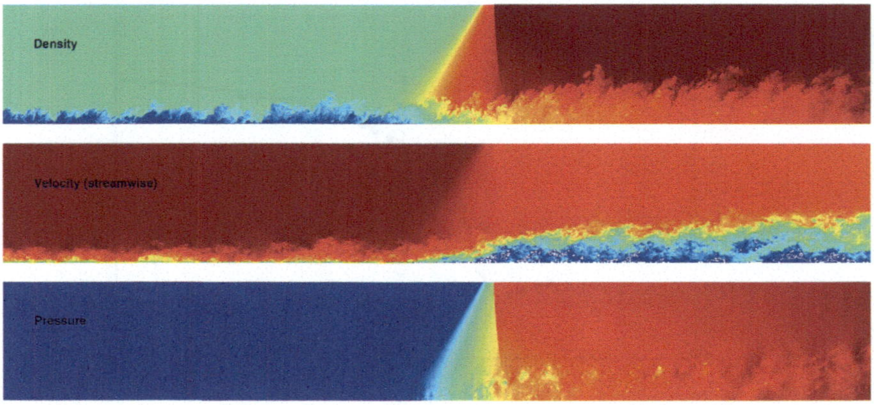

Fig. 48 Instantaneous fields of density, streamwise velocity and pressure in streamwise, wall-normal plane. The corresponding animation is available on the TFAST website

part of the boundary layer, separating boundary layer turbulence from the outer, essentially inviscid flow, and that become more convoluted past the interaction zone.

For a more complete representation of the flow, in Fig. 49 we report a three-dimensional instantaneous view of vortical structures educed with iso-surfaces of the swirling strength (i.e., the imaginary part of the intermediate eigenvalue of the velocity gradient tensor). The location of the normal shock is visible through contours of the pressure field, reported in a x-y plane slice. The figure allows to appreciate the full spatial evolution of the boundary layer, which undergoes transition past the

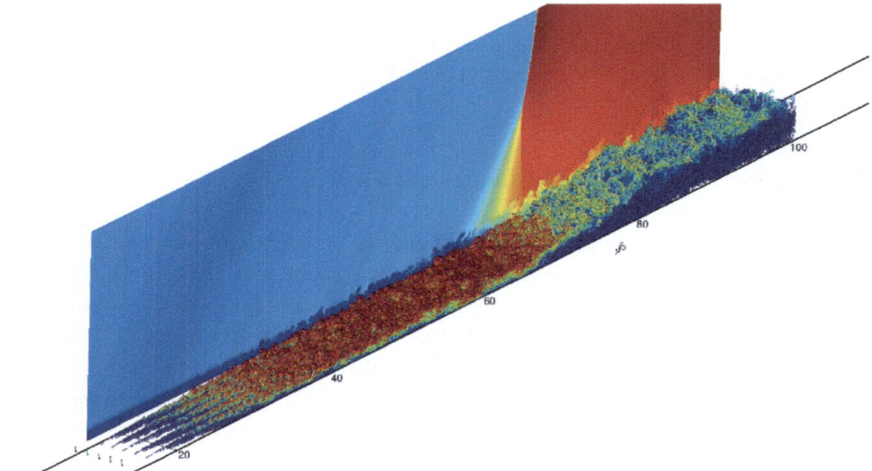

Fig. 49 Flow visualization for turbulent interaction at $M_s = 1.3$. Vortical structures are educed through iso-surfaces of the swirling strength. The pressure field is also shown in a x-y plane slice

VGs element. The most prominent feature is the shedding of hairpin vortices, which propagate downstream, past the roughness element, leading to flow breakdown. The generation and the evolution mechanisms of these structures have been widely discussed in the low-speed regime by Acarlar and Smith [13], Klebanoff et al. [14] and they are mainly related to the instability characteristics of the shear layer behind the obstacle.

2 Interaction Sensitivity to Transition Location

2.1 *IUSTI*

We present in this section an analysis of the influence of the position and the height of the perturbations on the transition to the turbulence of the interaction. All the experimental results have been summarized in order to define critical region of influence of upstream perturbations. From the various experiments achieved with different heights and unit Reynolds numbers it is possible to evaluate the critical size of the step able to generate turbulent conditions upstream and/or downstream from the interaction. In cases of turbulent upstream conditions, it has been found that the flow remains attached, whatever the imposed flow deviation considered. It has not been possible to obtained attached flow with no turbulent upstream boundary layer: all the other cases have been found separated with large aspect ratios of the interaction.

A Reynolds number based on the height of the step has been used to summarize these different results. It is based on the height of the step h, its location along the plate x and the maximum velocity seen by the step $U_h = U(y = h)$. This corresponds to an immersed Reynolds number. Then, Re_{hi} is defined as:

$$\mathrm{Re}_{hi} = \frac{\mathrm{U_h h}}{\upsilon_e} \tag{23}$$

In case of laminar boundary layers, the velocity profile can be considered as linear for $y/\delta < 0.5$. Therefore, U_h can be approximated by:

$$\frac{\mathrm{U_h}}{\mathrm{U_e}} \cong \frac{1}{\mathrm{k_1}} \frac{\mathrm{h}\sqrt{\mathrm{Re_u}}}{\sqrt{\mathrm{x}}} \tag{24}$$

with $k_1 = 4.2$ (see Fig. 21, chapter "WP-1 Reference Cases of Laminar and Turbulent Interactions") and the relation (22) can be expressed as:

$$\mathrm{Re_{hi}} = \frac{1}{\mathrm{k_1}} \mathrm{Re_u^{3/2}} \frac{\mathrm{h^2}}{\sqrt{\mathrm{x}}} \tag{25}$$

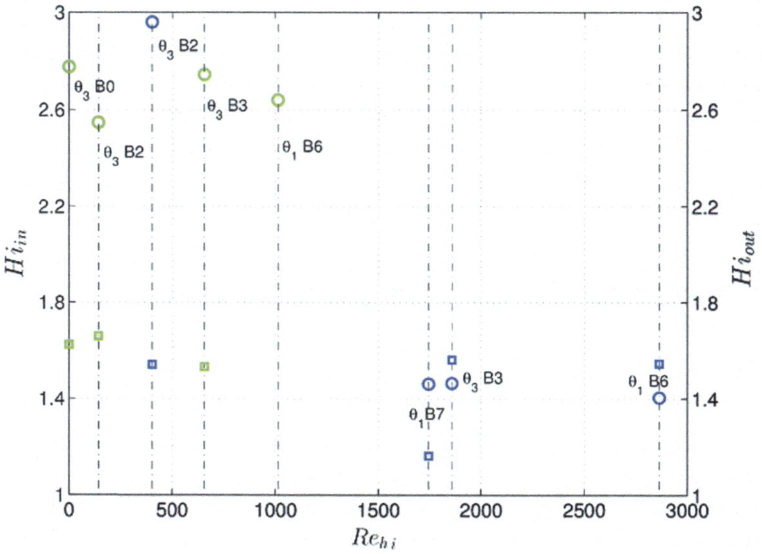

Fig. 50 $P_0 = 0.4$ atm

with Re_u the unit Reynolds number. The incompressible form factors of the upstream and downstream boundary layers are reported on Fig. 50 versus this parameter. Green and blue symbols correspond respectively to the $P_0 = 0.4$ atm and $P_0 = 0.8$ atm cases. The circles correspond to the upstream section and the square to the downstream one. Symbols are labeled with the corresponding flow deviation and tripping device. The form factors are plotted vs the immersed Reynolds number Re_{hi} and the natural case are reported for $\mathrm{Re}_{hi} = 0$. The largest form factors (about 2.7), correspond to the laminar conditions, when the turbulent conditions have form factors around 1.4. The downstream transitional states corresponds to the symbols labeled $\Theta_3 B_0$ and $\Theta_3 B_2$ for the $P_0 = 0.4$ atm cases: they are not clearly characterized from their form factors, only slightly larger than the other downstream turbulent cases. Different domains can be defined from these results:

- for experiments with values of immersed Reynolds number at least up to 140, laminar input conditions are observed and only a transitional boundary layer is developing downstream from the reattachment point
- for Reynolds number such as $400 \leq \mathrm{Re}_{hi} \leq 1000$, laminar input conditions are observed and a turbulent boundary layer is developing downstream from the reattachment point. The flow remains separated
- for Reynolds number such as $1750 \leq \mathrm{Re}_{hi}$, turbulent conditions are observed upstream and downstream from the interaction. The flow is attached whatever the imposed flow deviation.

These results are summarized on Fig. 51 for the two stagnation pressures $P_0 = 0.4$ atm and $P_0 = 0.8$ atm. The X axis is the longitudinal section of the flat plate and

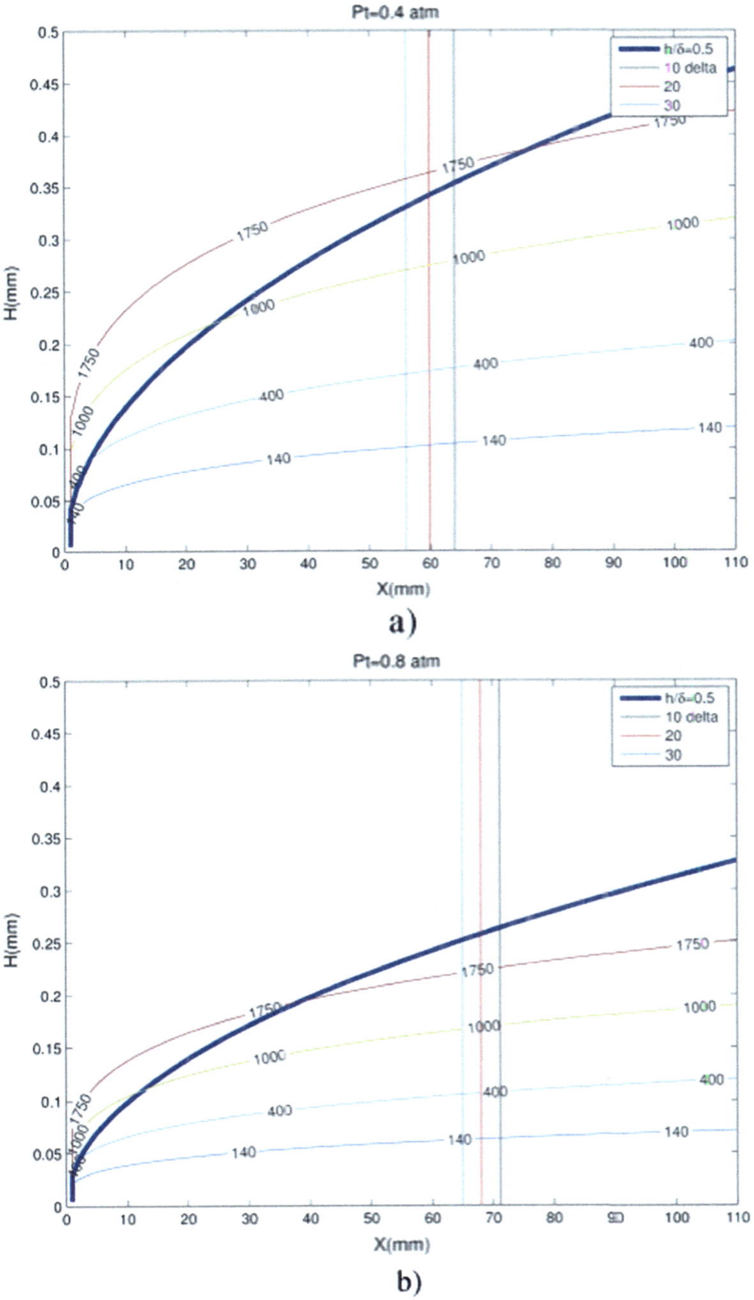

Fig. 51 Characteristic immersed Reynolds numbers. **a** $P_0 = 0.4$ atm; **b** $P_0 = 0.8$ atm

the Y axis is the height of the step in mm. The isovalues Reynolds number Re_{hi} curves are given for the values 140, 400, 1000 and 1750. Superimposed on the graphs is the curve corresponding to step heights equal to $0.5\delta_h$. This can be considered as an upper limit of the size of the steps. The 3 vertical lines show the section corresponding respectively to 10, 20 and 30 local boundary layer upstream from the location of the initial rise of pressure (compression waves).

These results are summarized in Fig. 51 for the two stagnation pressures $P_0 = 0.4$ atm and $P_0 = 0.8$ atm. The X axis is the longitudinal section of the flat plate and the Y axis is the height of the step in mm. The Reynolds number Re_{hi} isovalue curves are given for the values 140, 400, 1000 and 1750. Superimposed on the graphs is the curve corresponding to step heights equal to $0.5\delta_h$. This can be considered as an upper limit of the size of the steps. The 3 vertical lines show the section corresponding respectively to 10, 20 and 30 local boundary layer upstream from the location of the initial rise of pressure (compression waves). Therefore, these pictures give a first estimation of the possibilities (height and position) for control systems able to obtain:

- separated interactions, upstream laminar conditions, and downstream transitional conditions (region under the isoline 140)
- separated interactions, upstream laminar conditions, and downstream turbulent conditions (region between the isolines 400 and 1000)
- attached interactions, upstream turbulent conditions, and downstream turbulent conditions (region over the isoline 1750).
  The evolution of the unsteadiness inside along the interaction has been shown to follow these regions of immersed Reynolds numbers.
- unsteady interaction, involving low frequencies (if compared to the scales of the upstream boundary layer) developing along the mixing layer downstream of the separation point. Nevertheless, the dominant scales remain the unsteadiness coming from the upstream laminar boundary layer (region under the isoline 140)
- unsteady interaction, the dominant scales are the lower frequencies developing along the mixing layer (region between the isolines 400 and 1000)
- attached/turbulent interactions with no evidence of low frequency unsteadiness ($\mathrm{Re}_{hi} > 1700$).

From these figures it is clear that, with the actual set-up, the 0.4 atm case has limited possibilities to generate attached flow, it means upstream turbulent conditions. The 0.8 atm case has a larger efficient region, but requests quite small height (typically $y < 0.2$ mm). Of course, these results are based on steps experiments, and we can expect some differences for other perturbations such as Rod Vortex Generators, but the differences are not expected that large, and we should stay in vicinity of the regions highlighted on the figures.

2.2 ITAM

Passive and active methods of the laminar-turbulent transition control studied in ITAM (Sect. 1) have not led to a smooth displacement of the transition location. Therefore it was decided to investigate the sensitivity of the interaction zone to the transition location for the case of natural turbulization. Table 8 shows the main parameters of the experiments.

Figure 52 shows the results of schlieren visualization of the interaction zone. For the transitional case the size of the interaction region is less than for the laminar case. Decreasing of wedge angle results in slight shift of the position of the laminar-turbulent transition downstream. It is necessary to note that for zero pressure gradient BL (no shock) laminar-turbulent transition take place close to the interaction zone. Therefore, in this case the pressure gradient has little effect on the position of the transition. An increase of the Reynolds number significantly reduces the size of the separation region (not shown here).

The velocity fields measured at different wedge angles are shown in Fig. 53. It is interesting to note that in contrast to the laminar case, for the transitional one the separation shock wave was substantially weaker. Comparing the data obtained for similar size of the separation zone for the transition case (Fig. 53, $\beta = 3°$) and the laminar case (Fig. 53, $\beta = 2°$) it is clear to see that the reflected shock wave is well observed only for the laminar case. This flow feature can be explained by difference of unsteady flow in the interaction for different states of the incoming boundary layers. The periodic passage of turbulent spots through the interaction zone for the transitional case should lead to the suppression of the separation and the disappearance of the corresponding shock wave. After this the separation zone begins to return to its laminar state. As a result of this phenomenon the observed mean velocity fields are formed.

An important change of the flow was found in the wake, where the thickness of the turbulent boundary layer for the transitional case was much smaller than for the laminar or turbulent ones. This was confirmed by the data shown in the Fig. 65 (chapter "WP-1 Reference Cases of Laminar and Turbulent Interactions"). It is clear that weak growth of the momentum thickness for the transitional test case is accompanied by small level of pulsations in comparison with the laminar test case. In

Table 8 Parameters of experiments

M	P_0, bar	T_0 (K)	Re_1, 10^6 1/m	X_{imp} (mm)	β (°)	Re_{Ximp}, 10^3	δ* (mm)	
1.43	0.533	284.7	8.77	183	4	1600	–	Tran.
1.43	0.688	284.5	10.91	183	1,2, 3,4	2000	0.31	Tran.
1.43	0.841	285.6	13.27	183.5	4	2435	0.31	Tran.
1.43	0.976	286.5	15.32	183.5	1,2, 3,4	2810	0.24	Tran.

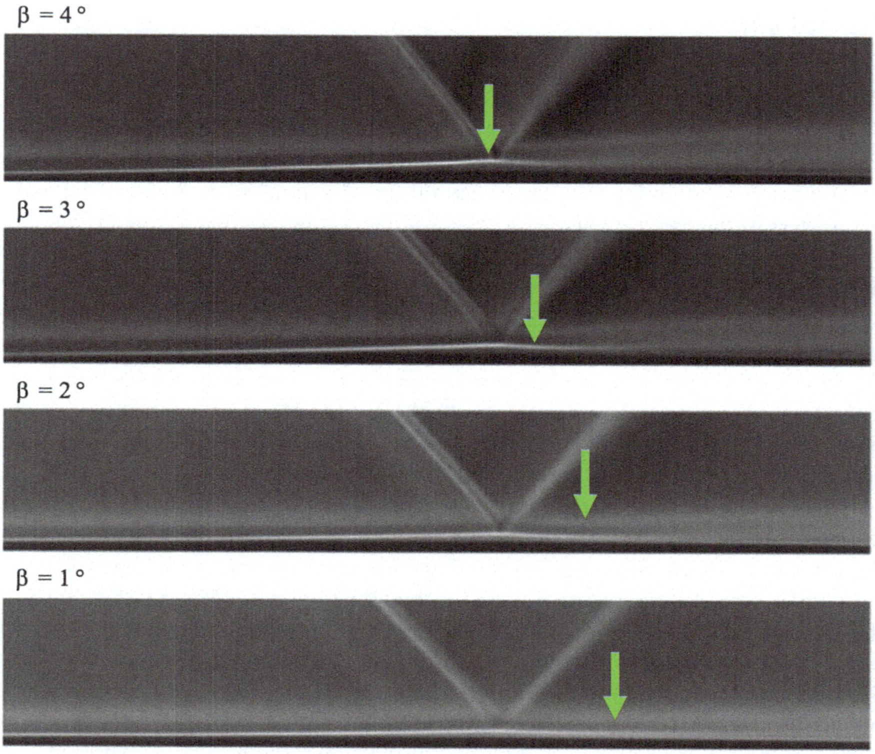

Fig. 52 Schlieren visualization (transitional case, L = 200 mm, Re_1 = 10.9e6 1/m)

Fig. 54 one can see the energy distribution for the first 20 modes estimated by POD analysis. The zones of measurements for all cases were approximately the same that allows to perform a quantitative comparison. The energy of structured oscillations of the flow for the laminar case significantly exceeds the values obtained for turbulent and transitional cases. Thus for laminar case the powerful coherent structures present in the interaction zone. It is most likely that these structures have a significant effect on mean flow and are responsible for the rapid growth of the thickness of the boundary layer in the wake. The presence of turbulent spots for the transitional case destroys the mechanism of generation of these structures.

In Fig. 55 one can see the distribution of PSD*(f)* f along the SWBLI in beginning (Fig. 55a) and ending (Fig. 55b) of the transition zone. For the transitional case in the incoming boundary layer on can see presence of a peak of pulsations with frequency of about 5–8 kHz, which may be related to processes of turbulent spot formation. Correlation analysis shows that these disturbances propagate downstream. Near the point of X_{imp} there is a sharp increase of the pulsations associated with the turbulization of the boundary layer. Further downstream we can see rapid increase of the

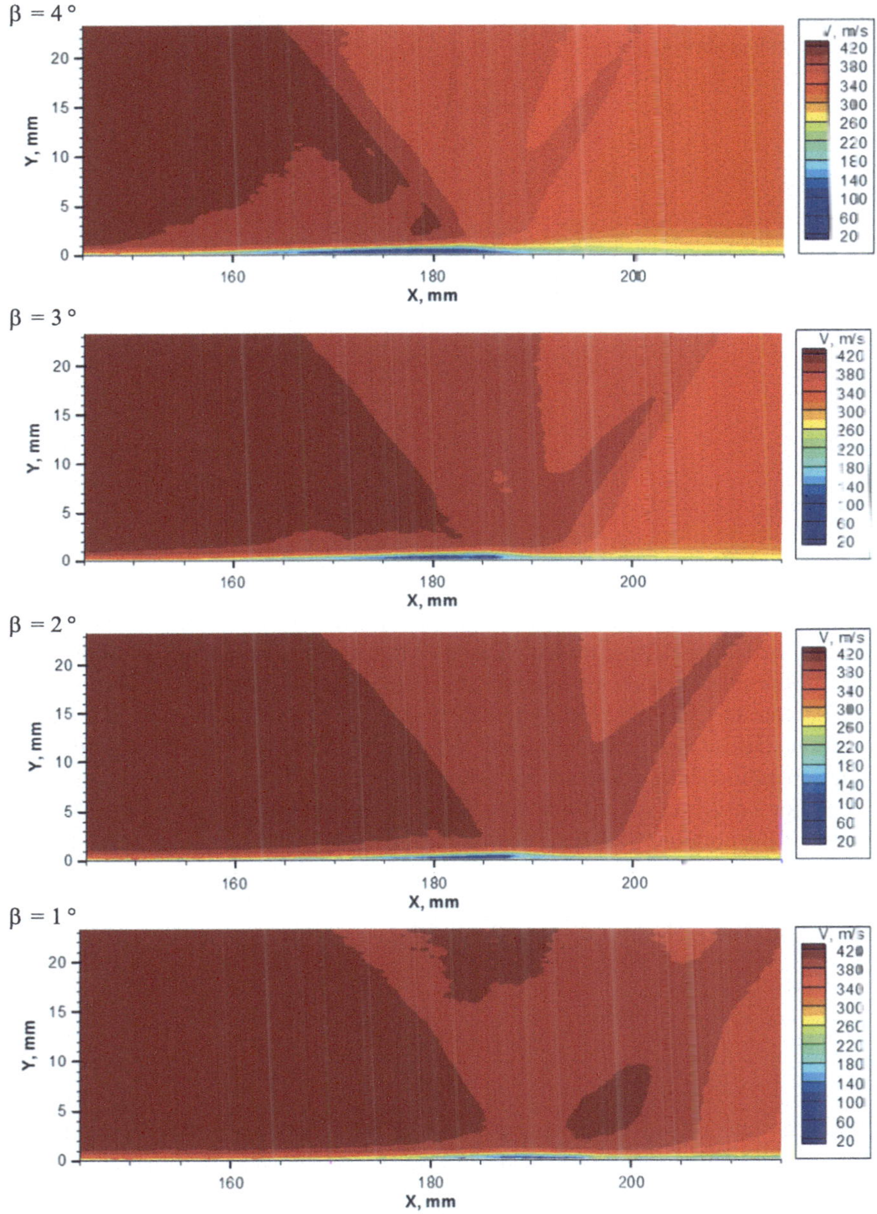

Fig. 53 Velocity fields (transitional case. L = 200 mm, Re_1 = 10.9e6 1/m)

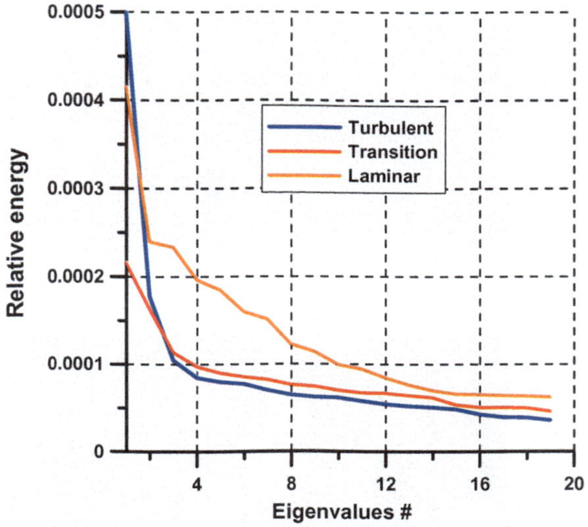

Fig. 54 Relative energy associated with mode #m

pulsations for all frequency band. This means the formation of well-developed turbulent boundary layer. Peak of pulsation at low frequency (about 1 kHz) is obviously due to some feedback process (oscillation of the shock wave and separation zone). From the correlation study it was concluded that disturbances up to 0.5 kHz are moving upstream in the separation region. The spectrogram obtained at the end of the transition zone is close to the turbulent case. However the low-frequency oscillations of the separation shock wave are not found here.

Data for a quantitative comparison can be obtained from the pressure fields. The method of the pressure fields calculation basing on PIV data is presented in [1]. For comparison two parameters are used: F_L and F_D. F_L corresponds to loss of lifting properties of a surface (for wing, compressor blades, etc.) due to the pressure rise in comparison with inviscid case and calculated as

$$F_L = \int_{X_{in}}^{X_{imp}} (P_{st} - P_{st_{in}})dx \tag{26}$$

F_D is an integral value of the total pressure losses in the shear layer due to the formation of SWBLI zone. It is calculated as difference between the integral pressure losses at inflow boundary and the value obtained in point X_{imp}:

$$F_D = \int_0^y P_{0_{in}}dy - \int_0^y P_0 dy \tag{27}$$

Fig. 55 Spectrogramm of wall pressure pulsations for laminar test case ($P_0 =$ 0.7 bar)

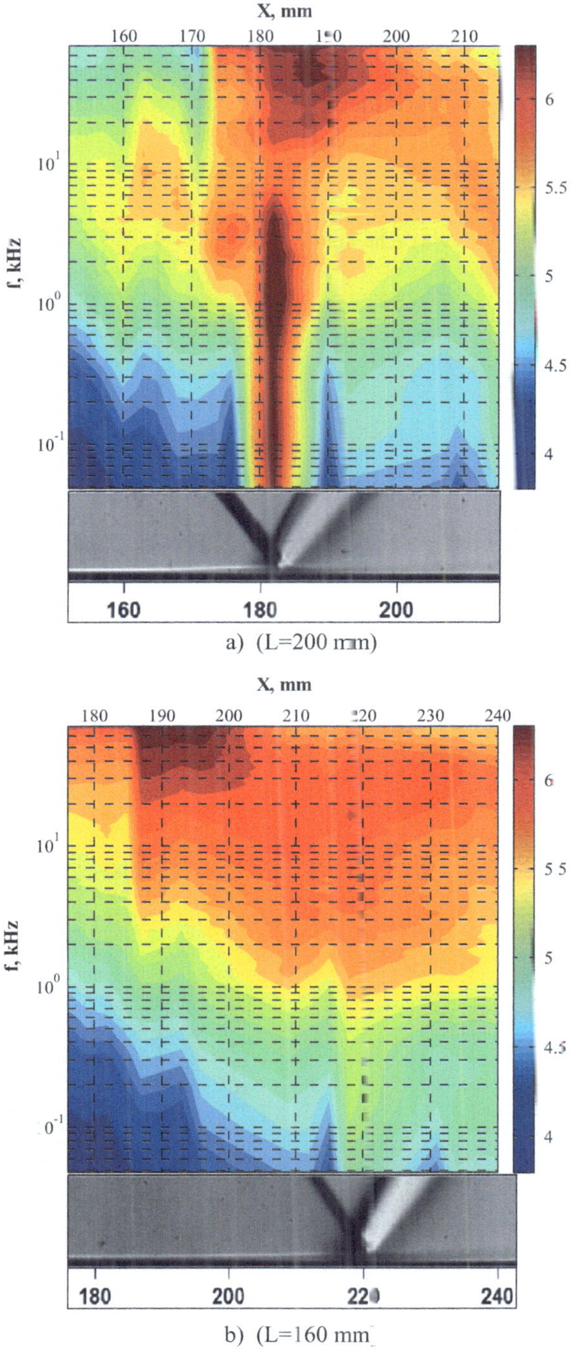

The production of the upstream dynamic pressure and upstream boundary layer displacement thickness was chosen to normalize the parameters F_L and F_D.

In Fig. 56 it is clear to see that for the dimensional and dimensionless plots there is approximately linear dependence of the stagnation pressure losses on the wall pressure, regardless of the inflow boundary layer state. Let's consider in detail the

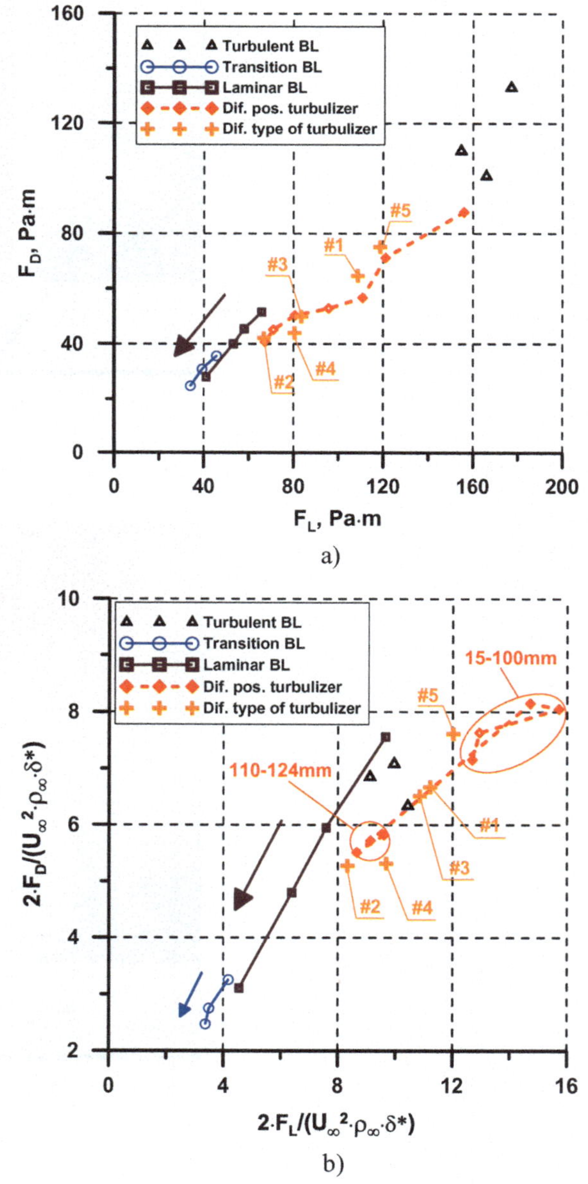

Fig. 56 Dimension (**a**) and dimensionless (**b**) values of total pressure loss vs the increase of wall pressure on the model

data presented in the Fig. 56b. The increase of unit Reynolds number (shown by arrow) improves the parameters of SWBLI for laminar and transitional cases. The values obtained for laminar and transitional cases lie on the same curve. This can be explained by the increase of inflow disturbances (up to the turbulent state of the inflow boundary layer) which make the boundary layer more resistant to adverse pressure gradient. However for the case of turbulent inflow boundary layer there is a sharp reduction of parameters of the interaction zone.

Unit Reynolds number has little effect on the dimensionless parameters F_L and F_D for the turbulent cases. This confirms the strong dependence of the SWBLI parameters on the level of and type of pulsations in the inflow boundary layer. It is interesting to note that the dimensionless parameters for the turbulent and laminar case at the low unit Reynolds number are approximately the same.

The turbulators producing laminar-turbulent transition upstream of the interaction zone, make the parameters of the zone slightly worse compared to the case of natural turbulization. As a result. we can make a preliminary conclusion that the best parameters of the SWBLI zone may be obtained when the inflow boundary layer is in the transitional state. Turbulent case has no advantages in comparison with laminar, therefore from this point of view, using of roughness for the laminar flow is not beneficial.

The obtained data show that the presence of turbulence spots (intermittency) can significantly improve the flow in the SWBLI zone, which is demonstrated in Sect. 1 for the example of active flow control by an electric discharge.

The physical mechanism of the development of nonstationary effects in the interaction zone for the transitional case raises many questions requiring an intensive study.

2.3 *ONERA-DAFE*

2.3.1 Comparison Between Cadcut and "ZZ" Tape Devices

Figure 57 shows a comparison of Schlieren visualizations from the reference case and the two controlled cases (Cadcut and ZZ100 tape) for the X_{a1} location on the flat plate. For the reference case (see Fig. 57a), the incoming laminar boundary layer is separated due to the impact of the shock-wave and the transition process appears in the interaction domain because vortex structures are visible only in the mixing layer downstream of the interaction region. The effect of both control devices (easily identified due the presence of compression and expansion waves, see Fig. 57b, c) is to eliminate the separation region and to rapidly trigger the transition. Indeed, vortex structures are now present far upstream of the interaction domain (behind the turbulators, more clearly visible for the ZZ100 tape device).

Figure 58 shows a comparison of oil flow visualizations from the reference case and the two controlled cases (Cadcut and ZZ100 tape) for the X_{a1} location on the flat plate. For the reference case, one notices the footprints of the shock pattern (I)

a) - reference case

b) - control case: ZZ100 tape

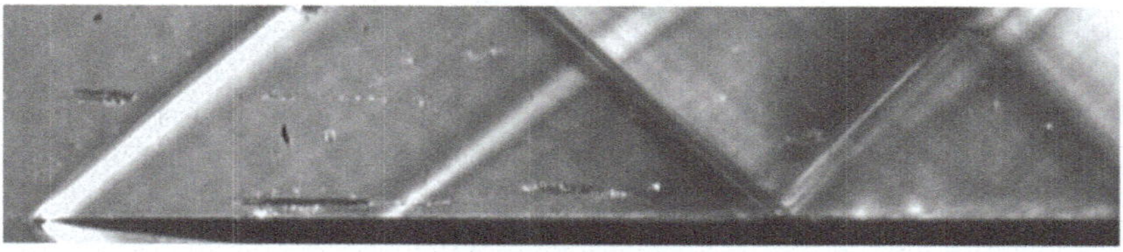

c) - control case: Cadcut

Fig. 57 Schlieren visualizations (devices at X_{d1} location)

and the waves emanating from the shock holder (R1 and R2). On the other hand, the separation line (D) is barely visible. For both controlled cases, flow topologies are very similar and one clearly sees the generation of vortices in the wake of the devices.

Figure 59 shows a comparison of temperature maps obtained by IR thermography measurements from the reference case and the two controlled cases (Cadcut and ZZ100 tape) for the X_{d1} location on the flat plate. While the temperature level increases in the interaction domain for the reference case, the temperature starts to increase just downstream of the devices (the ZZ100 tape is identified as a dark blue band, see Fig. 59b). This increasing is strengthened by the control with the Cadcut device (see on Fig. 59c the yellow spots in the vicinity of the device). This means that both devices forced the transition of the boundary layer. However, these temperature maps do not give a clear indication concerning the (no) existence of the separated region.

2.3.2 Effect of the Device Height

Figure 60 shows a comparison of oil flow visualizations from the "ZZ" tape device with the two selected heights (ZZ100 and ZZ200) for the X_{d1} location on the flat

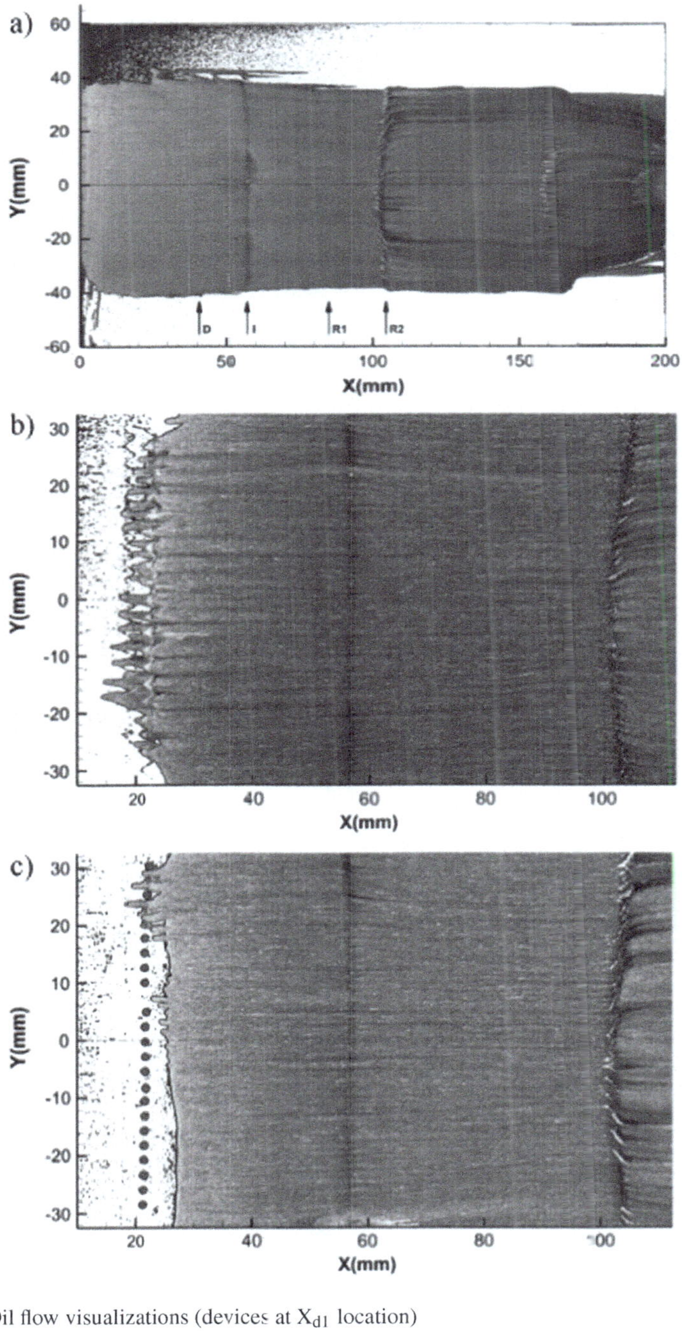

Fig. 58 Oil flow visualizations (devices at X_{d1} location)

Fig. 59 Temperature maps by IR thermography measurements (devices at X_{d1} location)

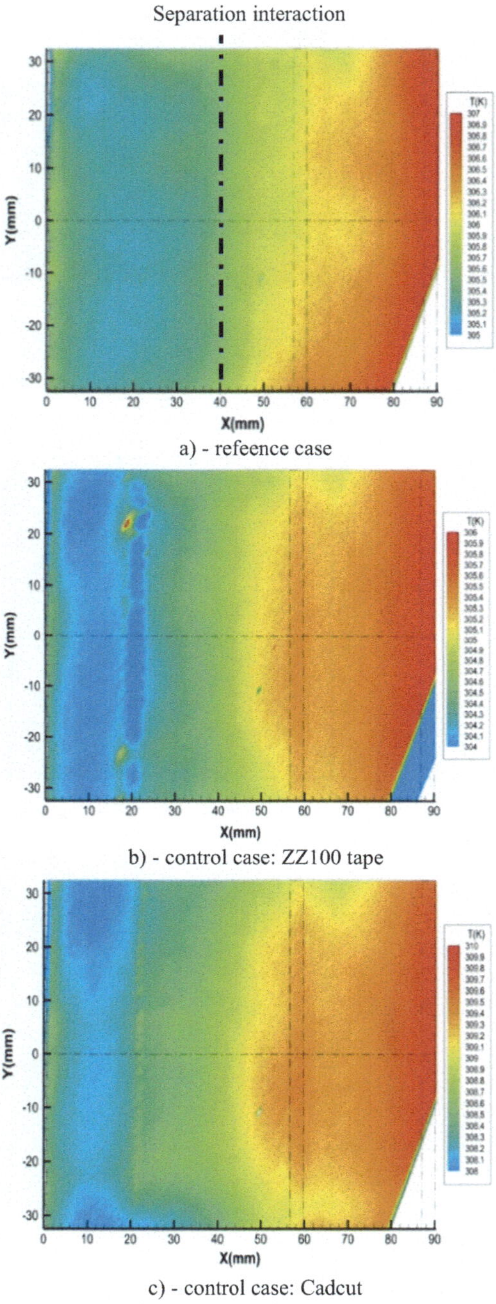

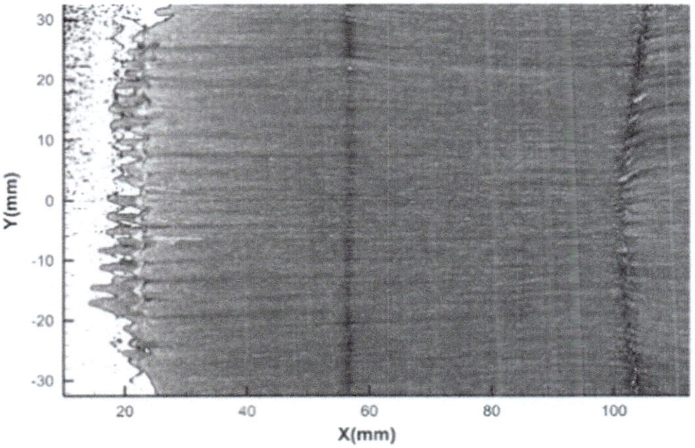

a) - control case: ZZ100 tape

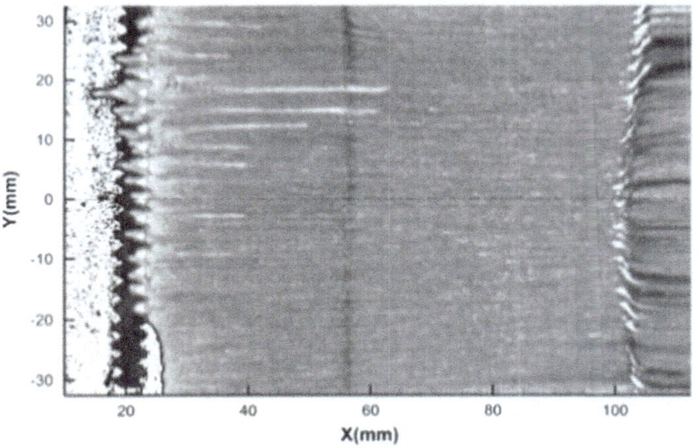

b) - control case: ZZ200 tape

Fig. 60 Oil flow visualizations (devices at X_{d1} location)

plate. Even if flow topologies are very similar, it seems that vortices generated in the wake of the highest device (ZZ200) are stronger.

Figure 61 shows a comparison of temperature maps obtained by IR thermography measurements from the "ZZ" tape device with the two selected heights (ZZ100 and ZZ200) for the X_{d1} location on the flat plate. While the temperature level starts to continuously increase just downstream of the ZZ100 tape device (see Fig. 61a), one notices a different evolution of temperature distribution behind the ZZ200 tape device: the temperature starts to increase but rapidly decreases to reach the level upstream of the turbulator and, then, increases again near the interaction region (see Fig. 61b). This (re-)laminar state of the boundary layer downstream of the turbulator

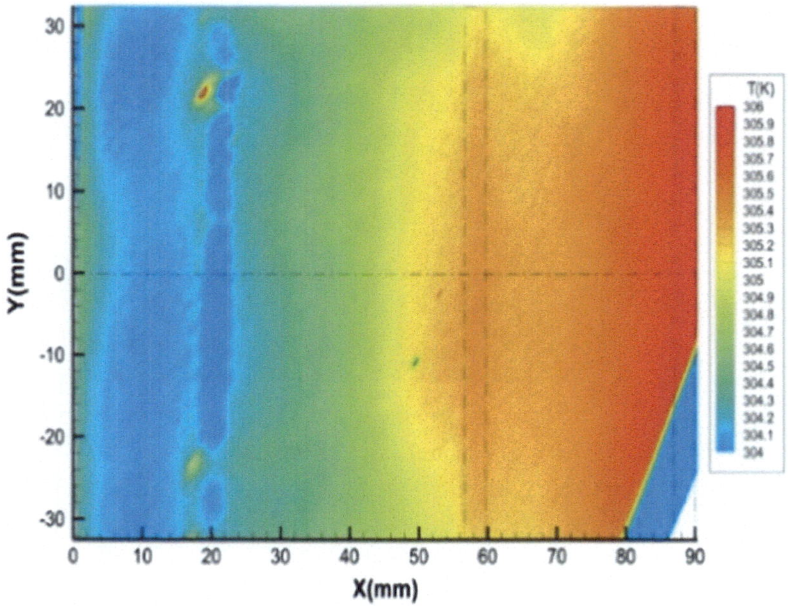

a) - control case: ZZ100 tape

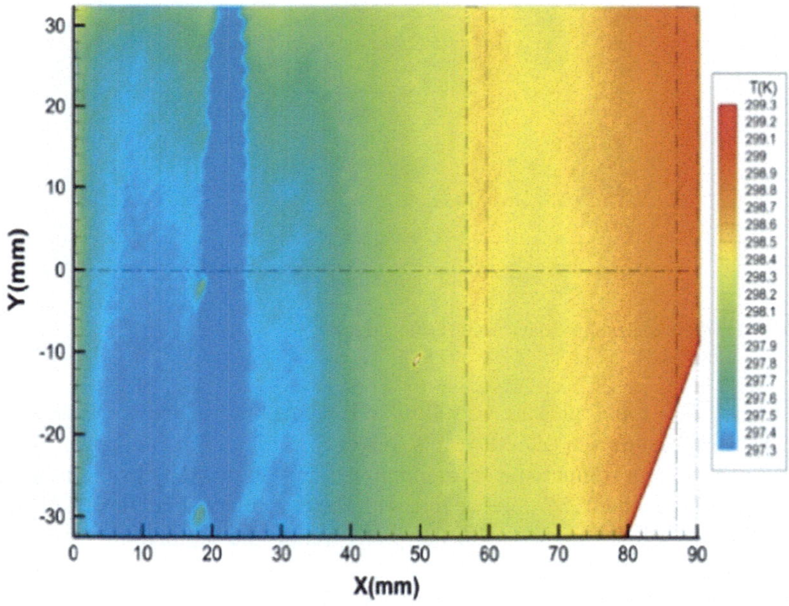

b) - control case: ZZ200 tape

Fig. 61 Temperature maps by IR thermography measurements (devices at X_{d1} location) is probably due to a too large height of the "ZZ" tape, which could lead to an "overtripping" of the boundary layer. Indeed, for this X_{d1} location of the devices, the choice of the ZZ100 tape is better because its height is closer to the value of the boundary layer displacement thickness at this distance from the flat plate leading edge.

2.3.3 Effect of the Device Location

Figure 62 shows a comparison of oil flow visualizations from the ZZ100 tape device for the two selected X_{d1} and X_{d2} locations on the flat plate. It seems that the turbulator in the upstream position (X_{d1}) is producing more intense vortices, but this does not modify the flow topology in the interaction region.

Figure 63 shows a comparison of temperature maps obtained by IR thermography measurements from the ZZ100 tape device for the two selected X_{d1} and X_{d2} locations on the flat plate. The position of the turbulator has a slight effect on the temperature increasing during the boundary layer transition process. This is due to the fact that both locations of the "ZZ" tape device are chosen upstream of the interaction domain and of the boundary layer separation line (for the reference case). One sees on Fig. 63 that the temperature starts to sharply rise into the interaction region, at $X > 40$ mm, for the two device locations.

2.3.4 Schlieren Visualizations and Fourier Modes Decomposition

Schlieren photography is used to gain a better understanding of the spatial distribution of the unsteadiness of the flow by performing Fourier analysis: FFT is the most important discrete transform used to perform Fourier analysis. In image processing, the samples can be the values of pixels along a row or column of a raster image. The time resolved image sequence can be written as $I(i, j, n)$ where I is the light intensity of each image point (or pixel). In this study, i and j vary from 1 to 464 and 1 to 360, respectively, and n defines the snapshot number and varies from 1 to $N = 60{,}000$. For a given position (i, j), the time series that represent the evolution of light intensity at a given pixel of the field can be written as $p_{ij}(n)$. Assuming a linear correspondence between light intensity and density gradient, one can compute the Fourier transform of $p_{ij}(n)$ using an FFT algorithm with Hanning window function, 60 blocks with 50% overlap of 2048 images. Due to the limited amount of data samples, the spectrum has a resolution of $f = 17$ Hz and does not present any peak.

Figure 64 shows a comparison of the flow density gradients obtained by Fourier modes decomposition from the reference case and the controlled case by ZZ100 tape for the X_{d1} location on the flat plate (see Fig. 57a, b, respectively). These pictures are obtained by a Phantom V710 camera and 60,000 images are recorded at 35 kHz frame rate. The three selected frequencies ($f = 1, 5$ and 10 kHz) are chosen to provide information on shock-wave unsteadiness and mixing layer instabilities. One notices the presence of a low-frequency (around 1 kHz) unsteadiness into the shock pattern

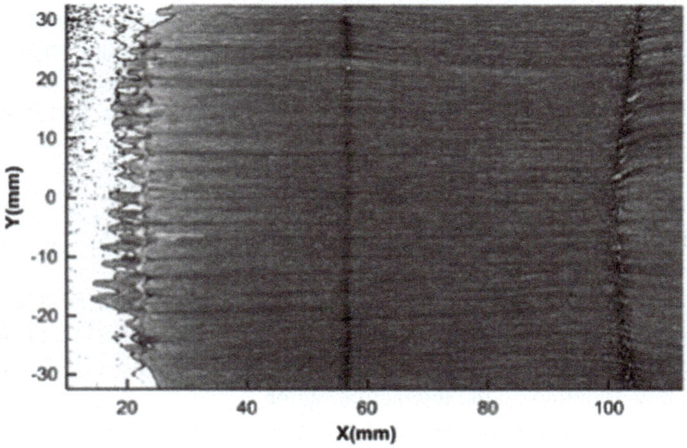

a) - control device at X_{d1} location

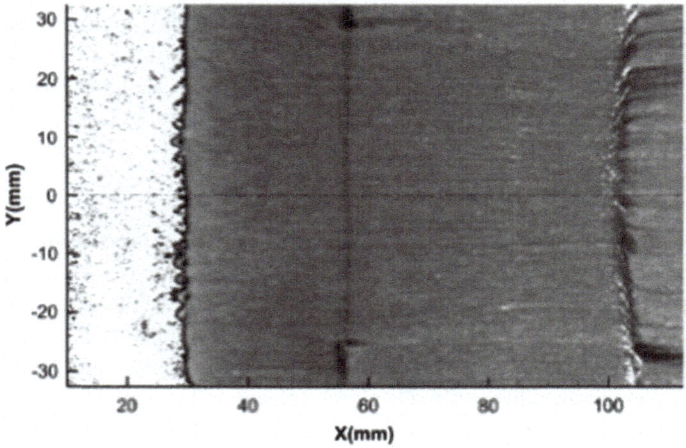

b) - control device at X_{d2} location

Fig. 62 Oil flow visualizations (for ZZ100 tape)

with a maximum of energy at the impact of the oblique shock on the boundary layer separation. So, the shock-waves seem to behave as a low-pass filter, displaying its maximum fluctuations at low frequencies. In the medium-frequency range (from 5 to 10 kHz), the circular structures can be related to the vortices generated by the Kelvin-Helmholtz instability, that are convected along the mixing layer in the downstream direction, behind the interaction region. However, these structures are already present upstream of the interaction domain when the control is applied (see Fig. 64b), which means that the "ZZ" tape device forced the transition and the boundary layer is turbulent upstream of the shock pattern.

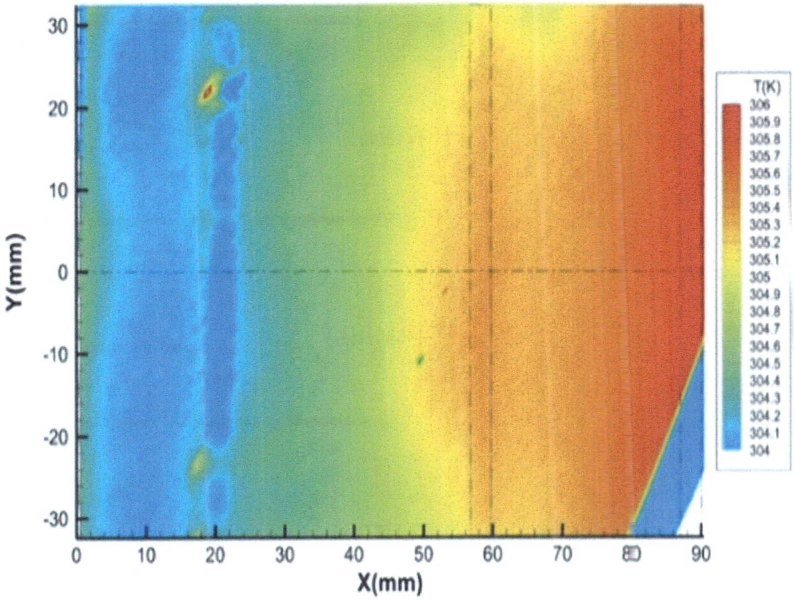

a) - control device at X_{d1} location

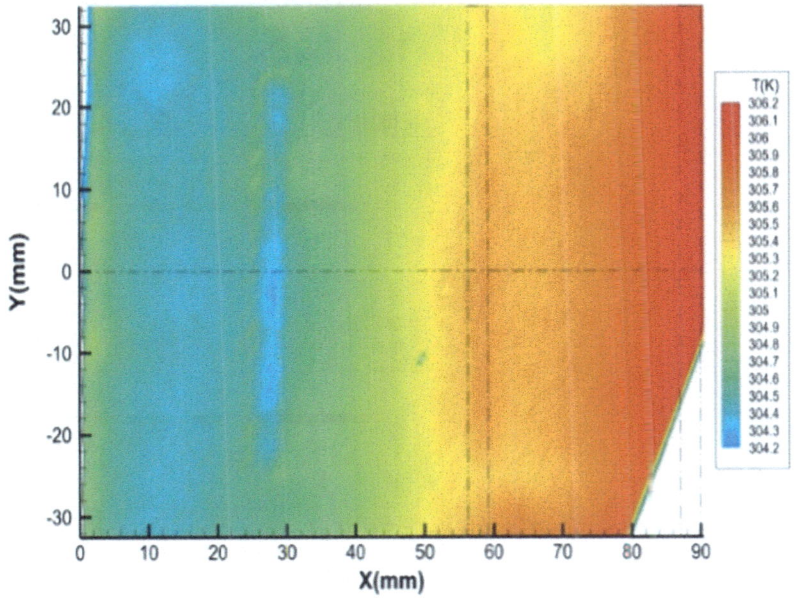

b) - control device at X_{d2} location

Fig. 63 Temperature maps by IR thermography measurements (for ZZ100 tape)

Fig. 64 Flow density gradients obtained by Fourier modes decomposition

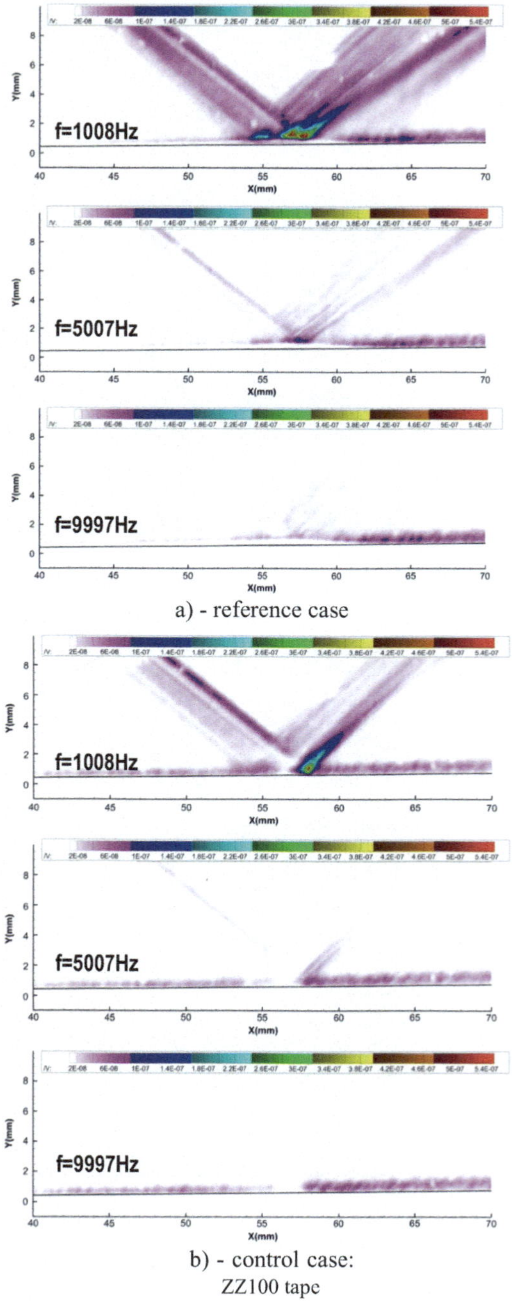

In our investigation the camera was equipped with a 12-bit colour depth sensor: the light intensity measured by each pixel is associated to a number between 0 and 4095. The available range is too small to describe at the same time both the strong density gradient due to the shock-wave and the small density variations in the mixing layer caused by the vortex shedding. For this reason, the image could present colour saturation on the shocks (black regions caused by the absence of light, see Fig. 57) or in the mixing layer (white region associated to too much light intensity). This problem can affect the results of the image processing, altering the Fourier modes by adding high-frequency energy in the saturated zone due to the signal truncation.

Finally, it was not possible to analyze flow unsteadiness for both very low-frequency (<500 Hz) and high-frequency (>15 kHz) modes.

2.3.5 Conclusions

Several parameters have been tested in order to optimize the effectiveness of control devices on the boundary layer transition, for a moderate oblique shock intensity configuration leading to a viscous interaction with separation. Both selected 3-D turbulators—a Cadcut device and a "ZZ" tape—forced the transition of the boundary layer and suppressed the separation region. The location of device with respect to the flat plate leading edge has a slight influence on the boundary layer transition process because the two tested positions of the turbulator are chosen far upstream of the interaction domain. Moreover, it is preferable to scale the height of device with the value of the boundary layer displacement thickness, where the control is applied, to avoid an "overtripping" of the boundary layer.

2.4 *TUD*

To study the effects of forced transition on a SWBLI, the oblique shock wave ($\vartheta = 3°$) is positioned at x = 51 mm, which under uncontrolled conditions corresponds to a strongly separated laminar interaction. The tripping devices are again installed and centred at x = 40 mm. Schlieren visualizations of the interaction were performed for the three tripping devices (Fig. 65), which may be compared to the uncontrolled interaction shown in Fig. 49 (chapter "WP-1 Reference Cases of Laminar and Turbulent Interactions"). The visualizations look very similar for the different tripping devices and suggest that flow separation has been eliminated in all cases. Notwithstanding the overall similarity in the three tripped cases, it may be observed that for the step the reflected shock region is slightly thicker for the other two trips. This might indicate a larger interaction region and a less effective tripping.

PIV measurements of the interaction region confirm that all trips indeed effectively remove the separation bubble. The PIV results will be further analysed in terms of the development of the integral boundary layer properties over the interaction.

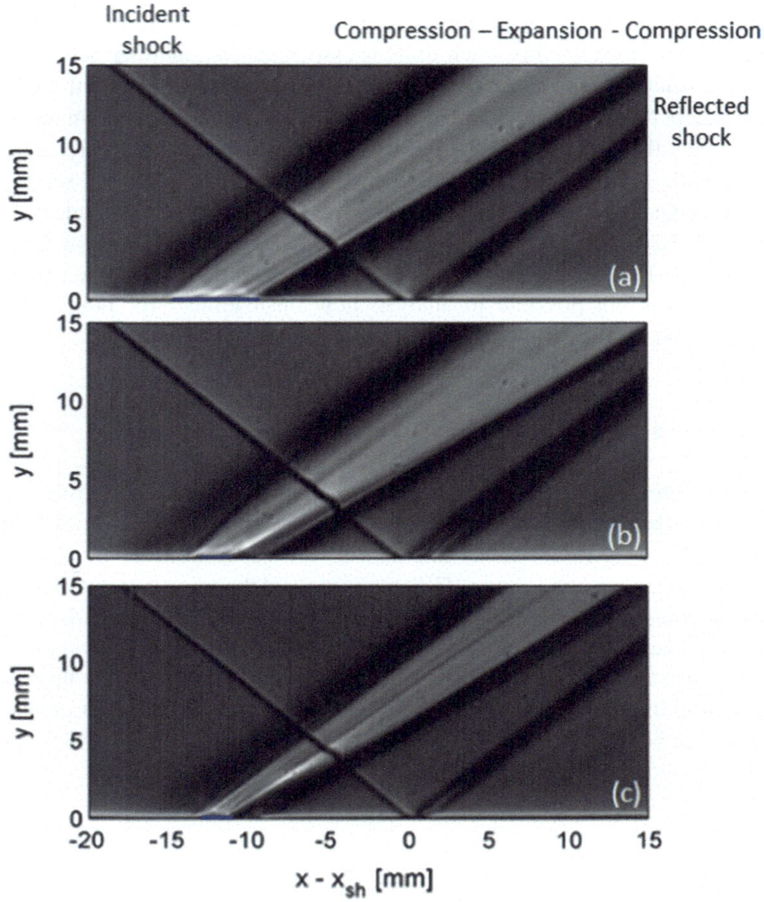

Fig. 65 Schlieren visualizations of the interaction for three tripping devices positioned 10 mm upstream of an impinging oblique shock wave ($x_{sh} = 51$ mm). **a** zig-zag strip; **b** step; **c** distributed roughness

First a comparison is made between the tripped boundary layer and the tripped interaction to assess the efficiency of the trip positioning on the interaction, as visualized by the incompressible shape factor (Fig. 66). Downstream of the trip the boundary layer quickly becomes turbulent, with an accompanying decrease of the shape factor, and at the shock impingement point (x = 51 mm) turbulent conditions have been established for all trips. Inside the interaction region the boundary layer velocity profile becomes less full, with the highest shape factors reached at the shock impingement location. For the step a maximum shape factor of 1.80 is reached in the interaction region, for the distributed roughness this is 1.72 while for the zig-zag strip a maximum shape factor of 1.61 is recorded. The flow downstream of the step is therefore concluded to be closest to separation in the interaction region, whereas

Fig. 66 Development of the incompressible shape factor H_i for the tripped boundary layer and tripped interaction: **a** zig-zag strip; **b** step; **c** distributed roughness

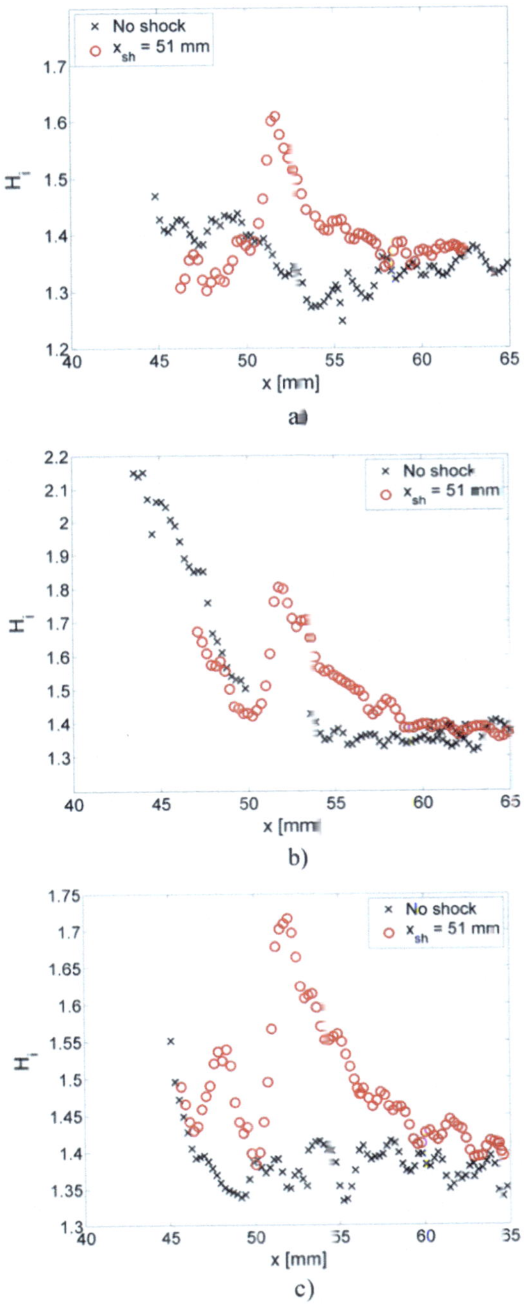

the flow downstream of the zig-zag strip is furthest removed from separation. The differences in shape factor are, however, relatively small.

Subsequently, a comparison is made between the tripped and untripped interactions (Fig. 67). For the untripped (laminar) interaction a large separation bubble is formed, which extends mostly upstream (ca. 8 mm) of the incident shock wave (see also Fig. 49, chapter "WP-1 Reference Cases of Laminar and Turbulent Interactions"). It was previously found that the boundary layer developing over the separation bubble remains in a quasi-laminar state up to the shock impingement location and transition only sets in after crossing the incident shock, resulting in a quick reattachment of the separated shear layer. This phenomenon is reflected in the development of the incompressible shape factor downstream of the interaction. The boundary layer for the untripped case recovers rapidly and within 5 mm of the shock wave it reaches the same level of fullness as recorded for the tripped interactions. Inside the interaction region the tripped cases display much lower values of the shape factor, consistent with the absence of a separation flow region.

From these results it is furthermore apparent that tripping the boundary layer upstream of the interaction results in a substantially thicker boundary layer downstream of the interaction. On average, tripping leads to an increase of the displacement thickness by 0.04 mm and an increase of the momentum thickness by 0.03 mm. In relative terms, this corresponds to a boundary layer that is ~50% thicker than the untripped boundary layer.

So, in summary it is confirmed that the transition control devices in the present configuration are able to remove the separation bubble, but do this at the price of having a substantially thicker boundary layer downstream of the interaction.

2.5 SOTON

2.5.1 Transition Location Effect on SWBLI

A quantitative comparison is made with the experiments conducted at the Institute of Applied Mechanics (ITAM) in Novosibirsk, Russia, for which large-eddy simulations (LES) simulations are performed due to the very large Reynolds number. The set of experiments by ITAM explores the effect of the unit Reynolds number Re_1 on the "perturbation zone" $\left(L^*_{PER}\right)$ for each type of interaction. By considering the flow quantities at the minimum distance from the wall that the particle image velocimetry (PIV) technology can accurately measure, the beginning of the perturbation zone is defined as the x-location where a 5% decrease of the streamwise velocity with respect to the inflow is recorded. A significant increase of the velocity indicates the end of the interaction. The experimental setup of ITAM for quantitative comparisons is summarised in Table 9.

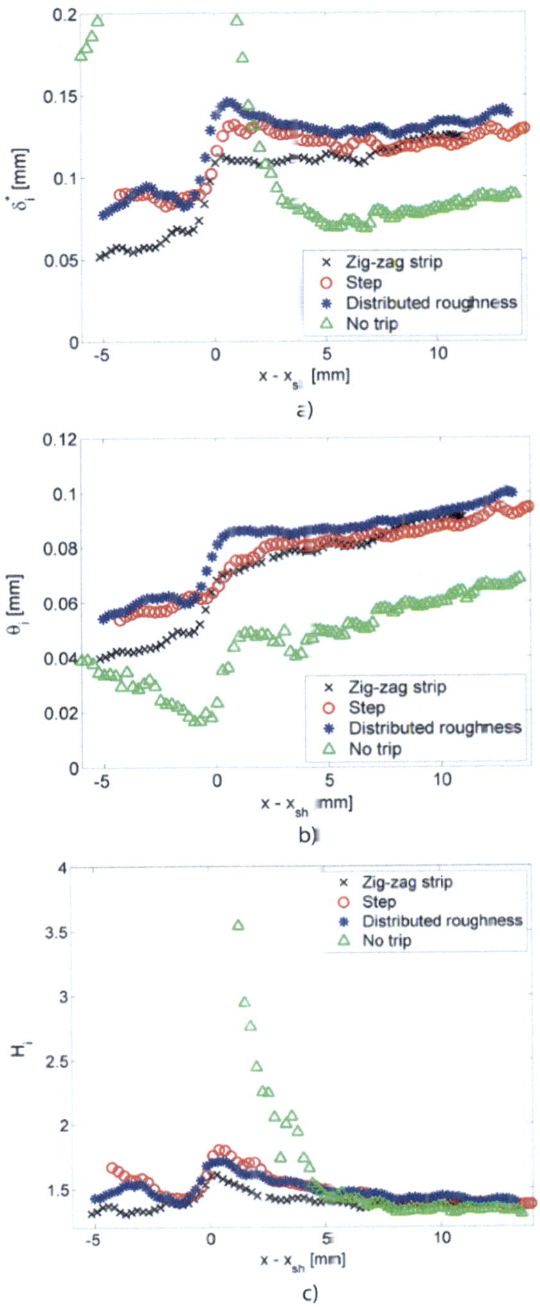

Fig. 67 Development of the boundary layer properties for the tripped and untripped interactions: **a** incompressible displacement thickness δ_i^*, **b** incompressible momentum thickness θ_i; **c** incompressible shape factor H_i

Table 9 ITAM experimental case selected for comparisons

M	$T_0^*(K)$	$P_0^*(bar)$	$\mathrm{Re}_1(m^{-1})$	x_{imp}^*(mm)	Re x_{imp}^*	$\theta(°)$	L_{PER}^*(mm)
1.4	289.1	0.65	$10.2 \cdot 10^6$	186.5	$1.77 \cdot 10^6$	3	20

2.5.2 Numerical Setup

The most relevant aspects of the numerical setup chosen to reproduce the experiments by ITAM are described here. Due to the large impingement Reynolds number, only LES calculations are carried out and the subgrid scale mixed-time scale (MTS) model is used. Due to computational resources limitations, only a transitional interaction. Details on the nature of the forcing, domain size and grid resolution are presented in the following sections. For consistency with the previous set of simulations the Mach number is M = 1.5, but nosignificant effects are expected since the difference between experimental and numerical pressure ratios p_3/p_1 (downstream of the reflected shock to upstream of the incident shock) is about 1%. The numerical inflow is placed at $x_0 = 0.113$ m, $\delta_{1,0} = 2.78 \times 10^{-4}$ m. With a unit Reynolds number $\mathrm{Re}_1 = 10.18 \times 10^6$ m^{-1}, the corresponding Reynolds number based on the displacement thickness at the inlet is $\mathrm{Re}_{\delta_{1,0}} = 2835.84$. Inflow profiles are given by a similarity solution using the Illingworth transformation. The angle of the shock generator is either $\theta = 3$ or 4 deg and the oblique shock is introduced by the Rankine-Hugoniot jump relations at the top boundary. For all the selected cases, the free-stream temperature is $T_\infty^* =$ 197.93 K and Sutherland's law is used to describe the variation of viscosity μ with temperature (Sutherland's constant temperature $TS^* = 110.4$ K). The integration time step is $\Delta t = 0.025$ for all simulations, that are run until statistical convergence.

A complete summary of the cases is reported in Table 10. If no shock is introduced ($\theta = 0°$) the solution is a zero-pressure gradient (ZPG) growing boundary-layer, that is used to understand the transition scenario in the absence of an interaction. A sensitivity analysis on the shock strength is also performed.

The grid distribution in the wall-normal direction is stretched and clusters about 30% of the grid points within the boundary-layer at the inlet. The grid is stretched in the streamwise direction following a 10th-order polynomial distribution whose derivatives are continuous up to 4th-order. The spatial step size Δx continuously decreases from the inflow up to the transition location, after which uniform grids are used. The grid resolution is $\Delta x^+ = 22$, $\Delta z^+ = 18$ and $\Delta y_{wall}^+ = 1.11$. Details of the domain size, normalised with the displacement thickness at the inlet, and grid

Table 10 Numerical simulation cases summary. All the simulations are run at M = 1.5 and $T_\infty^* =$ 197.93 K

Case	Facility	Simulation	$\mathrm{Re}_{\delta_{1,0}^*}$	θ (°)	$\mathrm{Re}_{x_{imp}^*} \times 10^6$	Interaction
IT-1	ITAM	LES-MTS	2835.84	0	–	–
IT-2	ITAM	LES-MTS	2835.84	4	1.77	Transitional
IT-3	ITAM	LES-MTS	2835.84	3	1.77	Transitional

Table 11 Numerical domain sizes and grid resolutions for each numerical simulation case

Case	$L_x \times L_y \times L_z$	$N_x \times N_y \times N_z$
IT-1	335 × 120 × 31	056 × 240 × 119
IT-2	335 × 120 × 31	056 × 240 × 119
IT-3	335 × 120 × 31	056 × 240 × 119

resolution for all interactions are reported in Table 11. The boundary conditions applied to the computational domain are no-slip and fixed temperature conditions (with temperature equal to the laminar adiabatic wall temperature) at the wall and time-dependent fixed inlet (where the modal forcing is applied). To minimise the reflection of waves into the domain, an integral characteristic method is applied to the top boundary and a standard characteristic boundary condition at the outflow.

To mimic the broadband disturbances in the wind tunnel, the modal forcing technique is used, including a large number of stable and unstable modes. For all cases, 42 eigenmodes are calculated for each combination of $\omega = 0.02: 0.02: 0.12$ and $\beta = -0.6: 0.2: 0.6$. Since pairs of oblique modes are selected to construct the broadband modal disturbances, random phases are added to each mode in order to avoid any symmetry of the breakdown.

2.5.3 ZPG Boundary-Layer

The LES-MTS model is used to study a ZPG boundary-layer forced with the broadband modal forcing and forcing amplitude $A_o = 0.03$ (corresponding to a turbulence intensity at the inflow $\rho_{rms} = 0.15\%$). The time- and span-averaged skin friction distribution (a) and contours of time-averaged skin friction (b) and instantaneous streamwise velocity (c) are reported in Fig. 68. The forcing amplitude has been tuned to have the shock impingement location set by the experiments approximately halfway between the beginning ($Re_x = 1.4 \times 10^6$) and end ($Re_x = 1.9 \times 10^6$) of the transition region. The transition scenario shows long streaky structures that precede the breakdown, as shown in Fig. 69.

2.5.4 Intermittency Calculation

The first and second time derivatives of the skin friction distribution are used to calculate the span averaged intermittency distribution in Fig. 7[illegible] (black solid line with error bars for the threshold f sensitivity). In agreement with the skin friction and shape factor distributions the region with $0 < \Gamma < 1$ extends between $1.4 \times 10^6 < Re_x < 1.9 \times 10^6$. The hyperbolic tangent (red solid line) fitted curve seems to better represent the intermittency distribution than the Narasimha [29] (blue solid line) approximation.

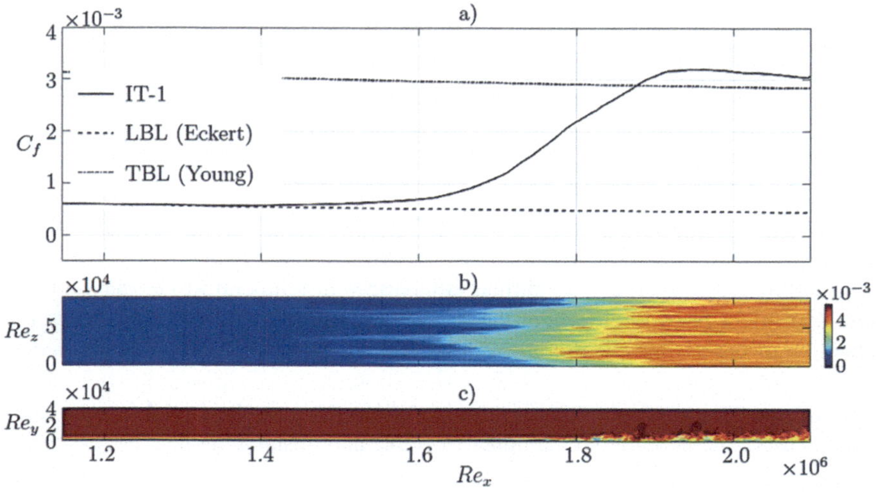

Fig. 68 **a** Time- and span-averaged skin friction distribution for the ZPG boundary-layer (black solid line) along with laminar (dashed-dotted black) and turbulent (dashed black) boundary-layer distributions by [9] and [10], respectively; contours of time-averaged skin friction (**b**) and instantaneous streamwise velocity (**c**). The extent of the transitional region is also confirmed by plotting the displacement thickness (**a**) and shape factor (**b**) distributions in Fig. 70, calculated with the compressible (solid black line) and incompressible (dashed black line) formulations

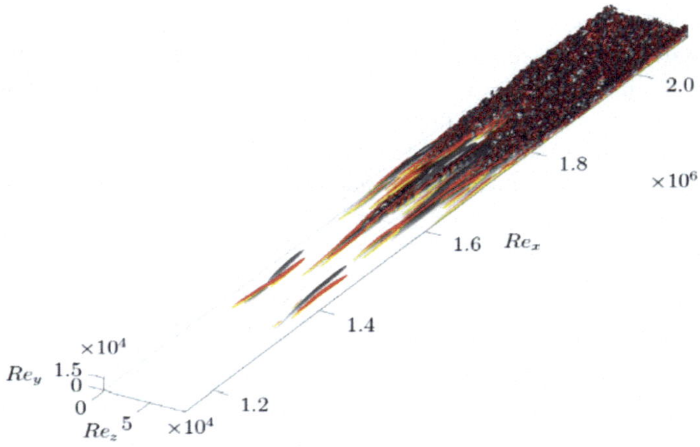

Fig. 69 Transition visualisation. Iso-surfaces of the wall-normal vorticity for the levels $\omega_y = +0.075$ (red) and $\omega_y = -0.075$ (black) and coloured with the streamwise velocity

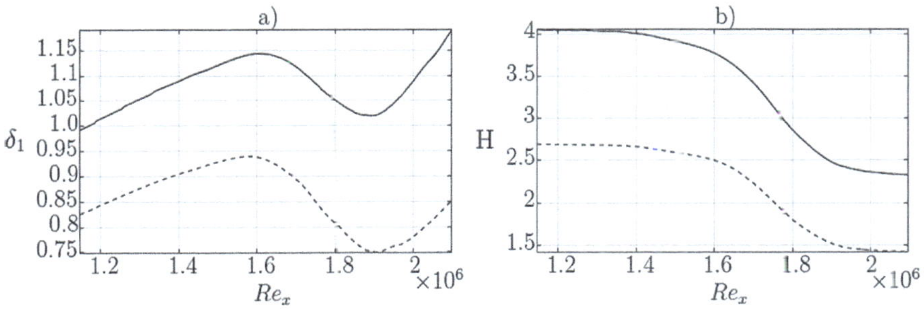

Fig. 70 Time- and span-averaged displacement thickness (**a**) and shape factor (**b**) distributions calculated with the incompressible (dashed lines) and compressible (solid lines) formulations

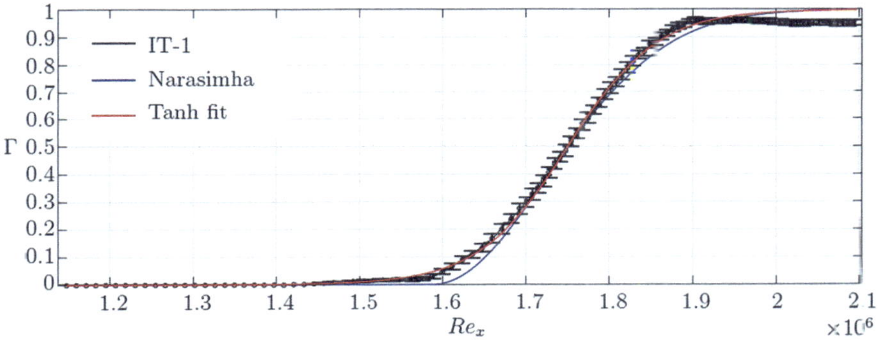

Fig. 71 DNS span averaged intermittency distribution (black solid line with error bars due to the threshold sensitivity) for the ZPG boundary-layer at $Re_{\delta_{1,0}} = 2835.84$ (case IT-1). Narasimha (blue solid line) and hyperbolic tangent (red solid line) fit curves are also plotted

2.5.5 SWBLI for Transitional Interactions

The transitional interaction at $Re_{x_{imp}} = 1.77 \times 10^6$ with $\theta = 4°$ (case IT-2) and $\theta = 3°$ (case IT-3) from the ITAM experiments are considered here for comparison. Figure 72 shows the skin friction (a), displacement thickness (b) and shape factor (c) distributions for the ZPG (solid black line), $\theta = 4°$ (blue solid line) and $\theta = 3°$ (red solid line) cases. Similarly to what happened for the previous setup at $Re_{\delta_{1,0}} = 1971.07$ (ONERA case), the presence of the shock causes breakdown to turbulence at the shock impingement location. The interaction size and boundary-layer thickness increases for increasing shock strength but the boundary-layer remains attached.

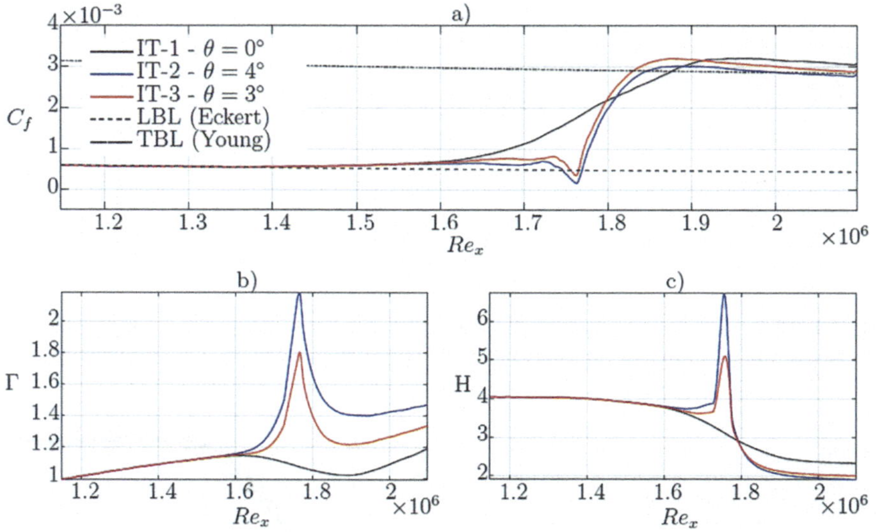

Fig. 72 Time- and span-averaged skin friction (**a**), displacement thickness (**b**) and shape factor (**c**) distributions for $\theta = 4°$ (blue solid line), $\theta = 3°$ (red solid line) and ZPG (black solid line—case IT-1). Laminar (dashed-dotted black) and turbulent (dashed black) boundary-layer distributions by [9] and [10], respectively

2.5.6 Intermittency Calculation

Figure 73 shows the span averaged intermittency distributions for the ZPG (black solid line), $\theta = 3°$ (red solid line) and $\theta = 4°$ (blue solid line) cases. Downstream

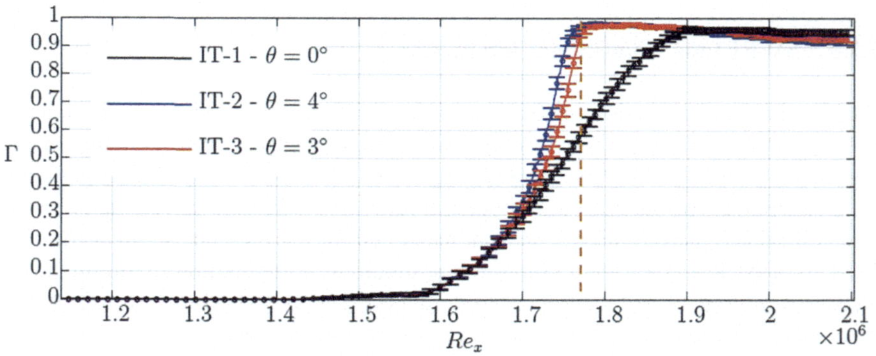

Fig. 73 Span averaged intermittency distributions (with error bars due to the threshold sensitivity) for the transitional case with $\theta = 4°$ (blue solid line—case IT-2) and $\theta = 3°$ (red solid line—case IT-3). The ZPG boundary-layer (black solid line—case IT-1) is also plotted. The vertical orange dashed line indicates the shock impingement location

of the interaction the boundary-layer becomes turbulent and the intermittency distribution deviates from the ZPG case and sharply jumps towards unity. Differently from the sensitivity study at lower Reynolds number, the effects of the shock strength on the intermittency distribution downstream of the interaction are negligible. The differences are more pronounced upstream of the impingement location ($Re_x imp = 1.77 \times 10^6$). The size of the interaction increases with the shock strength and the effects are visible for a larger upstream extent. Thus, the case with the weaker shock follows the ZPG solution for longer before deviating, however still reaching fully turbulent conditions at the impingement.

2.5.7 Experimental Comparisons

A comparison between the case at $\theta = 4°$ is done with the experimental PIV measurements performed by ITAM on a mean streamwise velocity x-y plane in Fig. 74. Although very qualitative, the comparison presents similar features: the boundary-layer visibly thickens after the shock and no recirculation region is detected. In order to compare the LES results with the experiments in terms of interaction length, a definition similar to the one provided by ITAM is used to calculate the perturbation zone. The beginning of the interaction region is taken as the location where the skin friction distribution deviates from the laminar solution by 5%. The interaction ends where a significant increase of the skin friction distribution is recorded. By following this definition, the perturbation length is LPER = 22 and 24 mm for $\theta = 3$ and $4°$, respectively, in reasonable agreement with the value measured at $\theta = 3°$ by ITAM (LPER = 20 mm), as reported in Table 9. However, the differences in the definitions of the interaction region still do not allow a satisfactory comparison. The lack of complete experimental measurements makes the search of more appropriate definitions currently out of scope, but the width of the peaks in the displacement thickness or shape factor distributions, or the streamwise extension where the intermittency is between 0 and 1, could be used for quantifying the interaction lengths.

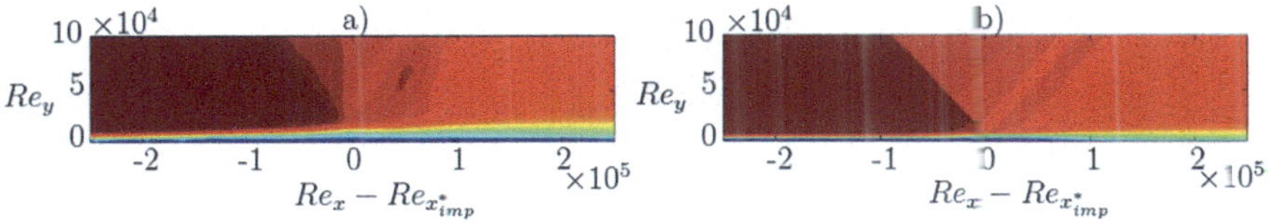

Fig. 74 Contours of instantaneous streamwise velocity from PIV measurements by ITAM (**a**) and LES-MTS (**b**). Levels between 0.0–1.0. (With permission of A. A. Sidorenko and P. A. Polivanov at the Institute of Theoretical and Applied Mechanics from Novosibirsk State University)

2.5.8 Reynolds Number Effect on the SWBLI

The present LES-MTS simulations for the ZPG and transitional SWBLI at $\theta = 4°$ are compared for the previous sets of calculations at $Re_{\delta_{1,0}} = 1971.07$ (ONERA setup). Figure 75 summarises the results in terms of time- and span-averaged skin friction (a), displacement thickness (b) and shape factor (c) distributions. The main features of the transitional interaction are very similar and the effect of the Reynolds number are small. In both cases, the boundary-layer becomes turbulent downstream of the shock impingement and proportionally thickens by a factor of 1.5. The minimum in the skin friction distribution is higher for higher Reynolds number but the interaction length is practically unchanged.

The span-averaged intermittency distributions in Fig. 77 also show very similar features. The effect of the interaction on the region upstream of the impingement is gradual with respect to the ZPG cases and the extent of the upstream boundary-layer portion that is affected by the shock is in agreement with the size of the interaction. Despite differences in the transition breakdown scenarios, these results suggest a scalability of the main features of the interaction with the Reynolds number. However,

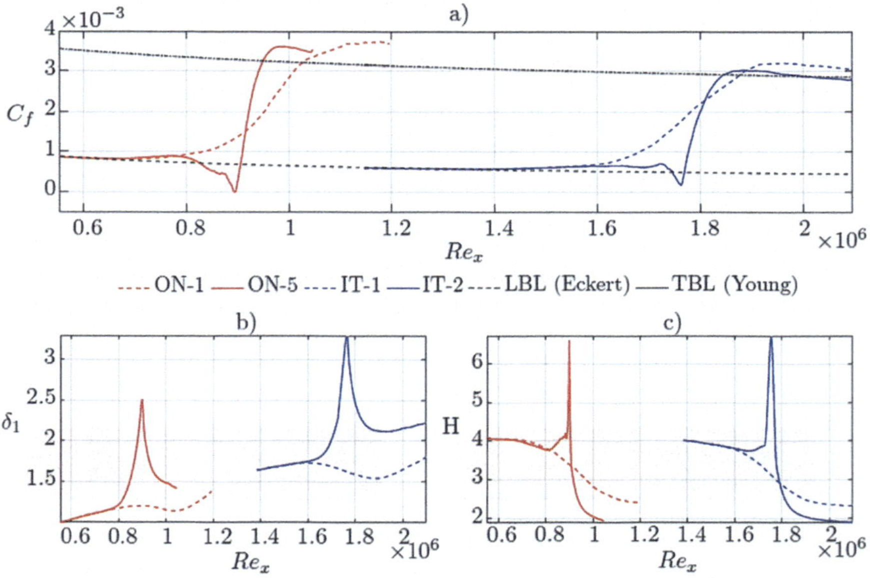

Fig. 75 Time- and span-averaged skin friction (**a**), displacement thickness (**b**) and shape factor (**c**) distributions for the ZPG (red dashed line—case ON-1) and $\theta = 4°$ (red solid line—case ON-5) cases at $Re_{\delta_{1,0}} = 1971.07$ and the ZPG (blue dashed line—case IT-1) and $\theta = 4°$ (blue solid line—case IT-2) cases at $Re_{\delta_{1,0}} = 2835.84$. Laminar (dashed-dotted black) and turbulent (dashed black) boundary-layer distributions by [9] and [10], respectively

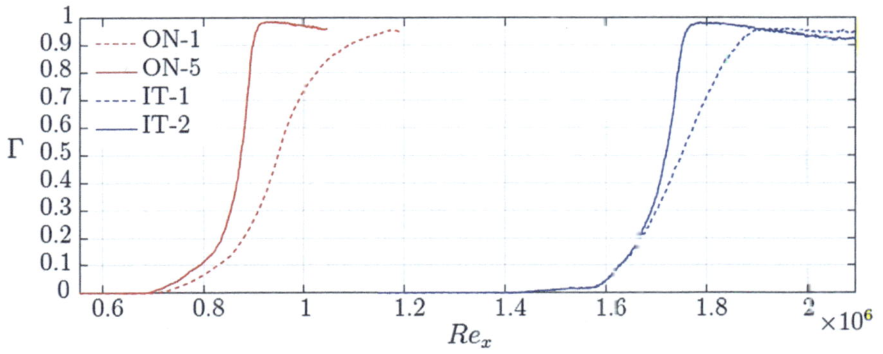

Fig. 76 Span averaged intermittency distributions for the ZPG (red dashed line—case ON-1) and $\theta = 4°$ (red solid line—case ON-5) cases at $\text{Re}_{\delta_{1,0}} = 1971.07$ and the ZPG (blue dashed line—case IT-1) and $\theta = 4°$ (blue solid line—case IT-2) cases at $\text{Re}_{\delta_{1,0}} = 2835.34$

Fig. 77 Coordinate system used to define unit vectors

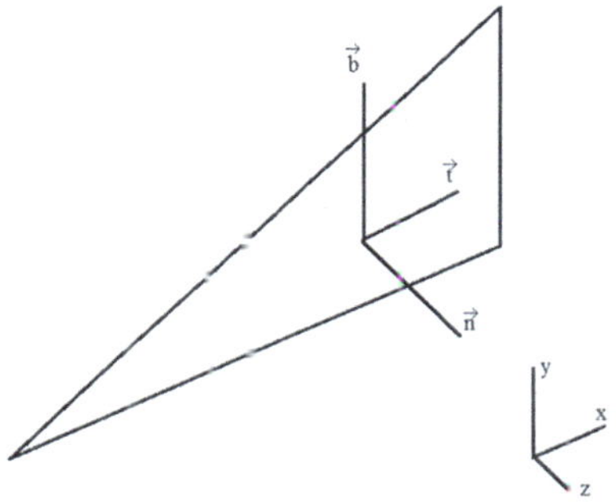

although not tested here, it is the authors' opinion that for the laminar cases the interaction region would significantly decrease in size for increasing shock impingement Reynolds number.

2.5.9 Conclusions

The transitional interaction is close to a marginal separation. The shock wave is strong and the boundary-layer becomes turbulent at the shock impingement location. Apart from boundary-layer thickness effects, similar results are obtained when the shock wave is weaker. For the same kind of interaction (transitional), the impingement Reynolds number has moderate effects apart from a larger boundary-layer thickening. Qualitative agreement is obtained with the experiments when LES calculations are compared with ITAM PIV measurements.

3 Physical Modelling of the Control Devices

3.1 IoA

Overview of BAY model of vortex generator

Implementation of numerical modelling of the effects of BAY model of thin-plate vortex generator

In BAY method developed by Bender et al. [30] the vortex generator (VG) is replaced by subdomain placed in original location of vortex generator, where the force distribution is applied. The force is introduced as momentum source by a source term in momentum balance equations. The vorticity is generated by force which depends on local velocity values and vortex generator's geometry and spatial orientation. The lifting force source term, L_i (3) acting at grid point i, is added to the governing discretized finite volume momentum (27) and energy (2) equations:

$$\Delta V_i \frac{\Delta \rho \overrightarrow{u_i}}{\Delta t} = \sum_j \overrightarrow{F_M} \Delta S + \overrightarrow{L_i} \tag{28}$$

$$\Delta V_i \frac{\Delta \rho \vec{E}}{\Delta t} = \sum_j \overrightarrow{F_E} \Delta S + \overrightarrow{u_i}\, \overrightarrow{L_i} \tag{29}$$

The source term $\overrightarrow{L_i}$ is a function of the lifting force caused by the vortex generator and is corrected for losses due to deviation of the flow from the vortex generator surface. F_M are inviscid and viscous fluxes in momentum equations. F_E are inviscid and viscous fluxes in energy equation. ΔS is the area of cell face *j*. The side force is approximated by Eq. (30).

$$\mathbf{L_i} = c_{VG} S_{VG} \frac{\Delta v_i}{v_m} \propto \rho u^2 \mathbf{l} \tag{30}$$

c_{VG} is relaxation parameter which controls the strength of the side force and consequently the intensity with which the local velocities align with the vortex generator, S_{VG} is plan parallel area of vortex generator, V_i is the volume of the cell where the force is calculated, V_m is the sum of volumes of cells where the force term is applied, α is the angle of local velocity $\mathbf{u}$ to the Vortex Generator, $\mathbf{l}$ is unit vector on which the side force acts.

The coordinate system is shown in the Fig. 77.

In the Fig. 77. the directions of x, y and z axes aren't parallel to t, b, n axes which represent vortex generator's vector due to Vortex Generator's angle of inclination and angle of attack. Vector **n** is the unit vector normal to vortex generator planform (36), **b** is the unit vector in the direction of the span of the VG (37) and **t** is unit vector tangential to the vortex generator planform and normal to **b** (38). The position of VG and its orientation are user inputs to User Defined Function in ANSYS Fluent commercial code.

Using a small angle approximation there is

$$\mathbf{l} = \frac{\mathbf{u}}{|\mathbf{u}|} \times \mathbf{b} \tag{31}$$

$$\propto \approx \sin\alpha = \cos\left(\frac{\pi}{2} - \alpha\right) = \frac{\mathbf{u} \cdot \mathbf{n}}{|\mathbf{u}|} \tag{32}$$

The final equation is multiplied by term (33)

$$\frac{\mathbf{u} \cdot \mathbf{t}}{|\mathbf{u}|} \tag{33}$$

which represents losses of side force due to high angles of attack The final expression for side force (34) according to Bender et al. is as follows:

$$\mathbf{L_i} = c_{VG} S_{VG} \frac{\Delta V_i}{V_m} \rho(\mathbf{u} \cdot \mathbf{n})(\mathbf{u} \times \mathbf{b})\left(\frac{\mathbf{u} \cdot \mathbf{t}}{|\mathbf{u}|}\right) \tag{34}$$

In the Eq. (35) **u** means local velocity vector

$$\mathbf{u} = [u, v, w] \tag{35}$$

Velocities u, v, w are the local velocities in the cells where momentum source is applied. Velocities are accessed by the $C_{U(c,t)}$, $C_{V(c,t)}$, $C_{W(c,t)}$ flow variable macros in the cells using User-Defined Functions. The unit vectors **n**, **b**, **t** in x, y and z directions are as follows

$$\mathbf{n} = \left[n_x, n_y, n_z\right] \tag{36}$$

$$\mathbf{b} = \left[b_x, b_y, b_z\right] \tag{37}$$

$$\mathbf{t} = \left[t_x, t_y, t_z\right] \tag{38}$$

$$\begin{aligned}
\mathbf{L_i} &= c_{VG} S_{VG} \frac{\Delta V_i}{V_m} \rho(\mathbf{u} \cdot \mathbf{n})(u \times b)\left(\frac{u \cdot t}{|u|}\right) \\
&= c_{VG} S_{VG} \frac{\Delta V_i}{V_m} \rho\left(u \cdot n_x + v \cdot n_y + w \cdot n_z\right) \\
&\quad \cdot \left(\left[v \cdot b_z - w \cdot b_y, w \cdot b_x - u \cdot b_z, u \cdot b_y - v \cdot b_x\right]\right) \cdot \left(\frac{u \cdot t_x - v \cdot t_y + w \cdot t_z}{|u|}\right)
\end{aligned} \tag{39}$$

The final expression for the side forces in cartesian coordinate system is as follows:

$$L_x = c_{VG} S_{VG} \frac{\Delta V_i}{V_m} \rho\left(u \cdot n_x + v \cdot n_y + w \cdot n_z\right)\left(v \cdot b_z - w \cdot b_y\right) \left(\frac{u \cdot t_x + v \cdot t_y + w \cdot t_z}{|\mathbf{u}|}\right) \quad (40)$$

$$L_y = c_{VG} S_{VG} \frac{\Delta V_i}{V_m} \rho(u \cdot n_x + v \cdot n_y + w \cdot n_z)(w \cdot b_x - u \cdot b_z)\left(\frac{u \cdot t_x + v \cdot t_y + w \cdot t_z}{|\mathbf{u}|}\right) \quad (41)$$

$$L_z = c_{VG} S_{VG} \frac{\Delta V_i}{V_m} \rho(u \cdot n_x + v \cdot n_y + w \cdot n_z)\left(u \cdot b_y - v \cdot b_x\right)\left(\frac{u \cdot t_x + v \cdot t_y + w \cdot t_z}{|\mathbf{u}|}\right) \quad (42)$$

The Eqs. (40), (41) and (42) are the equations directly implemented in the User-Defined Functions. After implementation of momentum sources as source terms in User-Defined Functions in ANSYS-FLUENT code the appropriateness of this implementation was checked. In order to simplify the equations and reduce computational cost and overcome the necessity of calculating the ΔV_i at every iteration it was assumed as 1. This simplification was taken into account during BAY model calibration.

3.1.1 BAY Model of Air-Jet Vortex Generator

In order to overcome difficulty with generating meshes of the air jet vortex generator at every investigated location the new source term model of air jet vortex generator was proposed. This model works by adding momentum source term to Reynolds-Averaged Navier-Stokes equations in Fluent commercial code. The model may be calibrated with respect to obtain the best circulation values agreement in location, where vortex is fully formed. The calibration with respect to circulation values is enough, because it is impossible to calibrate with one model constant more than one flow parameter, for example intermittency. Additionally, BAY model predicts in proper way other flow parameters such as turbulent kinetic energy and shear stresses as predicted by grid-resolved vortex generator if BAY model constant is calibrated with respect to circulation only, which will be shown in the plots.

3.1.2 Numerical Method and Flow Configuration

In the Fig. 78. is shown schematic of JVG orientation used in computations. The skew angle of a jet (the angle between main flow direction and the direction of jet blow) α is equal to 90° and the injection angle of a jet θ is equal to 45°.

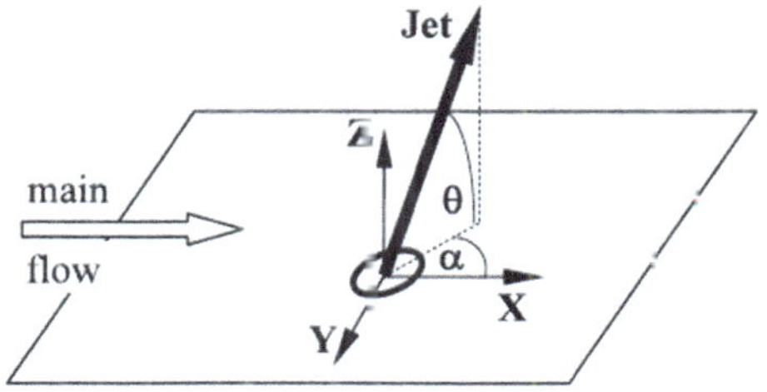

Fig. 78 Schematic of JVG orientation

3.1.3 Source Term Model of JVG—New Approach

A new ABAY model is a generalization of BAY model on JVG models. The name jBAY is reserved by Jirasek [30], who proposed a model of thin-plate vortex generator defined by grid points. A new approach to model JVG doesn't consider MFR_{jet}, because the mass flow rate in grid-resolved AJVG is regulated by total pressure at inlet of AJVG. The total pressure at jet inlet is set as in main flow. The new ABAY model takes into account the directional components of two unit vectors (**b** and **t**) of three (**n**, **b** and **t**) proposed originally by Bender et al. and assumes that vortex generator's angle of attack is equal to 1, because skew and injection angle of jet is just defined by unit vectors **b**, **t** orientation. The unit vectors values were used as presented in the Table 12 which correspond to vortex generator's orientation as mentioned above. In the Fig. 79 with the dashed edge is drawn the shape of modelled JVG.

Additional reason of doing assumption of $\alpha = 1$ is that the u velocity gradient in boundary layer is very high and it is difficult to control body force in this region. The volume of cell V_i is also assumed as 1 in order to reduce computational cost of calculating cell volume. The body force used in computations is (40). Cells are selected in the place of inlet of jet to the main flow and the height of selected cells is in described case the 1/10 of boundary layer height.

Table 12 Parameters of unit vectors used in model of JVG

Parameter	Value
n_x	0
n_y	0.5
n_z	0.5
t_x	−1
t_y	0
t_z	0
b_x	0
b_y	0.5
b_z	−0.5

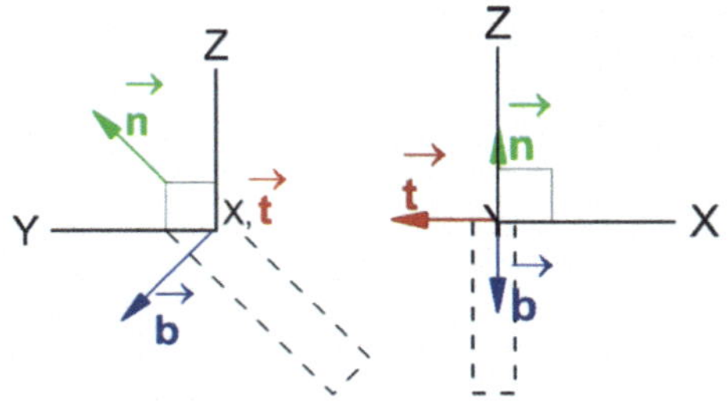

Fig. 79 Representation of unit vectors used in new model of JVG

$$\mathrm{L_i} = \mathrm{c_{VG} S_{VG}} \frac{1}{\mathrm{V_m}} \rho\alpha \left(\frac{\mathbf{u}}{|\mathbf{u}|} \times \mathbf{b} \right) \left(\frac{\mathbf{u} \cdot \mathbf{t}}{|\mathbf{u}|} \right) \tag{43}$$

S_{VG} is the area of the jet inlet with respect to the diameter of 1 mm. The V_m is the sum of volume of cells, where source term is applied, which is the area of the jet inlet multiplied by the height of model of vortex generator, which may be assumed as $H = 1/10$ of boundary layer thickness. The density ρ is calculated in every volume cell.

The cell is selected in the model of vortex generator if cell's centroid is in the region selected by constraints (44) and (45).

$$X^2 + Y^2 < D^2 \tag{44}$$

$$Z < H \tag{45}$$

where D is the diameter of jet inlet and H is the height of cells, where source term is applied.

In the Fig. 80. are presented the skin-friction coefficient distributions behind JVG modelled in the computational domain in shape of cuboid, which modelled the measurement chamber of wind-tunnel located in Institute of Fluid-Flow Machinery in Gdansk. In the Fig. 80. is presented the skin-friction coefficient at $X = 0.048$ m in the left plot and in the $X = 0.108$ m in the right plot compared with experimental data. The AJVG is located at $X = 0$ m. Good agreement of plots, especially in shape of distribution. In the region, where is the influence of the AJVG, the values are a bit different. This difference may be caused by different intensity of vortex calculated in CFD and measured in experiment, different location of vortex near wall, which influences the local velocity gradients and measurement error in experiment.

During numerical simulations the new model of JVG based on lifting-force theory was implemented to predict effects induced by grid-resolved JVG. It has been proven that new ABAY model of AJVG is is sufficiently precise to predict such flow parameters such as x-velocities, skin-friction coefficients and circulation as close as grid-resolved air-jet vortex generator under conditions that the BAY model c_{VG} is calibrated, the **b** and **t** vectors are defined properly and cells, where source term is applied are properly chosen. ABAY model is a good candidate model to predict flow

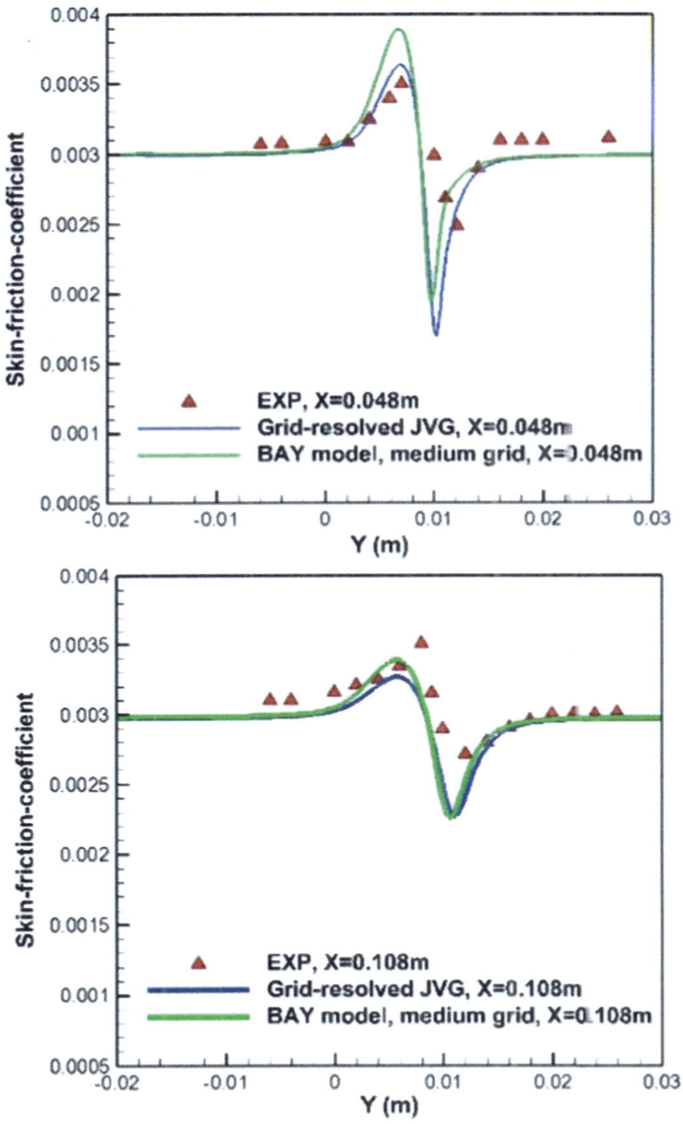

Fig. 80 Skin-friction coefficient plot on tunnel wall surface in the X = 0.048 m (left plot) and X = 0.108 m (right plot)

parameters as grid-resolved AJVG using the ANSYS Fluent solver with implementation by system of User Defined Functions. The disadvantage of the BAY model of AJVG is the fact that doesn't predict flow parameters in the close vicinity of AJVG hole (around 10 mm for a hole of diameter D = 1 mm) in the proper way due to different principle of work. The second disadvantage of the ABAY model is the fact, that it doesn't work in an asymptotic mode for large model constants C_{VG} as original BAY model. It implies the necessity of generation of mesh with grid-resolved Air Jet Vortex Generator (AJVG) and the necessity for calibration of the model constant. The effects in locations further from AJVG hole- from distance of 5 mm and higher the effects of AJVG are predicted in good agreement with grid-resolved AJVG. The advantage of BAY model is the fact, that during calculations it is possible to choose the cells, where source term is applied without need to generate new computational mesh. This model is also grid independent. The other advantage is the fact that using coarse mesh it is possible to obtain solution faster and with approximation error less than 10% in the distance to 5 cm and less than 5% in the distance to 10 cm. The best method to calibrate BAY model is to choose a slice far enough from AJVG—around 5 diameters of AJVG, because at location far enough from AJVG the vortices are fully formed and then using linear approximation method select proper c_{VG} constant.

References

1. P.A. Polivanov, A.A. Sidorenko, A.A. Maslov, Artificial turbulization of the supersonic boundary layer by dielectric barrier discharge, in *2014 29th Congress International Council of Aeronautical Sciences ICAS* (2014) pp. 1–10
2. R.H.M. Giepman, R. Louman, F.F.J. Schrijer, B.W. van Oudheusden, Experimental study into the effects of forced transition on a shock-wave/boundary-layer interaction. AIAA J. **54**(4), 1313–1325 (2016)
3. C.J. Gibbings, J.M.B. Ingham, D. Johnson, Flow in a Supersonic Jet expanding from a Convergent Nozzle (2002)
4. W. Shyy, B. Jayaraman, A. Andersson, Modeling of glow discharge-induced fluid dynamics. J. Appl. Phys. **92**(11), 6434–6443 (2002)
5. C.-C. Wang, S. Roy, Three-dimensional simulation of a microplasma pump. J. Phys. D. Appl. Phys. **42**(18), 185206 (2009)
6. M. Forte, J. Jolibois, J. Pons, E. Moreau, G. Touchard, M. Cazalens, Optimization of a dielectric barrier discharge actuator by stationary and non-stationary measurements of the induced flow velocity: application to airflow control. Exp. Fluids **43**(6), 917–928 (2007)
7. R.H.M. Giepman, M. Kotsonis, On the mechanical efficiency of dielectric barrier discharge plasma actuators. Appl. Phys. Lett. **98**(22) (2011)
8. B. Jayaraman, W. Shyy, Modeling of dielectric barrier discharge-induced fluid dynamics and heat transfer. Prog. Aerosp. Sci. **44**(3), 139–191 (2008)
9. E.R.G. Eckert, Engineering relations for friction and heat transfer to surfaces in high velocity flow. J. Aeronaut. Sci. **22**, 585–587 (1955)
10. A.D. Young, Boundary layers—AIAA education series. Blackwell Sci. (1989)
11. H.L. Dryden, Review of published data on the effect of roughness on transition from laminar to turbulent flow. J. Aeronaut. Sci. **20**(7), 477–482 (1953)
12. E. Reshotko, Is Retheta/Me a Meaningful Transition Criterion?. AIAA J. **45**(7), 1441–1443 (2007)

13. C.R. Smith, M.S. Acarlar, A study of hairpin vortices in a laminar boundary layer. Part 1. Hairpin vortices generated by hemisphere protuberances. J. Fluid Mech. **175**, 1–41 (1987)
14. K.P. Klebanoff, W. Cleveland, K.D. Tidstrom, On the evolution of a turbulent boundary layer induced by a three-dimensional roughness element.pdf. **237**, 101–187 (1992)
15. F.G. Ergin, E.B. White, Unsteady and transitional flows behind roughness elements. AIAA J. **44**(11), 2504–2514 (2006)
16. G. Groskopf, M.J. Kloker, O. Marxen, Bi-global secondary stability theory for high-speed boundary-layer flows, in *Proceedings of the 2008 Summer Program*, no. 1994 (2008) pp 55–72
17. M. Bernardini, S. Pirozzoli, P. Orlandi, Compressibility effects on roughness-induced boundary layer transition. Int. J. Heat Fluid Flow **35**, 45–51 (2012)
18. D.P. Rizzetta, M.R. Visbal, Direct numerical simulations of flow past an array of distributed roughness elements. AIAA J. **45**(8), 1967–1976 (2007)
19. E.R. Van Driest, C.B. Blumer, Boundary-layer transition at supersonic speeds—three-dimensional roughness effects (spheres). J. Aerosp. Sci. **29**(8), 909–916 (1962)
20. P. Balakumar, Receptivity of a supersonic boundary layer to acoustic disturbances. AIAA J. **47**(5), 1069–1078 (2009). https://doi.org/10.2514/1.33395
21. O. Marxen, G. Iaccarino, E.S.G. Shaqfeh, Disturbance evolution in a mach 4.8 boundary layer with two-dimensional roughness-induced separation and shock. J. Fluid Mech. **648**, 435–469 (2010)
22. J.A. Redford, N.D. Sandham, G.T. Roberts, Compressibility effects on boundary-layer transition induced by an isolated roughness element. AIAA J. **48**(12), 2818–2830 (2010)
23. S.P. Schneider, Effects of Roughness on Hypersonic Boundary-Layer Transition. AIAA J. **45**(2), 255–276 (2013)
24. T. Colonius, S.K. Lele, Computational aeroacoustics: progress on nonlinear problems of sound generation. Prog. Aerosp. Sci. **40**(6), 345–416 (2004)
25. J. Delery, J.G. Marvin, Shock wave - boundary layer interactions. AGARDograph **280** (1986)
26. S. Pirozzoli, Generalized conservative approximations of split convective derivative operators. J. Comput. Phys. **229**, 7180–7190 (2010). https://doi.org/10.1016/j.jcp.2010.06.006
27. S. Pirozzoli, F. Grasso, Direct numerical simulation of impinging shock wave/turbulent boundary layer interaction at M=2.25. Phys. Fluids. **18** (2006). https://doi.org/10.1063/1.2216989
28. Y. Na, P. Moin, Direct numerical simulation of a separated turbulent boundary layer. J. Fluid Mech. 0022-1120, 379 (1998)
29. R. Narasimha, The laminar-turbulent transition zone in the boundary layer. Prog. Aerospace Sci. **22**, 29–80 (1985)
30. A. Jirasek, Vortex-generator model and its application to flow control. J. Aircr. **42**(6), 1486–1491 (2005)
31. M. Choudhari, F. Li, C.-L. Chang, J. Edwards, M. Kegerise, R. King, Laminar-turbulent transition behind discrete roughness elements in a high-speed boundary layer, in *48th AIAA Aerospace Sciences Meeting Including the New Horizons Forum and Aerospace Exposition* (2010), pp. 1–21

Application Flow Cases

WP-3 Internal Flows—Compressors

Patrick Grothe, Pawel Flaszynski, Ryszard Szwaba, Michal Piotrowicz, Piotr Kaczynski, Benoit Tartinville, Charles Hirsch, and Alexander Hergt

Abstract In the case of a civil turbofan engine operating at particularly high altitudes the Reynolds number can drop by a factor of 4, when compared to the sea level values. The laminar boundary layer on the transonic compressor rotor blades interacts with shock waves and as a result a strong boundary layer separation forms. This can seriously affect the aero-engine performance and operation. One way to avoid strong separation is to ensure that the boundary layer upstream of the shock wave is turbulent. Forcing transition within the boundary layer can be achieved through the application of surface roughness patches. Although such passive control methods are already in use, the mechanism of the shock wave-laminar boundary layer interaction, and in particular the source of the strong shock unsteadiness are still not well understood. Furthermore, the benefits of boundary layer control obtained for

P. Grothe (✉)
Rolls-Royce Deutschland Ltd. & Co. KG, Blankenfelde-Mahlow, Germany
e-mail: Patrick.Grothe@Rolls-Royce.com; grothe.patrick@web.de

P. Flaszynski · R. Szwaba · M. Piotrowicz · P. Kaczynski
Institute of Fluid Flow Machinery, Polish Academy of Sciences, IMP PAN, Gdansk, Poland
e-mail: pflaszyn@imp.gda.pl

R. Szwaba
e-mail: rssz@imp.gda.pl

M. Piotrowicz
e-mail: mpiotrowicz@imp.gda.pl

P. Kaczynski
e-mail: pkaczynski@imp.gda.pl

B. Tartinville · C. Hirsch
Numeca International, Brussels, Belgium
e-mail: benoit.tartinville@numeca.be

C. Hirsch
e-mail: charles.hirsch@numeca.be

A. Hergt
Institute of Propulsion Technology German Aerospace Center (DLR), Cologne, Germany
e-mail: alexander.hergt@dlr.de

P. Doerffer et al. (eds.), *Transition Location Effect on Shock Wave Boundary Layer Interaction*, Notes on Numerical Fluid Mechanics and Multidisciplinary Design 144,
https://doi.org/10.1007/978-3-030-47461-4_4

Table 1 Geometrical and aerodynamic parameters of TFAST compressor cascade

Parameter	Unit	Value
Real Chord	mm	100
Stagger Angle	°	44.39
Pitch to Chord Ratio	–	0.6
Thickness to Chord Ratio	–	0.033
Blade Inlet Angle	°	50.91
Blade Exit Angle	°	33.22
Flow Inlet Angle	°	55.56
Deflection Angle	°	14.86
Inlet Mach	–	1.22
AVDR	–	1.203
Reynolds Number ALT	–	1.38×10^6
Turbulent Intensity	%	~1.0

low Reynolds numbers can turn into loss increase at the higher levels of Reynolds numbers. Another possibility of transition control is to use Vortex Generators driven by Air Jet (AJVG). In the compressor application the jets may be driven by the pressure difference between pressure and suction side of the blade. There are two effects which are present. The main effect is resulting from streamwise vortices generated on the blade suction side. The second effect is the suction of the boundary layer on the pressure side. The goal of Work Package 3 was to improve the understanding of the shock wave–laminar boundary layer interaction on the transonic compressor blade. This can potentially lead to successful new design solutions.

General Arrangement

In WP-3 the effect of laminar and turbulent SWBLI was investigated on a profile, representative for a section of a Fan Blade, defined by the industrial partner Rolls Royce-Deutschland (RRD). Boundary conditions of interest were chosen obtaining representative Reynolds numbers which occur at high altitude conditions. General geometrical and aerodynamic parameters are shown in Table 1. Upfront of the investigations, RRD did run 2D MISES simulations on the designed profile, showing that the boundary layer remains laminar until the shock.

Two facilities in which experiments were performed exist. The so called “Single passage test section” at IMP PAN was specifically adjusted to the required aerodynamic boundary conditions. At DLR a proper cascade consisting out of six aerofoils with main focus on the mid passage was used. The work split between the two facilities was defined in a way that the “Single passage test section” was used for wide range of parameter variations, whereas the DLR cascade was used obtaining more insight into the flow physics with usage of state of the art high fidelity measurement techniques.

Experiments performed at both facilities were supported by CFD simulations performed by NUMECA and RRD using their individual U/RANS flow solvers.

1 Single Passage Test Section

1.1 IMP PAN

Effect of the laminar-turbulent transition location on shock wave boundary layer interaction on the suction side of compressor profile was investigated in the IMP PAN transonic wind tunnel. For this purpose a single passage test section was designed [1] according to data for compressor cascade delivered by Rolls-Royce Deutschland.

A sketch of the test section is shown in Fig. 1. Reference conditions for the test section design are obtained from the numerical simulations for cascade configuration (Fig. 2), where the main flow features are shown: leading edge shock wave and its interaction with a boundary layer on the suction side of the next profile. The main parameters of cascade are shown in Table 1. One of the design criteria is the inlet Mach number and inflow uniformity which should be adequate to the real conditions. In order to maintain the required inflow conditions the nozzle designed during UFAST Project [2, 3] is used. Relative location of the profiles in the test section is the same as in the cascade in order to keep the similar flow structure. They are located in the nozzle at the required Mach number, M = 1.22.

A CAD view of the single passage compressor test section is shown in Fig. 3. Flow structure is investigated on the suction side (upper part) of the lower profile. The main shock waves are generated at leading edges of the profiles. The Reynolds number upstream of the profiles based on the blade chord (100 mm) is $Re = 1.4 \times 10^6$, what corresponds to the altitude aircraft conditions. Turbulence level in the nozzle upstream of the passage is 0.8%.

The outlet pressure and the shock wave location on the suction side of lower profile was controlled by the throat at the channel outlet (Fig. 3).

In case of test sections with linear compressor cascades, secondary flows and corner vortices play an important role and highly affect the Axial Velocity Density Ratio (AVDR) [4]. In such case, suction slots at sidewalls are applied to control and

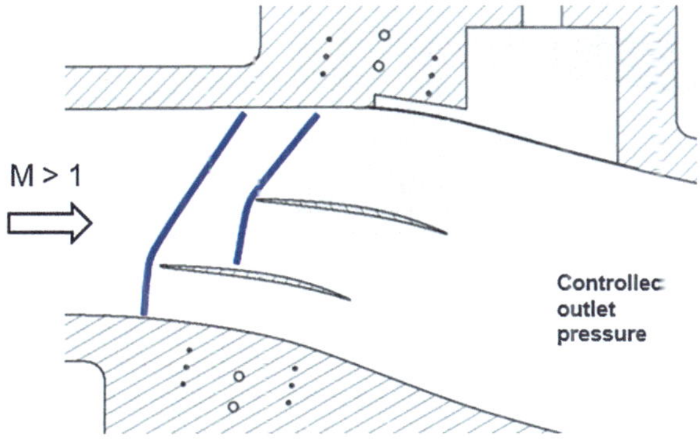

Fig. 1 Scheme of a single passage compressor test section

Fig. 2 Mach number in cascade

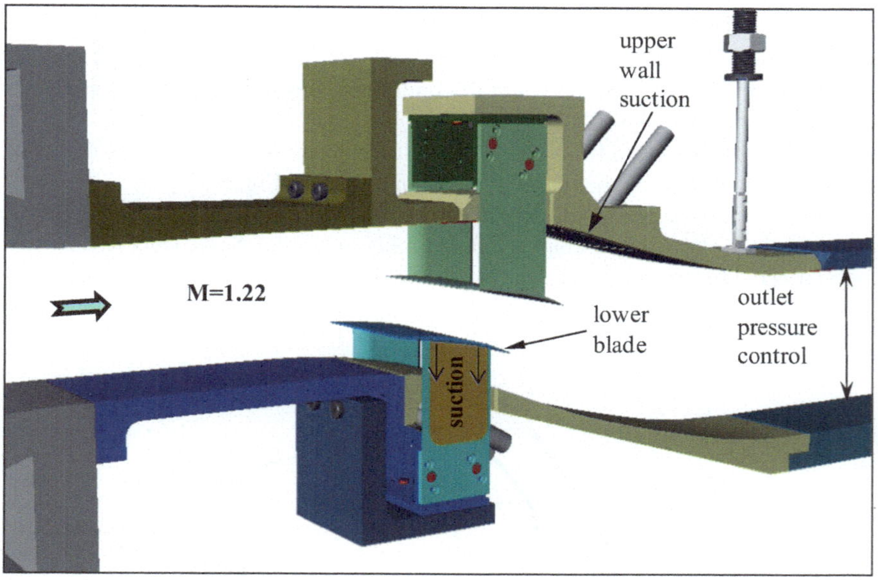

Fig. 3 View of test section

adjust flow conditions according to the cascade configuration. The details of suction slots designed and used in IMP PAN test section are described in [1]. The two suction systems are applied. One comprises two suction slots (black arrows) in supports of the lower profile at sidewalls (Fig. 3). These two slots are used for the proper AVDR control in the test section. Second system consists of suction on the upper and lower limiting walls. It allows to control the flow below the lower and above the upper one and finally to adjust the supersonic conditions upstream of profiles. Additionally, a perforated plate (perforation 5.2% and the holes diameter 0. 1 mm) is mounted in the upper wall. All suction slots are connected to the vacuum tank, so the flow is choked.

An effect of sidewall suction slots is shown (Figs. 4 and 5) by means of oil flow

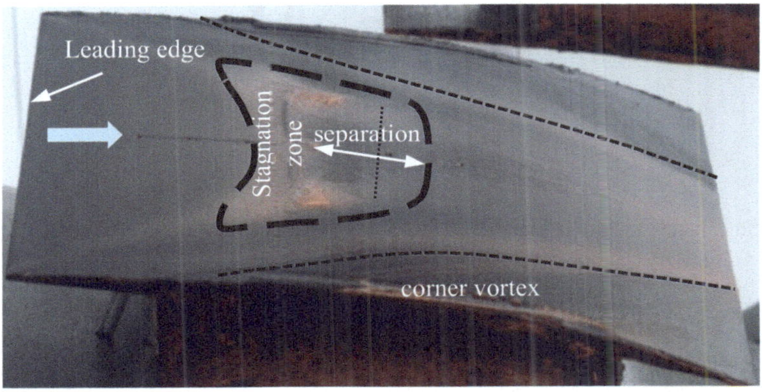

Fig. 4 Oil flow visualization—no sidewall suction

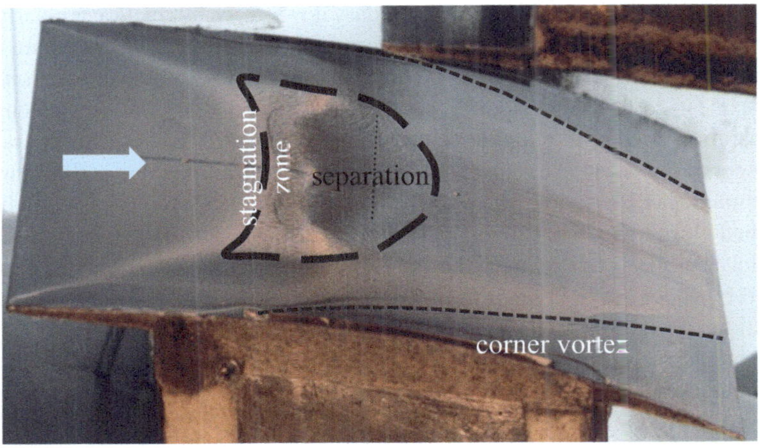

Fig. 5 Oil flow visualization—with sidewall suction

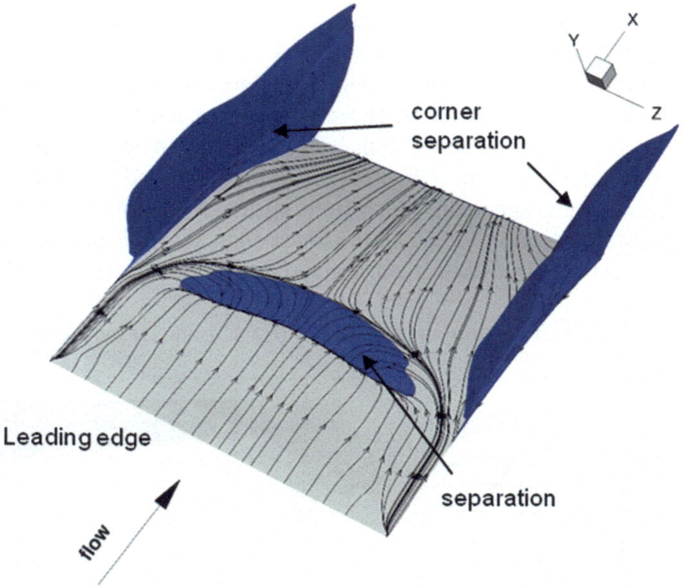

Fig. 6 Streamlines and iso-surface of negative X component of velocity (reversed flow)

visualization for the two flow cases, without and with sidewall suction. In Fig. 6, streamlines near the profile and iso-surfaces of negative X component of velocity predicted by numerical simulations are presented. Blue surfaces indicate existence of reversed flow. The effect of the shock induced separation is shown by zone in the middle of profile. Near the sidewalls, the corner separation is created at both sides leading to the contraction downstream of the shock wave. It can be noticed by the streamlines deflection as the effect of secondary flows. Such interaction is clearly seen on the oil flow visualization picture. The corner separation starts downstream of the shock wave due to interaction with the boundary layer. The corner interaction takes larger space in case without sidewall suction than in case with suction. The length of the separation zone in the middle of the channel is similar for both configurations, but the contraction downstream of the shock up to the trailing edge is different.

The main shock location is marked by a (dotted line). Within the zone indicated by dashed closed line, one can distinguish two subzones: stagnation and separation. The thicker oil layer upstream of the shock was created due to reduced skin friction induced by shock wave interaction with laminar boundary layer (stagnation zone). Separation zone is displayed by the oil streaks downstream of the stagnation zone.

The boundary layer development along the profile is dependent on the shock wave structure and pressure gradient on the profile. The shock waves topology in single passage test section is shown in schlieren picture (Fig. 7), while isentropic Mach number distribution on the suction side of lower profile is presented in Fig. 8. At the leading edge of the upper and lower profiles, shock waves are generated. The

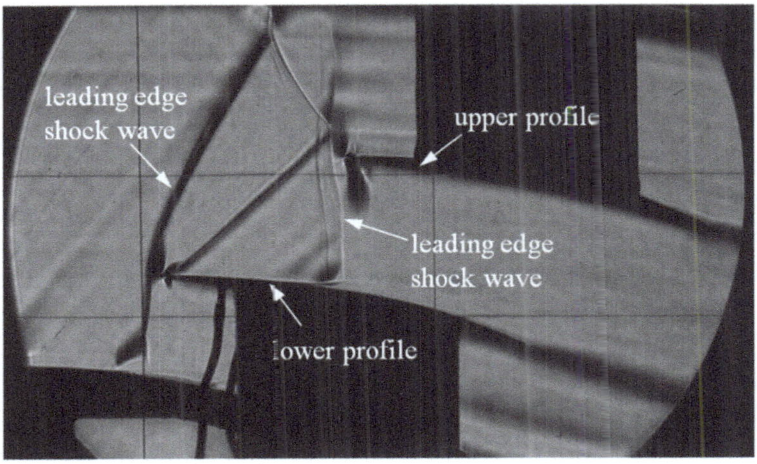

Fig. 7 Schlieren picture for reference case

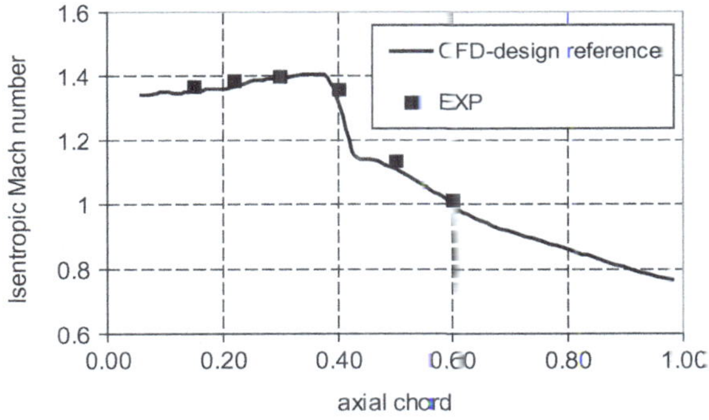

Fig. 8 Isentropic Mach number on suction side of profile

shock generated on the upper profile interacts with the boundary layer on the lower profile. Upstream of the shock, an oblique darker zone is visible, which indicates the interaction onset. Significant distance between the onset of the interaction and the main shock wave location is typical for shock wave interaction with laminar boundary layer.

Numerical results obtained for cascade configuration (by RRD), considered as the reference for the test section design, are compared with pressure tap measurements at middle span of the profile. Agreement of design assumptions and measurements proves that the flow structure existing in cascade is well reproduced by properly designed configuration of single passage test section.

Experimental and numerical investigations of the flow structure were carried out in two steps. The work done within the first one was focused on the reference case where shock wave interacts with boundary layer on suction side without tripping devices. In such case boundary layer upstream of the shock wave is laminar. As the next task, the tripping devices were applied on the suction side downstream of the leading edge and its effect on the shock wave boundary layer interaction was investigated. Three different methods were implemented to enforce laminar-turbulent transition upstream of the shock wave:

- tripping step
- roughness patches
- air jet vortex generators (AJVG) [5, 6].

Following measurements were carried out to investigate shock wave boundary layer interaction on compressor profile for cases without and with flow control devices:

- static pressure distribution at middle span of profile,
- schlieren visualization,
- high speed movie visualization to investigate normal shock wave oscillations,
- oil flow visualization,
- surface pressure measurements by means of Pressure Sensitive Paint (PSP),
- Temperature Sensitive Paint (TSP) to detect transition location on suction side of profile
- velocity measurement in the wake by means of Laser Doppler Anemometry (LDA).

1.1.1 Boundary Layer Tripped by Step

The step was mounted at three different positions on the profile: just downstream of the leading edge (LE), at 16% of chord length $X/C = 0.16$ and 22% of chord length $X/C = 0.22$. The height of the tripping step was varying in range from 0.05 mm to 0.4 mm, while the width was 5 mm or 10 mm. The main shock wave location for the investigated configuration is $X/C = 0.48$. Boundary layer thickness upstream of the shock wave estimated by means of numerical simulations is ~0.35 mm [7].

The location of the laminar-turbulent transition was detected by means of the Temperature Sensitive Paint (TSP), where the effect of local heat transfer and shear stress analogy is exploited. The results of temperature change on the suction side of the lower profile are presented for reference case and the case with transition control in Fig. 9. Due to the lack of the temperature paint calibration, the temperature change is shown only. The range of temperature variation is adequate to difference between the air stagnation temperature and profile wall temperature. In Fig. 9, one can notice that temperature is monotonically increasing from the leading edge up to the transition location upstream of the shock wave. It is the effect of decreasing shear stresses in this zone, so it indicates the highest shear stresses just downstream of the leading edge. The laminar-turbulent transition in boundary layer (sudden change of temperature) occurs at beginning of the separation zone location which corresponds to the location of the shock wave λ-foot front leg (see Fig. 5).

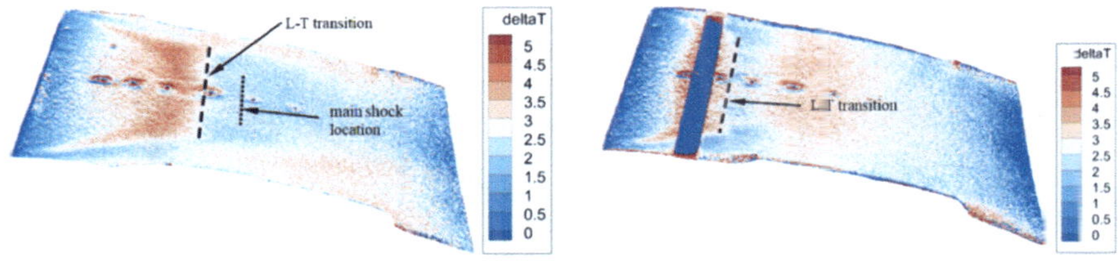

Fig. 9 Temperature Sensitive Paint—no tripping device (left), with tripping device (right)

Application of the microstep, as the tripping device (height 0.1 mm) leads to laminar-turbulent transition downstream of the step (Fig. 9-right). Upstream of the step, the zone of decreasing shear stresses is visible. It is the same situation as in the case without step, however the step embedded in the boundary layer influences on amplification of this effect. Downstream of the step, the low shear stress still exists, but further downstream the colour change indicates local increase of shear stress what allows for detection of laminar-turbulent transition location.

Further effect of the tripped boundary layer is observed downstream the profile. Velocity measurements in the wake (LDA) allow for comparison of investigated configurations and assessment of positive or negative effect on the aerodynamic performance. The location of the velocity traverse measurements downstream of the profile is shown in Fig. 10. The traverse is located in the middle of the channel, 20 mm downstream of the lower profile. The system of coordinates is located at the trailing edge (blue dot).

A comparison of the velocity profile for the clean case (*Ref-laminar*) and two configurations with step is shown in Fig. 11. The two cases selected are indicating high difference and opposite effect on the flow structure.

The step mounted closely downstream of the leading edge trips boundary layer early what creates a thicker boundary layer upstream of the shock wave. As the

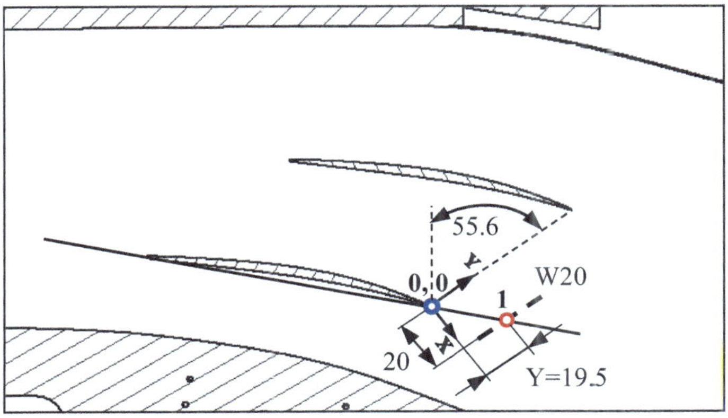

Fig. 10 Wake traverse location

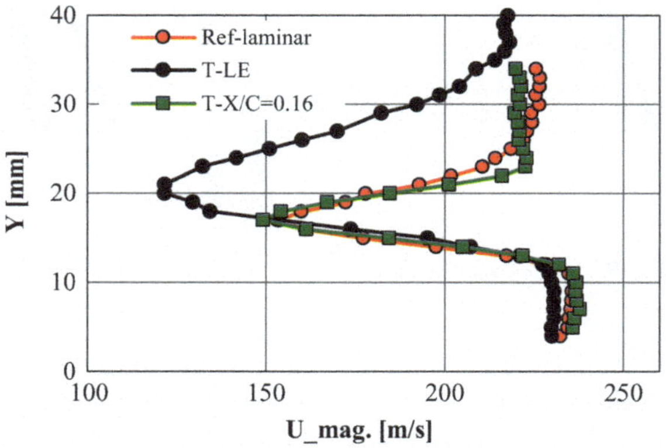

Fig. 11 Velocity in the wake

effect larger separation zone is created downstream of the shock wave and finally wider wake. If the step is moved downstream (x/c = 0.16), the boundary layer is also tripped, but shock interaction with turbulent boundary layer shows positive effect in the wake. In order to quantify obtained effects, integral parameters in the wake are computed for these cases. According to boundary layer analogy, thickness δ, displacement thickness δ* and momentum thickness δ** are evaluated for the upper part of the wake. As shown in Table 2, the same step located at two different positions on the suction side has significantly different effect, negative or positive.

The results of the velocity measurements and the wake parameters clearly show that too thick boundary layer upstream of the interaction zone (T-LE flow case) causes strong wake thickening. Improved results can be obtained if the location of the tripping device is properly adjusted. For the case T-X/C = 0.16 the wake thickness is reduced and integral parameters are lower than obtained for the reference case (laminar interaction).

Schlieren visualization recorded by the fast CCD camera and image analysis allows for investigations of shock unsteadiness. For each image (movie frame) one line (20 mm above profile) crossing the main shock wave is processed. The sampling frequency for all measurements was 2 kHz. Amplitude and RMS shock oscillations signal for the selected flow cases (Table 3) prove that they can be reduced if boundary layer is carefully tripped.

At the same time, numerical simulations have been carried out by means of Fine/Turbo Numeca. The structured mesh for this configuration is generated in IGG/Numeca and it consists of 18.6 × 10^6 cells. Explicit Algebraic Reynolds Stress

Table 2 Wake integral parameters

	Ref-laminar	T-LE	T-X/C = 0.16
δ (mm)	9.2	21.0	5.7
δ* (mm)	1.31	4.15	1.02
δ** (mm)	1.03	2.90	0.77

Table 3 RMS and amplitude of shock wave oscillations

	Ref-laminar	T-LE	T-X/C = 0.16
RMS (mm)	1.25	1.22	1.16
Amplitude (mm)	4.4	4.7	3.6

Model (EARSM) with transition modelling [8] was used. Ambient conditions according to the measurements (total pressure 101 kPa and total temperature 293 K) together with turbulence intensity 0.8% and turbulent viscosity ratio 10 are set at the inlet. At the outlet static pressure is adjusted in order to get the correct (as measured) shock wave location.

The effect of tripping device on isentropic Mach number distribution is shown in Fig. 12. At the step location, the disturbance is clearly visible and the minimum and maximum predicted numerical excitations go together with measured values. In case of transitional boundary layer, the shock wave is moved slightly downstream. The Mach number distribution downstream of the shock indicates very small difference between the cases.

Shock waves configuration for the reference case is already shown in schlieren picture (Fig. 7). Numerical simulations results are compared with schlieren picture by means of density gradient magnitude (Fig. 13). The main flow features such as the shock location on the profile, the shock generated on the leading edge and the shock reflection from the upper wall are well captured. Numerical prediction of the tripping step effect is shown in Fig. 13 (right). On the suction side, the step induces the compression and expansion waves. They are generated upstream and downstream of the step. The height of the λ-foot is very similar to the one obtained for the case without the tripping device.

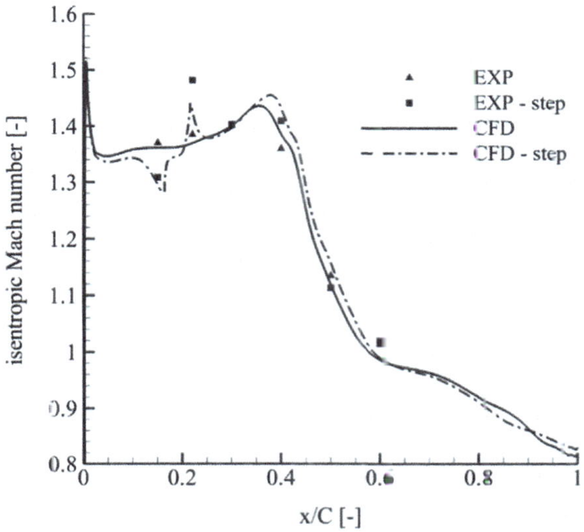

Fig. 12 Isentropic Mach number on the profile suction side

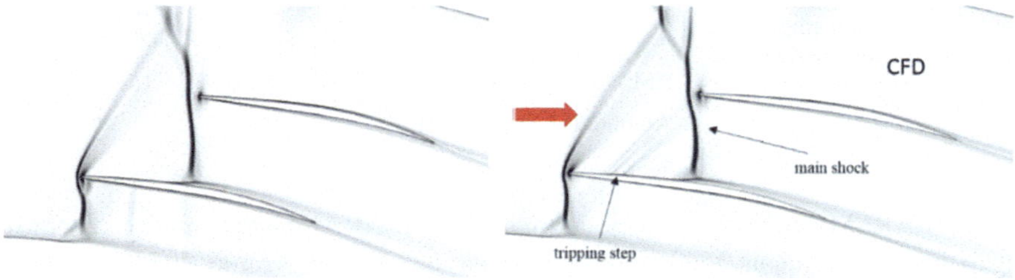

Fig. 13 Density gradient magnitude—no step (left), with step at X/C = 0.16 (right)

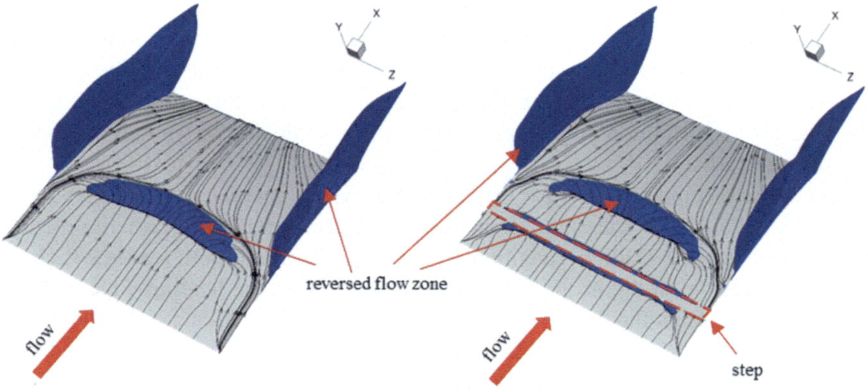

Fig. 14 Streamlines and iso-surface representing reversed flow

One of the main objectives of the transition location investigation is the effect on the shock induced boundary layer separation (Fig. 14). The reversed flow is represented by iso-surfaces of negative velocity component in X direction. There are three zones indicating backflow.

The two zones, located close to the sidewalls, visualize corner separation downstream of the shock wave and they are located in the profile and sidewall corner. Comparing two images, spanwise zones upstream and downstream of the step can be noticed. In case of configuration with microstep the separation downstream of the shock is slightly reduced. Although, such slight reduction is obtained, the positive effect shown by wake measurements is not properly predicted (Fig. 15). The difference is low, but the reason for that needs further investigation.

1.1.2 Boundary Layer Tripped by Roughness Patch

Boundary layer transition upstream of the shock wave on suction side of profile was also controlled by roughness patches. Roughness patch effect is investigated by means of sandpaper strips glued in groove (0.4 mm depth) on profile. The application

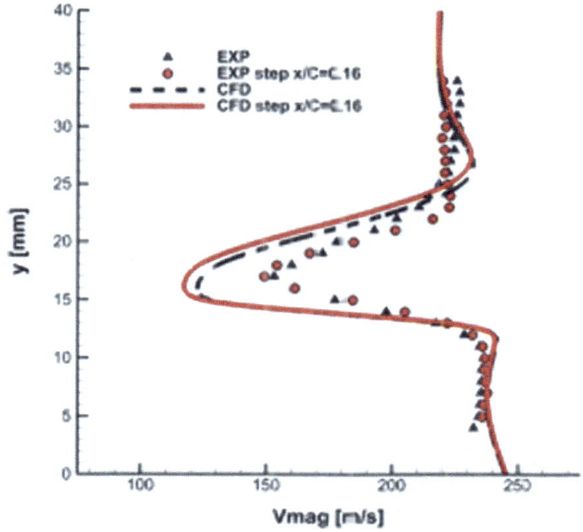

Fig. 15 Velocity in wake 20 mm downstream of profile

of roughness patches on the profile surface is a very important technical problem. In order to avoid steps on the suction surface allowing for roughness effect investigations, not coupled with step existence, the sandpaper strips have to be carefully mounted in a way that average level of roughness is approximately on the same level as smooth surface of the blade. Variation of investigated roughness and patch location is presented in Table 4.

An example of schlieren visualization for the flow with patch of Ra = 10 at X/C = 0.11 is shown in Fig. 16 (right), where Ra is an average roughness of the used rough strip. Roughness patch is a source of disturbance visible by weak waves at the starting and ending edges of roughness patch. Tripped boundary layer becomes turbulent and the interaction with shock wave has changed its character. The shock wave is bowed and the interaction zone is shorter than in laminar (reference) case. An important effect of applied roughness patch is significant reduction of separation downstream of the shock (Fig. 17) in comparison with the reference case (Fig. 5).

As it was already presented in the previous chapters for basic flow cases, boundary layer tripping to induce laminar-turbulent transition has to be done carefully. The

Table 4 Location and average roughness height (Ra)

Roughness strips locations and Ra (mm)		
X/C = 0.11	X/C = 0.16	X/C = 0.21
Strip width 10 mm	Strip width 5 mm	Strip width 5 mm
12	20	20
10	16	16
8	12	12
–	10	10

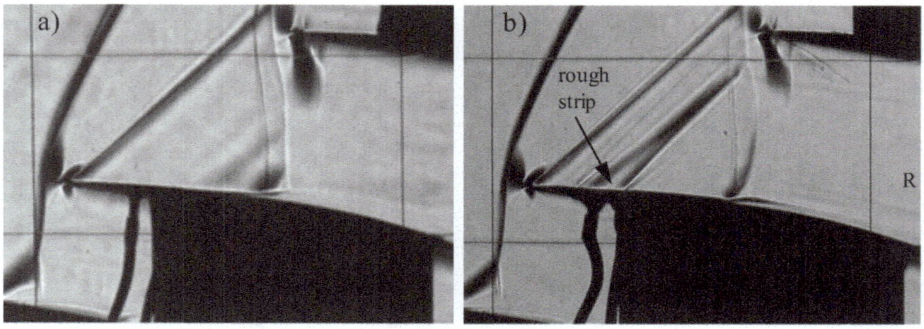

Fig. 16 Schlieren for reference case (left) and roughness patch Ra = 10 and X/C = 0.11 (right)

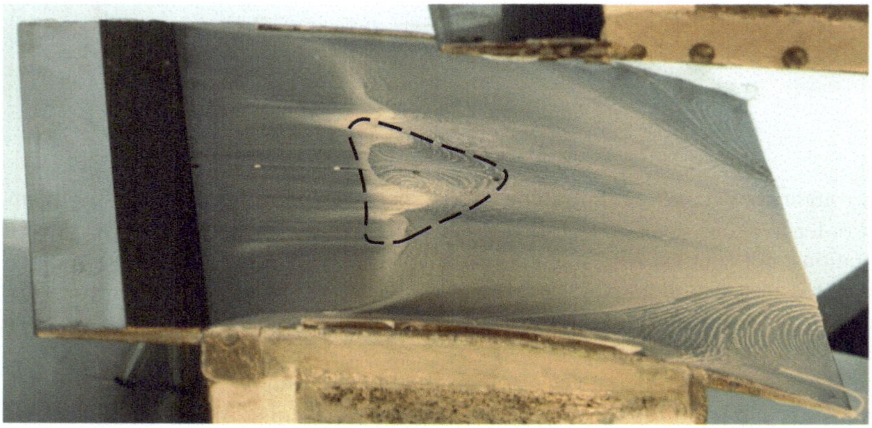

Fig. 17 Oil visualization for roughness patch Ra = 10 X/C = 0.11

same problem concerns roughness patches on compressor profile where the flow structure and aerodynamic performance can be improved if tripping device is applied, but it can be also deteriorated if "overtripping" is obtained and boundary layer is too thick.

In Fig. 18, the best and the worst measured effect on the wake is presented. In case of high roughness located far upstream of the shock wave, the shock wave interacts with thick boundary layer what leads to larger separation and wider wake. As the wake width is representative for aerodynamic performance of the profile, the best results obtained for particular configurations are shown in Fig. 19, where R—is a case with roughness patch and T—is a case with microstep. One can notice, that similar velocity distribution was measured for different locations of roughness patches and roughness height. These cases are compared with results shown above for boundary layer tripped by step. It means that similar effect is possible for different tripping methods. Quantitative comparison is presented by means of wake integral parameters (Table 5) computed for the upper part of the wake, which are considered

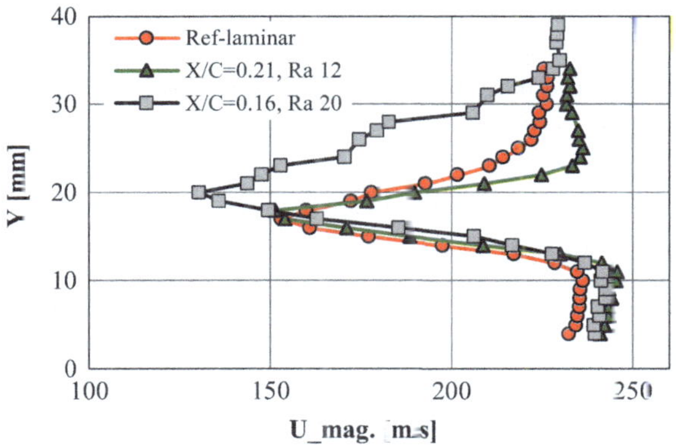

Fig. 18 Wake—velocity for selected cases

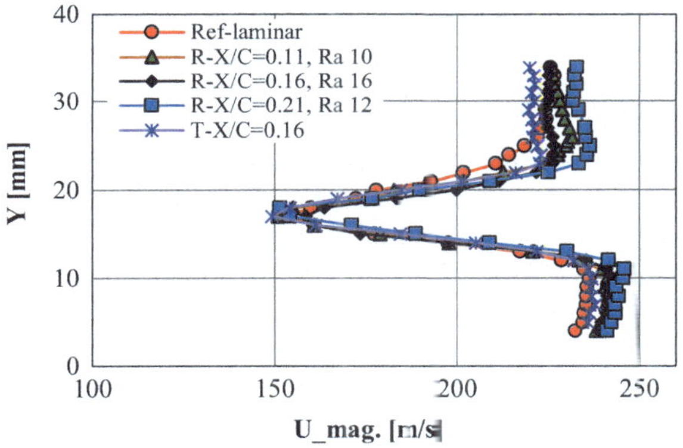

Fig. 19 Wake—velocity for the best flow configurations—roughness patch and step

Table 5 Wake integral parameters

	Reference	R-X/C = 0.11, Ra = 10	R-X/C = 0.16, Ra = 16	R-X/C = 0.21, Ra = 12	T-X/C = 0.16
δ (mm)	9.2	7.87	5.57	5.26	5.7
δ* (mm)	1.31	1.22	0.83	0.81	1.02
δ** (mm)	1.03	0.93	0.65	0.63	0.77

as the measure of tripping effect. The obtained values indicate that the most effective is the roughness patch located closer to the shock wave. Closer location requires higher roughness, but questions, how high and what is the limit for the positive effect, still needs further investigations.

As the conclusion one can state that the improvement of flow parameters downstream of shock wave boundary layer interaction requires moderate treatment of the boundary layer upstream of the interaction. Overtripping effect appears if boundary layer becomes too thick and leads to larger separation and deterioration of wake parameters or to higher shock oscillations. Positive effect was obtained for boundary layer tripped by roughness patch or step if the height and location is properly adjusted, although the height of the step is an order of magnitude larger than the roughness height.

1.2 NUMECA

The commercial package Fine^{TM}/Turbo is used for this study. The flow solver is a three-dimensional, density-based, structured, multi-block Navier-Stokes code using a finite volume method. Central-difference space discretization is employed for the spatial discretization with Jameson type artificial dissipation. A four-stage explicit Runge-Kutta scheme is applied for the temporal discretization. Multi-grid method, local time-stepping and implicit residual smoothing are used in order to speed-up the convergence. The S-BSL-EARSM turbulence model has been retained here together with the local correlation transition model (LCTM).

The mesh has been generated with Autogrid5^{TM}, it contains 22 million grid cells and the first inner grid-spacing has been fixed to 1 μm in order to reach a target y^+ of unity. The computational domain contains most of the details of the experimental test case as shown on Fig. 3. It starts upstream of the inlet nozzle, contains the two measuring blades, the slots and goes down to the end of the test section. A zoom of the mesh around the blades can be seen on Fig. 20. The lower blade is the instrumented one.

Boundary conditions are fixed based on the information provided by IMP. Inlet total quantities are fixed upstream of the nozzle. Total pressure is set to 101 kPa and total temperature is fixed to 293 K. The inlet flow is set to axial. The outlet static pressure is set to 76 kPa in order to reach a shock location in accordance to

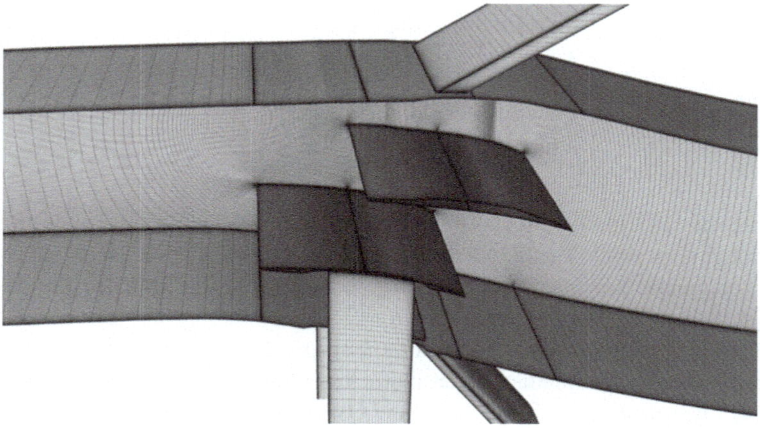

Fig. 20 Zoom of the mesh of the IMP test section

the experiment. Furthermore, three slots are located along end-walls. The objective of those slots is to suck air out of the main passage in order to be able to reach the target AVDR of 1.2 and upstream Mach number of 1.22. Slot 1 is located along the bottom end-wall below the lower blade. Slot 2 is made of two outlets. Each one is located at the corner between the lower blade suction side and the lateral end-wall. The mass flow that is extracted through this slot directly controls the blade suction side flow structure. Slot 3 is located along the upper wall. All the slots can be identified on Fig. 20. The back-pressure for slots 1, 2 and 3 is fixed to 40, 45 and 20 kPa, respectively. Knowing the importance of Slot 2 on the flow structure, IMP has also provided the measured flow rate through this slot. It is 8.44 g/s. Our numerical results lead to a flow rate of 8.51 g/s, i.e. rather close to the measured one. A sensitivity study has been performed on the outlet boundary condition for slot 2. It appears that decreasing the back-pressure by 5 kPa induces a modification of the limit-streamlines along the blade and of the pressure distribution at mid-span.

Figure 21 allows to compare the experimental and the numerical Schlierens of the flow at mid-span. From this figure it appears that numerical results are in an almost perfect match with the experimental ones. Indeed, the flow solver produces the upstream detached bow shock that starts at the leading edge of the lower blade. This shock interacts with the upper end-wall. Furthermore, the leading edge of the second blade induces a straight shock that impact the suction side of the lower blade with a lambda foot. It is likely to produce a laminar shock boundary layer interaction. Superimposing both images reveals that the computed straight shock is at the same location than the measured one.

The computed isentropic Mach number distribution at mid-span and along the suction side of the lower blade is displayed on Fig. 22. It is further compared to measured data along the blade surface (red dots). It appears that the flow solver produces a too high isentropic Mach number downstream of the shock boundary layer interaction. From previous sensitivity studies it appears that such a distribution is rather sensitive to the mass flow that is extracted through Slot 2. With a lower back-pressure, the pressure distribution in the downstream region will be closer to the

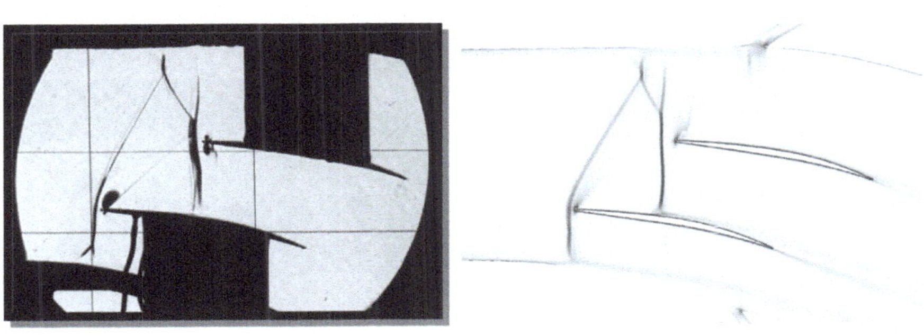

Fig. 21 Experimental (left panel) and numerical (right panel) Schlieren of the flow for IMP test section

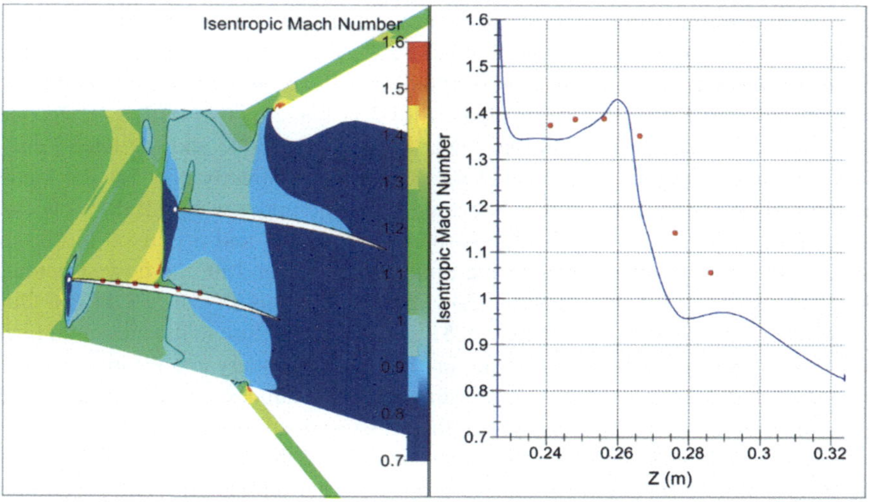

Fig. 22 Computed isentropic Mach number distribution at mid-span (left panel) and comparison along the blade suction side (right panel) between computed (blue line) and measured (red dots) data

experimental one, but the computed mass flow through Slot 2 will be over-estimated by more than 25%.

The flow through slot 2 has also an impact on the flow structure all along the blade suction side. This can be seen on Fig. 23. Indeed, slot 2 has a direct influence on the flow structure along the lateral end-walls and further downstream. Since the blade aspect ratio is rather small, it has also a direct influence on the tri-dimensional recirculation region that can be observed for both experimental and numerical results.

Such a recirculation is caused by the laminar shock boundary layer interaction. From this figure, it appears that the CFD results are in rather good match with experimental ones. The onset of recirculation is rather well predicted. But the stream-wise extend of this recirculation region is under-estimated. Furthermore, downstream

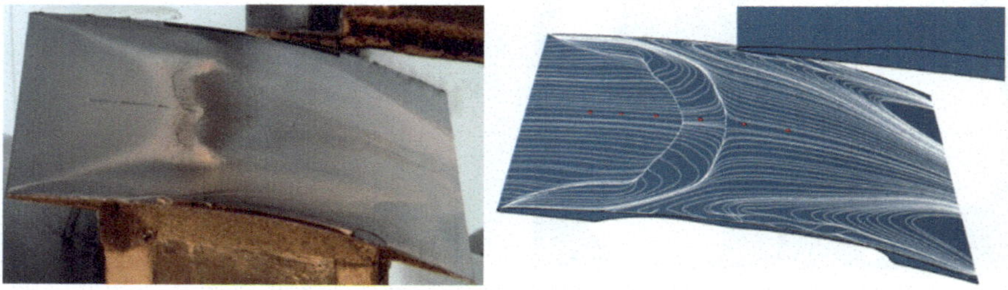

Fig. 23 Mesured (left panel) and computed (right panel) limit streamline along the blade suction side

corner separations originating either from the blade leading edge or from the slot 2 are very well captured by numerical results.

Figure 24 represents the measured and computed velocity profiles along three sections at mid span. One section is located 22 mm upstream of the blade leading edge whereas the two over are located 5 and 22 mm downstream of the blade trailing edges, respectively. The velocity magnitude distribution at mid-span is displayed on the left panel. It allows to have a more precise view of the exact measurement location. From this figure, it appears that the flow solver predicts in a rather accurate way the velocity distribution along these test sections. However a few discrepancies can be observed. Upstream of the blade, since the location of the computed bow shock does not exactly match the measured one, the velocity distribution is affected by this slight displacement of the shock (upper right profile). Nonetheless, the velocity magnitudes upstream and downstream of the shock are well predicted. 5 mm downstream of the blade trailing edge, the computed wake is slightly deeper than the measured one—however only a few experimental points are available. Nonetheless, the free stream velocity is correctly captured. For the most downstream velocity profile, again the computed wake is slightly too deep when compared to the experimental one.

To conclude, CFD simulations of the compressor blade have been performed for the IMP test case configuration. They reveal that the Fine^{TM}/Turbo flow solver together with S-BSL-EARSM turbulence model and LCTM is able to reproduce most of the flow features that have been observed experimentally. Indeed, a bow shock is present upstream of the lower blade. The upper blade produces a straight shock and induce a laminar shock boundary layer interaction along the lower blade suction side. This induces a three-dimensional separation along this blade that affects

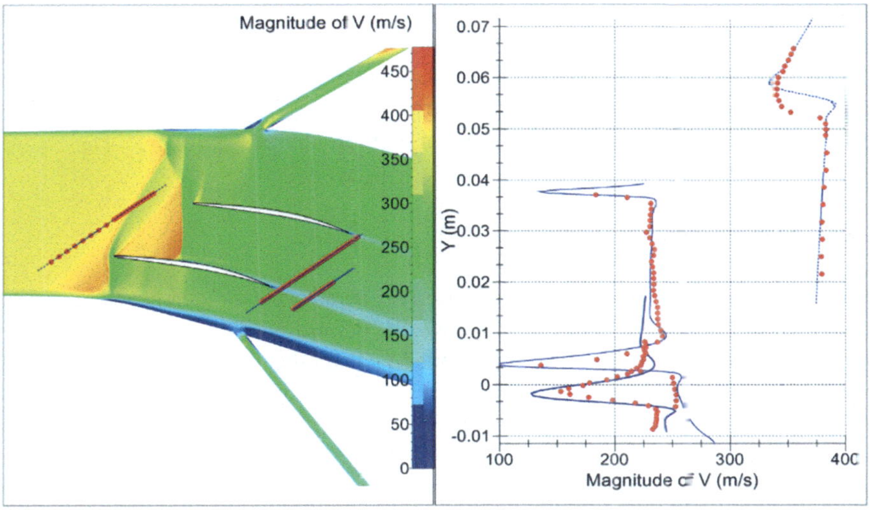

Fig. 24 Measured (red dots) and computed (blue lines) velocity profiles upstream and downstream of the blade for IMP test section

the downstream wake. Furthermore, a detailed comparison to available experimental data has also been performed. It points out the accuracy of the results produced with such a turbulence model.

2 Basic Compressor Cascade Flow

2.1 DLR

2.1.1 Test Facility and Experimental Setup

The experimental investigations on compressor cascade flow were performed in the Transonic Cascade Wind Tunnel at DLR in Cologne [9, 10] as shown in Fig. 25 (top). This tunnel is a closed loop, continuously running test facility with a variable nozzle, an upper transonic wall (Fig. 25, bottom) in order to reduce shock reflection, and a variable test section height. The air supply system of the wind tunnel enables an inlet Mach number up to 1.4 and a Mach number independent variation of the Reynolds number from 1×10^5 to 3×10^6.

The performed measurements were focused on the detection of the reference flow conditions of the cascade at laminar and turbulent shock boundary layer interaction. Figure 26 (left) shows the reference cascade which consists of six blades with 100 mm chord and 168 mm blade span. Due to the low turbulence level of the wind tunnel inflow the boundary layer typically is laminar on the blade suction side. In order to achieve the test conditions for a turbulent shock boundary layer interaction the turbulence level of the inflow has to be increased. For this purpose a turbulence grid has been placed upstream of the test section (based on the work of Hoheisel [11]). During the measurements boundary layer suction on the endwalls within the cascade passages are used. This is necessary to adjust the AVDR [12, 13] in order to achieve comparable operating points for the laminar and turbulent case.

2.1.2 Measurement Techniques

Different measurement techniques were used to characterize the steady and unsteady cascade flow. The right part of Fig. 26 shows the definition of the cascade parameter and measurement planes in detail. The profile Mach number distribution was detected by the measurement of the static pressure by static pressure taps on the pressure side of blade 4 and on the suction side of blade 3 (see Fig. 26). The passage between these two blades is defined as the investigation region of the cascade where homogenous inflow conditions have to be achieved during the measurements. In addition to the detection of the profile Mach number distribution at blade midspan the inflow angle distribution at midspan was measured by means of a L2F measurement system [14–16]. Figure 25 (bottom) shows a sketch of the wind tunnel test section with the

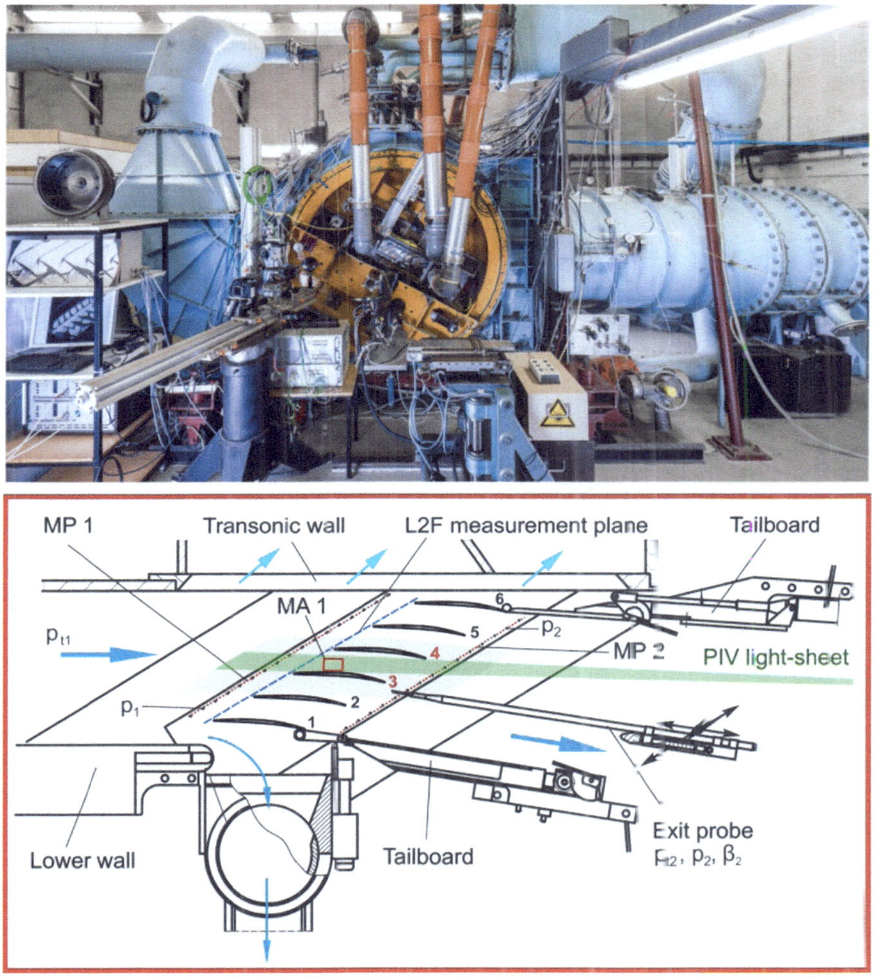

Fig. 25 DLR Transonic Cascade Wind Tunnel (top), sketch of the test section with the cascade (bottom)

cascade and the definition of the used measurement area for the PIV measurements. A detailed description regarding the PIV measurement setup are given in [17, 18].

The L2F measurement was performed at the L2F measurement plane (MP L2F) with a distance of 7 mm to the cascade inflow plane. The inflow Mach number is detected at measurement plane 1 (MP 1) by the measurement of the pitchwise static pressure distribution on the endwall at this plane as well as the measurement of the total pressure in the settling chamber. At measurement plane 2 (MP 2) the wake behind the cascade was measured by means of a 3-hole pressure probe and an additional static pressure probe. From the analysis of the wake measurements the

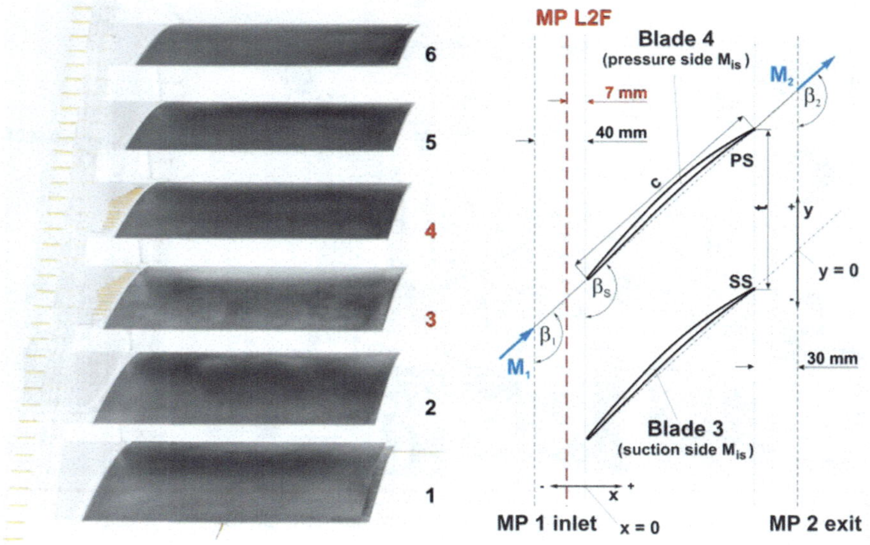

Fig. 26 Reference cascade with 6 blades for wind tunnel test (left), cascade parameter and definition of measurement planes (right)

pitchwise distributions of the losses, the outflow angle and the outflow velocity as well as the static pressure rise were achieved. In addition to the profile Mach number distribution liquid crystal measurements on the suction side of blade 4 were used to detect the boundary layer condition [19, 20]. The detailed analysis of the steady and unsteady shock behavior as well as the shock wave boundary layer interaction are based on Schlieren figures and PIV measurements. The Schlieren figures are used in order to characterize the shock structure and the PIV data are used for the detection of the shock position within the cascade and the measurement of the shock wave boundary layer interaction. Furthermore, a statistical analysis of the PIV data provides information of the shock movement range. Additionally, High Speed Shadowgraphy (HSS) is used to characterize the unsteady behavior of the shock fluctuation.

Figure 27 (top) shows the used setup for the PIV measurements. In this figure the light sheet within the cascade passage and on the left hand side the camera is visible. Additionally, the definition of the used PIV measurement areas (MA) is given in the lower part of this figure.

2.1.3 Results and Discussion

The measurement of the reference cascade includes two parts. The first part was the investigation of the uninfluenced cascades as a reference state. This means during the tests the incoming flow was characterized by low turbulence intensity of 0.5. In this

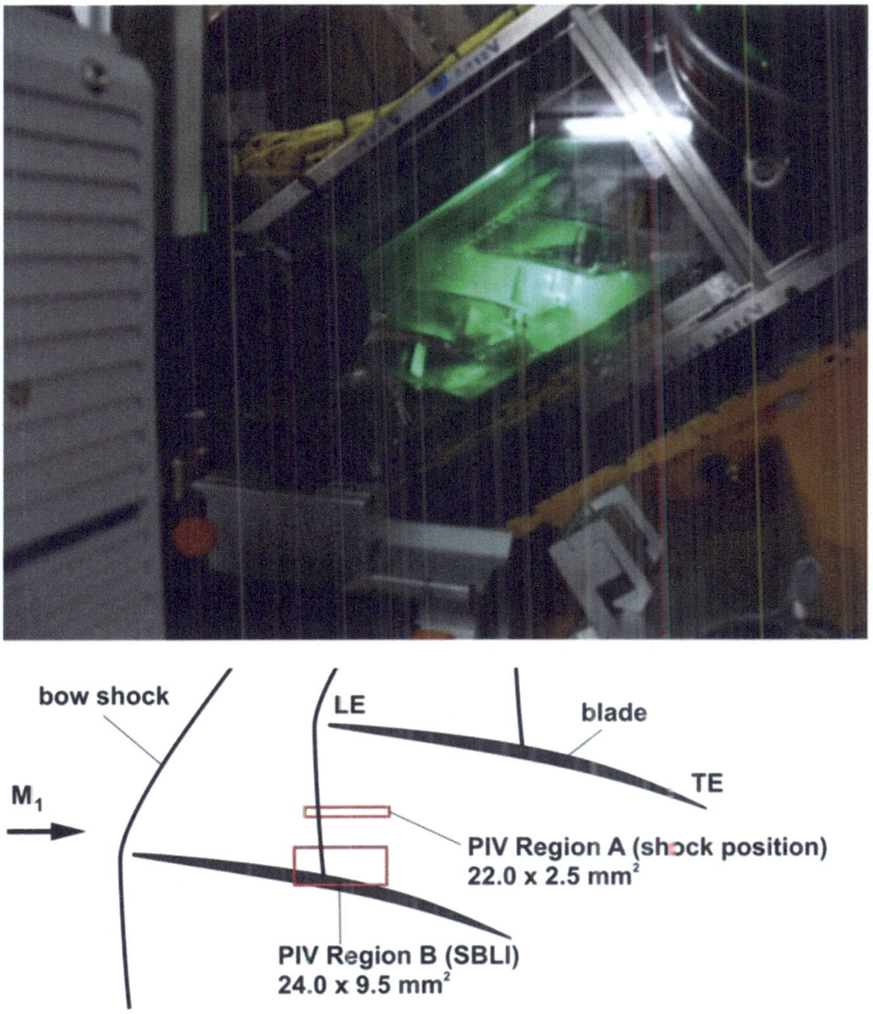

Fig. 27 Mounted cascade in the test section with PIV measurement setup (top), schematic figure of the shock system and location of the PIV measurement areas (bottom)

case the boundary layer on the blade surface should be laminar until the shock. In the further work this reference case will be named laminar. In addition to that (part two), the fully turbulent case was also investigated in order to be able to analyze the influence of the laminar and turbulent shock boundary layer interaction on the cascade performance in terms of losses and unsteadiness.

The detailed measurements of the cascade at laminar and turbulent case were performed at operating point 1. Figure 28 shows the measured inflow angle distribution for the both cases. The figure shows that the shape of the distribution as

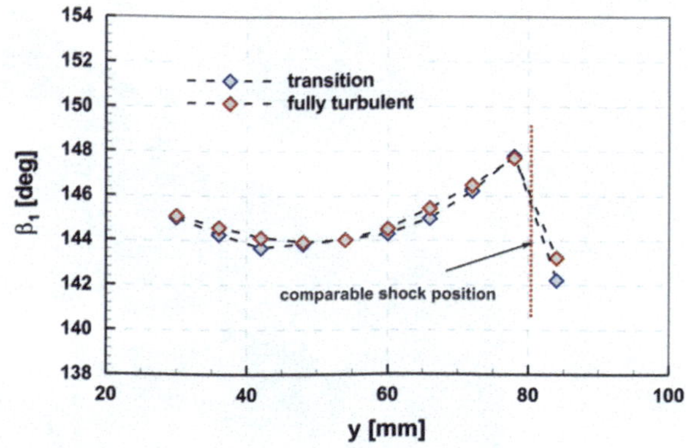

Fig. 28 Inflow angle distribution at MP L2F (cascade midspan)

Table 6 Measured averaged flow conditions of the laminar and fully turbulent reference case

	Laminar	Fully turbulent
M_1	1.21	1.21
Re_1 (10^6)	1.38	1.37
β_1 (°)	145.5	145.6
AVDR	1.21	1.20
β_2 (°)	130.0	130.1

well as the position of the bow shock is in a very good agreement. The averaged inflow angle of the cascade which was calculated by using conservation laws of fluid motions amounts to 145.5° for the laminar case and 145.6° for the turbulent case. The following Table 6 shows the measured flow conditions of both reference cases.

In Fig. 29 the Profile Mach number distribution of the laminar and fully turbulent case is shown. The mean outcome of this figure is that both cases have nearly the same

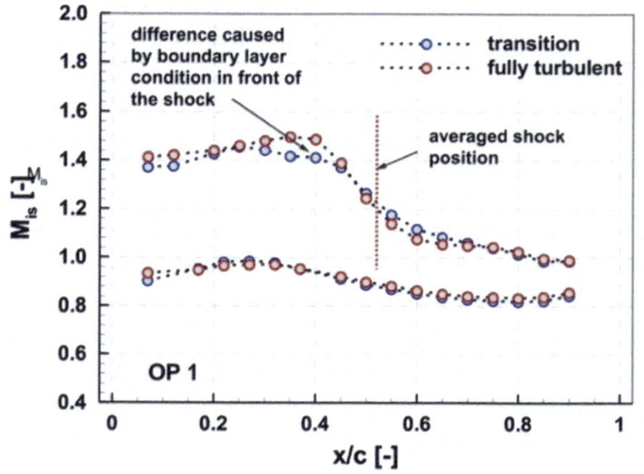

Fig. 29 Profile Mach number distribution at midspan (transition and fully turbulent)

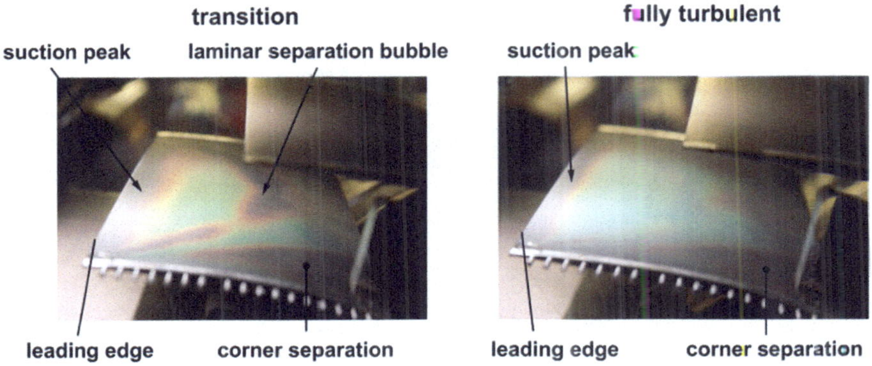

Fig. 30 Liquid crystal measurement on the blade suction side

shock position on the blade suction side. In addition to that, in the figure the difference between a laminar and a turbulent shock wave boundary layer interaction is visible. In front of the deceleration region, which is caused by the pressure rise over the shock, a Mach number peak is depicted for the fully turbulence case. In comparison to that, this Mach number peak disappears at the laminar case and furthermore the shock (deceleration region) is more smeared which means that the gradient of the Mach number decrease is lower for the laminar case.

Figure 30 shows the results of the liquid crystal measurement. From this figure the boundary layer behavior is observable. The used liquid crystal has a visible color range of 10° from 20° up to 30 °C. If the adiabatic wall temperature on the blade surface is outside this range the liquid crystal are black. Therefore, a black region close to the leading edge is shown in the figure which results from the suction peak and close to the endwalls the characteristically corner separation is visible. Furthermore, on the left hand side the separation bubble above the shock wave in the laminar case is shown in the figure. This separation bubble disappears at the fully turbulent case.

By the next step the differences of both reference cases will be investigated in more detail. Figure 31 shows the pitchwise loss distribution of the laminar (left) and the fully turbulent case (right). It is observable that the viscous losses represented by the wake are significantly reduced for the fully turbulent case. This means, that the shock interaction with a turbulent boundary layer produces lower losses than in the laminar case where a separation bubble occurs. Additionally, the distribution of the shock losses [21] is only slightly influenced for the turbulent case. The results in Fig. 31 also show that the shock loss level at the pressure side is lower than at the suction side which represents a not fully periodic outflow behind the cascade. This is an inherent effect at transonic and supersonic cascade wind tunnels and can only reduced by test adjustment but not fully avoided. A detailed explanation of this effect is given by Schreiber et al. [9].

The comparison of the Schlieren figures and the schematic diagram of the shock structure which are extracted from the Schlieren figures in Fig. 32 (laminar) and Fig. 33 (fully turbulent) shows the slightly differences of the shock structure. The

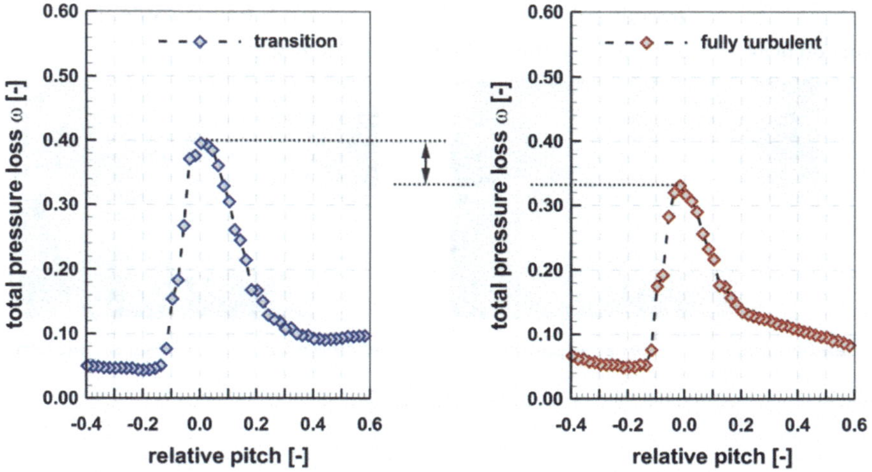

Fig. 31 Pitchwise distribution of the total pressure loss in the wake behind the cascade

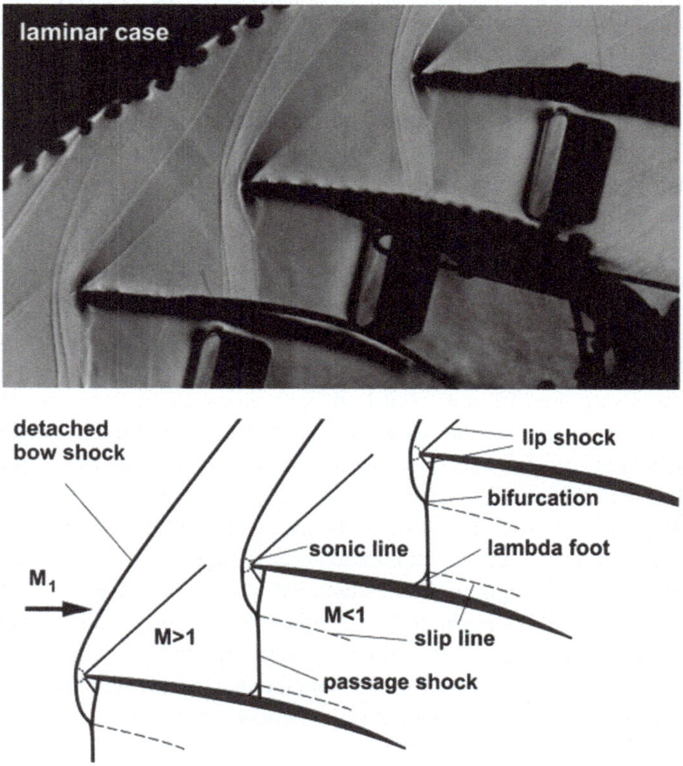

Fig. 32 Schlieren figure of the laminar reference cascade (top), schematic diagram of the shock structure (bottom)

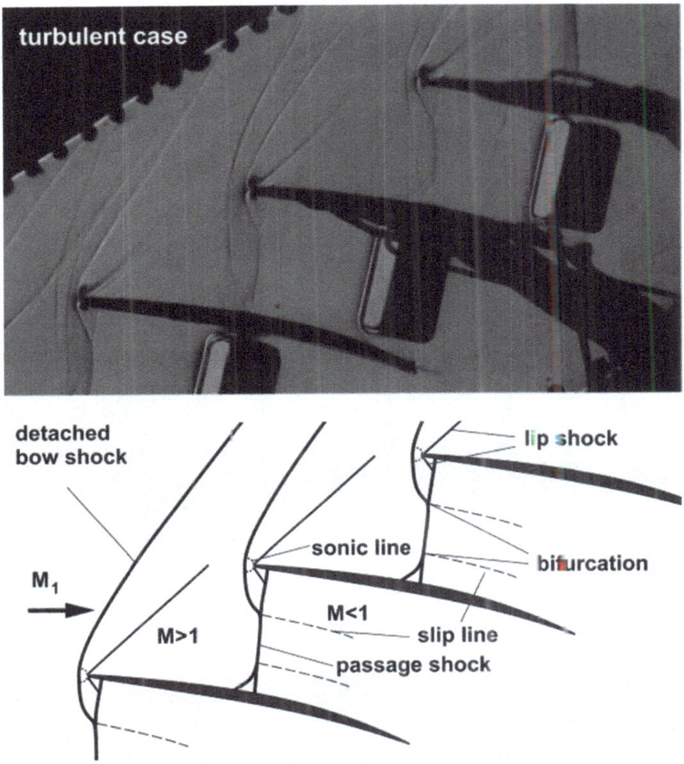

Fig. 33 Schlieren figure of the fully turbulence reference cascade (top), schematic diagram of the shock structure (bottom)

comparison of the both figures shows two main differences which are responsible for the different loss distributions in the wake. The lambda foot of the passage shock in the laminar case is an indication for the existence of a separation bubble underneath it, which is responsible for the higher viscous losses. In the fully turbulent case, the inclination of the passage shock is slightly decreased and at the moment of taking the picture also a bifurcation is visible above the suction side. These changes in the passage shock structure leads to the slight differences of the shock loss distribution.

In order to get more information about the shock movement and the behaviour of the laminar separation bubble PIV measurements were performed. The definition of the two used measurement areas is shown in Fig. 27. Area A is located within the cascade passage which allows the statistical analysis of the shock movement. Area B is located on the blade suction side were the shock boundary layer interaction take place.

During the measurements a significant movement of the passage shock was observable. Therefore, the analysis of the PIV measurements was divided in eight parts. At each part only the PIV images with the same shock position were used for a conditional averaging procedure. The results of this analysis process are shown in Fig. 34. At the front shock position (top) within this figure the separation length of the separation bubble on the blade surface amounts to 6 mm. If the shock position moves further downstream the separation length increases rapidly. At the rear shock position the separation length amounts to 18 mm. The increase of the separation length leads also to an increase of the boundary layer thickness behind the shock. The starting point of the separation bubble is approximately constant which means that this point is nearly unaffected by the shock position.

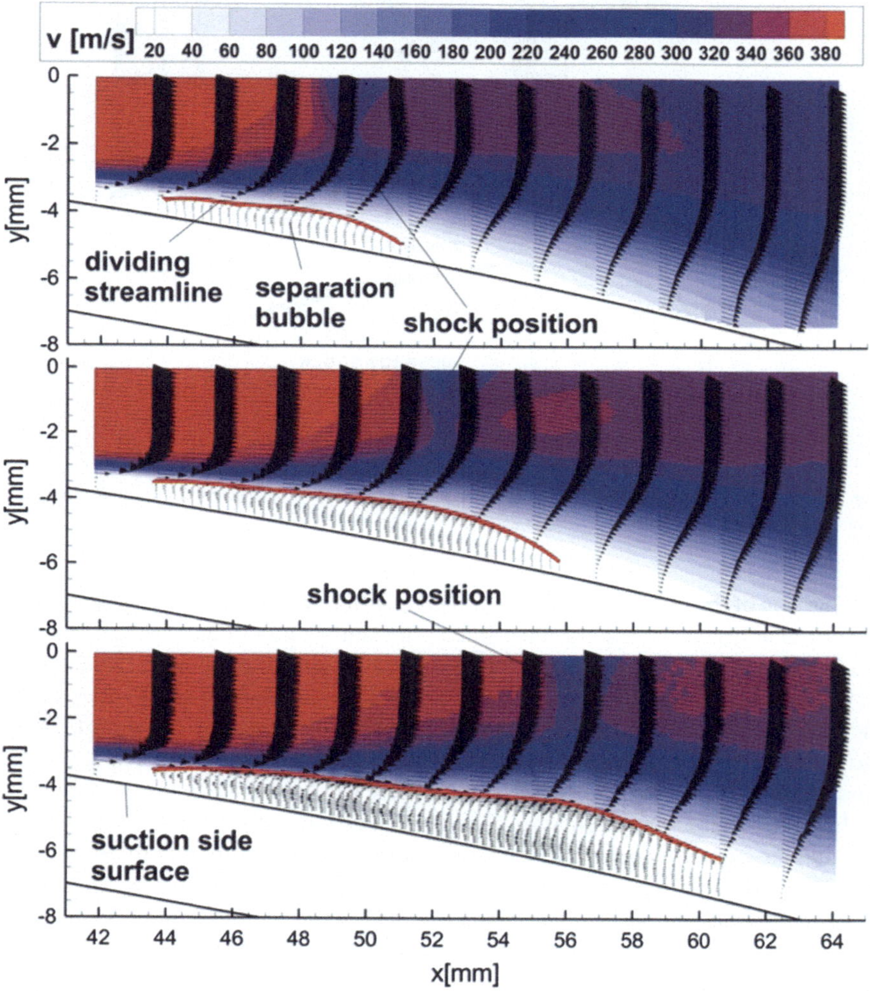

Fig. 34 Shock boundary layer interaction at different shock positions at the laminar reference cascade, conditional averaged PIV results, front shock position (top), mid shock position (middle), rear shock position (bottom)

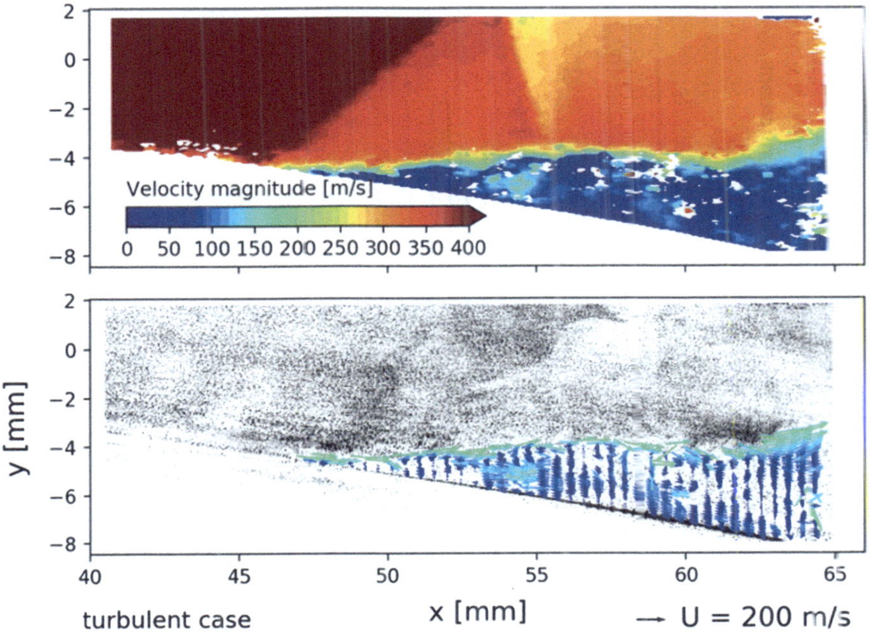

Fig. 35 Single PIV shot of the SBLI for the fully turbulent case (vectors at u > 200 m/s are clipped)

The analysis of the averaged PIV measurement results at the fully turbulent case shows also a shock movement but without separation bubble. If only the averaged data are analysed, however, a somewhat distorted picture of the real flow characteristics is obtained. The flow is extremely unsteady which also means that the flow condition on the blade suction side is changing. As an example Fig. 35 shows a large flow separation with strong oblique shock and a transitional separation bubble including reverse flow for a single PIV shot at the fully turbulent case. From this it can be deduced that a fully laminar or turbulent case does not really exist in this unsteady flow, but rather the laminar or turbulent boundary layer should only be labeled as a rather predominant flow condition.

The results of the statistical analysis of the PIV measurements are shown in Fig. 36 (laminar) and 37 (fully turbulent). The comparison of these results shows that in both cases the movement region of the shock is about 10 mm they remains uninfluenced by the incoming boundary layer condition. Only the mean shock position is changed from 52 mm at the laminar case to 55 mm at the fully turbulent case. This indicates a small change of the shock structure within the passage which was shown in the Schlieren figures. Nevertheless, the mean outcome here is that the movement range is uninfluenced in spite of the different boundary layer behavior.

In addition to the shock movement range the unsteady shock behavior was measured by means of a high speed shadowgraphy system. The resulting spectra of the laminar and turbulent case are shown in Fig. 38.

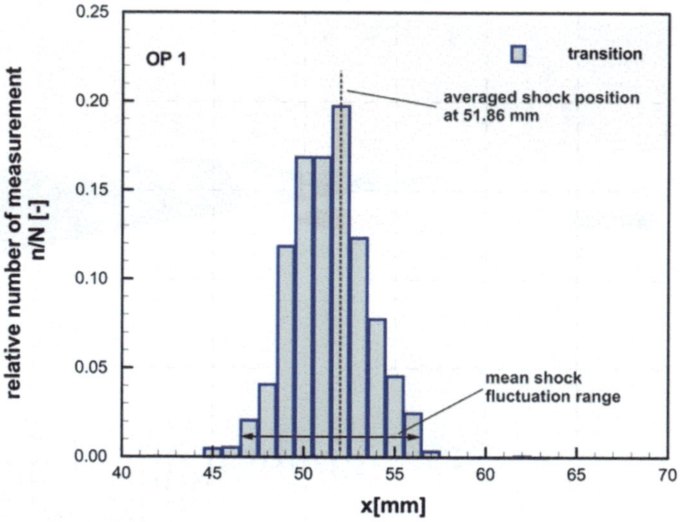

Fig. 36 Statistical analysis of the shock movement at PIV measurement area 1 (transition)

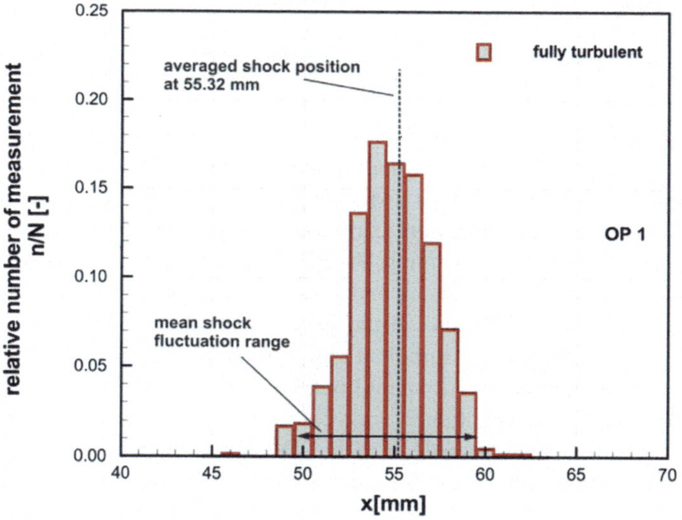

Fig. 37 Statistical analysis of the shock movement at PIV measurement area 1 (fully turbulent)

The spectra are generally characterized by broadband noise. In addition to that, the spectrum of the laminar case shows two dominant frequencies. The first frequency is at 1.15 kHz and the second at 2.3 kHz which is the first harmonic frequency. The source of the frequencies has to be found by further analysis. In addition to that, at the fully turbulent case these frequencies are disappeared. The further comparison

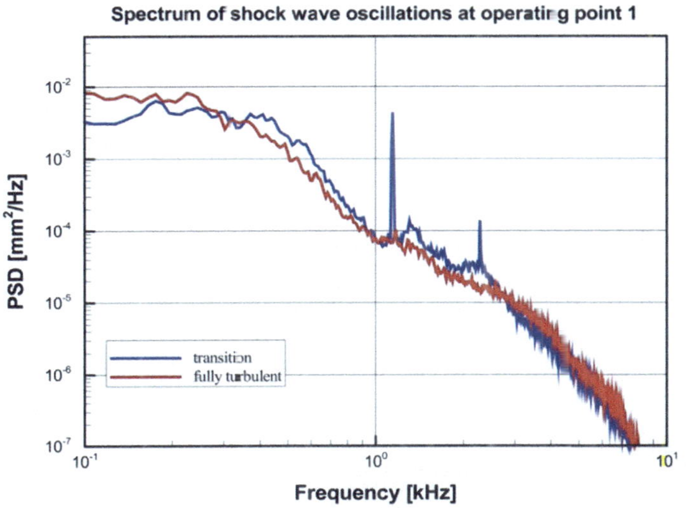

Fig. 38 Power density spectrum of the shock movement

of the spectra shows that the shock movement range of the turbulent case is larger than the laminar case at frequencies below 500 Hz and above 2.5 kHz. The higher squared amplitude at lower frequencies is equivalent with a much higher amplitude of the shock movement. From this behaviour it can be assumed that the forces acting the on the blade are also higher which lead to higher deflections of the blade in the fully turbulent case. Between 500 Hz and 2.5 kHz the shock movement range of the laminar case is larger.

The measurement of the boundary layer thickness on the blade suction side at 30% chord length which is a position in front of the shock and 75% chord length which is a position behind the shock shows that at the laminar case the boundary layer in front of the shock is smaller (<0.3 mm) than at the fully turbulent case (0.4 mm). But at the behind shock position the boundary layer thickness of the laminar case is higher (4.1 mm) than at the fully turbulent case (3.7 mm). This behaviour also indicates lower viscous losses as already described.

2.2 NUMECA

As for the single passage test section the commercial package Fine^{TM}/Turbo is used for the compressor cascade configuration. In order to validate our approach and to compare to the first experimental and numerical results a number of simulations with S-BSL-EARSM turbulence model has been performed first with and without LCTM transition model, with and without passage slots. Those simulations have been produced for sea-level and altitude configurations. For instance, measured and computed Schlierens of the fully turbulent flow for altitude configuration are provided on the Fig. 39.

However, the main objective was to produce transitional results with passage slots at altitude. Therefore, a series of simulations with S-BSL-EARSM, LCTM and with passage slots have been performed with a progressive increase of the flow rate through the slot in order to converge towards the target AVDR of 1.203. The main passage boundary conditions have also been adapted in order to reach the proper inlet flow angle and Mach number. For transitional configuration an AVDR down to 1.235 has been reached – it corresponds to an error in AVDR of 2.7%.

Pressure distribution along the blade surface is provided on Fig. 40. From this figure it appears that the comparison between the experimental and numerical is rather bad. Indeed, the numerical isentropic Mach number is always below the measured

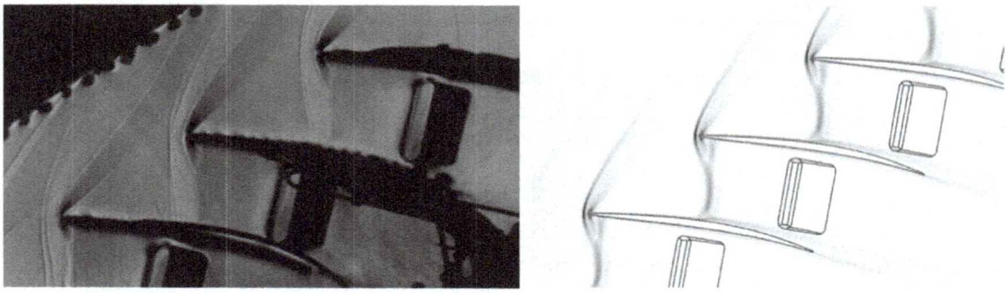

Fig. 39 Experimental run 95 (left panel) and numerical (right panel) Schlieren of the flow for fully turbulent configuration with passage slot

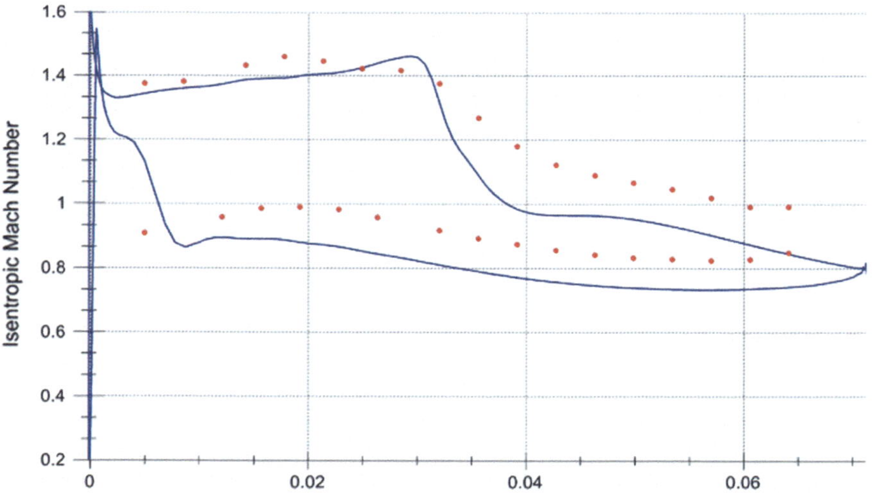

Fig. 40 Axial distribution of isentropic Mach number from experimental run 80 (red dots) and S-BSL-EARSM LCTM with passage slot (blue line)

one. Such a discrepancy along the pressure side and on the suction upstream of the shock reveals a potential difference in inlet flow angle or inlet Mach number between the experiment and the calculation and has already been identified for the fully turbulent configuration.

Since no experimental data of flow angle is available at measuring plane MP1, comparison will be performed at the L2F measuring plan, 7 mm upstream of the blade leading edge. The comparison between measured and computed inlet flow angle along this section is displayed on Fig. 41—two passages are measured. From this figure it is still obvious that the inlet flow condition is not the same for both passages. Indeed, a difference of more than 1° in inlet flow angle can be identified (measurement uncertainties in inlet flow angle are ±0.2°). Differences up to 0.1 in inlet Mach number are also produced between successive passages. Numerical results are closer to the experimental results from the left-most passage. But, this figure still exhibits significant differences in inlet flow angle and Mach number for this passage.

Experimental data are also available on the downstream plane MP2, 30 mm downstream of the blade trailing edge. Comparison of Mach number, flow angle are provided on Fig. 42. Only one passage has been measured and all the experimental data have been translated in blade-to-blade direction to match the numerical pitch-wise extend. From this figure, it appears that measuring data are not periodic. Large discrepancies in flow angle can be identified. This jump is much higher than the uncertainty in this measured data which is estimated to be ±0.3°. Therefore, a direct comparison between experimental and computed results is rather complex to conduct. When compared to the fully turbulent simulation, it appears that the transitional one produce a less pronounced downstream wake. Minimum Mach number within the wake goes from 0.5 to 0.54. The downstream flow angle is also about 1° lower in the transitional simulation. This is "partially" coherent with the experimental results.

Furthermore, comparison to measured downstream total pressure ratio is provided on Fig. 43. From this figure, it appears that the flow solver overestimates the pressure ratio within the passage. Contrary to the experimental results, the minimum pressure

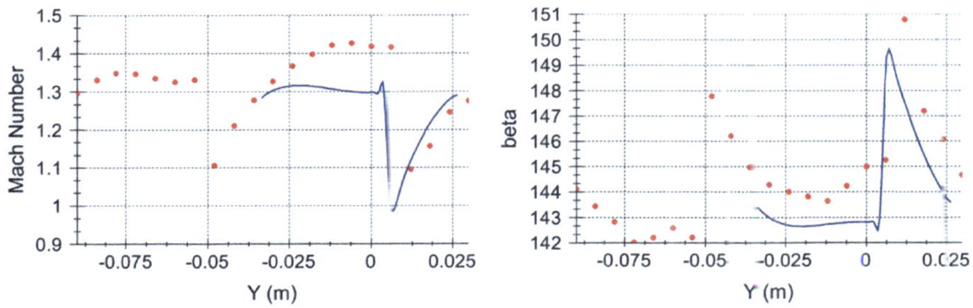

Fig. 41 Pitch-wise distribution of Mach number (left panel) and flow angle (right panel) along L2F measuring section from experimental run 80 (red dots) and S-BSL-EARSM LCTM with passage slot (blue line)

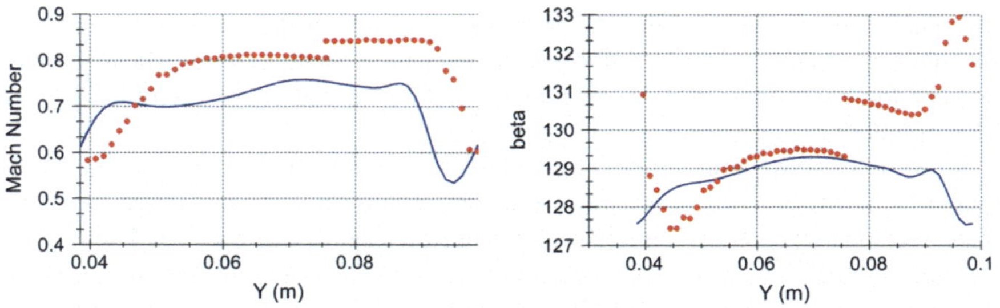

Fig. 42 Pitch-wise distribution of Mach number (left panel) and flow angle (right panel) along MP2 measuring section from experimental run 80 (red dots) and S-BSL-EARSM LCTM simulation with passage slot (blue line)

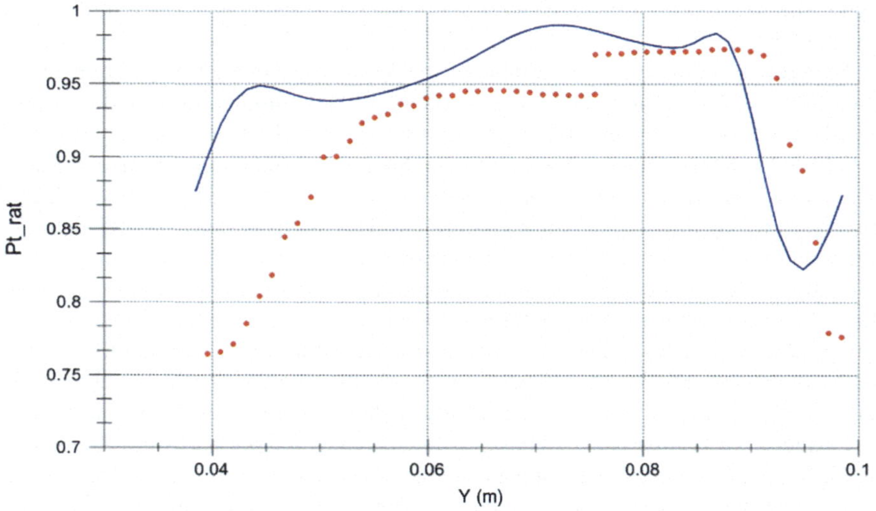

Fig. 43 Pitch-wise distribution of total pressure ratio along MP2 measuring section from experimental run 80 (red dots) and S-BSL-EARSM LCTM simulation with passage slot (blue line)

ratio has increased when going from a fully turbulent to a transitional configuration. This reveals that, at least from the CFD results, the laminar shock boundary layer interaction does not generate higher losses than the turbulent one. Indeed the boundary layer downstream of the shock is about 25% thinner and this has a direct impact on the wake and the losses. Such a result is coherent with results without any passage slot. For this configuration, transition leads to a global decrease in the losses at mid-span.

2.3 RRD

2.3.1 Introduction

RRD contribution to investigations in WP3 was related to numerical investigations of the cascade setup which was tested in the wind tunnel of DLR Cologne. The general profile and aerodynamic parameters of the cascade are shown in Table 1. For all numerical investigations the RRD in-house U/RANS flow solver HYDRA was utilised with k-ω Menter-SST turbulence model. Regarding transitional simulations the Langty & Menter correlation-based transition model was applied on top of the k-ω Menter-SST turbulence model.

The numerical setup is aimed to reproduce the measured data of the experiment which is performed from DLR in Cologne. In the experimental setup suction slots were specified to suck off the end wall boundary layer flow and control the end wall blockage by adjustment of the axial velocity density ratio (AVDR) to target AVDR = 1.2 (see Fig. 44).

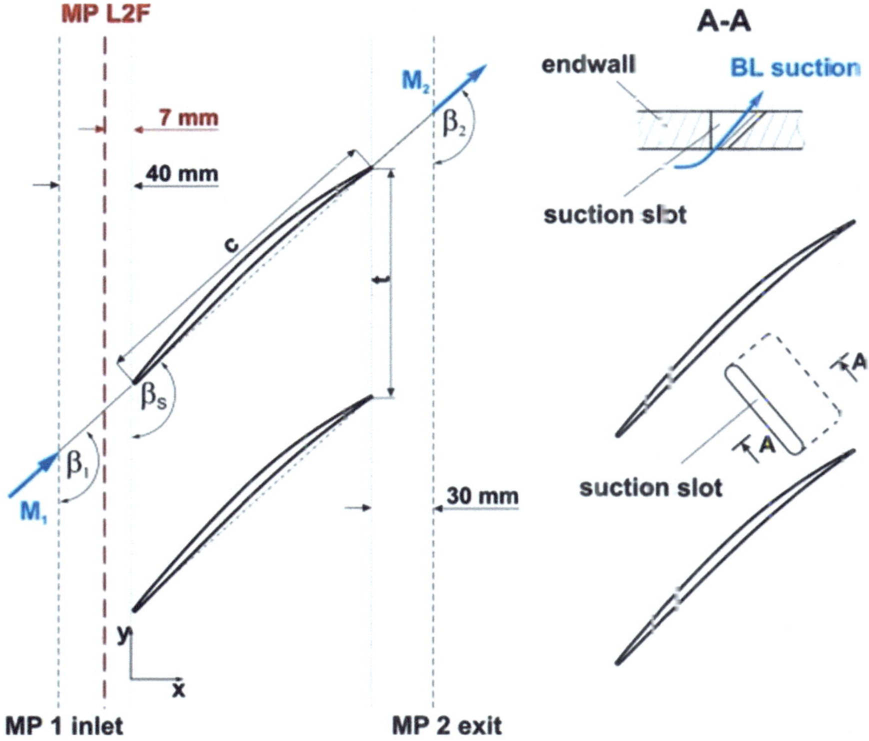

Fig. 44 DLR cascade test section

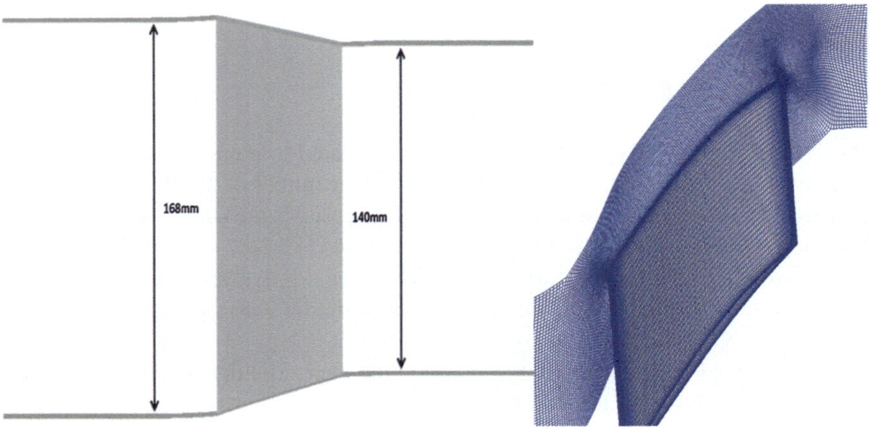

Fig. 45 Meridional view of simplified inviscid end wall model and mesh used for RRD TFAST investigations

Simulations on a one to one model of the DLR test cascade including suction slots lead for all investigated suction mass flows (up to choked slots) to higher AVDR than the target AVDR = 1.2. Hence end wall blockage could not be reduced as intended. Since several studies do indicate that k-ω Menter-SST turbulence model is overpredicting end wall blockage and furthermore the suction slot design and position was within DLR experience, it was decided to continue experiments with the chosen slot design and use a simplified geometrical model for numerical investigations of RRD.

Obtaining target AVDR of 1.2 a model with contracted inviscid end walls as it is shown in Fig. 45 was generated using AUTOGRID. The overall mesh has 2.3 million cells with y^+ values below unity.

2.3.2 Results

For reference investigations simulations were carried out at design inlet conditions (see Table 1). On top of those simulations, off design conditions were investigated, varying inlet Mach numbers and inlet angle.

At certain combination of inlet angle and inlet Mach number the state of the simulations changes from a steady state to a state with periodic unsteadiness. The unsteadiness is related to the coupling of the separation due to SWBLI and the separation occurring at the trailing edge of the aerofoil. In wing aerodynamics this phenomenon is known as buffeting.

Analysis at Design Point Conditions

Transitional and fully turbulent simulations at chosen design point of inlet angle $\alpha = 55.56°$ and inlet Mach number of $Ma = 1.22$ converge to a steady state solution.

Graphs characterizing the flow of the cascade are plotted from Figs. 46 to 49 where the solid black line is indicating the results achieved performing a fully turbulent simulation and the dashed line is indicating the results obtained including transition model.

Compared to the MISES Mach number distributions (fully turbulent used as reference by all WP-3 partners), HYDRA obtained results show adequate alignment for both transitional and fully turbulent simulations (see Fig. 46). Small offsets in shock position can be explained not meeting the inlet conditions perfectly when simulating both, MISES and HYDRA. Especially with MISES it is not straight forward obtaining converged solutions when having supersonic inlet conditions. The HYDRA obtained transitional profile shows the typical reduction of peak Mach number due to the laminar separation bubble occurring further upstream compared to the turbulent interaction. The proof that the interaction of the transitional simulation

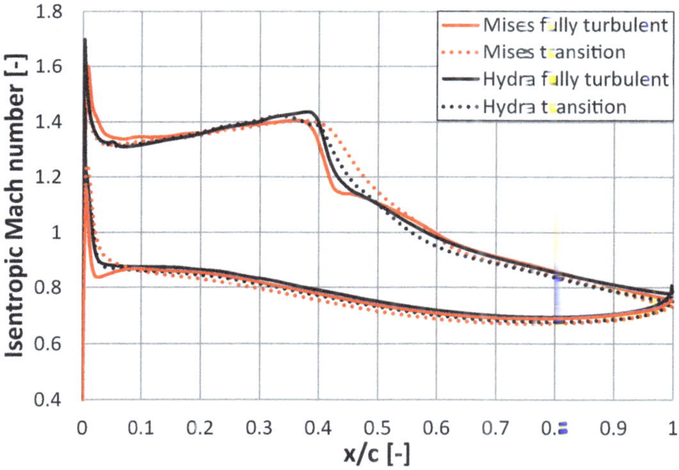

Fig. 46 Mach number profile comparison, Hydra versus Mises

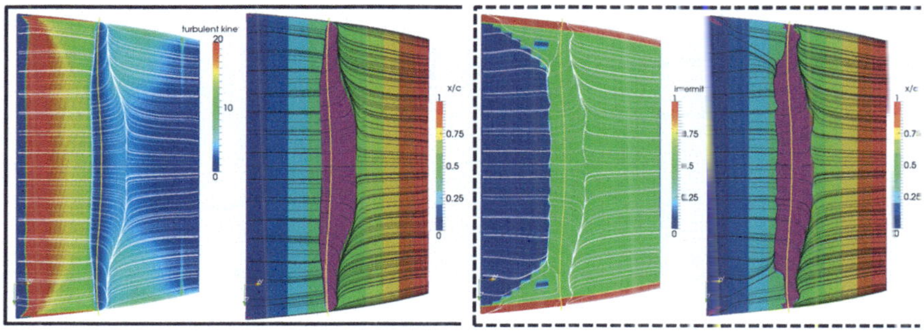

Fig. 47 Suction side streamlines for turbulent and transitional simulations at DP

with the shock is really a laminar interaction and transition is not already occurring upstream the shock can be seen when looking on intermittency on the suction side of the airfoil in Fig. 47. Important to note is that both turbulent and laminar SWBLI show a separation area due to SWBLI, which is highlighted via the pink contour which is clipped with negative axial velocity.

The laminar interaction of the transitional simulation is leading to an interaction of the boundary layer with the shock which is appearing further upstream compared to the interaction of the fully turbulent simulation. This is highlighted by the suction side skin friction coefficient in Fig. 48 and the suction side shape factor distribution in Fig. 49. Such behavior is expected from theory and also shown by oil flow visualization generated in IMP PAN experiments.

At mid height the generated profile loss is in the order of 6% lower for the transitional simulation relative to the fully turbulent simulation. Wake profiles shown in Fig. 48 show that the reduction of loss is appearing close to the suction side of the profile. The reduction of loss goes in line with higher skin friction coefficient downstream the shock boundary layer interaction zone. Opposite trend is depicted by the suction side shape factor which is plotted in Fig. 49. Although higher shape factor values for the laminar SWBLI indicate stronger separation, towards the trailing edge of the aerofoil the boundary shape factor is lower compared to the fully turbulent simulation.

Although the fully turbulent case is an artificial case, because in reality a certain portion of the chord will be always laminar before natural transition takes place, the numerical results go against the experimental results obtained by DLR Cologne, where thanks to higher inlet turbulence and hence turbulent SWBLI, losses could be reduced compared to the laminar SWBLI. On the other hand, the numerical simulations go in line with observations and explanations made by TU Delft on flat plate experiments. "Thanks to the further upstream located laminar separation

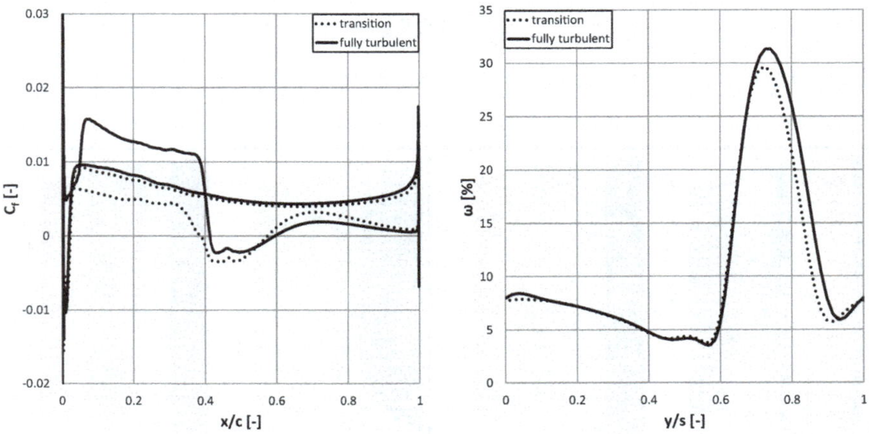

Fig. 48 Skin friction coefficients and total pressure wake loss for turbulent and transitional cases at DP

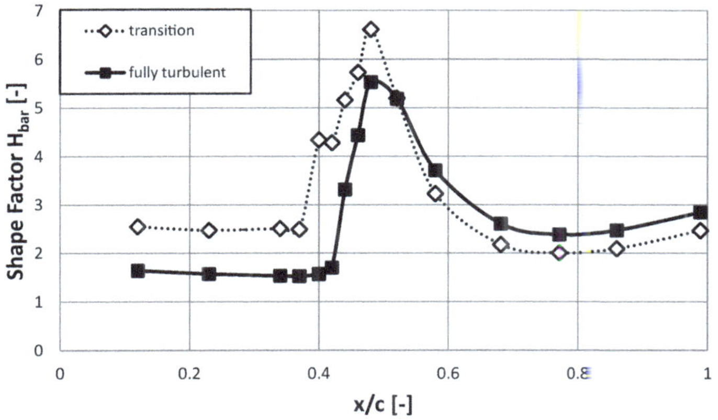

Fig. 49 Suction side boundary layer shape factor for fully turbulent and transitional simulations at DP

bubble isentropic compression waves exist which reduce the strength of the main shock, which is leading to milder interaction between shock and boundary layer, which finally leads to a reduction in loss compared to a turbulent SWBLI."

Analysis at Off-Design Conditions

Starting from design inlet conditions,

- inlet angle variation, retaining design inlet Mach number
- and inlet Mach number variation, retaining design inlet angle

were performed for both fully turbulent and transitional simulations. The loss polars for both variations are plotted in Fig. 50.

Even running all simulations using URANS, all plotted points converge to steady state solutions. The loss coefficient which is shown in those plots refers to the loss at mid span and is normalized to the dynamic pressure at the inlet. Being able to compare those results to the experimental results from DLR, the data is extracted at the same upstream (MP1) and downstream (MP2) position as it is done in the experiment (see Fig. 44).

Comparing polars it can be well noted that the transitional simulations predict lower losses compared to the fully turbulent simulations over the whole range of investigated inlet angles and inlet Mach numbers. The relative loss improvement of the transitional simulations remains almost constant when comparing same inlet angles and inlet Mach numbers. So, the conclusion given for the design point simulations, also holds for off-design conditions with different shock position and strength. Numerical simulations are indicating that laminar SWBLI does not harm the performance on the investigated profile.

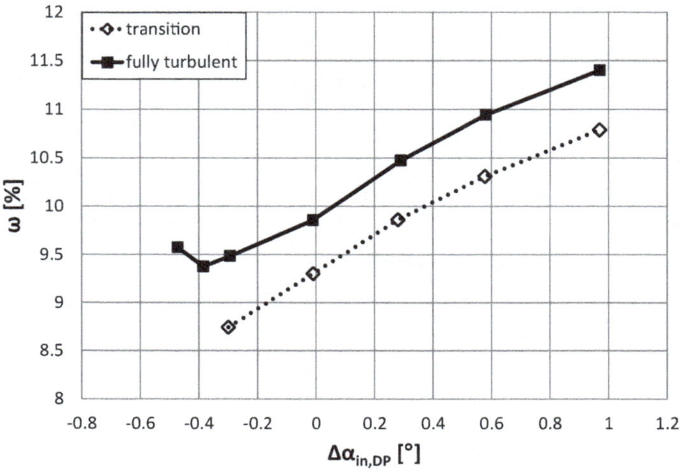

a) inlet angle variation at Ma=1.22

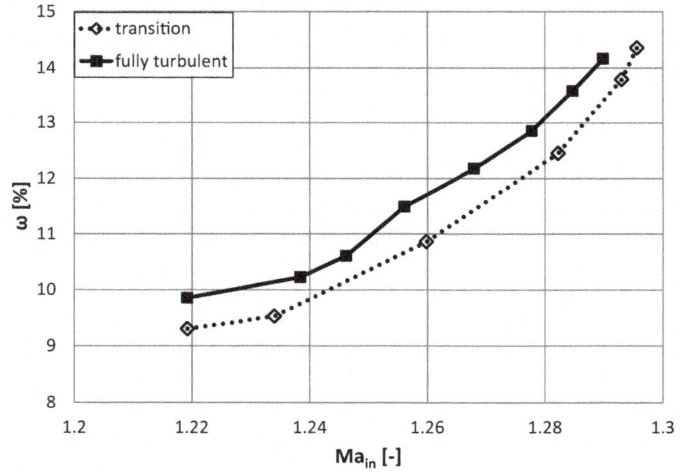

b) inlet Mach number variation at $\Delta\alpha in,DP = 0°$

Fig. 50 Loss polars varying inlet conditions

However, a conclusion if promoting transition upstream the shock can reduce losses cannot be drawn here, because the fully turbulent simulation is not representative for a real configuration with tripping device. In chapter "*Compressor cascade flow with transition control*" experiments and numerical simulations including tripping devices are described giving further insight on this subject.

At one degree higher inlet angle and slightly higher inlet Mach number compared to design intention, performed simulations predict a periodic unsteady behaviour.

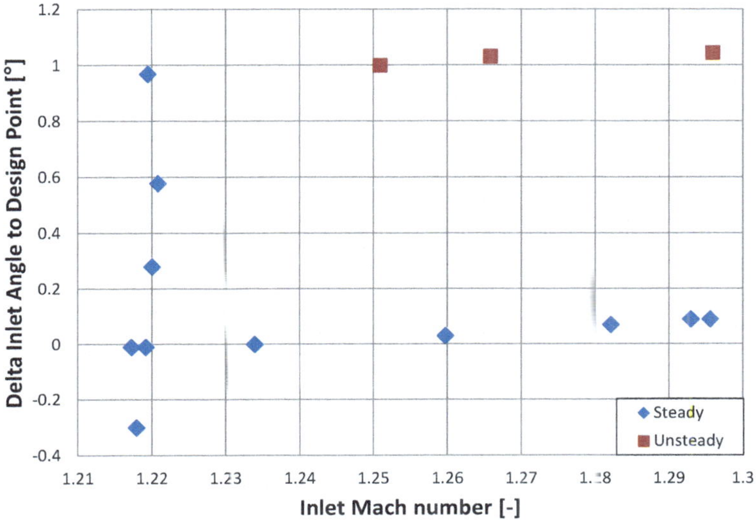

Fig. 51 Overview about performed unsteady transitional simulations

Figure 51 lists the matrix of investigated Mach number and inlet angle combinations. Time averaged profile Mach number distributions and suction side skin friction coefficient distributions are compared for cases with one degree higher inlet angle against the design point distributions in Fig. 52.

Amplitudes of the oscillations are indicated in Fig. 53 by the pressure force acting on the aerofoil. At design inlet Mach number of Ma = 1.22 the transitional simulation is still converging towards a steady state solution. Increasing the inlet Mach

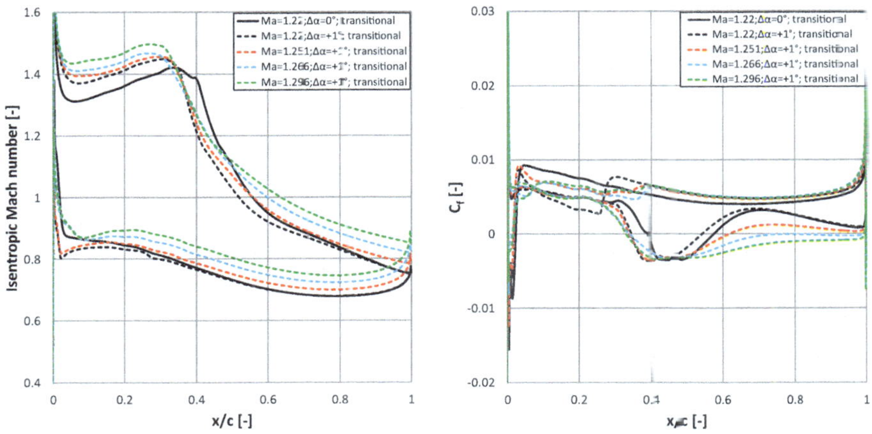

Fig. 52 Profile Mach number and skin friction coefficient for DP and $\Delta\alpha_{in,DP} = +1°$

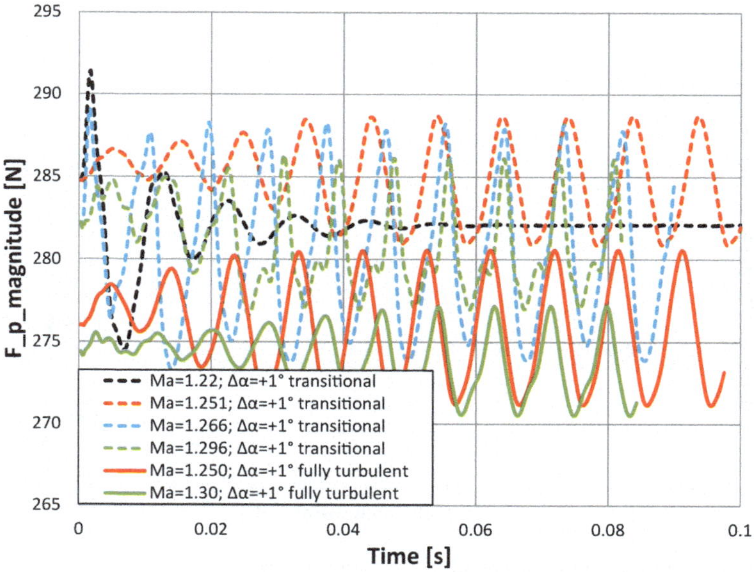

Fig. 53 Surface force monitor for chosen off design points of unsteady simulations

number, the other performed simulations clearly show a periodic unsteady behaviour, where highest amplitudes appear for the case with Ma = 1.266. Occuring frequencies increase when increasing the inlet Mach number from 100 Hz (Sr = 0.027) for inlet Mach number of Ma = 1.251 to 122 Hz (Sr = 0.032) for the case with inlet Mach number of Ma = 1.296. Fully turbulent simulations were also performed at those inlet conditions proving that the occuring unsteadiness is not related to the state of the boundary layer interacting with the shock. Depicted trends and magnitudes are similar compared to the traditional simulations.

The unsteady interaction is occuring when additional to separation due to SWBLI, separation at the trailing edge exists. This behaviour is similar compared to buffeting which is observed on transonic wings. For Ma = 1.251 small trailing edge separation is exists for 2 out of 12 time dependent solutions. Both separation zones are not yet merged for any timestep. The time averaged skin friction coefficient (see Fig. 52) is indicating that downstream the SWBLI the flow is attached to the suction side.

For an investigated inlet Mach number of Ma = 1.266 the time averaged skin friction coefficient is indicating that the boundary layer is on the edge being fully separated until the trailing edge. The separation due to SWBLI remains separated until the trailing edge for half of the time dependent solutions, and is reattaching for the others. Time dependent snap shots of separation contours are shown in Fig. 54 for 6 out of 12 time steps. The shock movement which is going in hand with the development of separation is shown in Fig. 55 via density gradient contours at midspan.

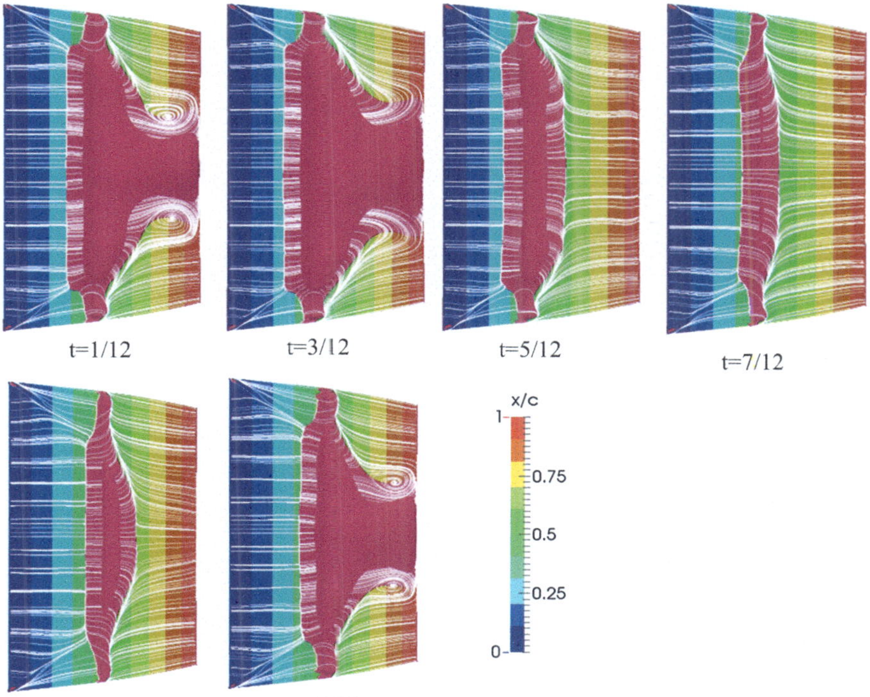

Fig. 54 Suction side streamlines and negative axial velocity contour for $\Delta\alpha_{in,DP} = +1°$ and $Ma_{in} = 1.266$

Increasing the Mach number further to Ma = 1.296 only 3 out of 12 time dependent solutions show a reattachment of the shock induced separated boundary layer. Hence the amplitude of the shock movement is reduced (see Fig. 53). Increasing the inlet Mach number even further unsteadiness and shock movement disappear due to the fact that the flow remains fully separated from shock beyond trailing edge for all time steps.

2.3.3 Conclusions

Fully turbulent and transitional steady state CFD simulations are performed on a cascade section with supersonic inlet conditions. A wide range of different inlet Mach numbers and inlet angles was analysed. At design inlet conditions both, transitional and fully turbulent simulations show similar flow structures but the transitional SWBLI is leading to lower total pressure losses compared to the fully turbulent simulation. For the transitional simulation, transition is promoted by the shock itself. Increasing inlet angle and inlet Mach number transitional results show over the whole

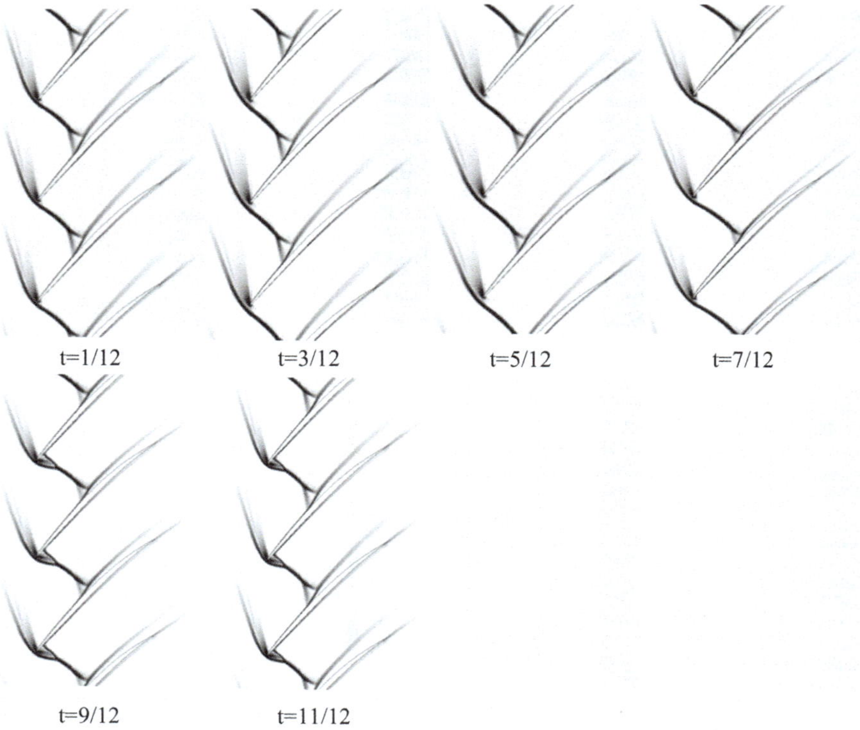

Fig. 55 Density gradient—numerical *Schlieren* for $\Delta\alpha_{in,DP} = +1^{\circ}$ and $Ma_{in} = 1.266$

range of the investigated variations lower losses compared to the fully turbulent simulations when comparing similar inlet conditions. At a certain aerodynamic loading while increasing inlet angle and inlet Mach number periodic oscillations appear for both fully turbulent and transitional simulations. Those oscillations go in line with the appearance of trailing edge separation. At certain loading the SWBLI induced separation and the trailing edge separation are merging for some time steps. Increasing the loading further oscillations disappear due to the fact that the boundary layer remains fully separated from the shock position towards trailing edge for all time steps. The observed phenomenon is very similar to buffeting which is observed on transonic wings.

3 Compressor Cascade Flow with Transition Control

3.1 DLR

3.1.1 Experimental Investigation with AJVG and Roughness Patch

The general experimental setup for the investigation with AJVG and roughness patch is unchanged and can be taken from the chapter about the investigations at the reference cascade. Also the measurement technique used and the definition of the measurement planes are unchanged.

3.1.2 Configuration of the AJVG Cascade

The investigated AJVG configuration is based on preliminary tests at Szewalski Institute of Fluid-Flow Machinery Polish Academy of Sciences (IMP PAN) which are described above. The results of the investigations show that the AJVG to be used has a diameter D of 0.4 mm, a skew angle of 75°. and a pitch angle of 30° and the distance ratio L/D should be 15. Figure 56 shows the configurations of the used AJVG on the blades suction side of each cascade blade. The jet for the AJVG is not produced by external compressed air supply. Rather, in the configuration investigated, a hole from the pressure side to the suction side of the blade is used to produce the jet. The diameter of the inlet hole on the blade pressure side is also 0.4 mm, with a pitch

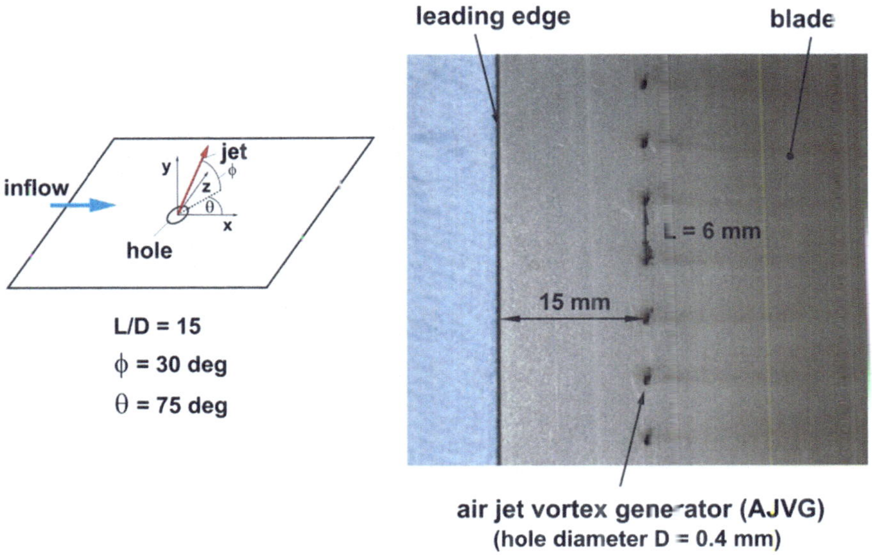

Fig. 56 Blade configuration with AJVG

angle of 30°. The distance of the AJVG from the blade leading edge also results from the studies carried out at the IMP PAN. It is designed that the distance between AJVG and the area of the shock boundary layer interaction should be minimized but it is ensured that the transition in any case occurs in front of the area of the shock boundary layer interaction.

3.1.3 Configuration of the Roughness Patch Cascade

For the determination of an optimal surface roughness patch (SRP) configuration for the measurements in the transonic cascade, also preliminary investigations were carried out at the IMP PAN. The aim of the single test section investigations was to determine the required roughness parameters Ra and Rz as well as the position and necessary extension of the patch to force the transition in front of the shock boundary layer interaction region. However, the turbulent length of the boundary layer in front of the shock should be as short as possible to avoid unnecessary additional losses.

Figure 57 shows the configurations of the used roughness patch on the blades. It starts at 11.1% of the chord length and has a width of 10 mm. The roughness values used are Ra = 10 mm and Rz = 35 mm.

Fig. 57 Blade configuration with roughness patch

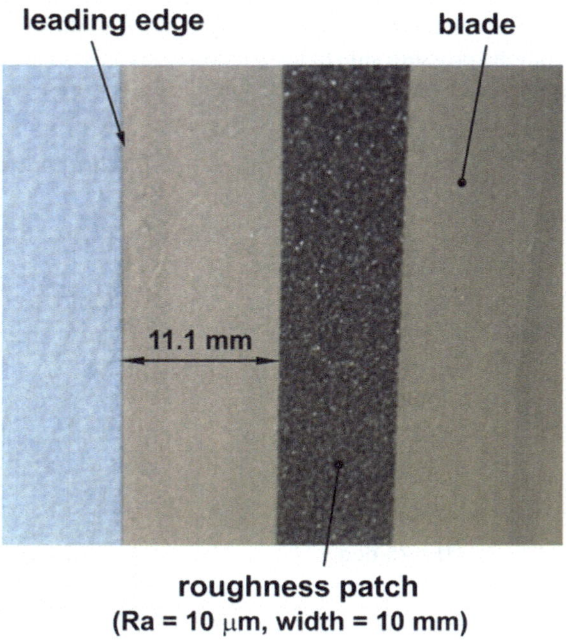

3.1.4 Results and Discussion

The detailed measurements of the cascades with boundary layer control were performed at operating point 1. Figures 58 and 59 show the measured inflow angle distribution of the AJVG and SRP configuration. The averaged inflow angle of the both cascades which was calculated by using conservation laws of fluid motions amounts to 145.5° for AJVG case and 145.6° for the SRP case. The comparison with the two reference cases (laminar and fully turbulent) of the cascade shows that similar inflow conditions were achieved at the measurements. Table 7 shows the measured flow conditions of the investigated AJVG and SRP configuration.

In Figs. 60 and 61 the profile Mach number distribution of the AJVG and SRP case is shown. In the figures it becomes apparent, that the suction side shape of the Mach number distributions is similar as for the fully turbulent cascade. In front of the deceleration region, which is caused by the pressure rise over the shock, a Mach number peak is depicted for these cases. For the cascades with AJVG and SRP the transition occurs already before the shock, so that in these two cases a turbulent shock boundary layer interaction takes place. This confirms the effectiveness of the

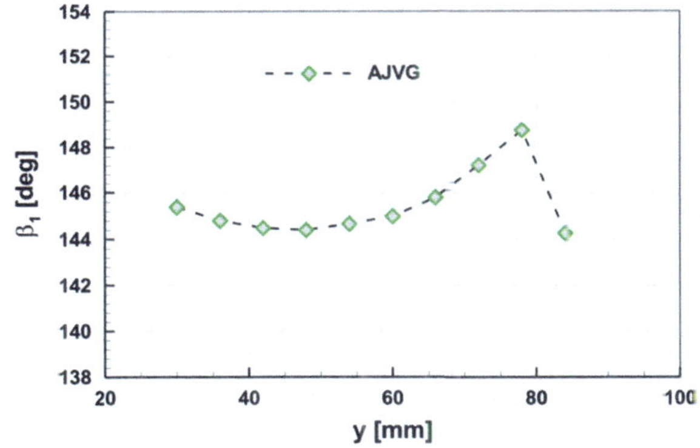

Fig. 58 Inflow angle distribution of AJVG cascade at MP L2F (cascade midspan)

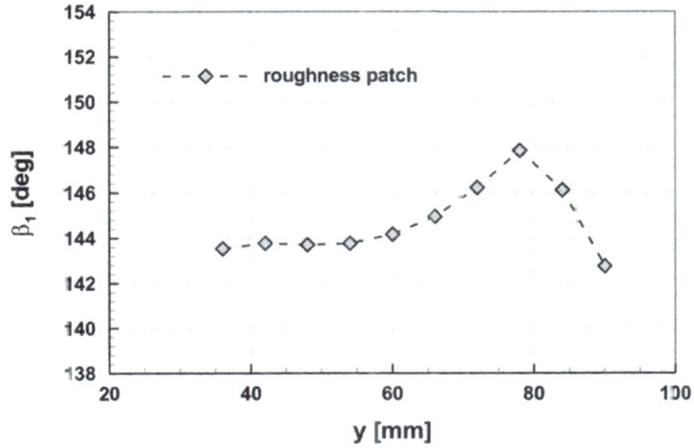

Fig. 59 Inflow angle distribution of the SRP cascade at MP L2F (cascade midspan)

Table 7 Measured averaged flow conditions of the AJVG and SRP configuration

	AJVG	SRP
M_1	1.20	1.21
Re_1 (10^6)	1.41	1.40
β_1 (°)	145.5	145.6
AVDR	1.21	1.22
β_2 (°)	129.9	129.6

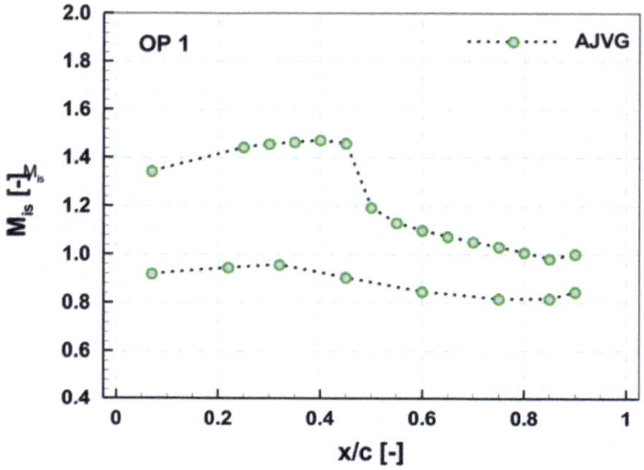

Fig. 60 Profile Mach number distribution at midspan (AJVG)

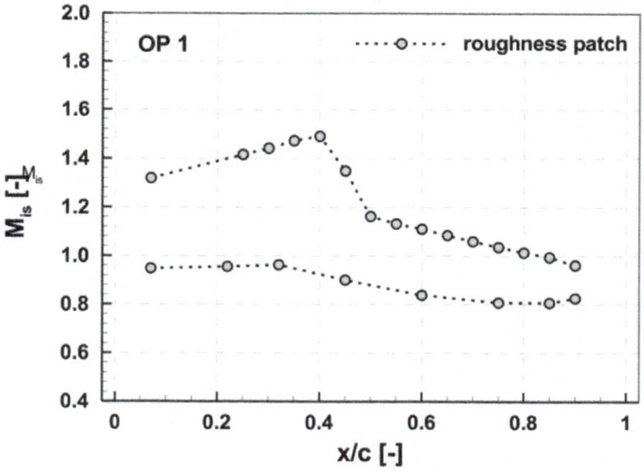

Fig. 61 Profile Mach number distribution at midspan (roughness patch)

investigated methods with regards to the enforcement of the transition before the shock.

Figures 62 and 63 show the comparison of the pitch dependent total pressure loss distribution of the laminar reference cascade and the cascade using AJVG as well as SRP. In Fig. 62 it can be observed that on the one hand the shock losses close to

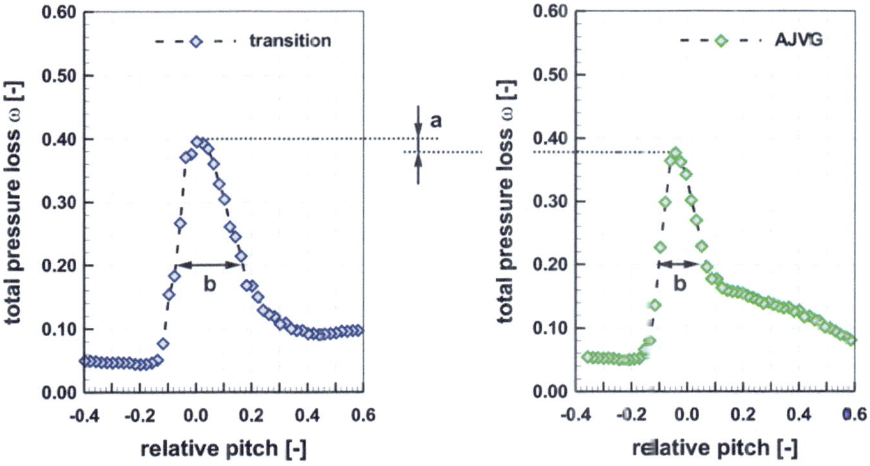

Fig. 62 Pitchwise distribution of the total pressure loss in the wake behind the AJVG cascade

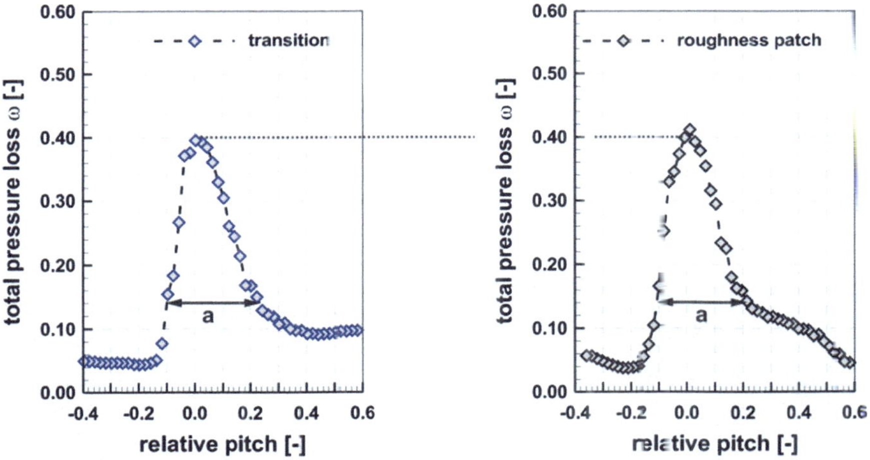

Fig. 63 Pitchwise distribution of the total pressure loss in the wake behind the SRP cascade

the blade suction side (y/t = 0.2 to 0.5) are higher for the case with AJVG than in the reference case. On the other hand, the viscous losses represented by the wake behind the blade trailing edge are significantly reduced for the AJVG case. The wake depth and width is reduced which is marked by b in Fig. 62. This means that the use of the AJVG results in lower viscous losses than in the laminar case, where a laminar separation bubble exists (mass flow averaged total pressure loss coefficient in the laminar case is 0.148 and in the turbulent case 0.133, corresponds to a loss difference of 10%). With regard to the additional shock losses, it can be stated that this is probably due to a change in the shock structure within the cascade and a higher Mach number in front of the shock caused by the changed boundary layer condition compared to the laminar case. Nevertheless, the additional shock losses lead to the fact that the total loss reduction is lower than in the fully turbulent reference case. However, from these results it can be stated that the AJVG are very efficient to reduce the viscous losses on the blade suction side by forcing the transition in front of the shock. This means, that only a small amount of additional losses are added to the viscous losses within the boundary layer by this control method. The mass flow averaged total pressure loss coefficient of the AJVG case is 0.141, which corresponds to a loss difference of 5% compared to the laminar case. This also confirms the advantage of the AJVG in a transonic application as a flow control method instead of conventional vortex generators.

With regard to the resulting loss behavior for the SRP cascade compared to the reference case, it can be observed in Fig. 63 that the differences in wake losses behind the trailing edge of the blades are not so significant. However, the shock losses at the blade suction side region are significantly reduced. From these results it can be deduced that the use of SRP as a flow control method on the blades shows an ambivalent influence on the loss behaviour. This results from the additional losses caused by the SRP. In this case, the measured boundary layer thickness at 30% of the chord length is 50% higher than in measurements with the turbulent reference cascade and the AJVG cascade. When using the SRP, the loss reduction is essentially due to the reduction of the shock losses, which also suggests that the shock structure is significantly changed. The mass flow averaged total pressure loss coefficient of the SRP case is also 0.141.

To verify the assumption of modified shock structures at the AJVG and SRP cascade Schlieren images are used. Figures 64 and 65 show the Schlieren images for

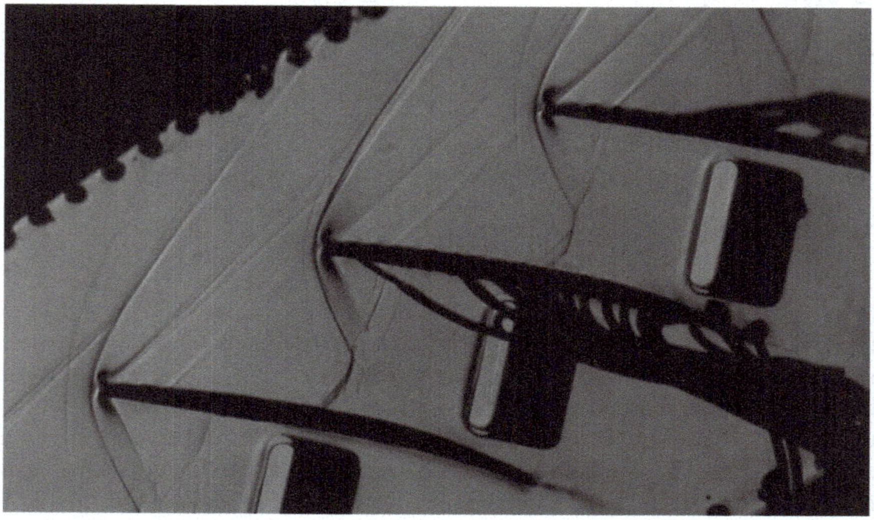

Fig. 64 Schlieren figure of the cascade with AJVG

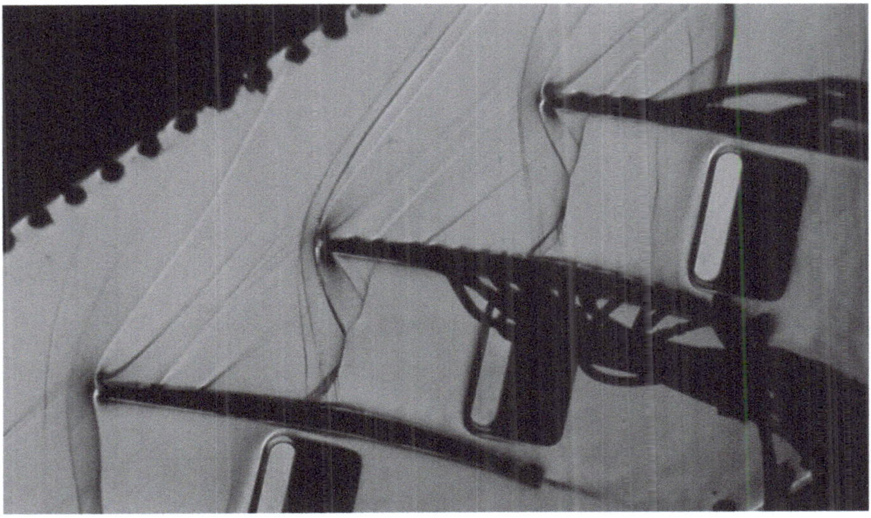

Fig. 65 Schlieren figure of the cascade with roughness patch

the AJVG and SRP cascade. At the investigated operating point of the cascade, the bow shock and the passage shock meet each other at a bifurcation approximately in the middle of the passage. This is a main difference of the AJVG and SRP cascade compared to the laminar reference cascade. Because for the laminar reference the bifurcation point is closer to the blade pressure side. This leads to the fact that in this range of displacement (between 50 and 75% pitch) the deceleration of the flow into the subsonic region is not only effected by a nearly normal shock, as in the reference case. Instead there is an upstream deceleration by an oblique shock. Rather, deceleration occurs first by an oblique shock above the bifurcation and then by the nearly normal passage shock. Due to the reduced inflow velocity to the passage shock there are less shock losses, which is reflected in the measured wake distribution in Figs. 62 and 63.

In order to get more information about the shock movement and the behavior of the shock boundary layer interaction on the blade suction side PIV measurements were performed [22]. During the measurements a significant movement of the passage shock at the AJVG and SRP case was observable. The statistical analysis of shock movement at the AJVG case is shown in Fig. 66. It is observable that the movement region of the shock is about 10 mm. This is the same results as for the both reference cases (laminar and fully turbulent), which means that the shock movement range remains uninfluenced by the used boundary layer control method. Only the mean shock position is different from the laminar (52 mm) and fully turbulent (55 mm) cases and amounts to 53.5 mm. This indicates a small change of the shock structure within the passage which was visible in the Schlieren figures. Nevertheless, the mean outcome here is that the movement range is again uninfluenced in spite of the different boundary layer behavior.

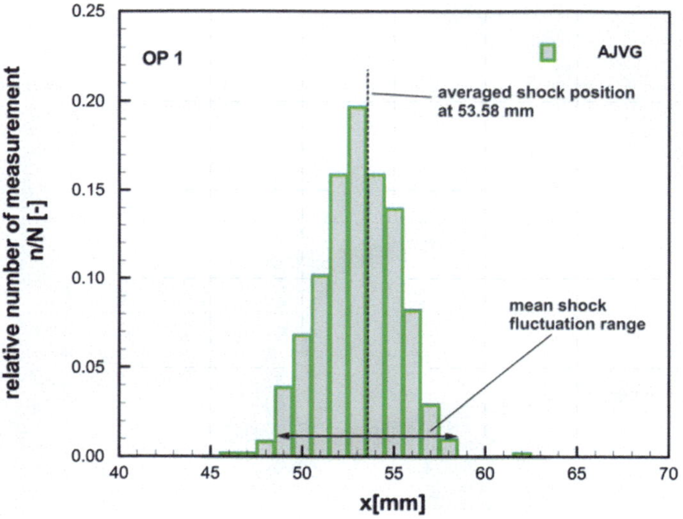

Fig. 66 Statistical analysis of the shock movement of the AJVG cascade at PIV measurement area A

The statistical analysis for the SRP cascade is shown in Fig. 67. In this figure it is observable that the movement region of the shock is only about 8 mm. Compared to the other investigated cases a reduction of the movement region is achieved.

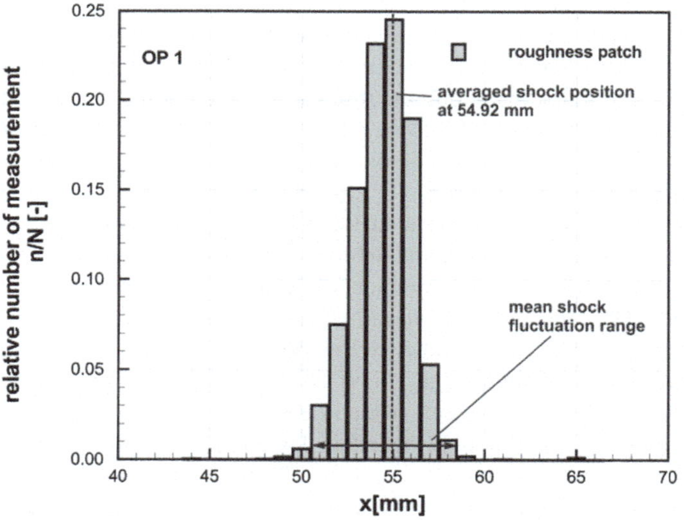

Fig. 67 Statistical analysis of the shock movement of the SRP cascade at PIV measurement area A

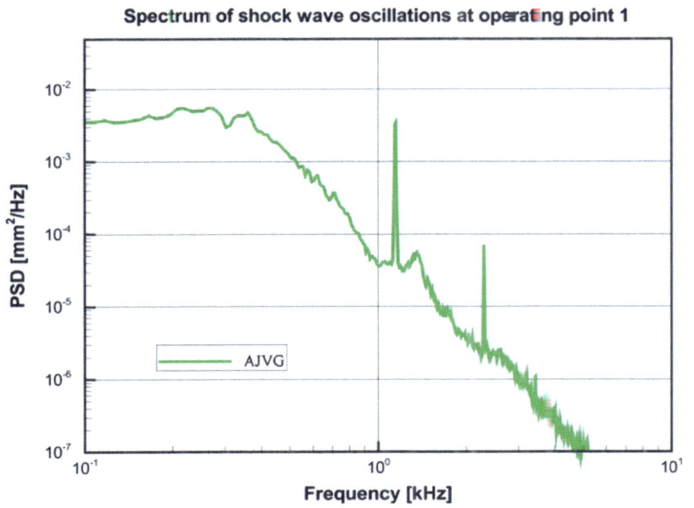

Fig. 68 Power density spectra of the shock wave fluctuation

This means that the shock movement range is significantly influenced by SRP as a boundary layer control method. Also the mean shock position is different from the laminar (52 mm) and fully turbulent (55 mm) cases and amounts to 54.9 mm (Fig. 67).

In order to assess the influence of the pre shock boundary layer condition on the unsteady shock behavior high-speed shadowgraphy measurements were performed. Figure 68 shows the power density spectrum of the AJVG case and Fig. 69 shows the same spectrum for the roughness patch case. The y-axis represents the shock movement per Hertz, which means that the high values are represent wide shock movement and the lower values represent small shock movement. The spectra are generally characterized by broadband noise. In addition to that, the spectrum of the AJVG case show two dominant frequencies. The first frequency is at 1.14 kHz and the second at 2.28 kHz which corresponds to the first harmonic frequency. The source of these frequencies can be a periodic separation on the throttle behind the test section or a periodic excitation of the shock oscillation by the pressure fluctuations in the wake of the blade. An exact determination is not possible, because no unsteady wake measurements were carried out. A further comparison of the spectra shows that the shock movement squared amplitude of the flow control cases is lower than in the laminar reference at frequencies above 800 Hz. At frequencies below 200 Hz only the SRP case shows significantly lower amplitudes. The lower squared amplitude at lower frequencies is equivalent to a much lower amplitude of the shock movement. From this behaviour it can be assumed that the forces acting on the blade are also lower which should lead to lower displacement of the blade. However, it can be concluded that the shock oscillation is reduced by the use of AJVG and SRP and that the SRP reduces the oscillation especially at low frequencies below 200 Hz.

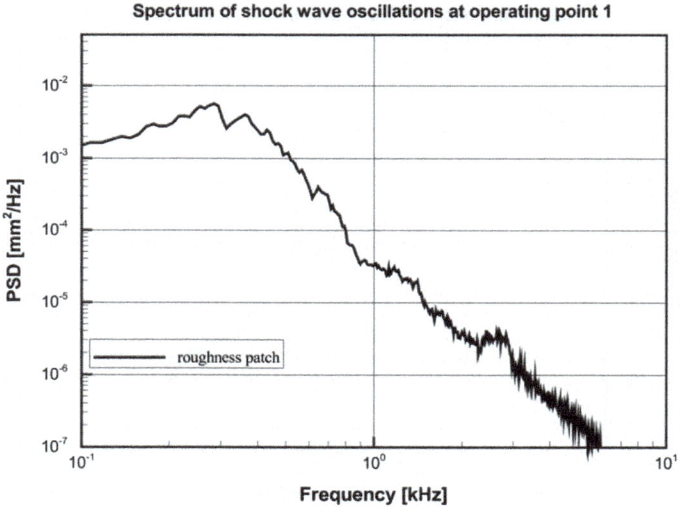

Fig. 69 Power density spectra of the shock wave fluctuation

3.2 *NUMECA*

A first simulation has been performed with a rough element corresponding to the experimental set-up of DLR. This set-up is based on the sensitivity study already performed by IMP. A 10 mm wide rough element with a centreline average roughness of 10.2 μm has been selected. Its axial location is set to 11.1%. The measured roughness distribution for IMP and DLR experiments are displayed on Fig. 70. It appears that both are rather consistent. Furthermore, those distributions are also consistent with the one of the experiment of Acharya et al. [23]. According to this paper, the ratio of equivalent sand-grain roughness to the centreline average roughness should be equal to 6. Such a value has been retained for the present investigation.

For this reference simulation a series of preliminary simulations has been performed in order to reach the target flow condition—mainly inlet Mach number and AVDR.

According to the results from former simulations of the flow over a flat plate, one of the major effects of the rough patch should be to promote laminar to turbulent transition upstream of the interaction. In order to identify such an effect and its impact of the flow solution the current simulation should been compared to fully turbulent and transitional simulation already obtained in the scope of this project. Therefore, we will use, as references, two simulations from the basic cascade configuration. The former one is the fully turbulent run performed at the target flow condition, and the second one is the transitional simulation with a slightly too high AVDR when compared to target value. Both have been computed using S-BSL-EARSM turbulence model. The limit streamlines on the blade suction side are displayed on Fig. 71. Furthermore, for the transitional runs, the green line represents the location

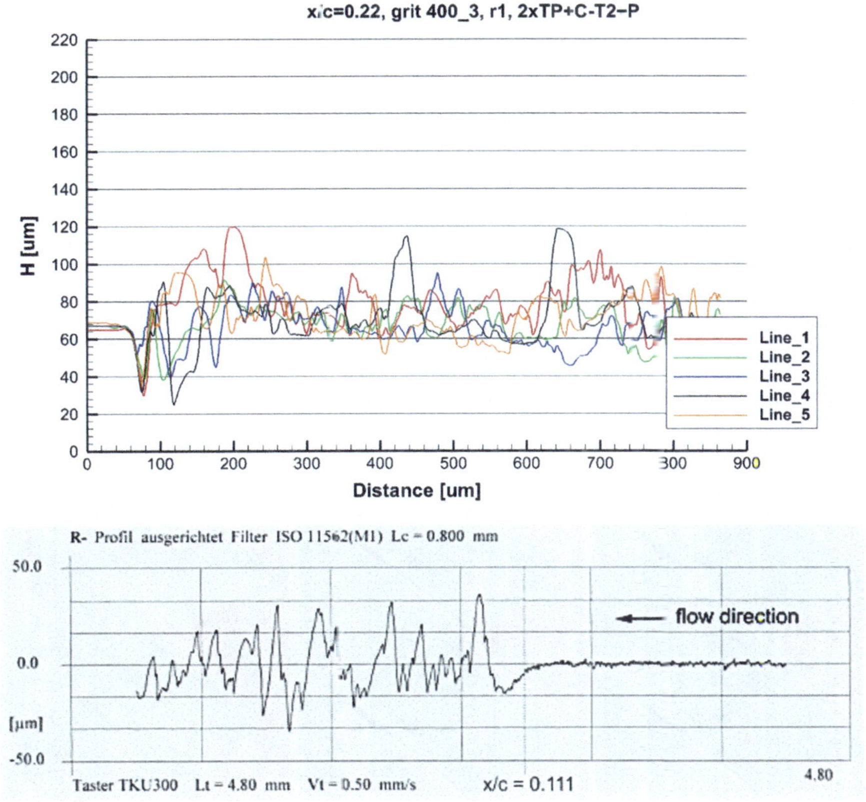

Fig. 70 Measured roughness distribution along IMP (top) and DLR (bottom) roughness patches

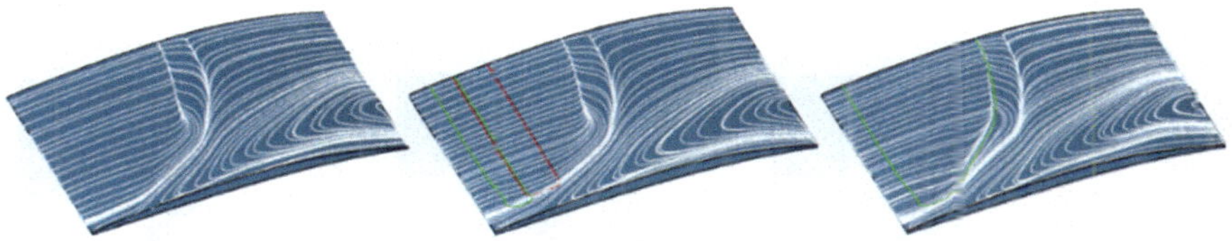

Fig. 71 Limit streamlines along the blade suction side from S-BSL-EARSM turbulence model in fully turbulent (left panel), transitional with rough patch (central panel), and transitional (right panel) modes. The green line indicates laminar to turbulent transition, and the red one corresponds to the limits of the rough patch

of laminar to turbulent transition 0.15 mm away from the wall. From this figure it appears that the rough patch succeeds in promoting laminar to turbulent transition. Indeed, the flow becomes fully turbulent as soon as it reached the edge of the rough element. Therefore, the shock wave boundary layer interaction is rather similar to the fully turbulent one as depicted on Fig. 71. Indeed, the tri-dimensional recirculating structure associated with the shock wave boundary layer interaction appears to be rather sensitive to the upstream boundary layer thickness and turbulence intensity. It should be mentioned that the rough element induces a slightly thicker boundary layer upstream of the shock, when compared to the fully turbulent configuration.

Such a turbulent interaction has a direct impact on the boundary layer downstream of the shock as can be shown from Fig. 72. Indeed, the turbulent interaction produces a thicker boundary layer along the rear part of the blade suction side. This has a direct impact on the thickness of the downstream wake. For this test case the turbulent interaction produces a thicker wake than the laminar one.

The distribution of pressure loss coefficient along the downstream measuring plane MP2 is displayed on Fig. 73. Total pressure difference has been normalised by inlet dynamic pressure. Again, fully turbulent and rough configurations are rather close to each other. Losses are higher than the one produced by the transitional run with smooth walls. Furthermore, as already mentioned, at mid-span, the downstream

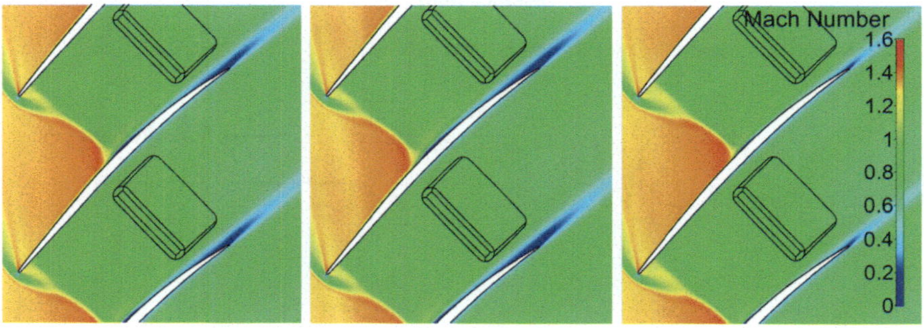

Fig. 72 Mach number distribution from S-BSL-EARSM turbulence model in fully turbulent mode (left panel), transitional mode with rough patch (central panel), and transitional mode (right panel)

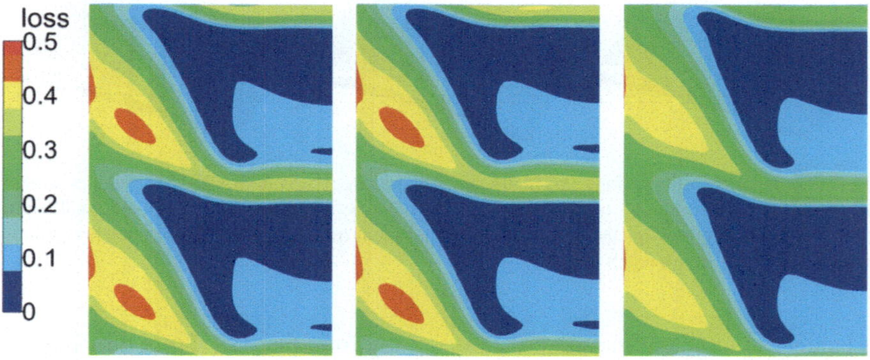

Fig. 73 Total pressure loss coefficient at measuring plane MP2 from S-BSL-EARSM turbulence model in fully turbulent mode (left panel), transitional mode with rough patch (central panel), and transitional mode (right panel)

Table 8 Test matrix. The green marker corresponds to the experimental set-up from DLR

	Centreline average roughness Ra (μm)				
Axial location (%)	8	10.2	12	16	20
11.1	●	● (green)	●	●	●
16		●			

wake is thinner in the latter simulation and the losses are reduced. The passage vortex is also strongly impacted.

In order to further investigate more precisely the impact of the rough element onto the shock boundary layer interaction, a sensitivity study has been performed. For such a study the location and the height of the roughness element has been modified. The idea is to mimic the sensitivity study already performed by IMP for their experimental device. The test matrix is summarized on Table 8.

The green marker corresponds to the experimental setting from DLR. All the simulations have been performed so has to reach the target flow conditions. As for the IMP experiment, for the most downstream location the wide of the rough patch is only 5 mm. It should be mentioned that all the simulations were conducted in a steady mode. The convergence was check by monitoring the mass flow error which was lower than 0.05%. Furthermore, no variations higher than 0.1% were observed for inlet Mach and AVDR over the last hundreds iterations.

The investigated roughness heights correspond to a transitionally rough surface with k_s^+ ranging from 10 to 24. Therefore, those patches are likely to promote laminar to turbulent transition. Total pressure losses increase with the height of the rough element as depicted on Fig. 74 (circles). Indeed such an increase is almost linear with the centreline roughness height. Above a certain threshold value of about 9 μm, the pressure losses are higher than the losses computed by the fully turbulent simulation. From this figure it is also clear that the roughness element corresponding to Ra = 20 μm is over-tripping the boundary layer and induces a loss increase of half a percent with respect to the fully turbulent configuration.

A simulation has also been conducted with a rough patch located slightly downstream. Since the wide of the patch has also been halved it corresponds only to the latter portion of the original rough patch. It centreline roughness height is again 10.2 μm. The slot mass flow rate has been adapted to reach an AVDR of 1.214. The pressure losses for this simulation are represented by a square on Fig. 74. It appears that putting more downstream the rough patch leads to a decrease in the pressure losses. Indeed, such a rough patch configuration behaves in a similar way than the previous configuration with a Ra = 8 μm. Turbulent kinetic energy profiles upstream of the interaction are rather similar and therefore, losses are also similar. This confirms the importance of the level of turbulence intensity upstream of the shock in order to control the shock boundary layer interaction and therefore the downstream losses.

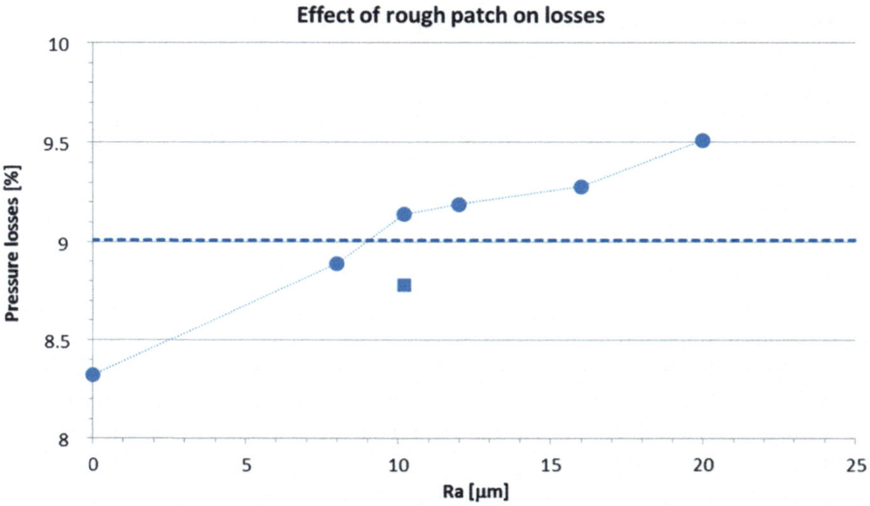

Fig. 74 Computed pressure losses between MP2 and MP1 at mid-span as a function of roughness high coefficient. The smooth case corresponds to a zero roughness. The pressure losses from the fully turbulent simulation is represented by the thick dashed line

3.3 RRD

3.3.1 Introduction

On top of reference simulations which are presented in former RRD chapter, simulations including features with intent of tripping the boundary layer were performed. First a roughness patch is modelled on the suction side of the aerofoil. The second feature applies a source term model on the suction side and mimics the presence of AJVGs (air jet vortex generators). The aim of this study is to investigate if promoting transition upstream the shock which then leads to a turbulent SWBLI can reduce pressure losses compared to the laminar SWBLI.

All transitional reference simulations varying inlet Mach number and inlet angle which were performed on the simplified model (see Fig. 45) are plotted in Fig. 51. The blue points characterize those inlet conditions which lead in unsteady transitional simulations to steady converged solutions. The red points indicate inlet conditions, which lead to periodic unsteady oscillations.

The roughness patch which is investigated in here performed simulations is at same location as it was tested in the experiment in the cascade at DLR. The investigated patch is applied on the suction side of the aerofoil between $x/c = 0.11$ and $x/c = 0.21$. In Fig. 75 the patch is highlighted by a grey pattern. In that figure, main shock (yellow line) and separation zone (pink contour) are also shown, extracted from the baseline transitional simulation with design inlet conditions. The patch location was chosen from a parametric study which was carried out in experiments at IMP PAN. In those

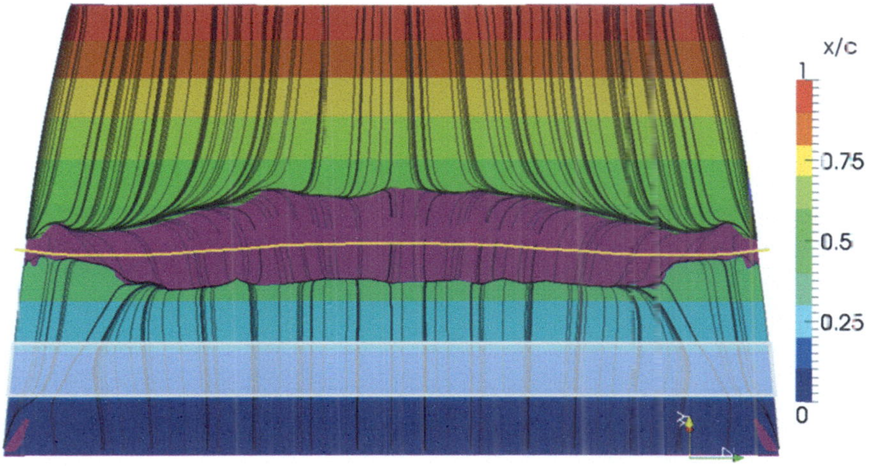

Fig. 75 Suction side view highlighting the location of roughness patch in respect to shock and separation location

experiments roughness heights were varied for different patch locations and widths. The here presented roughness patch was chosen yielding the best compromise in robust tripping of the boundary layer, reduction of wake losses and reduction of unsteady amplitudes due to shock movement. The roughness patch with roughness height of Ra=10 μm finally tested in DLR cascade, was chosen yielding the best compromise in robust tripping of the boundary layer, reduction of wake losses and reduction of unsteady amplitudes due to shock movement.

The roughness model which is applied in RRD flow solver is based on the ideas from Aupoix [24] and Eca and Hoekstra [25]. This model is artificially producing turbulence by tweaking k and ω equations in k-ω Menter-SST turbulence model. To prescribe the roughness heights using such model, a certain sand grain roughness ks needs to be specified. The chosen sand grain roughness will not match the measured values of Ra or Rz one to one. Bons [26] is summarizing a bunch of correction equations, different authors use to correct sand grain roughness to measured values of Ra and/or Rz.

The AJVGs investigated here by means of numerical simulations are at same location as they were tested in the experiment in the cascade at DLR. The used source terms shown in Fig. 76 are applied on the suction side of the aerofoil at $x/c = 0.15$. The location of the injection holes is chosen to be in line with tested roughness patches. AJVGs tested at IMP PAN being connected from pressure to suction side have injection holes with a diameter of $D = 0.3$ mm. Due to manufacturing constraints the holes which were drilled into the surface of the DLR aerofoils have a diameter of roughly $D = 0.5$ mm. The angular definition of the injected air is prescribed by IMP PAN. In stream wise direction the holes have an inclination angle of $\alpha = 65°$ against the suction side of the aerofoil. The tangential angle of the injection holes

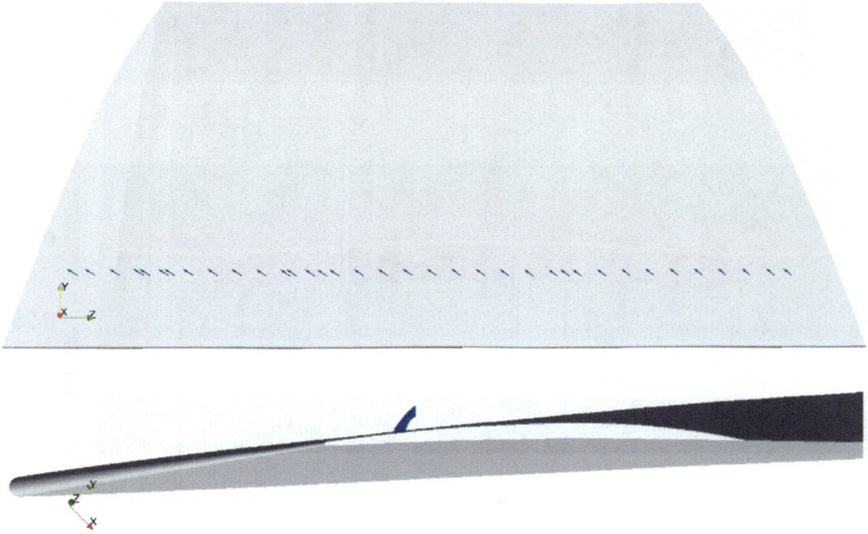

Fig. 76 Geometrical views of source term model on suction side representing AJVGs

is manufactured to be $\beta = 33°$ measured against the suction side of the aerofoil. In numerical simulations the AJVG jet velocity is derived from the pressure to suction side pressure gradient obtained from the baseline transitional simulation.

3.3.2 Design Point Results

Results obtained from performed steady state simulations with roughness patch and AJVGs at design inlet conditions of $\alpha = 55.56°$ and Ma = 1.22 are analysed and compared against transitional and fully turbulent simulations introduced in Sect. 2.3.

Investigation of Roughness Patch

On the suction side of the reference AUTOGRID mesh a roughness patch is specified from x/c = 0.11 to x/c = 0.21 (see Fig. 75). At design inlet conditions several roughness levels prescribed as sand grain roughness ks were simulated.

Pressure losses for design inlet conditions are shown in Fig. 77. The highlighted point at roughness height of ks = 0 μm is extracted from the reference transitional simulation. In addition to that the solid black line is indicating the pressure loss of the fully turbulent simulation. Pressure losses are normalized to the dynamic pressure at the inlet at mid height.

Increasing the roughness level to ks = 50 μm leads to a slight drop in pressure loss can be observed. However this drop is rather small with relative pressure loss reduction of one percent. Moreover suction side intermittency is showing that for

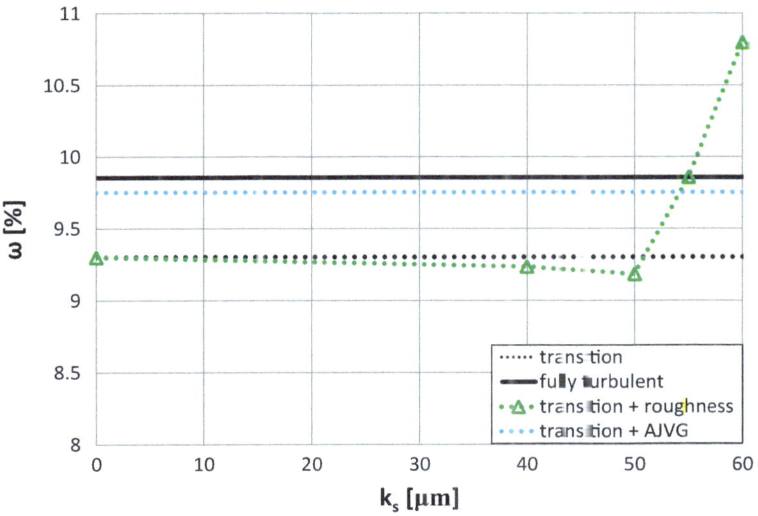

Fig. 77 Pressure loss comparison at midspan for fully turbulent, transitional, transitional + roughness (several heights) and AJVG simulations at DP

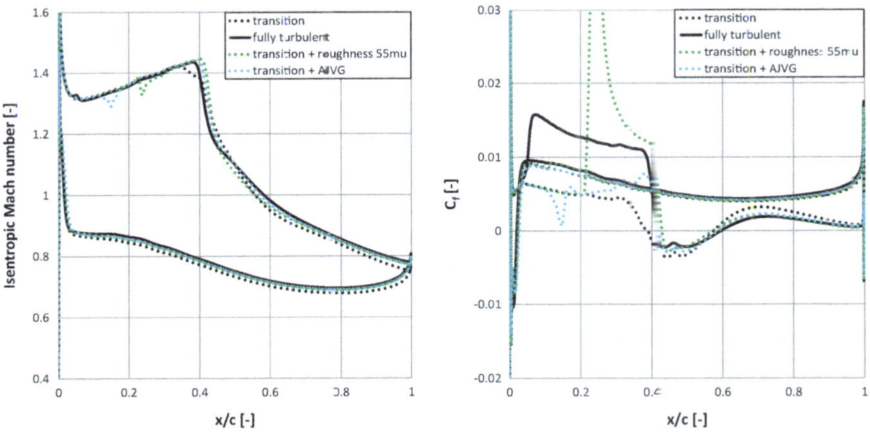

Fig. 78 Profile Mach number and skin friction coefficient for fully turbulent, transitional, transitional + roughness (55 μ) and AJVG simulations at DP

such roughness level the boundary layer remains laminar until the shock position (see Fig. 80).

Successful promotion of transition upstream of the shock occurs for a sand grain roughness of ks = 55 μm and higher. At such level of roughness height pressure losses are increased compared to the reference transitional case and line up with the loss level of the fully turbulent simulation. For that case profile Mach number distribution and

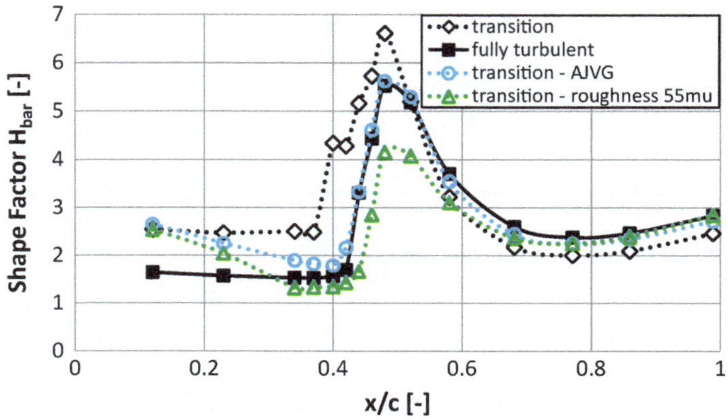

Fig. 79 Boundary layer parameters for fully turbulent, transitional, transitional + roughness (55 μ) and AJVG simulations at DP

suction side skin friction coefficient are compared in Fig. 78 against the distributions extracted from transitional and fully turbulent simulations. Comparing suction side shape factor distributions, it can be noticed that tripping the boundary layer leads to a qualitative improvement resulting in a smaller shock induced boundary layer separation length. However towards the trailing edge the shape factor is rising again to a similar level as obtained in the fully turbulent simulation.

For the over tripped case with a roughness height of ks = 60 μm pressure losses are even further increased compared to the simulation with roughness height of ks = 55 μm.

It might be of interest to investigate further points with roughness heights in-between ks = 50 μm and ks = 55 μm, figuring out the optimum in respect to pressure loss reduction. On the other hand, the benefit is expected being small and more important not be related to successful promotion of transition upstream the shock. Moreover, currently it cannot be concluded if the small drop in pressure loss applying a roughness patch whilst still getting a laminar SWBLI has a physical background or is purely numerically driven. It is assumed that the loss is reduced thanks to a laminar boundary layer which has slightly higher momentum resulting from turbulence production induced by the roughness patch and is hence more resistant against the adverse pressure gradient induced by the shock.

Investigation of AJVGs Using Source Term Model

On the suction side of the reference AUTOGRID mesh a source term model is applied to mimic the presence of an injected jet induced by AJVGs. In Fig. 76 two different lateral views of the aerofoil are shown including the injecting source term jets represented by blue arrows. Those arrows indicate the jet directions and the length by how far the jets are penetrating into the main flow. Due to the mesh resolution which is needed for comparability with the reference simulations no discrete jets are obtained from the performed simulations.

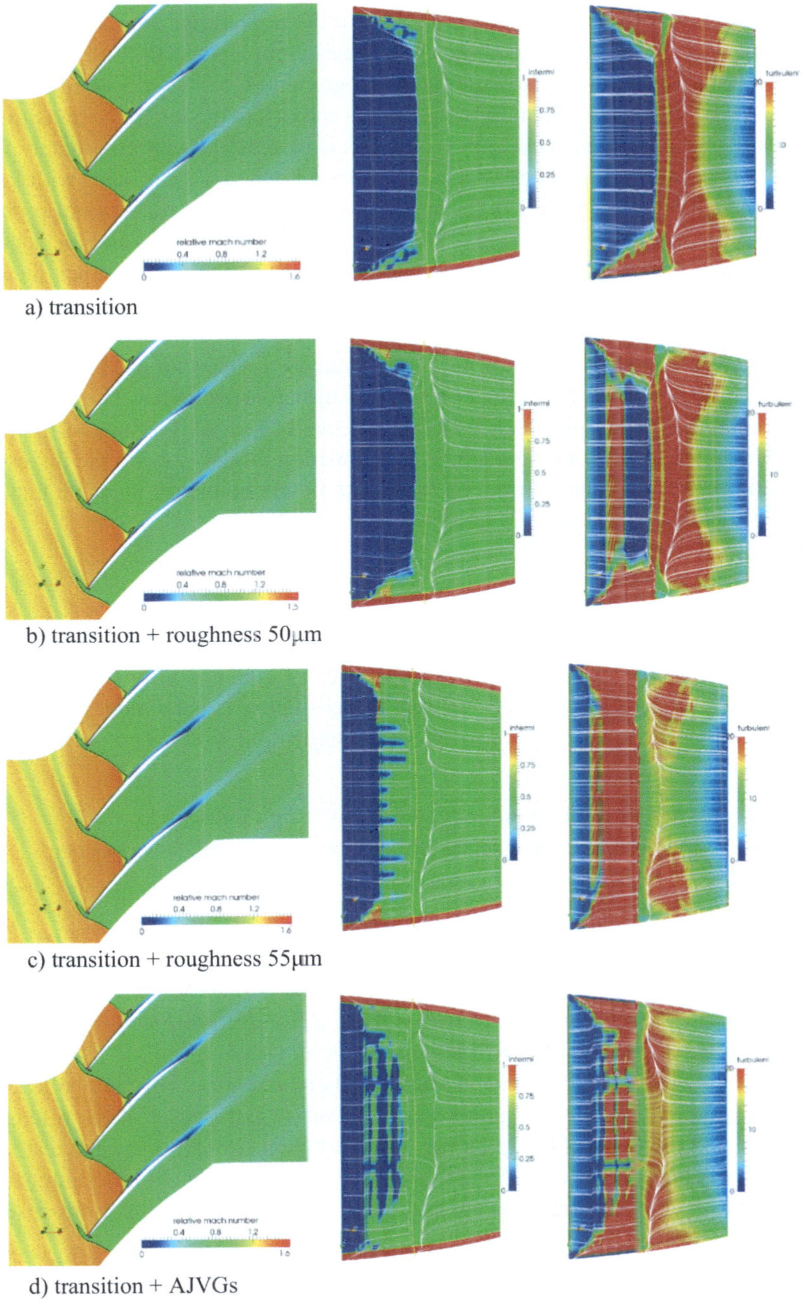

Fig. 80 Mach number contours at midspan + suction side streamlines (intermittency and turb. kinetic energy) for transitional, transitional + roughness and AJVG simulations at DP

At design inlet conditions the pressure loss of the transitional simulation including AJVGs via source term modelling is reduced compared to the fully turbulent simulation and is therefore also slightly lower compared to the successful tripped roughness case with ks = 55 μm (see Fig. 77). Compared to the reference transitional case on the other hand the relative pressure loss is increased by approximately four percent. Compared to the roughness patch tripped boundary layer case with ks = 55 μm, the source term model is leading to a more moderate transitional process as it is shown by the suction side skin friction coefficient in Fig. 78 and the suction side shape factor distribution in Fig. 79. However downstream of the shock there is hardly any difference observed compared to the fully turbulent case for both, suction side skin friction coefficient and suction side shape factor.

The intermittency plot shown in Fig. 80 is indicating that with used source term model intermittency is switching twice back from turbulent to laminar. This behaviour cannot be seen as being physical. It is rather related to stability issues of the flow solver which appears due to the fact that the transition model is using the fluxes for calculation of intermittency. In the end the state of the boundary layer interacting with the shock is of interest.

3.3.3 Off—Design Results

For baseline transitional simulations various different inlet conditions were investigated (see Fig. 51). Thanks to the fact that both, roughness patch and source term model are capable to promote transition, off-design simulations were performed at different-inlet conditions:

- Source term model: Mach number variation at design inlet angle
- Roughness patch: Mach number variation at one degree higher incidence.

Investigation of AJVGs Using Source Term Model

Same source term model as used for design inlet conditions with same jet velocity is used for simulations at design inlet angle and variation of inlet Mach number. For all investigated inlet Mach numbers, transition is successfully promoted upstream the shock and hence a turbulent SWBLI is present for all cases.

As shown in Fig. 81 increasing inlet Mach number the source term model is leading to a higher gradients of the loss polar compared to both, the polar of the reference transitional and the polar of the reference fully turbulent simulations. Hence at inlet Mach numbers beyond Ma = 1.25 the transitional simulations including source term model lead to even higher losses compared to the fully turbulent simulations at same inlet Mach numbers.

Investigation of Roughness Patch

The roughness height which is chosen for inlet Mach number variation at one degree higher inlet angle has a sand grain roughness of ks = 55 μm. Such sand grain roughness successfully promotes transition at design inlet conditions.

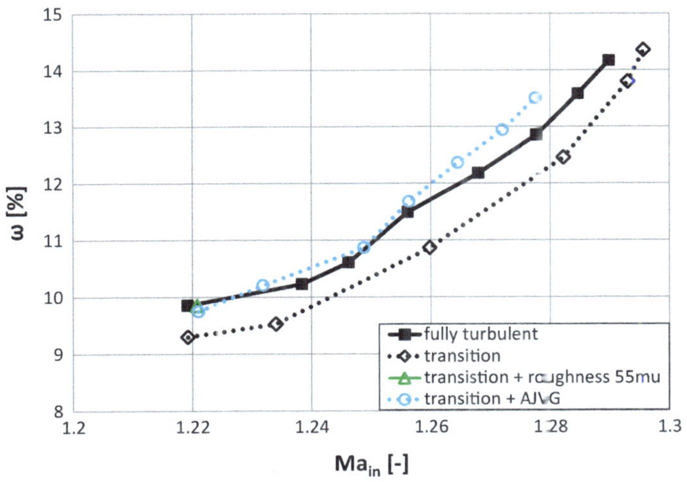

Fig. 81 Loss polar for fully turbulent, transitional and transitional + AJVG simulations for inlet Mach number variation at $\Delta\alpha_{in,DP} = 0°$

Time average pressure losses for here performed roughness height simulations with ks = 55 μm are compared to pressure losses of pure transitional and fully turbulent simulations in Fig. 82. It is noted that losses of transitional and fully turbulent simulations are aligning at Ma = 1.266. At that inlet condition, separation due to SWBLI and TE separation are merging (see Fig. 54). Anyhow, for simulations including roughness patch losses are reduced compared to pure transitional simulations. The relative pressure loss reduction by use of the roughness patch is approximately 1.5 percent.

The reason why simulations including roughness patch lead to lower losses compared to the reference transitional simulations can be probably found in the upstream shock laminar boundary layer interaction with slightly higher momentum thanks to production of turbulence induced by the roughness patch (similar to ks = 50 μm at design inlet conditions). The objective of promoting transition upstream the shock is not achieved with same roughness height which is capable to trip the boundary layer at design inlet conditions. On the other hand, at a modelled roughness height of ks = 60 μm the boundary layer also gets over tripped and pressure losses are increased considerably compared to the pure transitional and fully turbulent cases. Considering the design point results regarding use of roughness promoting transition upstream the shock, it is expected that also at off-design conditions no loss reduction with successful promotion of transition upstream the shock is present.

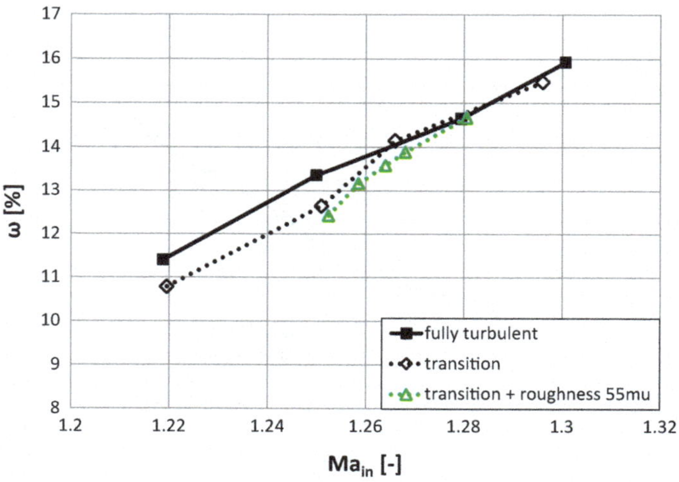

Fig. 82 Pressure losses at midspan for fully turbulent, transitional and transitional + roughness (55 μ) simulations at $\Delta\alpha_{in,DP} = +1°$ (time average)

3.3.4 Conclusions

Both, roughness model and source term model which are implemented in RRD flow solver are capable to promote transition in a physical manor. Although intermittency contours are indicating that transition is occurring immediately, extracted shape factor distributions show that a certain length is required to finish transition.

Compared to the reference transitional simulation only small reduction in pressure loss is observed for simulations including roughness patch where transition is not yet promoted upstream the shock. It is not clarified if the loss reduction, possibly resulting from higher momentum in the laminar boundary layer is physically correct or is just a numerical artefact. For a roughness height which is leading to a reasonable tripped boundary layer, the pressure loss is increased compared to the reference transitional simulation. Although the shock induced separation size is considerably reduced for the transition promoted case, towards the trailing edge the shape factor is increased to the level of the fully turbulent simulation, resulting to a pressure loss lining up with the loss of the fully turbulent simulation. Increasing the roughness height further leads to considerably higher pressure losses is due to over tripping.

The utilized source term model approach does not allow discrete AJVGs due to the fact that the reference mesh has to be used for consistency to former performed simulations. However, mild transition promotion with only small boundary layer thickening is obtained by use of the source term model. Similar to the roughness induced transition, also the source term model induced transition is decreasing the shock induced separation length, but is leading to a pressure loss increase compared to the reference transitional simulation.

None of the here investigated methods promoting an early transition onset leads to a reduction in losses compared to the case with laminar SWBLI. A measure indicating that trend is the suction side shape factor at the TE which is for all cases with turbulent SWBLI higher compared to the case with laminar SWBLI. The beneficial effect of a laminar SWBLI in numerical simulations is the presence of isentropic compression waves above the laminar separation bubble leading to a reduction of peak Mach number and hence a reduction in strength of the main shock.

References

1. P. Flaszynski, P. Doerffer, R. Szwaba, P. Kaczynski, M. Piotrowicz, Shock wave boundary layer interaction on suction side of compressor profile in single passage test section. J. Therm. Sci. **24**(6), 510–515 (2015)
2. P. Doerffer, C. Hirsch, J.-P. Dussauge, H. Babinsky, G.N. Barakos, *Unsteady Effects of Shock Wave Induced Separation* (2011)
3. P. Doerffer, J. Telega, Flow control effect on unsteadiness of shock wave induced separation. J. Therm. Sci. **22**(6), 511–516 (2013)
4. J.P. Gostelow, Cascade aerodynamics (1984), ss. 270 p: ill, 23 cm
5. P. Flaszynski, R. Szwaba, Experimental and numerical analysis of a streamwise vortex generator for subsonic flow, January (2006)
6. R. Szwaba, Comparison of the influence of different air-jet vortex generators on the separation region. Aerosp. Sci. Technol. **15**(1), 45–52 (2011)
7. P. Flaszynski, P. Doerffer, R. Szwaba, M. Piotrowicz, P. Kaczynski, Laminar-turbulent transition tripped by step on transonic compressor profile. J. Therm. Sci. **27**(1), 1–7 (2018)
8. F.R. Menter, A.V. Garbaruk, Y. Egorov, Explicit algebraic reynolds stress models for anisotropic wall-bounded flows, in *EUCASS—3rd European Conference for Aerospace Sciences, July 6–9th 2009* (2009)
9. H.A. Schreiber, H. Starken, W. Steinert, Transonic and supersonic cascades. AGARDOgraph – Adv. Methods Cascade Test., AGARD AG **328**, 35–59 (1993)
10. W. Steinert, R. Fuchs, H. Starken, H, Inlet flow angle determination of transonic compressor cascade. ASME J. Turbomach. **114**(3), 487–493 (1992)
11. R. Kiock, G. Laskowski, H. Hoheisel, *Die Erzeugung höherer Turbulenzgrade in der Meßstrecke des Hochgeschwindigkeits-Gitterwindkanals, Braunschweig, zur Simulation turbomaschinenähnlicher Bedingungen*. DLR Forschungsbericht (FB82-25, 1982)
12. H.A. Schreiber, Comparison between flows in cascades and rotors in the transonic range. Lect Ser. Transonic Blade-to-Blade Flows Axial Turbomach., von Karman Inst. Fluid Dyn. (1976)
13. H.A. Schreiber, H. Starken, On the definition of the axial velocity density ratio in theoretical and experimental cascade investigation. Symp. Measuring Tech. Transonic Supersonic Flow Cascades Turbomachines (1981)
14. P. Schimming, Experimental Investigation of Supersonic Inflow of Compressor Cascade by the Laser-2-Focus Method. Symp. Measuring Tech. Transonic Supersonic Cascade Flow (1976)
15. R. Schodl, A laser-two-focus (L2F) velocimeter for automatic flow vector measurements in the rotating components of turbomachines. ASME J. Fluids Eng. **102**(4), 412–419 (1980)
16. R. Schodl, Laser two focus techniques. VKI Lect. Ser. 1989–05 Meas. Tech. Aerodyn. (1989)
17. J. Klinner, A. Hergt, C. Willert, Experimental Investigation of the transonic flow around the leading edge of an eroded fan airfoil. Exp. Fluids **55**(9) (2014)
18. A. Hergt, J. Klinner, J. Wellner, C. Willert, S. Grund, W. Steinert, M. Beversdorff, The present challenge of transonic compressor blade design. ASME J. Turbomach. **141**(9) (2019)
19. D.J. Mee, T.W. Walton, S.B. Harrison, T. Jones, *A Comparison of Liquid Crystal Techniques for Transition Detection*. No. AIAA-91-0062 in 29th Aerospace Science Meeting, AIAA (1991)

20. W. Steinert, H. Starken, Off-design transition and separation behavior of a CDA cascade. ASME J. Turbomach. **118**(2), 204–210 (1996)
21. H.A. Schreiber, Experimental Investigation on Shock Losses of Transonic and Supersonic Compressor Cascades. No. AGARD-CP-401 in *Conference Proceedings for Transonic and Supersonic Phenomena in Turbomachines*, AGARD, pp. 11–1–11–15 (1986)
22. J. Klinner, A. Hergt, S. Grund, C. Willert, Experimental investigation of shock-induced separation and flow control in a transonic compressor cascade. Exp. Fluids **60**(6) (2019)
23. M. Acharya, J. Bornstein, M.P. Escudier, Turbulent boundary layers on rough surfaces. Exp. Fluids **4**, 33–47 (1986)
24. B. Aupoix, A general strategy to extend turbulence models to rough surfaces: application to Smith's k-L model. J. Fluids Eng. **129**(10), 1245 (2007)
25. L. Eça, M. Hoekstra, Numerical aspects of including wall roughness effects in the SST k-ω eddy-viscosity turbulence model. Comput. Fluids **40**(1), 299–314 (2011)
26. J.P. Bons, A review of surface roughness effects in gas turbines. J. Turbomach. **132**(2), 21004 (2010)

WP-4 Internal Flows—Turbine

Anna Petersen, Piotr Doerffer, Pawel Flaszynski, Ryszard Szwaba, Michal Piotrowicz, Piotr Kaczynski, Benoît Tartinville, and Charles Hirsch

Abstract Modern HP turbine stages consist of highly loaded aerofoils, including transonic and even supersonic flow regions. In terms of stators a normal shock wave in the passage throat chokes the flow, stabilizing the flow conditions at the operating point. In addition to these flow phenomena, the strong acceleration along the early suction side leads to a relaminarisation of the flow, which in turn has a strong impact on the size of the shock induced separation bubble. Finally, the injection of film coolant via rows of cooling holes further influences the boundary layer state. As heat transfer and film cooling effectiveness are of crucial importance in high pressure turbines, an in-depth understanding of transition mechanisms is needed for a competitive design. The main objective was to study transition location effects (from natural transition to fully turbulent) on separation size, shock structure and unsteadiness. The transition interaction with cooling flow was a real challenge for the TFAST project. This research delivered new knowledge, which is crucial for further improvement of HP turbine stages. As transition control devices AJVG and disturbance strip were used.

1 Introduction

Modern HP turbine stages consist of highly loaded aerofoils, including transonic and even supersonic flow regions. In terms of stators a normal shock wave in the passage throat chokes the flow, stabilizing the flow conditions at the operating point. In addition to these flow phenomena, the strong acceleration along the early suction side leads to a relaminarisation of the flow, which in turn has a strong impact on the

A. Petersen (✉)
DLR, Bunsenstraße 10, 37073 Göttingen, Germany
e-mail: Anna.Petersen@dlr.de

P. Doerffer · P. Flaszynski · R. Szwaba · M. Piotrowicz · P. Kaczynski
Institute of Fluid Flow Machinery, Polish Academy of Sciences, IMP PAN, Gdansk, Poland

B. Tartinville · C. Hirsch
NUMECA, Chaussée de la Hulpe 187-189, 1170 Bruxelles, Belgium

P. Doerffer et al. (eds.), *Transition Location Effect on Shock Wave Boundary Layer Interaction*, Notes on Numerical Fluid Mechanics and Multidisciplinary Design 144,
https://doi.org/10.1007/978-3-030-47461-4_5

size of the shock induced separation bubble. Finally, the injection of film coolant via rows of cooling holes further influences the boundary layer state. As heat transfer and film cooling effectiveness are of crucial importance in high pressure turbines, an in-depth understanding of transition mechanisms is needed for a competitive design. The main objective was to study transition location effects (from natural transition to fully turbulent) on separation size, shock structure and unsteadiness. The transition interaction with cooling flow was a real challenge for the TFAST project. This research delivered new knowledge, which is crucial for further improvement of HP turbine stages. As transition control devices AJVG and disturbance strip were used.

2 Turbine Blade Suction Side Model

2.1 IMP PAN

For the purpose of investigations of the shock wave boundary layer interaction, a single passage test section in wind tunnel of IMP PAN is designed. The main objective of the investigation is the shock wave boundary layer interaction on the suction side of the gas turbine profile. In such situation the middle passage has to be designed in order to reproduce the pressure distribution on suction side, as the one existing in the cascade. The upper profile plays important role in creation of required pressure distribution and the wake downstream.

General view of the test section is shown in Fig. 1. Upstream of the profiles, the inlet part is designed in order to maintain Mach number ~0.13. The profiles and cascade configuration are defined by TFAST Partner (Rolls-Royce Deutschland). Downstream of the profiles, the movable upper wall enabling the outlet pressure adjustment is mounted. The shape and the location of the upper limiting wall influences on the Mach number distribution on suction side, which is an important parameter required for the defined conditions. The outlet Mach number for the defined cascade configuration is 1.05.

An advantage of the simplified test section at IMP PAN is the measurement accessibility. The nozzle guide vane configuration allows for investigations of:

- basic flow without cooling holes—location of natural transition
- effect of film cooling on transition location and shock wave boundary layer interaction
- shock wave interaction with turbulent boundary layer obtained by means of tripping devices.

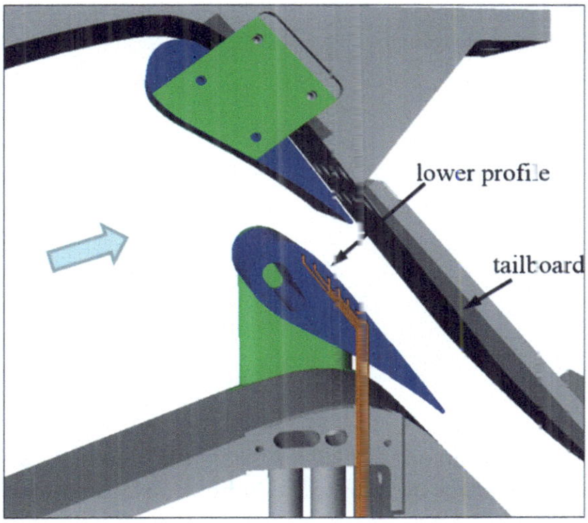

Fig. 1 Test section

The cascade configuration, including 2 rows of cooling holes, is defined as follows: profile chord 112.5 mm, axial chord 52 mm, profile span 100 mm, pitch to axial chord ~2, cooling hole diameter 1.35 mm (both rows), hole distance to hole diameter 2.5.

Effect of the film cooling on transition and shock interaction wave boundary layer interaction was investigated for various pressure ratio (coolant/main flow) 1.05, 1.1 and 1.15. Additionally, the two tripping devices were also investigated: step (strip stuck in spanwise direction) and jet vortex generator. Experimental investigations have been done only for the step. The strip centre was located on the blade at the $x/c_{ax} = 0.55$ (Fig. 2). The height of the step was 50 μm and its width 4 mm. The location of main shock wave in the blades passage was at $x/c_{ax} = 0.65$.

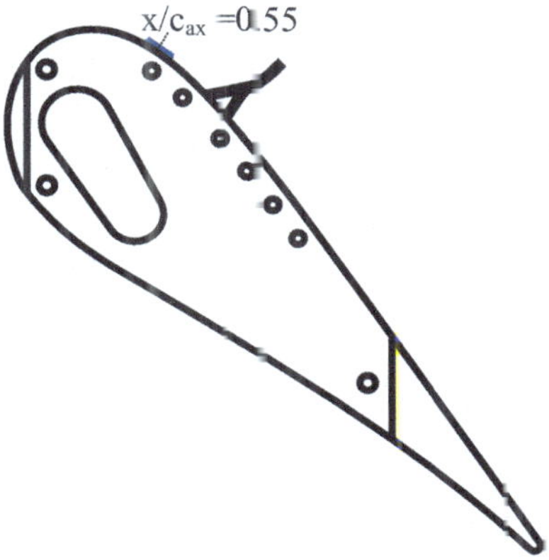

Fig. 2 Locations of the tripping step on the blade

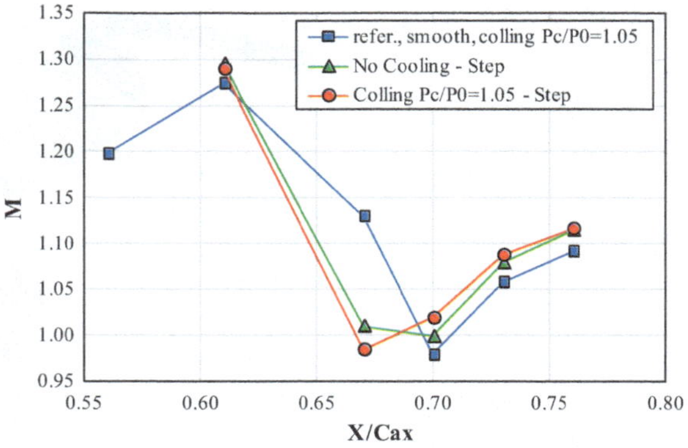

Fig. 3 Mach number distribution along on the blade

In this chapter, selected results for film cooling pressure ratio 1.05 with or without tripping device are presented. In Fig. 3, Mach number on suction side of the profile is shown for 3 cases: (a) with film cooling without step, (b) without film cooling with step, (c) with film cooling with step. The configuration (a) is considered as the reference case.

One can notice that pressure rise (Mach number decrease) on the shock wave takes place on shorter distance for the case with tripping device (step) located upstream of the interaction (step), than for the reference configuration. If no tripping device is applied, there is no natural transition upstream of the shock wave. The pressure jump zone is more extended in streamwise direction if shock wave interacts with laminar boundary layer and the two shocks exist, following one after another. Shock wave structure is shown in schlieren picture (Fig. 4). If tripping device is applied and transition is induced upstream of the shock, pressure jump zone is shorter. In these cases, the λ-foot is very small (Figs. 5 and 6), so the shock wave interacts with boundary layer on a much shorter distance than in case of laminar boundary layer.

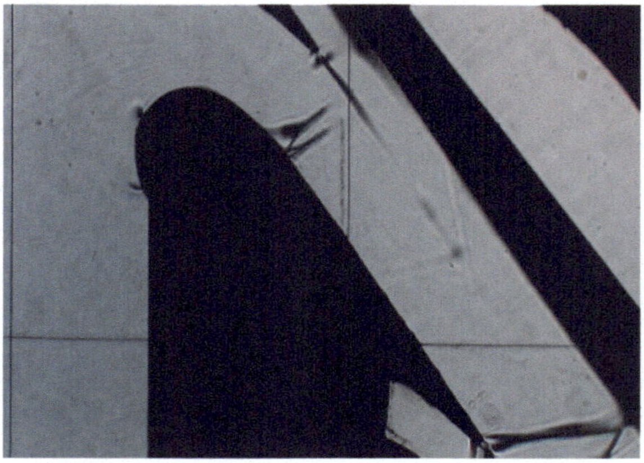

Fig. 4 Schlieren picture for the reference case—with cooling, $P_c/P_0 = 1.05$

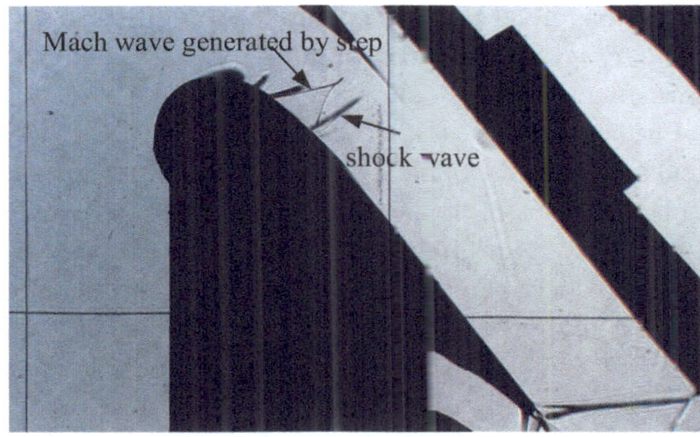

Fig. 5 Schlieren picture—with step and w/o cooling

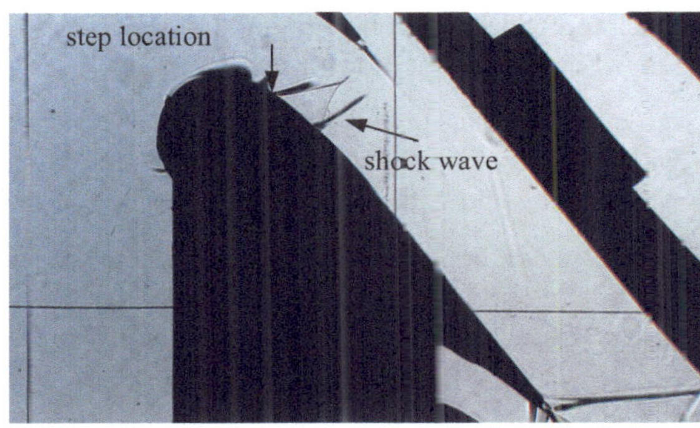

Fig. 6 Schlieren picture—with the step and with film cooling

A typical effect of the shock wave boundary layer interaction is a separation downstream of the interaction. In Fig. 7, oil visualization is shown for the two cases: without and with step. For not tripped configuration, the separation zone spreads uniformly between both lateral walls of the wind tunnel, what indicates that corner flows are very weak. The separation region starts at the location of front λ-foot shock and its length is about 9 mm.

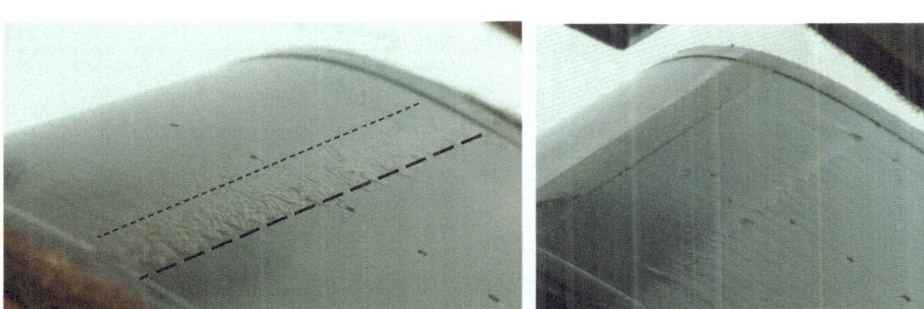

Fig. 7 Oil visualization—with film cooling/without step (left), with film cooling/with step (right)

Positive effect of the step application upstream of the shock wave and tripping boundary layer is also shown in Fig. 7 (right). No evidence of separation is noticed downstream of the shock wave.

Further downstream, close to the trailing edge ($x/c_{ax} = 0.94$), velocity measurements (by Laser Doppler Anemometry) were carried out at 50% span to compare boundary layer for the investigated cases (Fig. 8). In Table 1, the integral parameters (δ-boundary layer thickness, δ*-displacement thickness, δ**-momentum thickness, H_{12}-shape parameter) for boundary layer profiles are also presented. All of them are very similar, what indicates that boundary layer transition induced by the step reduces locally separation size and does not increase viscous losses downstream of the interaction.

Another investigated tripping device was Air Jet Vortex Generator (numerical simulations only). Numerical simulations are carried out by means of Fine/Turbo Numeca for the test section configuration shown in Fig. 9. The structured mesh was generated in IGG/Numeca for 3 cases:

- reference without film cooling and Air Jet Vortex Generators (no FC-no AJVG)
- with cooling and without Air Jet Vortex Generators (FC-no AJVG)
- with cooling and with Air Jet Vortex Generators (FC-AJVG).

Boundary conditions are set according to the measurement data: total temperature, total pressure and turbulence intensity. In case of cooling holes, the total pressure at the inlet is adjusted according to the required pressure ratio. At the outlet, static

Fig. 8 Velocity profile (boundary layer) at $x/c_{ax} = 0.94$

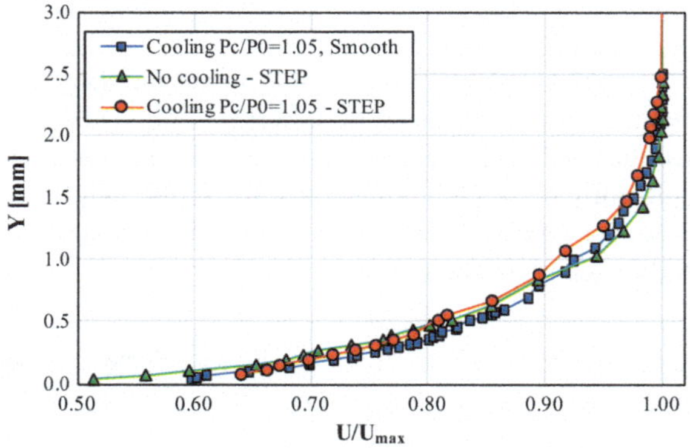

Table 1 Boundary layer integral parameters at $x/c_{ax} = 0.94$

	Reference, cooling	No cooling, step	Cooling, step
δ [mm]	1.80	1.60	1.98
δ* [mm]	0.20	0.21	0.21
δ** [mm]	0.16	0.16	0.17
H_{12}	1.25	1.33	1.24

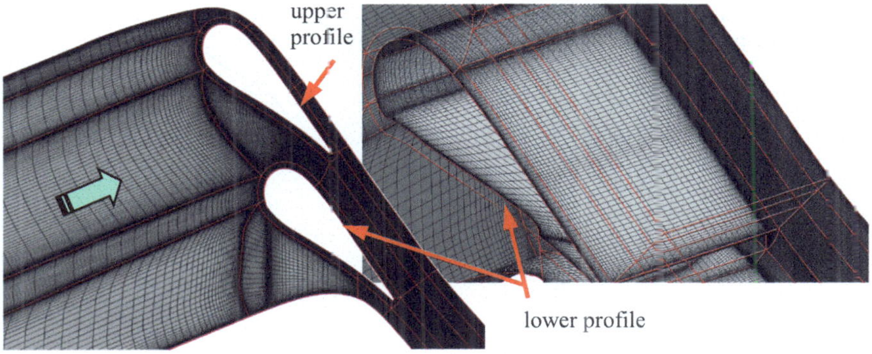

Fig. 9 General view of the computational mesh

pressure is adjusted in order to obtain outlet Mach number 1.05. The coolant flow is considered to investigate the effect on the boundary layer development and transition location only and thermal effects are not investigated. The total pressure at the cooling holes inlet is related to the required pressure ratio 1.05, while at the AJVG inlet, the total conditions are the same as in the main flow. Spatial discretization using 2nd order central difference scheme with scalar artificial dissipation formulated by Jameson et al. [1] is applied. The results presented in the paper are obtained for Explicit Algebraic Reynolds Stress Model (EARSM) turbulence model including transition prediction [2].

As the first step, the size of the separation downstream of the shock is compared for the reference case. In Fig. 10, oil flow visualization and computed streamlines close to the blade surface are presented. Separation and reattachment lines are marked on the oil flow picture. In case of numerical simulations, the separation is visualized by iso-surface of negative streamwise velocity component. One can notice, that separation size is under predicted by numerical simulations. The measured and calculated separation length is 9 mm and 7 mm, respectively.

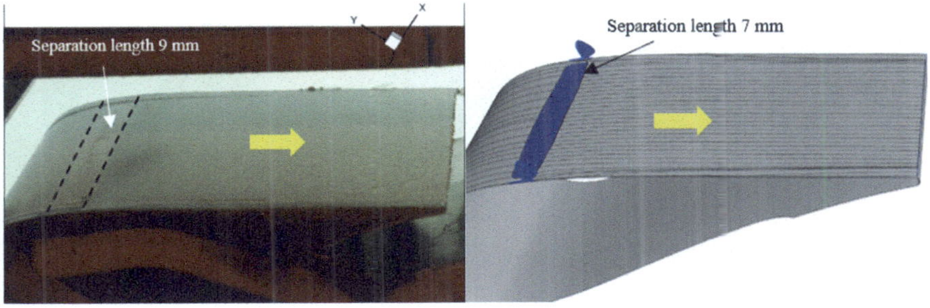

Fig. 10 Oil flow visualization (EXP—top) and streamlines and iso-surface of negative velocity component (CFD—bottom)—no FC—no AJVG

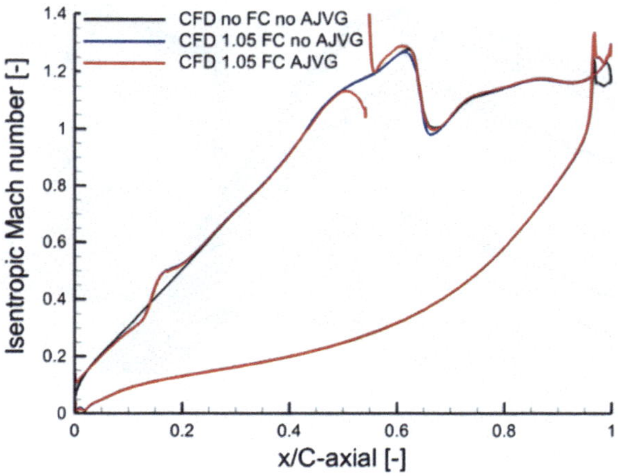

Fig. 11 Isentropic Mach number at midspan—FC and AJVG effect

The comparison of the isentropic Mach number at the middle section of profile for 3 cases (with/without FC and AJVG) is shown in Fig. 11. The first difference can be noticed downstream of the leading edge, at location 15% of axial chord, where the pressure distribution is disturbed by the 2nd row of cooling holes. Further downstream, at 55% of axial chord, the existence of jets influences on the local pressure distribution. Finally, cooling flow and generated streamwise vortices lead to the boundary layer modification upstream of the shock wave and the interaction.

The existence of film cooling influences on the boundary layer evolution on the suction side and the shock structure (Fig. 12). As shown in numerical schlieren (density gradient magnitude) pictures, the λ-foot size is lower than for clean case, which can be considered as the interaction of laminar boundary layer. The laminar type of interaction is indicated by fact that at such relatively low Mach number upstream of the shock ($M = 1.26$), the large λ-foot appears. The application of jet vortex generators does not influence meaningfully on the λ-foot and no difference, in comparison with "FC-no AJVG" case is observed. The AJVG effect is seen upstream

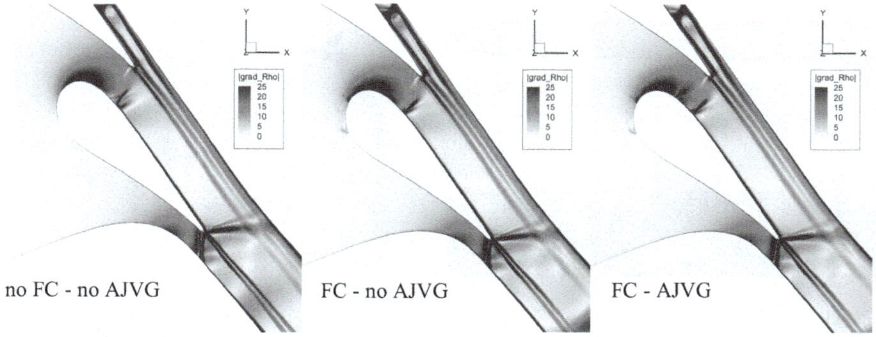

Fig. 12 Shock wave structure

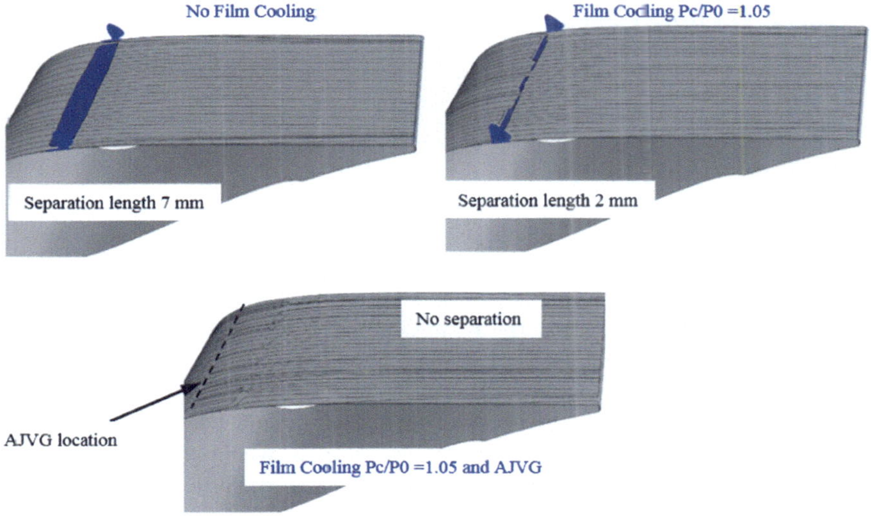

Fig. 13 Streamlines and iso-surface of negative velocity component

of the shock wave, where the disturbance generated by jets influences on the pressure distribution.

The upstream boundary layer influenced by film cooling and generated streamwise vortices affects the λ-foot size and the separation length downstream of the shock wave. In Fig. 13, streamlines close to the profile surface and the iso-surface of negative streamwise velocity component are shown. As presented above, the separation length for the case without film cooling is 7 mm. Existence of film cooling at the pressure ratio 1.05 causes reduction of the separation size to 2 mm. The positive effect is obtained by influence of film cooling on the boundary layer and the increased shear stresses upstream of the shock wave. The application of jet vortex generators leads to further boundary layer modification by creation of 3D flow structures upstream of the shock. As the consequence, the separation is fully reduced and the flow is attached. It is very positive effect which should be also investigated experimentally. Anyway, the predicted effect of the generated vortices on the shock wave boundary layer interaction is consistent with the other results [3].

3 Basic Turbine Cascade Flow

3.1 DLR

3.1.1 Experimental Set-up

Test facility

The measurements are performed at the Wind Tunnel for Straight Cascades (EGG) at the German Aerospace Centre (DLR) in Göttingen. The wind tunnel operates in a closed circuit. It is driven by a 3.7 MW compressor, which achieves a maximal pressure ratio of 14 and a volumetric flow of 60 to 230,000 m^3/h.

Figure 14a shows a schematic view of the closed circuit. On the right side the picture shows the Wind Tunnel for Straight Cascades. The whole system is evacuated at the start of a test day. After achieving a certain pressure in the system the compressor can be started. Subsequently, the system will be filled up with dried air. Additional drying is done during operation. It is possible to control the pressure and temperature in a certain range at the inlet of the test section. The air for the cooling flow is taken downstream of the pressure air supply.

Figure 14b shows the test section of the EGG. The air is accelerated through an adjustable duct into the test section. The cascade is installed on a rotatable disc, which allows the adjustment of the inlet angle. Adjustable walls limit the height of the cascade. The maximum height is 400 mm. The channel width is fixed at 125 mm. Downstream of the test section the flow passes a plenum and then a diffuser. The diffuser is adjustable and controls the chamber pressure and thus the turbine exit Mach number. All measurements are performed at an exit Mach number of Ma = 1.05.

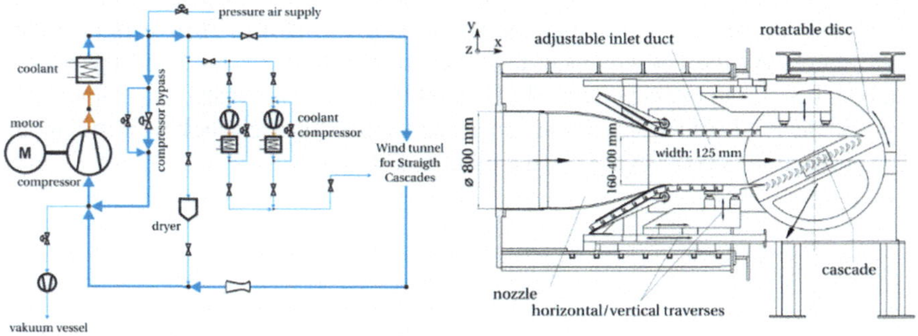

a) Schematic view of closed circuit b) Test section

Fig. 14 Wind tunnel for straight cascades

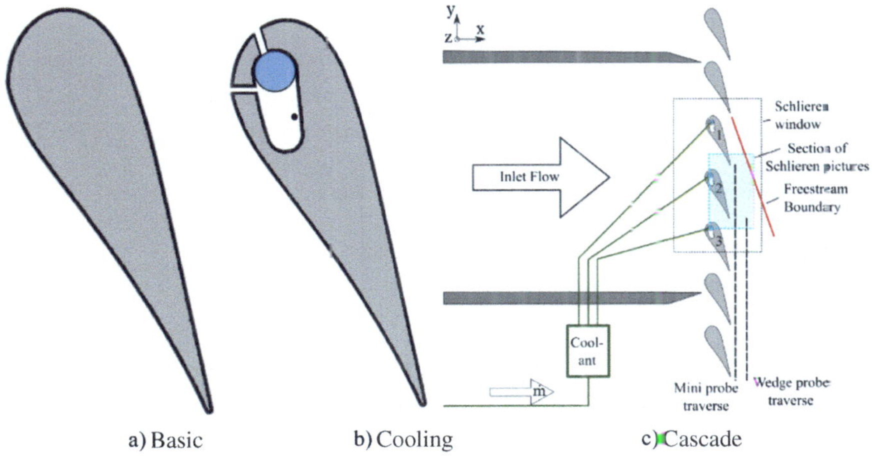

Fig. 15 TFAST profile and cascade

The coolant supply delivers air into a distribution plenum. In this plenum is installed a thermocouple, which gives the temperature of the coolant flow. The distance of the distribution plenum to the blades is approximately 2 m. Three blades are connected to the cooling supply (Fig. 15c).

Cascade and air cooling geometry

Two types of high-pressure stator blades are investigated. The profile contour is the same for every blade type. The blowing ratio of the cooling blade will be adjusted. Rolls-Royce Deutschland (RRD) performed the profile design. Table 2 lists the geometric cascade parameters.

RRD also delivered specifications of the blade with applied film cooling. The cooling blade is the second blade type. Both blade types are shown in Fig. 15a, b. The basic blade has no cavity and no film cooling holes. The cooling blade has one cavity and two rows of cooling holes. In Fig. 15b, the mass flow supply area is highlighted in blue. The supply pipe has a minimum diameter of $d = 7.48$ mm. A black dot in the cavity highlights the location of a pressure probe. At the leading edge of the blade, the lower row is named “Row 1”. The suction side row is designated as “Row 2”.

Figure 15c shows a sketch of the test section and the TFAST cascade. Due to a small stagger angle and a long chord, five blades remain in full flow. The center blade with number ② is denoted as the measurement blade. All measurement techniques are applied to this blade. A black dashed line in Fig. 15c shows the approximate position

Table 2 Geometric parameters of the cascade

Chord length [mm]	75	Throat [mm]	14.56
Pitch [mm]	68.2	Blade height [mm]	125
Axial chord [mm]	34.66	Stagger angle [°]	31.72

Table 3 Geometric quantities of coolant rows and holes

Coolant row	Row 1	Row 2	Coolant row	Row 1	Row 2
Hole diameter [mm]	0.9	0.9	Coolant area [10^{-4} m^2]	1.17	1.17
Hole distance/Hole diameter [–]	2.5	2.5	Hole angle [°] (bitang.)	90	16.25
Hole length/Hole diameter [–]	≈5	≈5	Hole angle [°] (axial)	0	0
Number of holes [–]	46	46	Radial hole angle [°]	0	0
Span of row [mm]	101.24	101.24			

Table 4 Coolant flow configurations

c_m [%]	Cavity pressure [kPa]	Mass flow [g/s]	Discharge coefficient [–]
0	–	–	–
1.1	120.7	13.3	0.48
1.4	126.5	17.1	0.49

of the Schlieren window. A blue dashed line frames the section of the Schlieren pictures. A red solid line highlights the freestream boundary. The measurement blade is chosen to minimize the influence of the free stream and is located inside the Schlieren window.

Cooling holes are applied to the measurement blade and two adjacent blades to ensure periodic flow conditions. The form of the cooling holes is cylindrical. The distribution plenum of the coolant supply is shown below in Fig. 15c. The geometric specifications of the coolant rows are listed in Table 3.

The ratio hole length/hole diameter l/d is designed as suggested in [4]. The tube flow should be fully developed at a ratio of five. The tuning of the cooling parameters is done by the cavity pressure in the measurement blade. The specifications are listed in Table 4. The table lists the total cavity pressure, the mass flow of the supply pipe at the wake measurements and the discharge coefficient. The average total temperature in the distribution plenum had an amount of $T = 297$ K.

Metrology

A 1D hot-wire-anemometer and a 3 hole probe are applied for the measurement of the inlet flow conditions. For the investigation of the boundary layer, infrared thermography is used. The flow field is visualized by Schlieren technique and multi-hole pressure probes. Additionally, the measurement blade is equipped with 20 pressure taps. At the coolant blade type, a hot-film element is installed on the measurement blade ②. Each measurement technique has its own test set-up. For each test set-up a separate measurement is performed.

Figure 16 shows the blade with normalized geometric dimensions. For the normalization the length of the blade is taken. The positions of the pressure taps and the

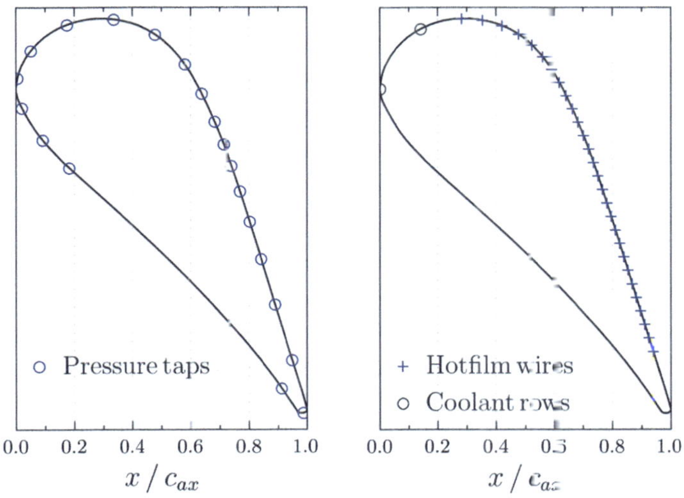

Fig. 16 TFAST Profile, position of pressure taps, coolant rows and hot-film wires

hot-film wires are shown. In addition, the positions of the film cooling holes at the blade with the hot-film sensors are depicted.

Inlet measurements

The inlet measurements are applied to determine the boundary layer thickness at one sidewall of the wind tunnel. The turbulence intensity at the inlet plane of the test section is investigated by hot-wire anemometry. To measure the pressure distribution a three-hole cobra probe is used. Both probes are moved spanwise between one wall and mid-span by a traverse mechanism. The 1D hot-wire probe Dantec 55P11 is applied. It uses a 5 μm tungsten wire and operates with the StreamLine 90N10 Frame. The turbulence intensity (*Tu*) is determined by the following equation:

$$Tu = \frac{\frac{1}{n}\sum_{i=1}^{n} \overline{u'^2_i}}{\overline{u}}. \quad (1)$$

For each measurement a windowing of 1000 values is applied. The resulting turbulence intensities are averaged afterwards.

Infrared thermography

The infrared measurements of the basic case are performed with a Nikon Laird S270A infrared camera. A Stirling circle cools a PtSi Schottky-Barrier IR CCD sensor It's resolution is 475 × 442 pixels. The sensitivity of the camera is in the range of 3 to 5 μm. The frame rate can resolve 30 Hz in an observable temperature range of −20 to 250 °C. The accuracy of the measurement lies within ±2 K. At the cooling case, a VarioCAM HD infrared camera is applied with a resolution of 640 × 480 pixels. It uses an uncooled Microbolometer-FPA (pitch 25 μm) in a spectral range of 7.5–14 μm. The highest frame rate reaches 60 Hz in full range. The thermal resolution at 303 K is better than 0.03 K and the accuracy is ±1 K. The calibration is performed by reference of the camera and of a calibrated pyrometer at blackbody radiators (distance: 0.5 m).

The infrared camera is installed downstream of the measurement blade. The material of the blades is plastic (Polyetheretherketon) due to its low thermoconductivity (0.27 W/(m K)). Markers are engraved close to both sidewalls to relate the infrared pixels and the spatial resolution. For this purpose, a polynomial function is found by means of polynomial regression of the sixth degree between the markers and the image pixels. During the measurement, the test section inlet temperature is changed. Most measurements are made with a positive temperature gradient over the time of the channel inlet flow. In a few cases, a negative temperature gradient has been used to shorten the operating time of the wind tunnel. As the temperature of the flow at the test section inlet changes, infrared images are taken at the maximum frame rate of each camera. The heating rate for the film-cooling blade is 0.1 K/s. With the basic blade, the heating rate is lower.

For post-processing, a dimensionless T_{rel} temperature is defined:

$$T_{rel} = \frac{T_{cold}}{T_{warm}}. \tag{2}$$

Two infrared images are divided from one another for each measurement. The time span between both recordings is t = 20 s for the film cooling blade. To compare the surface temperatures of different operating points in a graph, the relative infrared images T_{rel} are averaged around the midsection:

$$\overline{T_{rel}} = \frac{1}{n}\sum_{i=1}^{n} T_{rel,i}. \tag{3}$$

A Savitzky-Golay filter [5] with window size 11 and polynomial grade 1 is applied to the result of Eq. 3. The induced temperature difference through the inlet temperature of the wind tunnel is not comparable. Therefore, $\overline{T_{rel}}$ is normalized with the mean over the first n = 10 pixels in the flow direction:

$$\overline{T_{rel,n}} = \frac{\overline{T_{rel}}}{\frac{1}{n}\sum_{i=1}^{n}\overline{T_{rel,i}}}. \tag{4}$$

This normalization is also used for data calculated with Eq. 2:

$$\overline{T_{rel,n}} = \frac{\overline{T_{rel}}}{\frac{1}{n}\sum_{i=1}^{n}\overline{T_{rel,i}}}. \tag{5}$$

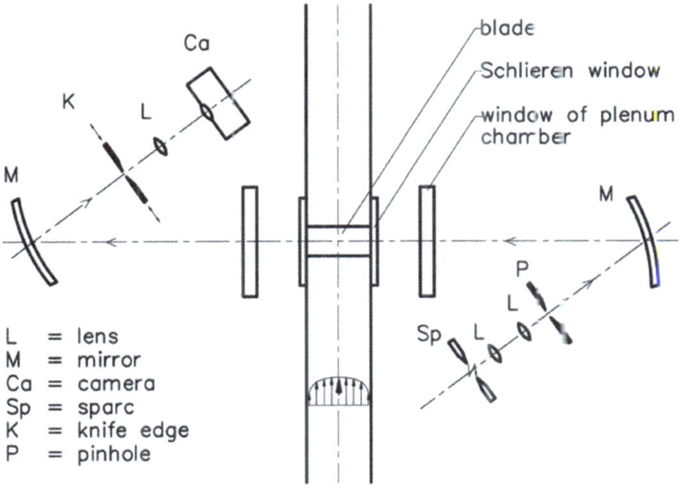

Fig. 17 Schlieren arrangement [6]

Schlieren visualization

The arrangement of the Schlieren visualization at the Wind tunnel for Straight Cascades is conventionally and shown in Fig. 17. The light source is a flash lamp of the company Nanolite (rare gas: Argon). The flash pulses with a length of 20 μs. The camera is a Canon EOS 7D with a CMOS sensor. With a resolution of 2496 × 3420 pixels. An ISO speed of 160 and a shutter speed of 1/400 are used.

Flow field pressure measurements

A blade equipped with 20 pressure taps is applied for the measurement of the surface pressure distribution at the basic blade. The exit flow conditions are investigated with two multi-hole probes. With a relative axial distance of 11.54% to the blade trailing edge plane, a 4-hole Mini probe is deployed. 51.91% behind the trailing edge plane a 3-hole Wedge probe is installed. The denotation of the wake traverses is depicted in Fig. 15c. In Fig. 18 a sketch of both probe heads is shown. Further information about the probe and its calibrations are described in [7].

The traverse is discretized by a step size of 1 mm. Figure 15c shows the start positions for the wake positioning of each probe. The measured inhomogeneous data are recalculated to the homogeneous data by applying the equations of [8] according to [9]. The thermodynamic loss (ξ_{th}) is calculated according to [10]:

$$\xi_{th} = 1 - \frac{\left[1 - \left(\frac{p_2}{p_{02}}\right)^{\frac{\kappa-1}{\kappa}}\right]\left(1 + \sum_{i=0}^{n} c_{m,i}\right)}{\left[1 - \left(\frac{p_2}{p_{01}}\right)^{\frac{\kappa-1}{\kappa}}\right] + \sum_{i=0}^{n} c_{m,i}\left\{\left[\left(1 - \frac{p_2}{p_{0,jet3}}\right)^{\frac{\kappa-1}{\kappa}}\right] c_{m,i}\right\}}. \quad (6)$$

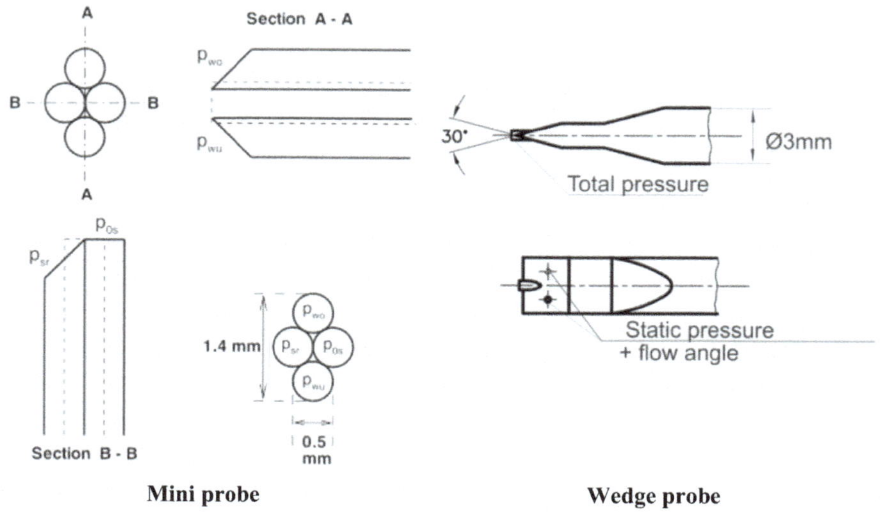

Fig. 18 Multi-hole probes [7]

The thermodynamic loss coefficient, the circumferential angle, and the Mach number are shown relative to the reference case:

$$\xi_{th}^{*} = \frac{\xi_{th}}{\xi_{th,ref}};\ \alpha_{th}^{*} = \frac{\alpha_{th}}{\alpha_{th,ref}};\ Ma_{th}^{*} = \frac{Ma_{th}}{Ma_{th,ref}}. \tag{7}$$

The given losses are the homogeneous data around the wake of the measurement blade (② in Fig. 15c). The homogenization reaches half pitch in each direction of the measurement blade. The pressure data are gained by a PSI 8400 from Pressure Systems. The sensors are silicon piezoresistive pressure sensors and the total error is ±0.25%.

Mass flow measurement

The mass flow is measured at one pipe of the coolant supply. The pipe leads to the distribution plenum in the test section. The mass flow is determined considering the standards of DIN EN ISO 5167-1/2 (2003). For the temperature measurement, a K-Type thermocouple is applied. The mineral insulated thermocouple has a diameter of 6 mm. The connected multimeter (3706) and screw terminal (3724-ST) are manufactured by the company Keithley Instruments Inc. The Co-Junction-Compensation (CJC) is integrated. The pressure data are gauged by a PSI 9010 from Pressure Systems. To calculate the mass flow coefficient, it is primarily necessary to calculate the passage mass flow:

$$\dot{m}_h = \frac{h_c}{h}\left(\frac{2}{\kappa+1}\right)^{\frac{\kappa+1}{2(\kappa-1)}} \sqrt{\frac{\kappa}{R}}\frac{p_{01}}{\sqrt{T}}A. \tag{8}$$

The area A covers the channel width where the coolant holes ejecting the coolant mass flow $\dot{m}_c$. The mass flow of each blade $\dot{m}_c$ is a third of the measured mass flow because three blades are connected to the pipe. The mass flow coefficient c_m is determined by the following equation:

$$c_m = \frac{\dot{m}_c}{\dot{m}_h}. \tag{9}$$

Hot-film element

On one cooled blade a 28-nickel wire hot-film element has been installed. The spatial dimensions are restricted through the coolant rows and the trailing edge. The hot-film element is manufactured by Tao Systems. The leads have a height of 12.7 µm and the nickel wire of 0.2 µm. The polyamide film is 50 µm thick. The film is glued by a 130 µm thick glue tape in a 0.2 mm deep pocket on the blade suction side. The pocket edges are filled with filler to ensure a smooth transition to the blade.

The length of each hot-film sensor is 1.45 mm and the width 0.1 mm. The sensors are equally spaced to each other with a distance of 2.5 mm. At the edge of the hot-film element, the leads are countersunk into the blade. Inside the blade, the cables are soldered on the leads. All cables exit the blade on one side. They are connected outside of the test section to a multiplexer. It is manufactured especially for this application at the DLR. The multiplexer has four output channels and is connected to four modules of the Dantec StreamLine 90N10 Frame. Analog signals of the hot-film are digitalized with an A/D converter and send to a computer.

The ground wire of the hot-film is summarized to one. At other cases, it is possible to measure a few hot-film-wires parallel. In this case, the mutual influence is too high. All hot-film-wires are measured at different times. For the analysis, the fluctuating voltages are calculated:

$$U' = U - \overline{U}. \tag{10}$$

The root mean square of the voltages is calculated by:

$$U'_{rms} = \sqrt{\frac{1}{n}\sum_{i=1}^{n} U'^2_i} \tag{11}$$

Additionally, the skewness data are evaluated. These are computed by the python package 'scipy'.

3.1.2 Results and Discussion

Inlet measurements

Two measurement techniques are applied for the inlet measurements: Hot-wire anemometry and a three-hole pressure probe. The location for the inlet measurements is 80 mm upstream of the blade leading edge for both measurements. The results of the measurements are depicted in Figs. 19 and 20 for the hot-wire and the pressure probe respectively. The data are plotted over the relative channel width z/z_{EGG}. The measurement devices traversed from one wall to the midsection.

In Fig. 19, the turbulence intensity of the hot-wire data is shown. Furthermore, the standard variance of the turbulence intensity is plotted for each measurement. At $z/z_{EGG} = 0.05$ a peak in the turbulence intensity appears. The wall friction can

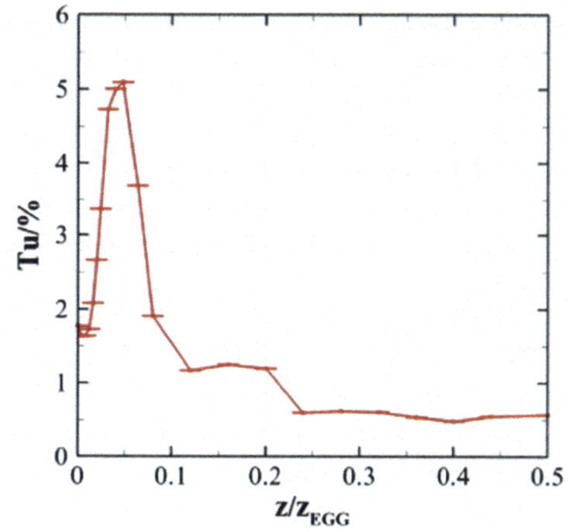

Fig. 19 Hot-wire measurements: Turbulence intensity

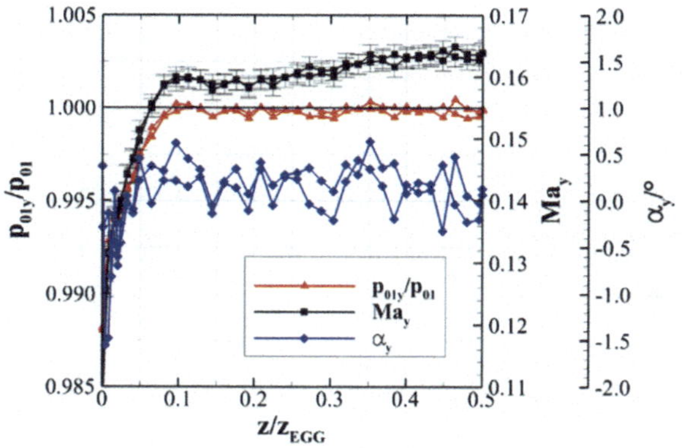

Fig. 20 Pressure probe measurements: Pressure distribution

explain the decrease at the wall. Inside the channel, the turbulence intensity decreases until it reaches approximately $Tu = 0.5\%$ at the center of the wind tunnel.

In Fig. 20 the normalized pressure data p_{01y}/p_{01} and the probe Mach number Ma_y with its errors are depicted. The probe is a 3-hole probe and is able to measure one angle. At this application, the angle variation in y-direction is measured, the alpha angle α_y. The pressure data p_{01y}/p_{01} show the wall boundary layer limit at $z/z_{EGG} = 0.1$. Outside the boundary layer, the averaged inflow Mach number is $Ma_y = 0.16$. The velocity is increased slightly towards the center of the wind tunnel. The results for the angle are close to zero outside the wall boundary layer.

A Fast Fourier Transformation is applied to the fluctuating voltage signal of the hot-wire sensors. Figure 21 shows the power spectra. Four measurement locations are displayed. Three are taken of the wall boundary layer and the fourth shows the center of the wind tunnel. Between $f = 15$–30 kHz different peaks appear depending on the location. They emerge the strongest at $z/z_{EGG} = 0.5$. One possible explanation is thought to be the blade passing frequency of the compressor. This can be excluded because the blade passing frequency is far beneath 1 kHz. Another reason may be vortex shedding of the rods of the hot-wire probe. The brackets of the hot-wire probe have a diameter of 0.5 mm. Applying a Strouhal number of 0.2, a vortex shedding frequency of approximately 20 kHz is the result. Therefore, the vortex shedding is a suitable explanation for the peaks in the power spectra.

Flow field measurements

In Fig. 23 the results of the Schlieren technique are shown. Figure 23a shows the basic case. In Fig. 23b–d three different mass flow rates of the cooling blade are depicted. The alignment of the knife-edge is designated in each picture.

All Schlieren pictures in Fig. 23 show a nearly similar shock structure. At the passage, an oblique shock is generated at the trailing edge of the blade. This incident shock is reflected on the suction side. Upstream indicates a separation shock the beginning of a separation (S). At the reattachment of the separation bubble (R),

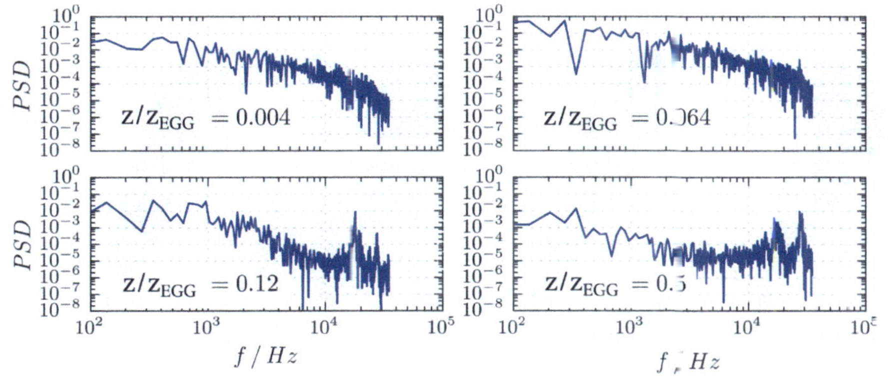

Fig. 21 Power spectra of hot-wire data

an oblique reattachment shock originates. Babinsky and Harvey [11] describes this shock structure and an according sketch is shown in Fig. 22.

In Fig. 23a, b, the separation shock reflects normal at the upper wake. The reflection is oblique if cooling flow is activated. Downstream of the shock system at the separation bubble a normal shock is shown. Another normal shock appears at the trailing edge and emerges weaker than the other one upstream. These normal shocks are unsteady. This fact is revealed by the comparison of different pictures.

Figure 23b shows a similar shock structure. The apparent differences might be induced by the unsteadiness of the flow field. In Fig. 23c, a difference in the shock structure appears at the separation shock. It's initial point is closer to the incidence shock impingement point. This distance is even smaller in Fig. 23d. Additionally, the reflection at the wake is further downstream in (c) and (d) than in the cases without cooling flow. This leads to an oblique reflection. When the coolant mass flow is applied, the exit Mach number is kept constant. The coolant flow inserts additional mass upstream of the smallest cross section. This leads to a higher effective Mach number upstream of the shock. Hence, the shock appears oblique. The reflection of the incidence oblique shock is an expansion wave. This is an indication for a separation bubble in Fig. 23. The reflections of the separation and the reattachment shock propagate downstream until the appearance of the normal shock terminates

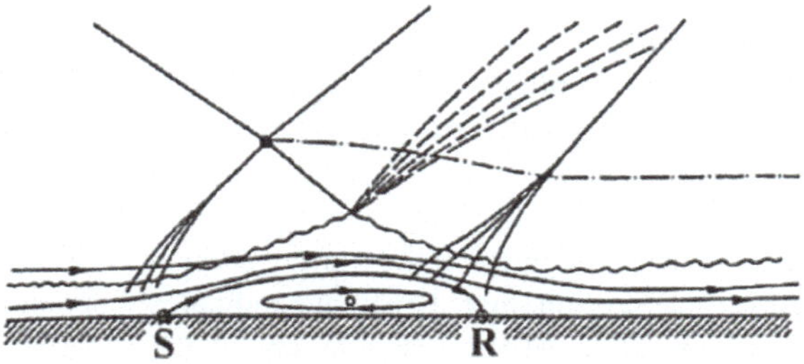

Fig. 22 Shock system with separation bubble [11]

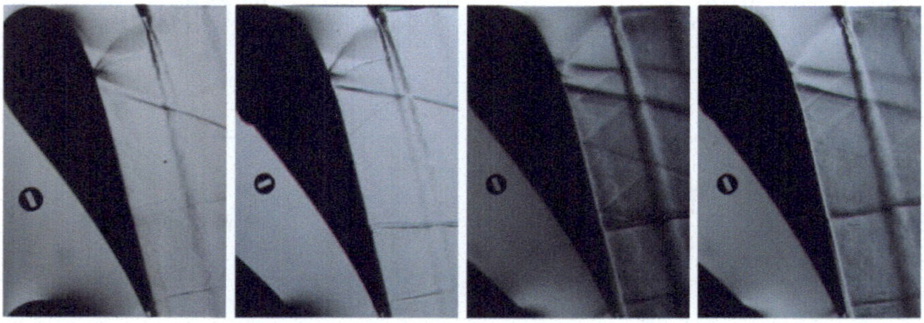

a) Basic b) Cooling, $c_m = 0\%$ c) Cooling, $c_m = 1.1\%$ d) Cooling, $c_m = 1.4\%$

Fig. 23 Schlieren technique

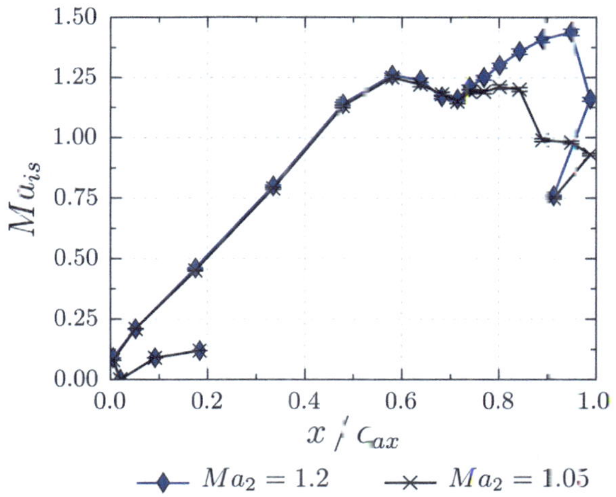

Fig. 24 Isentropic Mach number, Basic case

them. At the cooling blade (Fig. 23b–d) an increase of the boundary layer thickness downstream of the transition is visible. The thickness of the boundary layer increases distinctly with the normal shock. The second normal shock at the trailing edge has no influence on the boundary layer thickness. Figure 24 shows the isentropic Mach number. There are two exit Mach numbers shown. The error bars due to errors of the PSI system are denoted as well. The results of exit Mach number 1.05 show a good agreement with Fig. 23a. Downstream of the suction peak is the flow decelerated by the incidence shock and the re- and attachment shocks of the separation bubble. The strong normal shock appears between $x/c_{ax} = 0.85$ and 0.9. The next weaker normal shock is detected between $x/c_{ax} = 0.95$ and 1. All shocks detected by the Schlieren technique are confirmed by the pressure distribution.

$Ma_2 = 1.2$ is depicted to show the development of the normal shock at $x/c_{ax} = 0.95$. The numerical simulations of the partners show the appearance of a normal shock close to $x/c_{ax} = 1$ at $Ma_2 = 1.05$. In the experiments at $Ma_2 = 1.2$, the strong normal shock which is normal at Fig. 23a at $x/c_{ax} = 0.85$ and 0.9, moved to the trailing edge. At $Ma_2 = 1.2$ it is an oblique shock due to the higher exit Mach number.

Figure 25 shows the results of the Wedge probe at the coolant case. The probe is installed downstream of the Mini probe. The results of the Mini probe are depicted in Fig. 26. The relative wake traverse positions $(y - y_0)/t$ refer to the numbers shown in Fig. 15c. The starting position for the Mini probe (Fig. 26) is set to $(y - y_0)/t = 1$ (downstream the trailing edge of the upper blade). The starting position of the Wedge probe is one pitch lower. The results of the cooling case are shown solely because at the measurements of the basic blade the humidity was too high. Due to sparse time and cost resources, the repetition of the measurements was not undertaken. Figure 25 shows the relation of probe pressure p_{02y} to inlet pressure p_{01}, the total pressure ratio. The results of the probe Mach number Ma_{2y} are shown below. At the bottom of Fig. 25, the circumferential angle α_{2y} is depicted. The mixing of the flow is further advanced than for the Mini probe.

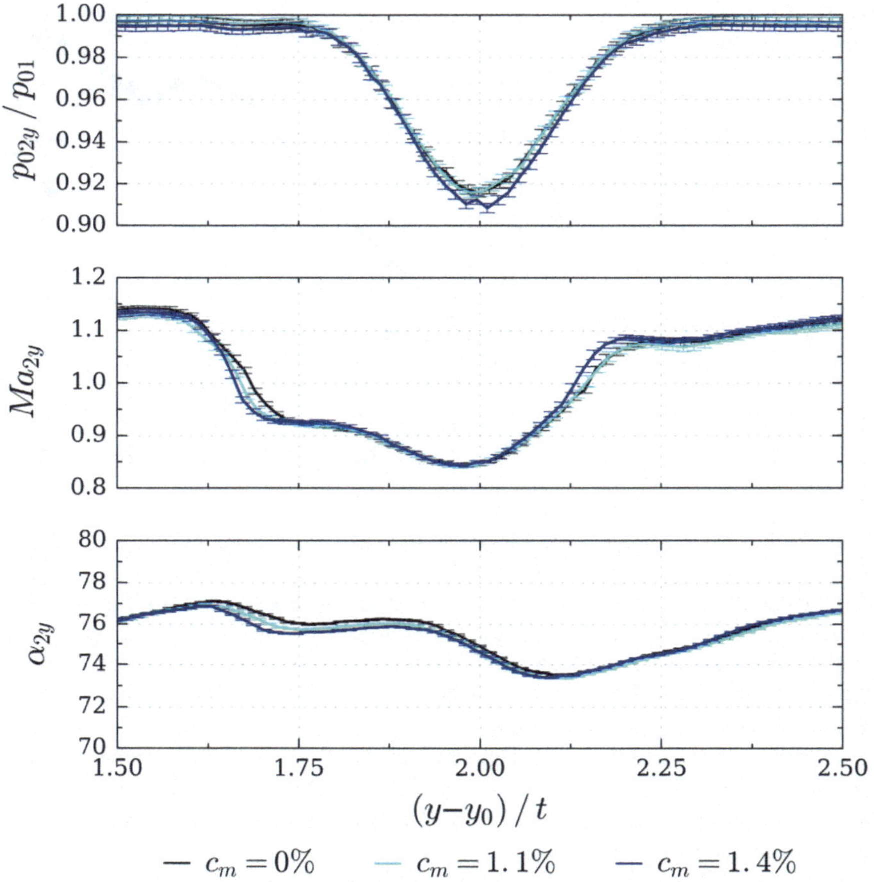

Fig. 25 Wake traverse of cooling blade, Wedge probe

In Fig. 25 the total pressure ratio peak at the wake is slightly distinct at $c_m = 1.4\%$ than at the other two. At Ma_{2y} the normal shock is detected at $(y - y_0)/t = 1.7$. It appears further upstream with increased coolant flow. On the pressure side of the blade at $(y - y_0)/t = 2.3$ the flow is locally decelerated. This is the reflected incidence shock at the separation bubble. It appears at $c_m = 1.4\%$ closer to the trailing edge. The higher coolant mass flow causes an increased upstream Mach number and the angle of the shock is raised. Hence, it appears closer to the trailing edge. The higher mass flow at $c_m = 1.4\%$ has also an effect on the normal shock at $(y - y_0)/t = 1.7$. Due to the higher mass flow the normal shock appears further upstream. At the circumferential angle α_{2y} a slight difference between the cases at the position of the normal shock ($(y - y_0)/t = 1.7$) is detected.

Figure 26 shows the results of the Mini probe traverses. An extra hole enables the measurement of the radial angle β_{2y}. Due to a small probe head a distance to the

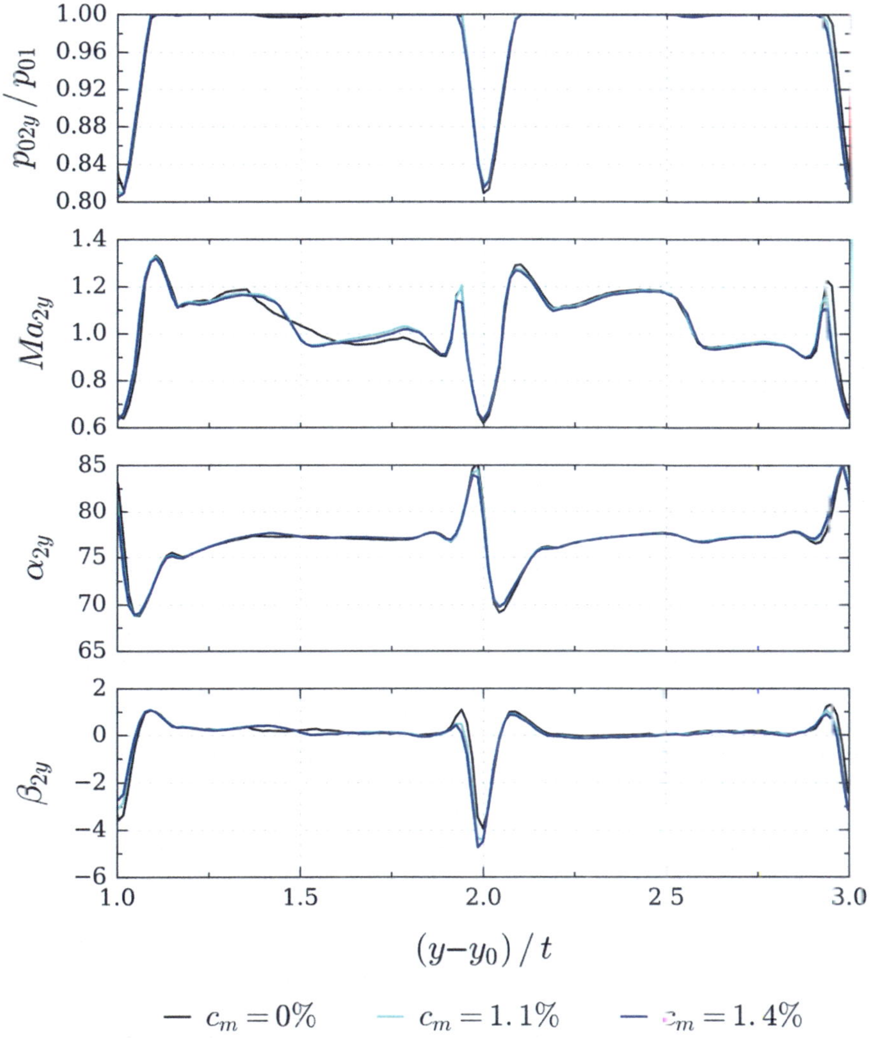

Fig. 26 Wake traverse of cooling blade, Mini probe

blades of 4 mm can be reached. The mixing of the flow is not far progressed at this position and the flow characteristics are more pronounced. The total pressure ratio shows sharper wakes than at the Wedge probe, with a value of $p_{02y}/p_{01} \approx 0.8$. The normal shock appears around $(y - y_0)/t = 1.5$ and is more spread due to the other wall distance of the Mini probe. The next weaker normal shock at the trailing edge is visible at $(y - y_0)/t = 1.9$.

The circumferential angle is $\alpha_{2y} \approx 77°$ at the passage. Solely at the wake, the α_{2y} results show minor differences. At the wake edges, the case without film cooling has

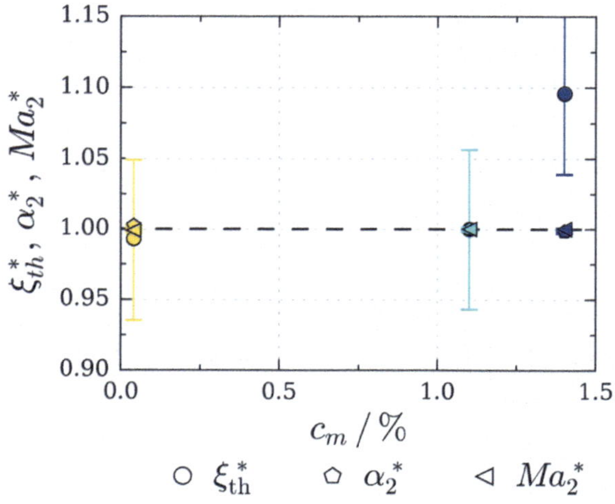

Fig. 27 Losses, exit angle, mixed out Mach number of Wedge probe measurements

angles that are more distinct. It's wake is sharper also. This case shows a slightly higher radial angle β_{2y} at the wakes suction side. At the wake the flow undergoes a radial angle change with one dominant side.

In Fig. 25 the inhomogeneous results of the Wedge probe are shown. These results are not post processed, except for the determination of the values with the help of the probe calibration. Averaging over one pitch according to [9] is called homogenization. In Fig. 27 homogenized results of the Wedge probe are shown. The homogenization is performed for the thermodynamic loss ξ_{th}. In the calculation of the loss the coolant mass flow (Eq. 6) is regarded. Additionally, the circumferential angle α_2 and the Mach number Ma_2 are shown. Those are referenced by Eq. 7. The reference values are of the film cooling at $c_m = 1.1\%$ and have an amount as follows: $\xi_{th,ref} = 6.013$, $\alpha_{th,ref} = 74.46°$ and $Ma_{th,ref} = 0.931$. The position of the data is $(y - y_0)/t = 2$ (Fig. 15c).

The thermodynamic loss ξ_{th} increases with a raised coolant mass flow. The result at $c_m = 1.4\%$ protrudes by its high loss. Figure 25 indicates two reasons for the higher loss at this mass flow rate. First, at the passage the total pressure ratio is smaller. This is caused by higher losses due to stronger shocks. The other reason is the smaller total pressure ratio p_{02y}/p_{01} at the wake. This is caused by the higher portion of turbulent boundary layer upstream of the separation bubble. The homogenized circumferential angle and Mach number show neglectable minor deviations.

Transition detection techniques

The temperature distribution of the basic and the cooling blade is shown in Fig. 28. The figures are post processed according to Eq. 5. In Fig. 28a, a minor influence of the markers at the sidewall is visible above ①. It is limited on a region close to the sidewalls. A high temperature gradient $T_{rel,n}$ indicates a high heat transfer and vice versa. At the basic blade, the separation bubble at ① appears at $x/c_{ax} \approx 0.63$. Downstream at regions of low heat transfer ③ alternate with regions of high heat transfer ②. The latter are turbulent wedges. At ③, the boundary layer sustains the treatment of the shock-wave boundary layer interaction and remains laminar. The normal shock is situated at $x/c_{ax} \approx 0.85$ ④.

At the coolant blade, no turbulent wedges are visible. Without the coolant flow the boundary layer is turbulent downstream of ⑤ and the normal shock appears at position ⑥. With coolant flow ejection, the heat transfer differences are minor. The

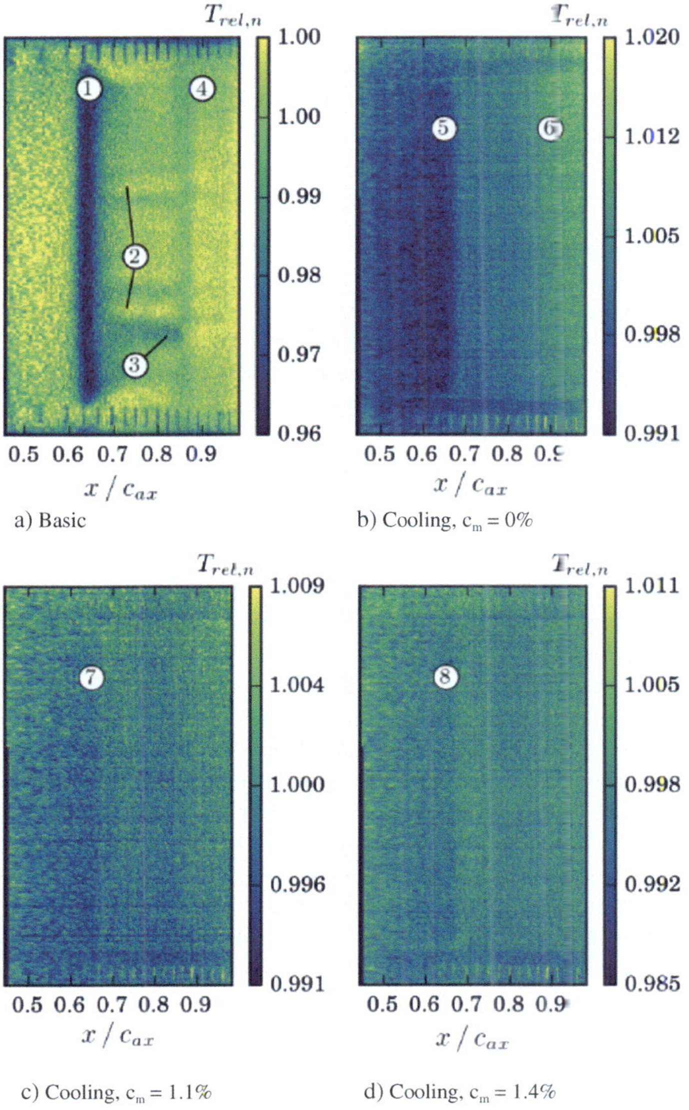

Fig. 28 Temperature distributions on suction side

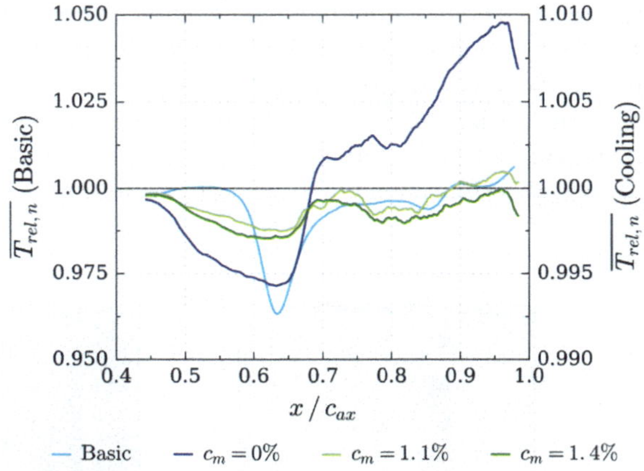

Fig. 29 Averaged temperature distributions of all cases and Mach number distribution

reattachment of separation is highlighted by ⑦ and ⑧. The normal shock is not visible any more.

The results of Fig. 29 are averaged in spanwise direction (Eq. 4). Different infrared cameras are taken for the basic case and for the cooling case. The cameras measure at different wavelengths. The temperature step while the measurement between both cases is different. Those are the reasons for other level of $\overline{T_{rel,n}}$. The results of each blade are shown on its own axis.

At the basic case, the onset of separation is clearly visible around $x/c_{ax} \approx 0.57$. The basic case separation reattaches *at* $x/c_{ax} \approx 0.67$ at the increase of $\overline{T_{rel,n}}$. The reattachment at the cooling case is located at the same position. The onset of separation at this blade type cannot be defined that clear. A shock-induced transition at the separation bubble appears at the cooling case. A significant reduction of the separation size is not revealed. The strong normal shock at $x/c_{ax} \approx 0.88$ is only visible at cases without any active cooling flow.

The next figures show the results of the hot-film measurements. In Fig. 30, 10 ms of the fluctuating voltages are shown. The data of each sensor are recorded successively. The laminar part of the boundary layer reaches up to $x/c_{ax} \approx 0.56$. Hence, the onset of separation cannot be detected by the infrared image of the coolant flow. With increasing coolant mass flow, the turbulent disturbances of the boundary layer rise. At every case, the boundary layer is fully turbulent at $x/c_{ax} \approx 0.72$. Around $x/c_{ax} \approx 0.88$, the normal shock causes higher fluctuations.

In the intermitting region, the separation bubble will be analyzed by a detailed view in Fig. 31. The whole region with intermitting flow is superimposed by fluctuations of the shock wave boundary layer interaction. Without coolant mass flow (Fig. 31a), disturbances are visible at $x/c_{ax} \approx 0.64$. Those are amplified at the next hot-film sensor downstream. A fully intermitting flow can be seen at the sensor situated at $x/c_{ax} \approx 0.68$. At the next sensor, minor peaks are still visible. A fluctuating separation bubble causes these peaks. The coolant flow enhances the amplitudes.

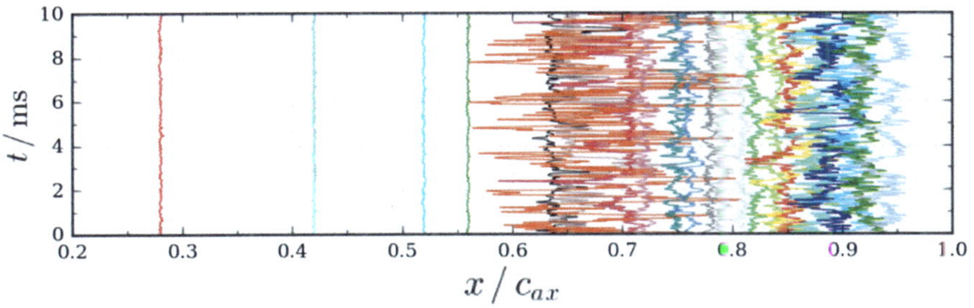

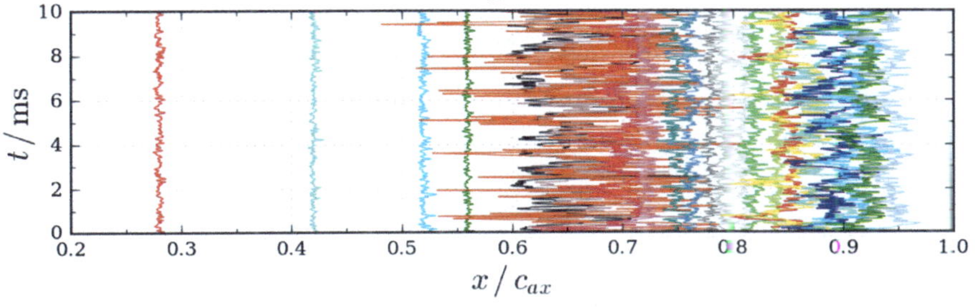

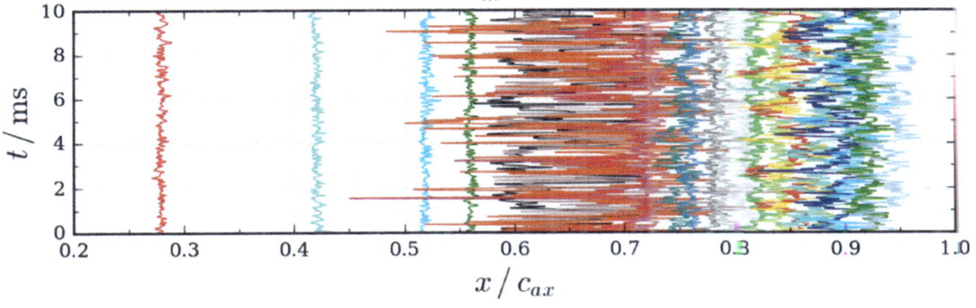

Fig. 30 Time signals of fluctuating hot-film voltages

The voltage fluctuations U' of each wire are evaluated to achieve the root mean square U'_{rms} (Eq. 11) and skewness U'_S. The results are shown in Fig. 32. In the laminar region, the higher coolant mass flow is visible by increased values of U'_{rms} and U'_s. According to recent investigations [12] slightly positive skewness data can be interpreted as a predominantly laminar boundary layer with occasional turbulent components.

A peak of the rms data is found for all three cases at the same position. The amount of U'_{rms} is different for each case. This indicates that the genuine peak of each case appears close by. At the same position of the U'_{rms} peak, the intermittency reaches approximately 50 percent [12]. The skewness data show an inflection point after

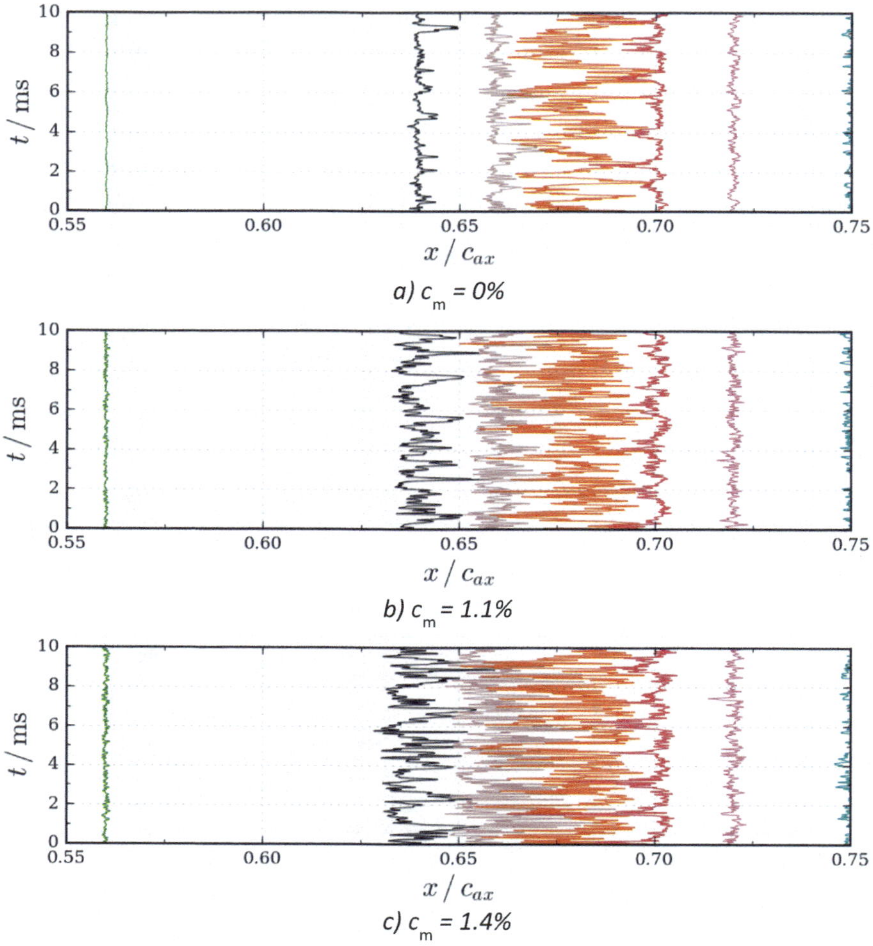

Fig. 31 Time signals of fluctuating hot-film voltages, Section of Separation bubble

reaching a positive peak. Further downstream, the boundary layer starts to become turbulent indicated by the negative peak of the skewness data.

The transition process is completed at $x/c_{ax} \approx 0.68$. The indicators are the negative peak of U'_S and a higher level of U'_{rms} downstream of the separation bubble compared to the laminar region. The normal shock at $x/c_{ax} \approx 0.9$ causes another negative peak of the skewness and a rise of the rms data.

Summary and conclusions

At transonic flow conditions, experimental research is carried out at high-pressure turbine stator blades in a straight cascade to determine the aerodynamic behavior. Special emphasis is laid on the detection of the transition location. Its influence on the shock wave boundary layer interaction is the main research object in the TFAST

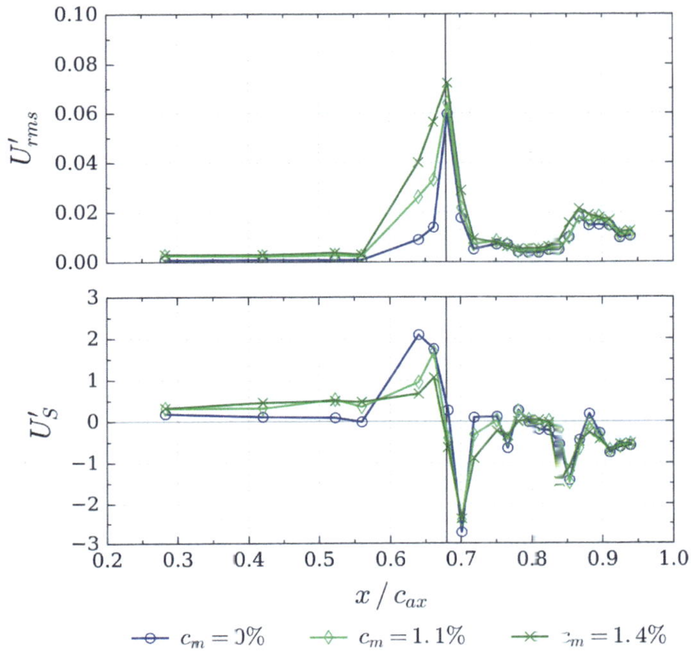

Fig. 32 Rms and Skewness of fluctuating hot-film voltages

project. The main investigation assessment for a turbine is the aerodynamic loss. In this study, the focus is on these criteria. Two cascade set-ups are applied, one without coolant and the other with coolant flow with different coolant mass flow ratios. The experiments are performed in the Wind Tunnel for Straight Cascades (EGG) in the German Aerospace Centre (DLR), Göttingen. RollsRoyce Deutschland (RRD) has designed the profile geometry. The blade profile has to sustain high loads and the boundary layer is stabilized by high acceleration.

The inlet parameters of the blade leading edge are investigated with a hot-wire anemometer and a pressure probe. One of the solid blades without coolant ejection is equipped with pressure taps to obtain the surface pressure distribution. Further information about the flow field is obtained by the application of Schlieren technique and wake traverse measurements. The transition location is detected with an infrared camera and a stripe of 28 hot-film elements on the suction side of a cooled blade. All measurements are performed at zero incidence angle and the transonic exit Mach number $Ma_2 = 1.05$. The exit Reynolds number is $Re_2 \approx 1.3 \times 10^6$. The coolant efficiency is neglected in this investigation. Therefore, the temperature of the coolant flow is set equal to the inlet temperature.

To investigate an actively cooled turbine environment, the cascade is equipped with three cooling blades. Each blade has two film coolant rows. One coolant row is located at the stagnation point and the second in the acceleration region on the suction side. Measurements of the cooled cascade without coolant flow are performed to investigate the influence of the holes itself on the transition location. RRD specified the coolant pressure to 120.7 kPa, which gives a coolant mass flow ratio of 1.1%. Additionally, a higher coolant mass flow of 1.4% is applied to investigate the coolant flow influence on the transition location.

Subsequently, the specific research findings of this investigation are summarized:

Shock structure
An incidence shock at the passage causes a separation bubble at all cases. This indicates the reflected expansion wave as well as a reattachment shock. Downstream of the passage a normal shock at $x/c_{ax} \approx 0.88$ appears. This shock terminates the reflections of the upstream shock system. At the trailing edge a second weaker normal shock emerges. The normal shocks are oscillating.

Separation
Next to the Schlieren technique, the temperature distribution obtained by the infrared camera reveals a separation bubble caused by the incidence shock at the passage. The introduction of coolant flow does not change the reattachment position. The onset of the separation cannot be determined that specific, as it would be necessary. A significant shortening of the separation bubble cannot be detected.

Transition location
At the basic blade without any coolant flow, the incidence shock at the passage induces a partial transition of the boundary layer state. Nevertheless, this does not happen for the boundary layer over the whole span. The temperature distribution reveals turbulent wedges. These are not observed at the cooled blade. At the cooled blade, a separation-induced transition could be observed at all three different mass flow rates. Additionally, an increase of turbulent disturbances upstream of the separation is detected.

Turbine loss coefficient
The losses rise with the increase of coolant mass flow. The higher portion of turbulent boundary layer upstream of the separation bubble causes this.

4 Turbine Cascade Flow with Transition Control

4.1 DLR

4.1.1 Experimental Set-up

In this work package, three types of high-pressure stator blades are investigated. The profile contour is the same for every blade type. The blowing type and ratio will be adjusted. This section comprises investigations about a blade having two film-cooling rows and one row of air jet vortex generators (AJVG's).

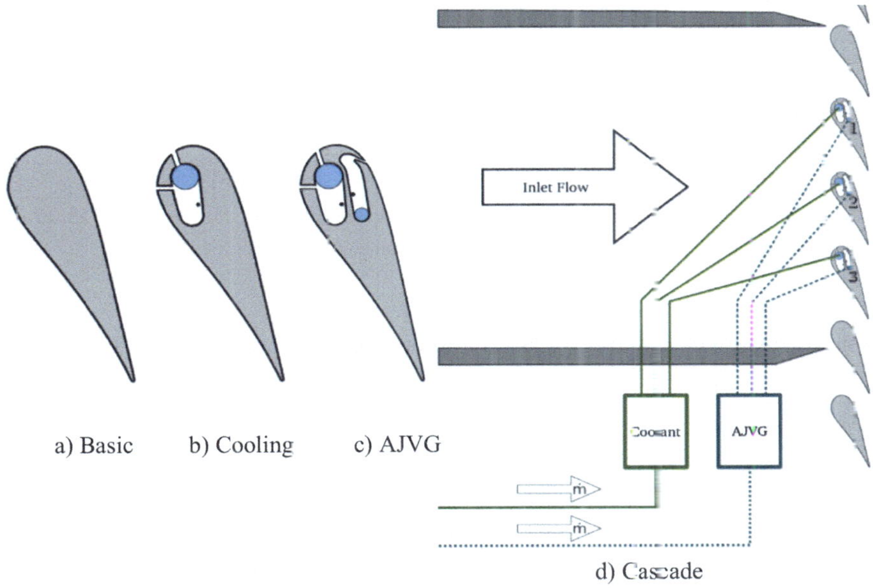

Fig. 33 TFAST profile and cascade

The basic and cooling configurations are shown in Fig. 33a, b. The mass flow supply area is highlighted in blue for each blade. A black dot in the cavity highlights the location of a pressure probe. It is located at the mid of the cavity and the measurement is used to regulate the cavity pressure.

The last blade type has air jet vortex generators (AJVG) additional to the coolant orifices (Fig. 33c). IMP PAN has designed the AJVG's. The design is based on findings in [13], [3] and [14].

Figure 33d shows a sketch of the test section and the TFAST cascade. Due to the small stagger angle and a long chord, five blades remain in full flow. The numbers designate the wake traverse measurements to the blades. Blade number ② in Fig. 33d is the measurement blade. Later explanations about the metrology will refer to this designation. The coolant supply delivers air through one pipe. It ends in the test section in a distribution plenum. A thermocouple is placed in the plenum, which gives the temperature of the coolant flow. The length of the pipes from the distribution plenum to the blade is approximately 2 m. Three blades are connected to the cooling supply (Fig. 33d). This includes the measurement blade and both blades adjacent to it. An extra pipe supplies the AJVG's. The supply set-up is the same as for the cooling cavity. The AJVG specifications are listed in Table 5.

The AJVG parameters are adjusted by the cavity pressure in the measurement blade (number ②). The coolant flow is fixed at $c_m = 1.1\%$. Table 6 shows the flow supply data for the AJVG cavity. The supply temperature has an average value of 296 K.

Table 5 AJVG configuration

Hole diameter [mm]	0.3	Number of holes in spanwise direction [–]	18
Theta angle [°]	35	Space between wall and AJVG holes [mm]	18.5 and 19.6
Alpha angle [°]	65	Area of AJVG cavity [mm^2]	77.89
Pitch [mm]	4.5	Cavity pressure [kPa]	80, 115, 130
Relative exit hole position (x/c_{ax}) [%]	54.4	Cavity temperature [K]	300
Length of AJVG hole [mm]	7.3	Inlet Pressure [kPa]	115
Number of holes in spanwise direction [–]	18	Number of holes in spanwise direction [–]	18

Table 6 AJVG flow configurations

c_m [%]	Cavity pressure [kPa]	Mass flow (one blade) [g/s]	Discharge coefficient [–]
0.08	80	0.25	0.71
0.13	115	0.41	0.83
0.15	130	0.45	0.83

The AJVG cavity pressure is chosen by the relation of the cavity pressure to static and total pressure of the flow at the AJVG exit. The lowest mass flow rate is intended to be right between the static and total pressure at the orifice outlet. The intermediate cavity pressure is set to be equal to the total pressure and the highest setting is chosen well beyond the total pressure.

Metrology

The approaches for Schlieren visualization, Wake traverse measurements, and mass flow are similar to the blades described in Sect. 2. The normalized geometric dimensions of the blade are shown in Fig. 16.

Infrared thermography

The AJVG blade has two cavities. Due to the higher geometrical complexity, the production tolerances with plastic material as applied at the other blade types are not reachable. Metal is used for the AJVG blade material. It has a high thermal conductivity but the production tolerances are realizable. To isolate the suction side a coating application is employed. For the coating, a 0.12 mm deep pocket on the blade suction side surface is inserted with plastic material. The pocket starts downstream of the AJVG holes and reaches the trailing edge of the blade. The borders are outlined with a dotted rectangle in Fig. 34. Additionally, an image of this blade is shown in Fig. 35. The evaluable region reaches from 59% to 96% relative axial chord. The infrared camera and the evaluation procedure are identical applied as to the film-cooling blade.

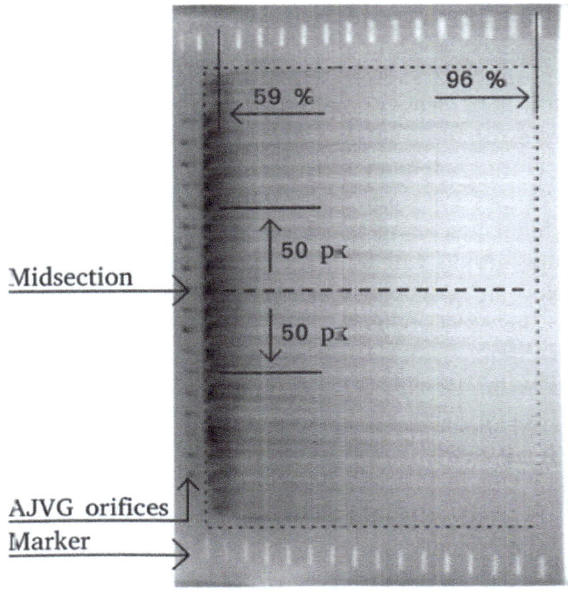

Fig. 34 AJVG Infrared image explanations

Fig. 35 Image of AJVG Infrared Blade

4.1.2 Results and Discussion

Flow field measurements

In Fig. 36, the Schlieren images are shown. Next to the results of the AJVG measurements, a reference Schlieren picture is depicted. It is taken from the cooling case at $c_m = 1.1\%$ (from Fig. 23a). The alignment of the knife-edges required for the Schlieren technique is shown in each picture. The upper row of Schlieren images shows results with a knife-edge parallel to the blade chord. The images of the row below are taken with a knife-edge adjusted across the blade chord.

The AJVG bores produce weak shocks upstream of the throat. In streamwise direction, a shock at the beginning and at the end of the orifice is visible in each flow case. Downstream, an oblique incidence shock of the upper blade appears. Its reflection does not look like an expansion wave. This indicates the non-existence of a separation bubble. However, the existence of a weak separation bubble is still possible. Other indicators for this theory are the absence of a compression wave upstream of the impingement point and a reattachment shock. There are slight changes of

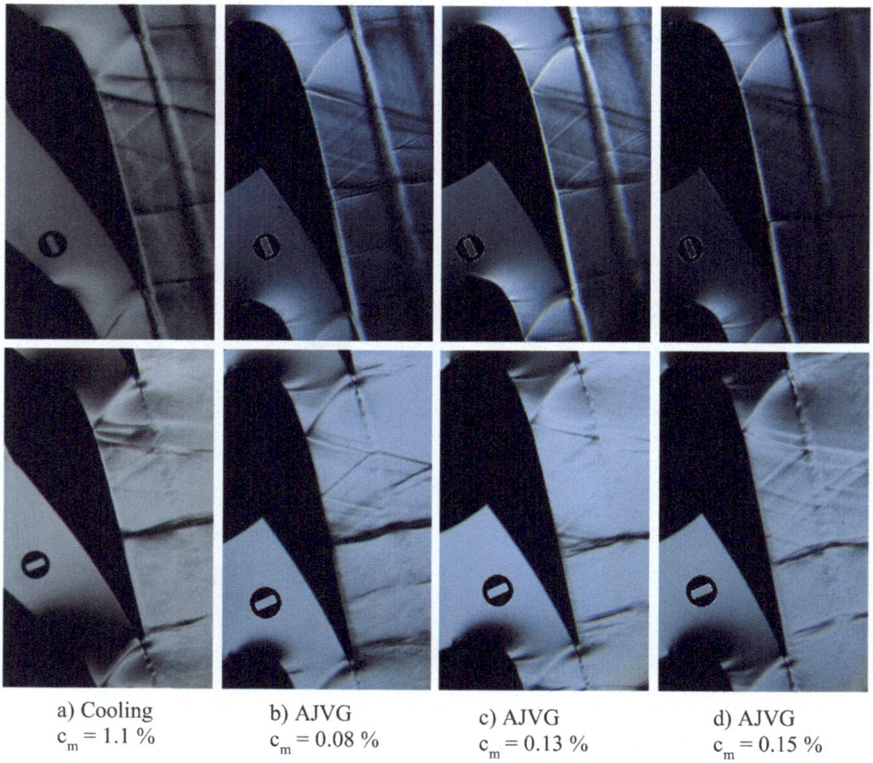

Fig. 36 Results of Schlieren technique

contrast level in the Schlieren images where the compression and reattachment wave would be expected. This indicates a weak appearance of those shocks. The reflection of the incidence shock produces downstream reflections itself until a strong normal shock stops them. This strong normal shock shows a highly unsteady behavior. At the trailing edge of the blade a second weaker normal shock appears.

Downstream of the AJVG bores, a thin turbulent boundary layer is visible especially at the pictures with the knife-edge being parallel to the blade chord. The boundary layer thickness rises through the incident shock downstream of the passage and again through the strong normal shock closer to the trailing edge. The effect of the AJVG in all cases is those of a turbulator because the boundary layer is turbulent downstream of the AJVG's. All visible differences of the strong, normal shock between the AJVG cases are caused by the temporal fluctuation. The general flow structure is not dependent from the coolant mass flow.

Compared to Fig. 36a the clearest difference is the diminishing of the separation bubble. The cooling blade shows the appearance of a compression shock upstream of the separation bubble and a reattachment shock. It does not show a turbulent boundary layer upstream of the separation. This indicates, downstream to the compression wave, a laminar boundary layer. Downstream the passage, similar effects are observed: Thickening of the boundary layer through shocks, appearance of a strong and unsteady normal shock and a second weaker shock at the trailing edge.

Figures 37 and 38 show the results of the probe measurements. The wake measurements are undertaken at the midsection with the wedge probe. Figure 37 shows a total pressure ratio (p_{02y}/p_{01}), a Mach number (Ma_{2y}) and a circumferential angle (α_{2y}). The abscissa refers to the trailing edges of the blades and the numbers are outlined in Fig. 15c. The reference case of the cooling blade at $c_{m,\ Cooling} = 1.1\%$ is depicted as well.

At the total pressure ratio, a clear difference at the wake between cooling and AJVG blade can be seen. The total pressure loss at the AJVG is higher. The wake profile at $c_m = 0.08\%$ has the broadest expansion.

At the Mach number distribution, a difference in the position of the strong normal shock at $(y - y_0)/t \approx 1.7$ is visible. This shock is located on the blade at $x/c_{ax} \approx 0.9$ (at the basic case). With increasing AJVG mass flow, the normal shock is shifted downstream. Except for the both higher mass flows, the amount of mass flow seems to be too small to relocate this shock. The reflected incidence shock of the cooling blade appears at $(y - y_0)/t \approx 2.3$. The reflection appears downstream at $(y - y_0)/t \approx 2.35$ for $c_m = 0.08\%$ and even more downstream for the higher AJVG mass flow ratios. Hence, the angle of the oblique shock is raised with increased AJVG mass flow. The mass flow injection of the AJVG's is located upstream of the incidence shock. Therefore, the mass flow increases the Mach number upstream of the incidence shock and influences its angle alignment. The results for the circumferential angle α_{2y} have an averaged amount of $\alpha_{2y} \approx 75^{\circ}$. The deviations are neglectable.

Figure 38 shows the wake traverses of the Mini probe. The traverse is located closer to the trailing edge. The velocity gradients are higher in comparison to the location of the Wedge probe measurement. The wake in the total pressure ratio graph is two times deeper and narrower. The wake shape does not differ significantly

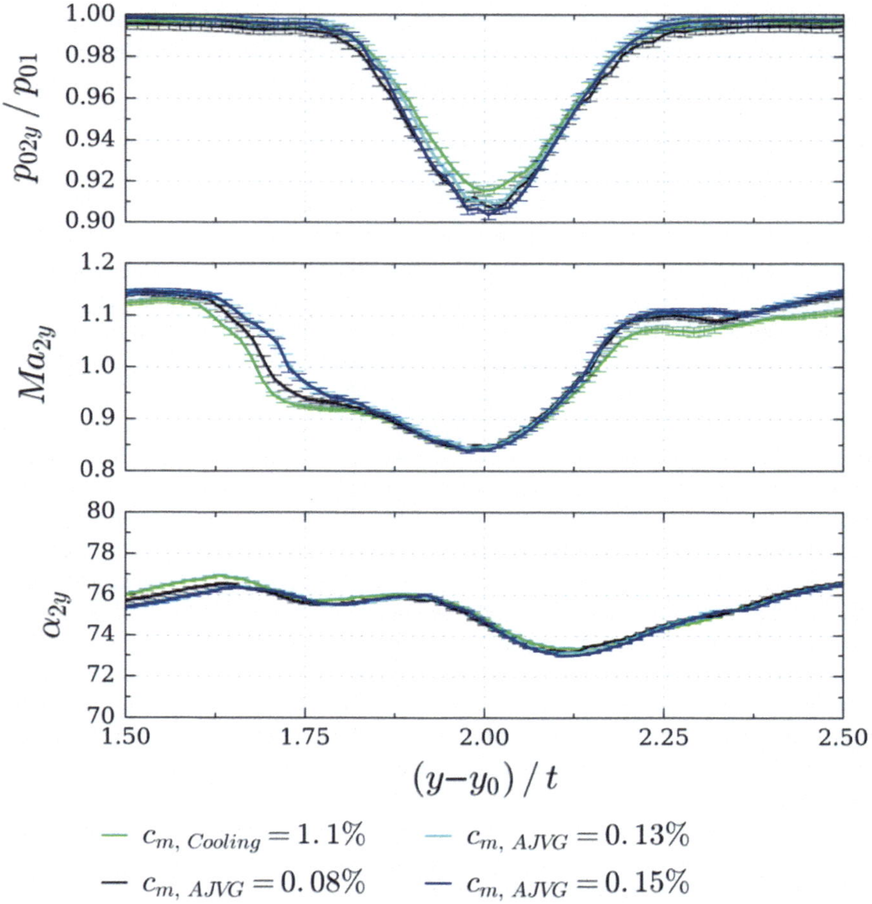

Fig. 37 Wake traverse, Wedge probe

between the AJVG cases. The strong normal shock appears at the cooling case further upstream as it is measured with the Wedge probe. The circumferential angle shows deflections at the edges of the wakes in various directions. This is an effect caused by the application of a probe. While one hole is already influenced by the wake, the other one is still in the passage flow. This gives a fictitious angle change at the wake edges.

The Mini probe is able to determine the radial angle β_{2y} as well. The probe shows a strong deflection in one direction at the wake. The deflection is caused by the installation process of the probe.

In Fig. 39 the homogeneous thermodynamic losses ξ_{th}, the homogeneous exit circumferential angle α_2 and the homogeneous exit Mach number Ma_2 are depicted.

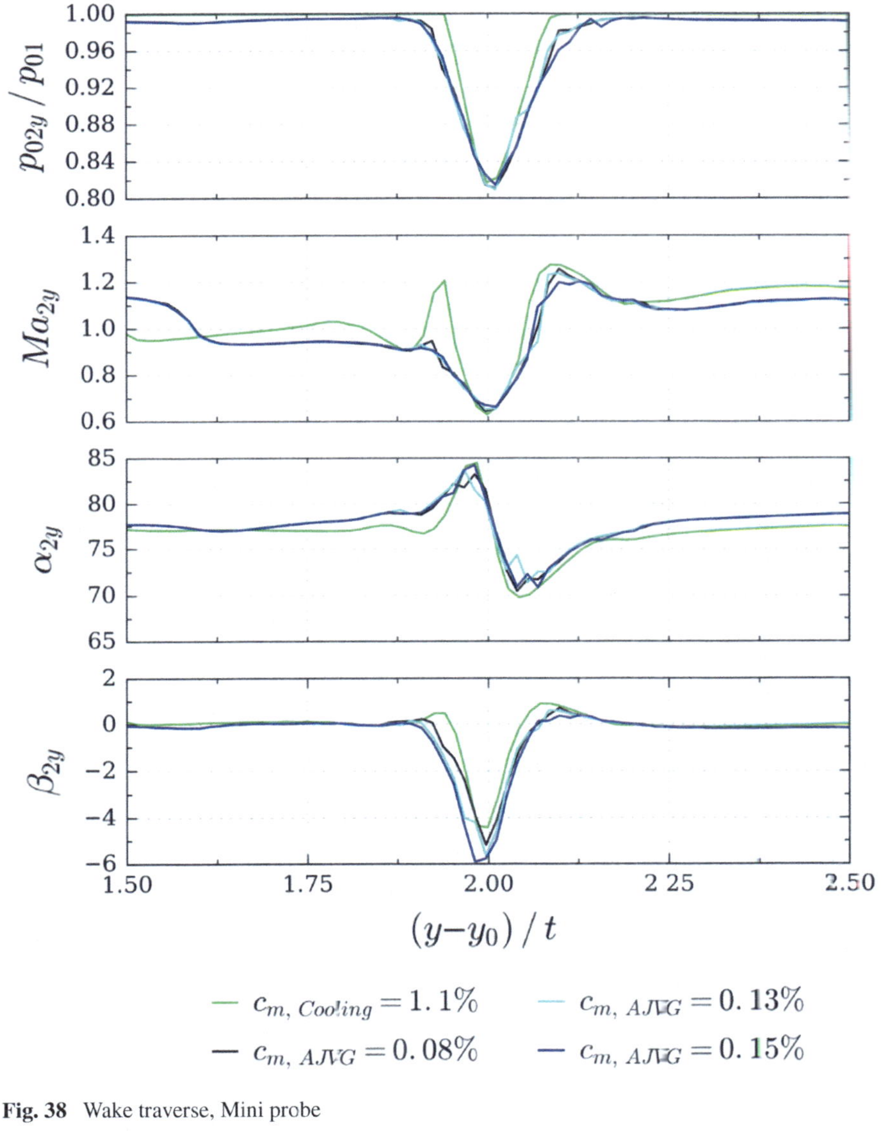

Fig. 38 Wake traverse, Mini probe

The abscissa shows the AJVG mass flow ratio. The results are referenced to the homogeneous results of the cooling blade at $c_{m,Cooling} = 1.1\%$.

The results of Mach number and circumferential angle do not show a significant difference to the reference case at all mass flow ratios.

There are significant variations of the thermodynamic loss. Those do not have a plausible development respectively to the AJVG mass flow. There is no plausible explanation, why the losses do not show a trend line. The errors of the PSI system

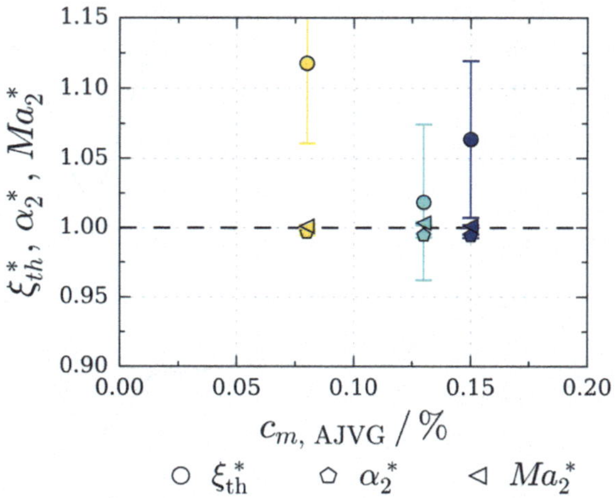

Fig. 39 Evaluated probe data

have, in contrast to the Mach number and the circumferential angle, a huge influence on the result. Hence, it cannot be excluded that the variations of the thermodynamic losses ξ_{th} are within the error fluctuations of the PSI system.

Nevertheless, it can be claimed that the vortex generators in comparison to the cooling case increase the loss coefficient.

Infrared technique

This chapter shows results of the wall temperature distribution measured by an infrared camera. The flow in Fig. 40 goes from left to right. Explanations are outlined in Fig. 34. The visibility of the AJVG flow increases with the AJVG mass flow. The images in Fig. 40 do not show a ratio of two different images. Hence, information about transition is not visible. The division of two infrared images causes that the static temperature will be removed. Hence, the static temperature is still included in Fig. 40. Shocks are visible. The reattachment is visible at $x/c_{ax} \approx 0.65$ as well as the strong normal shock at $x/c_{ax} \approx 0.85$ by an increase of the temperature.

Downstream the AJVG holes on the coating, small triangles are visible. Here the flow is accelerated locally. This indicates the decreased wall temperature. The triangles with higher temperature between the AJVG holes are the regions where the AJVG mass flow is still not affecting the wall temperature distribution.

The temperature gradients are depicted in Fig. 41. Downstream of the vortex generators the wakes of the AJVG flow are visible. Those are stronger at the two higher mass flow rates. An increase in temperature appears at the normal shock ($x/c_{ax} \approx 0.9$). The images are averaged in spanwise direction and shown in Fig. 42 next to the cooling case.

The transition is triggered by the incidence shock in the cooling case. The boundary layer is turbulent downstream of $x/c_{ax} \approx 0.65$. The step of rise in heat transfer is not visible in the AJVG results. This is a confirmation of the analysis of the Schlieren

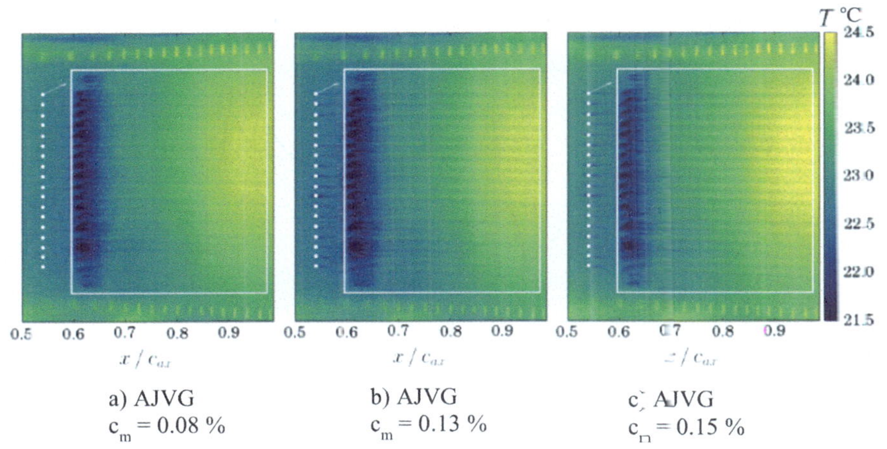

Fig. 40 Temperature distribution at 300 K inlet temperature

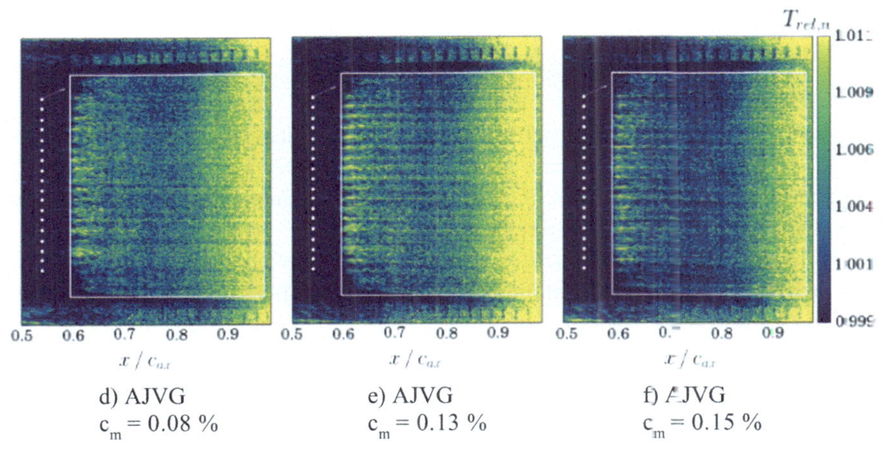

Fig. 41 Temperature distribution at 300 K inlet temperature

technique. It is noticed that the boundary layer turns turbulent through the vortex generators. The infrared results confirm this outcome. The thickening of the boundary layer causes the rise in temperature gradient at $x/c_{ax} \approx 0.9$.

Summary and conclusions

At transonic flow conditions, experimental research is carried out at high-pressure turbine stator blades in a straight cascade. The aerodynamic behavior is investigated with emphasis on the detection of the transition location. The main investigation assessment for a turbine is the loss coefficient. In this work package, a special attention is paid on these criteria. A cascade set-up with air jet vortex generators (AJVG)

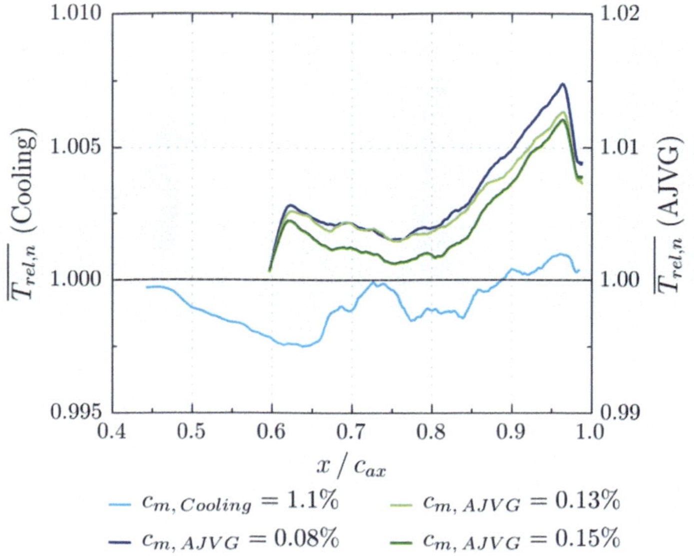

Fig. 42 Temperature distribution

is applied. The experiments are performed at the Wind Tunnel for Straight Cascades (EGG) at the German Aerospace Centre (DLR), Göttingen. The profile geometry is designed by Rolls-Royce Deutschland (RRD) to sustain high loads and to stabilize the boundary layer upstream of a region of separation.

Information about the flow field is obtained by the application of Schlieren technique and wake traverse measurements. An infrared camera detects the transition location. All measurements are performed at a zero incidence angle and at an exit Mach number $Ma_2 = 1.05$. The exit Reynolds number is $Re_2 \approx 1.3 \times 10^6$.

Three blades are equipped with coolant and AJVG cavities. Each coolant cavity has two coolant rows. One exits at the stagnation point and the second one in the acceleration region on the suction side. The coolant mass flow ratio is set to $c_m = 1.1\%$. The design of the AJVG's has been undertaken by IMP PAN. The AJVG mass flow ratio will be varied between 0.08%, 0.13%, and 0.15%.

Subsequently the specific research findings of this investigation are summarized:

Shock structure

The AJVG holes produce weak shocks. An oblique incident shock from the trailing edge of the upper blade reflects at the suction side of the blade but an expansion wave is not clearly visible. These shocks reflect downstream between wake and surface until a strong, unsteady, normal shock stops them. A second weaker normal shock at the trailing edge can be seen. There are no visible differences between the three mass flow ratios in the shock structure.

Separation
The separation is clearly diminished by the AJVG's. The Schlieren images show only weak indications for the presence of a separation bubble.

Transition location
The boundary layer turns turbulent at the location of the AJVG orifice in every case. Hence, their function is those of a turbulator. The boundary layer thickens downstream of every occurring shock.

Turbine loss coefficient
The AJVG produce higher losses compared to the cooling case and extra shocks at its orifices. It's amount overweight's probably the profits by diminishing the separation bubble.

4.2 NUMECA

The commercial package Fine™/Turbo is used for this study. The flow solver is a three-dimensional, density-based, structured, multi-block Navier-Stokes code using a finite volume method. Central-difference space discretization is employed for the spatial discretization with Jameson type artificial dissipation. A four-stage explicit Runge-Kutta scheme is applied for the temporal discretization. Multi-grid method, local time-stepping and implicit residual smoothing are used in order to speed-up the convergence. Both the SST and the S-BSL-EARSM turbulence model has been retained here together with the local correlation transition model (LCTM).

First simulations on the turbine cascade have been performed without the two lines of holes. Both SST and S-BSL-EARSM turbulence models are used without and with laminar-to-turbulent transition activated. The isentropic Mach number distribution along the blade surface obtained from those simulations at mid span is displayed on Fig. 43. It appears that most of the differences are obtained within the interaction region. Differences are more pronounced when going from a fully turbulent model to a transition module—whatever turbulence model is used. Indeed, for fully turbulent simulations the flow remains attached, whereas flow separation occurs when transition module is activated. This corresponds to a smoother pressure gradient in the interaction region. Furthermore, when looking more deeply at the recirculation region, S-BSL-EARSM and SST produce rather different results. Indeed, the recirculation zone is more bi-dimensional with the simplest turbulence model.

The comparison with experimental data reveals that all the models are able to reproduce the pressure distribution along the blade upstream of the interaction region. The accordance is less satisfactory downstream of the interaction. Experiment shows a plateau at an isentropic Mach number of 1.2, whereas it is only 1.15 from CFD results. One important finding is that none of those models is able to produce the second shock that appears in the experiment at about 85% stream-wise position

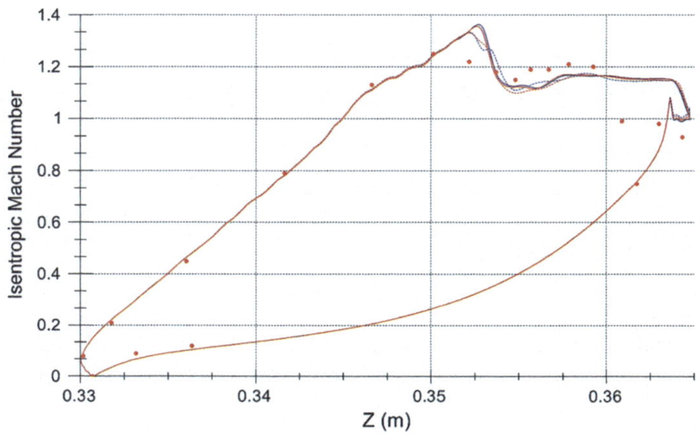

Fig. 43 Isentropic Mach number distribution along the blade at mid span obtained from SST turbulence model (red lines) and from S-BSL-EARSM turbulence model (blue lines) in fully turbulent mode (continuous lines) and with transition model (dashed lines). Red dots correspond to experimental data

on the suction side of the blade. Even if a large number of simulations (including sensitivity to geometrical uncertainties) have been performed, the exact reason for this discrepancy remains unknown and its potential impact on the interaction is also unknown.

Two series of wake traverse data are available from the experimental data for the basic configuration. Mini probe and wedge probe are located 4 mm and 18 mm downstream of the blade trailing edge, respectively. It should be mentioned that, because of the rather close location to the blade surface for the 1.4 mm wide mini probe, the uncertainties associated to this probe is rather important. Comparison between measured and computed total pressure ratio, Mach number and flow angle for those two sections are provided on Fig. 44.

From this figure it appears that the flow solver produces a rather good prediction of the total pressure ratio just downstream of the blade. Indeed, for the clean blade, the minimum total pressure within the wake, 4 mm downstream of the blade trailing edge, is 0.85 from both experiment and CFD results. Further downstream, the flow solver produces a slightly too thick wake and overestimate the minimum pressure ratio. The computed flow angle is over-estimated by about 2° when compared to the measured one. Furthermore, one can clearly notice that in experimental and numerical results, the normal shock that appears over the last part of the blade suction side is not at the same location—further upstream for the experiment. This has a direct impact on flow angle and Mach number predictions downstream of the blade. One should also notice that the flow angle measured by the mini probe does not reflect any shock pattern downstream for both measured passages but produce large variations of flow direction within the wake (that can also be noticed from experimental data of radial flow angle).

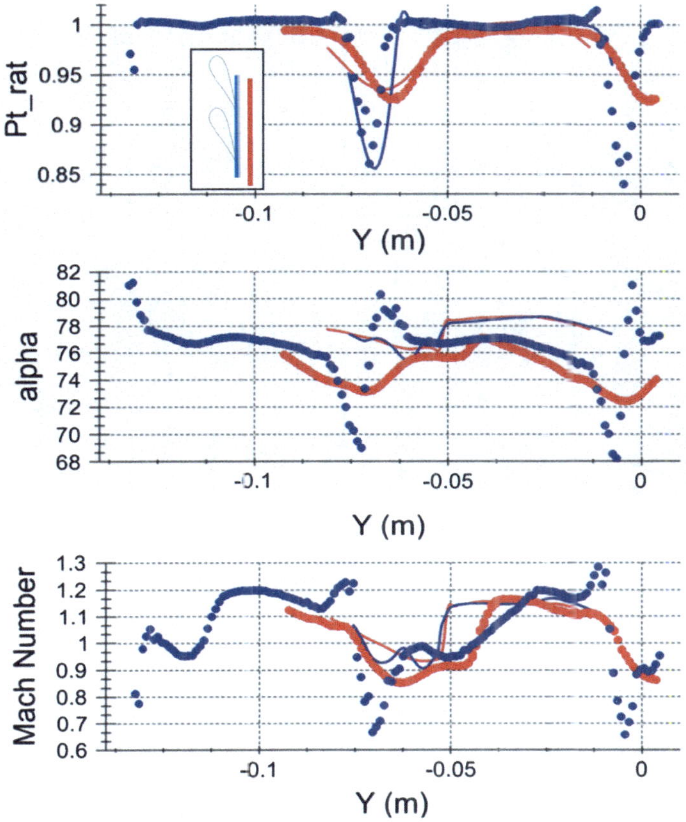

Fig. 44 Experimental (symbols) and computed S-BSL-EARSM with LCTM (lines) total pressure ratio, flow angle [deg], and Mach number downstream of the blade trailing edge from mini probe (blue) and wedge probe (red) at mid-span for basic configuration

Cooling holes are considered as source terms along the blade surface. The cooling flow rate and its total temperature have been provided by DLR Göttingen—only the configuration with $P_{cav}/P_{01} = 1.05$ has been considered hereafter. Simulations on the cooled blade have been conducted with S-BSL-EARSM turbulence model with taking into account laminar to turbulence transition.

The flow rate through those two rows is imposed in order to match inputs provided by DLR. It is 4.44 g/s for both lines of holes. The flow rate through each line of holes has been distributed in order to reach a hole outlet total pressure close to the measured one, i.e., 120.7 kPa. Therefore, the flow rates through line 1 and line 2 are 1 g/s and 3.4 g/s, respectively. The cooling hole total temperature is set to 300 K. This simulation corresponds to the experiment with a mass flow coefficient of 1.1%—denoted $c_m = 1.1\%$.

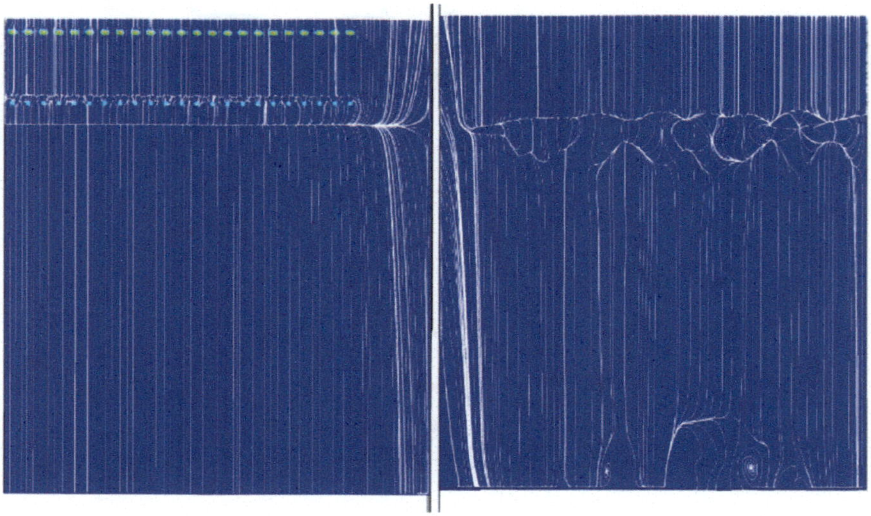

Fig. 45 Limit streamlines along the turbine blade obtained from S-BSL-EARSM turbulence model with transition for the cooling configuration. View from upstream on the left and from downstream on the right

The computed limit-streamline along the blade surface are displayed on Fig. 45. It appears that the presence of the first row of cooling hole has an impact on the flow structure around the leading edge. Indeed, the injection direction is opposite to the flow direction. This induces a strong tri-dimensional recirculation pattern in this region. Line 2 being further downstream has much less impact on the limit-streamlines.

Downstream wake traverse data are also available from the experimental device. Comparison between measured and computed total pressure ratio, flow angle, and Mach number for the two measured downstream sections are provided on Fig. 46. It appears that the experiment show a significant discrepancy between the cooled and the un-cooled configuration. The former leads to a reduction of the observed total pressure ratio within the wake for both probes. Experimental data already reveals significant differences in the wake traverse measurements between the clean blade configuration and the one with cooling holes but without cooling flow. Numerical results show only minor differences between the two configurations with and without cooling. Therefore, cooling plays a minor role in term of downstream wake. This is also, somewhat, coherent with experimental outcomes, since adding flow rate through cooling hole does not drastically modify the downstream wake profiles. Therefore, Numerical results produce a rather important over-estimation of the minimum total pressure with the blade wake for both traverses. As already mentioned for the clean blade, the different of location of the normal shock over the last portion of the blade suction side leads to rather important discrepancies in term of Mach number and flow angle distribution downstream of the blade.

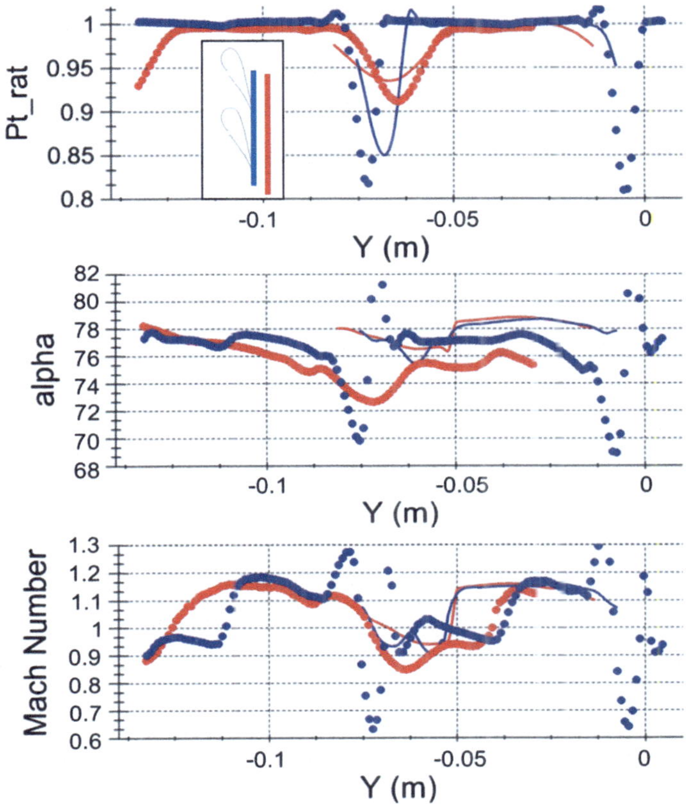

Fig. 46 Figure B4. Experimental (symbols) and computed S-BSL-EARSM with LCTM (lines) total pressure ratio, flow angle [deg], and Mach number downstream of the blade trailing edge from mini probe (blue) and wedge probe (red) at mid-span for cooling configuration

In order to identify the exact role of cooling holes on the laminar-to-turbulent transition, a deeper analysis and comparison between the basic and the cooling configurations is performed hereafter. Since most of the experimental data are provided at mid-span, the first comparison will be performed along the symmetry plane. Figure 47 is a composite graph showing the numerical Schlieren and the intermitency at mid-span for both configurations. From this figure it appears that the two lines of cooling holes induce a thicker laminar boundary layer upstream of the interaction region. The location of the start of the separation region is identical for both simulations but the cooled configuration presents a larger recirculation bubble (9 mm instead of 6 mm). The impact of the cooling rows on the recirculation region has already been highlighted by hot-film measurements. This has an impact on the location of the rear shock that is associated to the reattachment of the boundary layer as can be seen from Fig. 47. For both simulations, the boundary layer is not fully turbulent within

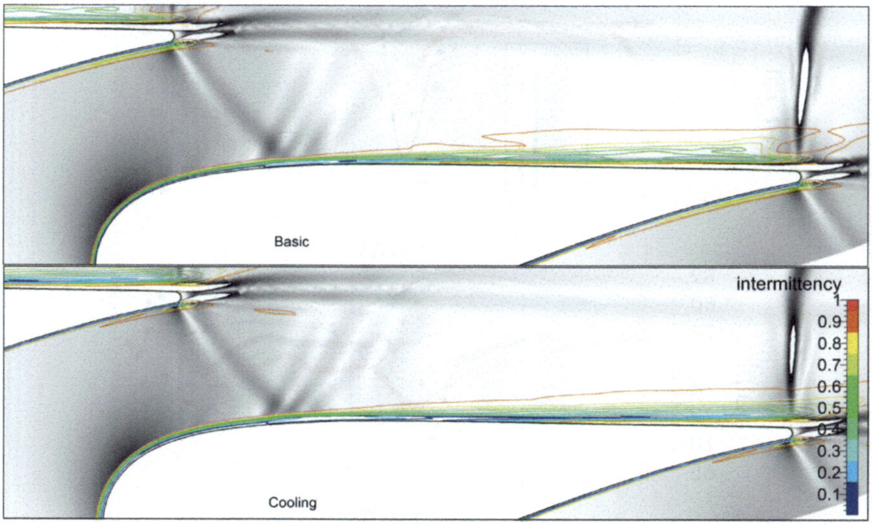

Fig. 47 Numerical Schlieren obtained by S-BSL-EARSM with transition, together with iso-lines of intermittency at mid-span for cooled and basic configurations

the recirculation zone. Indeed intermittency is about 30% in this region. Laminar-to-turbulent transition appears downstream of the recirculation zone and seems to be delayed for the cooled configuration.

However, such a pattern is only valid at mid-span. The recirculation region is not uniform all along the span. Zone of larger recirculation can be identified together with span-wise portion with almost no recirculation—mainly close to the end-walls. Since transition onset appears downstream of recirculation region, one can infer a significant span-wise variation in intermittency distribution.

In order to investigate the effect of AJVGs on the control of laminar to turbulent transition, simulations have been performed for the three different AJVG mass flow coefficients as summarised in Table 7. For the experimental location, each one corresponds to a different experimental AJVG cavity total pressure, 80, 115, and 130 kPa, from low to high mass flow coefficient. With respect to the inlet total pressure, this corresponds to a total pressure ratio of 0.7, 1, and 1.13, respectively. Mass flow for each row of AJVG has also been provided by the DLR Göttingen and will be used as an input for the cooling module of the flow solver. Produced numerical results

Table 7 Test matrix

AVJG mass flow coefficient			
Axial location	0.08%	0.13%	0.15%
55% (experiment)	•	•	•
50%	•	•	•
60%	•	•	•

will be compared to available experimental data. Furthermore, a sensitivity study has been carried on. Based on the outcomes of Work-Package 2, it appears that the efficiency of AJVG to promote laminar-to-turbulent transition is dependant of the control device location upstream of the shock-boundary layer interaction. A certain delay exists between the location of the AJVG and the onset of transition. This has also been observed by Herbst et al. [15]. Therefore, the sensitivity study is mainly based on the location of the AJVG. The test matrix is summarised in Table 7.

In order to investigate the effect of the AJVG, those simulations will be compared to already available simulations without AJVGs that have been presented before.

First simulations have been performed with the experimental set-up of the AJVGs. They correspond to an AJVG axial location of about 55% with three different cavity total pressures, i.e., three different flow rates through AJVGs. The simulations have all been conducted with S-BSL-EARSM turbulence model together with LCTM. They have all been converged to reach a mass flow error lower than 0.02%.

From those simulations it appears that the presence of the AJVGs does not significantly modify the shape and extend of the recirculation region downstream of the incident oblique shock. Indeed, none of those simulations succeed in promoting laminar-to-turbulent transition upstream of the interaction zone. Therefore, the interaction is still laminar and produces a laminar separation bubble.

The experimental data provide measurement of the entropy losses along the wedge probe at mid span. Following Petersen (2015) and in accordance with [16] entropy losses ε_s are computed following the equation:

$$\xi_s = 1 - \frac{1 - \left(\frac{p_2}{p_{t2}}\right)^{\frac{\gamma-1}{\gamma}}}{1 - \left(\frac{p_2}{p_{t1}}\right)^{\frac{\gamma-1}{\gamma}}},$$

where p stands for pressure, t denotes total quantity. *1* corresponds to the inlet plane which total pressure is fixed to 115 kPa, and *2* corresponds to the wedge plane or traverse. From our numerical results, entropy losses are computed in two different ways. On the one hand, the averaging process is done along the wedge traverse, i.e. only at mid-span. On the second hand, it is done along the full plane corresponding to wedge axial coordinate. Because of end-walls the second method produces slightly higher entropy losses. Static pressure comes from a surface averaging, whereas total pressure is flux averaged. Such an averaging process is consistent with best practices in turbomachinery industry. Differences in measured and computed entropy losses can also be explained by the presence of a normal shock in the experiment at approximately 85% axial position. Such a shock should increase the losses.

Distribution of entropy losses as a function of the AJVG mass flow coefficient is displayed on Fig. 48. The blue symbols correspond to the computation at mid-span only, whereas the red symbols correspond to computation of the losses along the surface located at the wedge probe axial position. It appears that the increase in mass flow through AJVG does not directly induce an increase in entropy losses. The maximum loss is computed for the intermediate mass flow coefficient and further

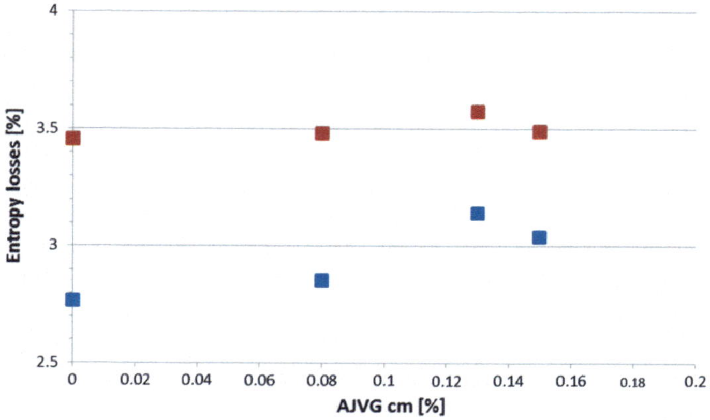

Fig. 48 Computed entropy losses 18 mm downstream of the blade trailing edge as a function of AJVG mass flow coefficient at mid-span (blue symbols) and over the full surface (red symbols)

increasing the flow rate leads to a decrease in losses. This finding is consistent with available experimental data. Furthermore, it reflects the observed deeper downstream wake produced at this intermediate flow coefficient. The difference between the maximum entropy loss at intermediate flow rate and the one for the simulation without AJVG is about 0.4%. It is 0.6% from the preliminary experimental data. Averaging over the surface instead of along the mid-span line does not drastically modify our findings. Because of the presence of secondary losses along the end-walls, differences between simulations are reduced but the same trend can be observed with a maximum loss at intermediate flow rate.

As already mentioned, a sensitivity study to AJVG definition has also been performed. For such purely numerical experiments no measured data are available. As for the previous simulations, they have all been conducted with S-BSL-EARSM turbulence model together with LCTM. It should be mentioned that the mass flow coefficient and the flow angle with respect to the blade surface have been maintained identical to the reference simulations.

The entropy losses computed for all the simulations are summarised on Fig. 49. On the left panel only the simulations without AJVG are reported. They summarise the impact of the two rows of cooling holes and the effect of having a fully turbulent or a laminar boundary layer upstream of the interaction. From this part of the figure, one can notice that using a transition model leads to a drastic decrease in entropy losses, and consequently an increase in efficiency. The difference is about 0.5% and is almost identical with and without the two rows of cooling holes. Furthermore, the effect of the two rows of holes on the losses is relatively minor. By looking only at mid-span traverse, one can conclude that it induces a tiny decrease in losses for the transitional simulation. This is consistent with the experimental findings that reveal a decrease in local losses. However, by looking at the losses based on surface averaging, it appears that the two rows of cooling induce a tiny increase in global losses of 0.1%.

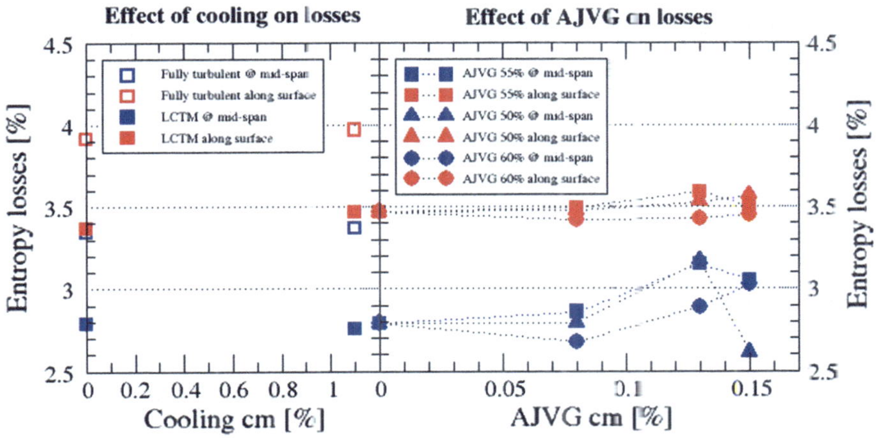

Fig. 49 Computed entropy losses 18 mm downstream of the blade trailing edge as a function of cooling mass flow coefficient (left panel), and AJVG mass flow coefficient and location (right panel). Blue symbols are computed at mid-span, whereas red ones are averaged over the downstream surface

The effect of AJVG mass flow for the experimental set-up has already been described above. By moving the AJVG location upstream or downstream one can observe variations in the computed losses. At the lower mass flow coefficient, the differences are rather small and the computed losses are almost similar to the one coming from the simulation without AJVGs. The maximum difference is only 0.2% on the losses computed at mid-span. Progressively increasing the flow rate leads to higher differences in losses at mid-span. This reflects the tri-dimensional nature of the flow. Indeed, wake of the cooling holes and the AJVGs interacts together. They produce a rather complex structure of vortices at the shock-boundary layer interaction and further downstream. Therefore, the losses computed at mid-span drastically depend on this complex inter-play. The effects on the losses averaged over the downstream surface are much less dependent on the AJVG mass flow coefficient.

For the reference and upstream locations, none of the simulations is able to modify the shock boundary layer interaction. Upstream of the shock the boundary layer remains laminar. Shock boundary layer interaction produces a series of recirculation along the blade span. Turbulence starts to growth at the separation onset. However, two of those simulations produce a distinct scenario. Indeed, at the two highest mass flow coefficients and at the most downstream AJVG location, the shock boundary layer interaction produces an almost continuous separation in the span-wise direction. The maximum extension of the recirculation zone has been drastically reduced and the maximum thickness of the separation zone has also been reduced—from 0.3 mm to 0.1 mm. It appears that coherent turbulent spots are produced downstream of the recirculation zone. Therefore, the intensity of the shock boundary layer interaction has been reduced so that a partially laminar boundary layer is present downstream of the shock. This has a direct impact on the viscous forces acting on the blade surface and potentially on the heat exchange and the life-time of such a high pressure turbine blade.

References

1. A. Jameson, W. Schmidt, D. Gmbh, E. Turkel, T. Aviv, Numerical solution of the euler equations by finite volume methods using Runge-Kutta time-stepping schemes, in *14th Fluid and Plasma Dynamics Conference* (1981)
2. F. R. Menter, A. V. Garbaruk, Y. Egorov, Explicit algebraic reynolds stress models for anisotropic wall-bounded flows. Prog. Flight Phys., 89–104 (2012)
3. P. Flaszynski, P. Doerffer, R. Szwaba, Numerical simulations of transonic flow with film cooling and jet vortex generators, in *9th European Conference on Turbomachinery Fluid Dynamics and Thermodynamics. ETC 2011—Conference Proceedings*, vol. 2, pp. 1615–1625 (2011)
4. K. Jung, Mehrreihige Filmkühlung an gekrümmten Oberflächen. PhD, TU Darmstadt, Darmstadt (2001)
5. A. Savitzky, M.J.E. Golay, Smoothing and differentiation of data by simplified least squares procedures. Anal. Chem. **36**(8), 1627–1639 (1964)
6. F. Kost, P. Giess, Experimental turbine research at DLR Goettingen. J. Gas Turbine Soc. Jpn. **32**(6) (2004)
7. F. Kost, The behaviour of probes in transonic flow fields of turbomachinery, in *8th European Conference on Turbomachinery Fluid Dynamics and Thermodynamics. ETC 2009—Conference Proceedings* (2009)
8. J. Amecke, Auswertung von Nachlaufmessungen an ebenen Schaufelgittern. AVA (1967)
9. M. Pianko, F. Wazelt, Propulsion and energetics panel working group 14 on suitable averaging techniques in non-uniform internal flows (1983)
10. R. Kiock, H. Hoheisel, H. J. Dietrichs, A.T. Holmes, The boundary layer behaviour of an advanced gas turbine rotor blade under the influence of simulated film cooling, in *AGARD-CP 390, Heat Transfer and Cooling in Gas Turbines* (1985)
11. H. Babinsky, J.K. Harvey (ed.), *Shock wave-boundary-layer interactions* (Cambridge University Press, Cambridge, UK, 2011)
12. R.J. Howell, Boundary layer development in the BR710 and BR715 LP turbines—the implementation of high-lift and ultra-high-lift concepts. J. Turbomach. **124**(3), 385–392 (2002)
13. R. Szwaba, Influence of air-jet vortex generator diameter on separation region. J. Therm. Sci. **22**(4), 294–303 (2013)
14. R. Szwaba, P. Doerffer, Separation control by air-jet vortex generators on turbine blades, in *Proceedings of the 9th European Turbomachinery Conference* (2011)
15. F. Herbst, A. Fiala, J. Seume, Modeling vortex generating jet-induced transition in low-pressure turbines, J. Turbomach. **136**(7), 071005 (2014)
16. S. Dykas, W. Wróblewski, D. Machalica, Numerical analysis of the losses in unsteady flow through turbine stage. Open J. Fluid Dyn. **3**(4), 252–260 (2013)

WP-5 External Flows—Wing

Flavien Billard, Todd Davidson, Holger Babinsky, Robert Placek, Marek Miller, Paweł Ruchała, Wit Stryczniewicz, Tomasz Kwiatkowski, Wieńczysław Stalewski, Janusz Sznajder, Sara Kuprianowicz, Matteo Bernardini, Sergio Pirozzoli, George Barakos, George Zografakis, Benoit Tartinville, Charles Hirsch, Damien Szubert, Marianna Braza, Ioannis Asproulias, Nikos Simiriotis, Jean-Baptiste Tô, and Yannick Hoarau

Abstract Study of transition location effect (from natural transition to fully turbulent) on separation size, shock structure and unsteadiness was the focus of this WP. Boundary layer tripping (by wire or roughness) and flow control devices (VG) were used for boundary layer transition induction. Although this type of flow field had been studied widely in the past, there remains considerable uncertainty on the effects of transition on transonic aerofoil performance. In particular it is not known how close to the shock location transition has to occur to avoid detrimental effects associated with laminar shock-induced separation. Furthermore, it was unclear how best to provoke transition on an airfoil featuring significant laminar flow and how close to

---

F. Billard (✉)
Aircraft Engineering and Loads Division, Dassault-Aviation, 78, quai Marcel Dassault, 92512 Saint-Cloud, France
e-mail: flavien.Billard@dassault-aviation.com

T. Davidson · H. Babinsky
Department of Engineering, University of Cambridge, Trumpington Street, Cambridge CB2 1PZ, UK

R. Placek · M. Miller · P. Ruchała · W. Stryczniewicz · T. Kwiatkowski · W. Stalewski · J. Sznajder · S. Kuprianowicz
Aerodynamic Department, Institute of Aviation, Al. Krakowska 110/114, Warsaw, Poland

M. Bernardini · S. Pirozzoli
Dept. of Mechanical and Aerospace Engineering, Sapienza University of Rome, Via Eudossiana 18, 00184 Roma, Italy

G. Barakos · G. Zografakis
School of Engineering, University of Glasgow, University Avenue, Glasgow G128QQ, UK

B. Tartinville · C. Hirsch
NUMECA, Chaussée de la Hulpe 187-189, 1170 Bruxelles, Belgium

D. Szubert · M. Braza · I. Asproulias · N. Simiriotis · J.-B. Tô
UMR CNRS/INPT N° 5502, Institut de Mécanique des Fluides de Toulouse (IMFT), Av. du prof Camille Soula, 31400 Toulouse, France

Y. Hoarau
Département de mécanique ICUBE Laboratory, Université de Strasbourg, 2 Rue Boussingault, 67000 Strasbourg, France

P. Doerffer et al. (eds.), *Transition Location Effect on Shock Wave Boundary Layer Interaction*, Notes on Numerical Fluid Mechanics and Multidisciplinary Design 144,
https://doi.org/10.1007/978-3-030-47461-4_6

the shock this needs to be performed. Finally, current CFD methods are particularly challenged by such transitional flows. In this work package some of the findings from the basic research performed in other WPs was applied. Specialized large-scale transonic wind tunnels running cost is very high therefore using such facilities is not appropriate for upstream research programs such as TFAST. Therefore we have used existing wind tunnels within our consortium. One of these is a transonic test section at UCAM where laminar and transitional profiles were studied previously at Reynolds numbers up to 2 million (based on chord length). This wind tunnel allowed basic investigations of the transition location effects on a shock induced separation and unsteadiness for a relatively large number of parameters. A larger wind tunnel at Institute of Aviation in Warsaw was used, which enabled the investigation of a much larger aspect ratio profile. In this facility it was possible to measure a whole force polar up to and including the buffet boundary. The research was carried out for the natural b/l transition location as well as different methods of tripping.

1 Transonic Flow Over Laminar Wing

1.1 DAAV

1.1.1 Introduction

WP5 is devoted to the study of transition location effect on separation size, shock structure and unsteadiness. In particular it should address how close to the shock location transition has to occur to avoid detrimental effects, if any, associated with laminar shock-induced separation, and what is the best way to trigger the transition.

To initiate the WP5 work, DAAV has designed a 2D laminar airfoil for the transonic tests in the UCAM and IoA wind tunnels. The main objective in this design is to guarantee a laminar region from the leading edge to the shock location. By lack of information and calibration of the consortium wind tunnel facilities, the N factor limit is unknown and had to be chosen a priori. The value of 3 was retained. (In flight, the limit is 8 for Tollmien-Schlichting instabilities). Roughly, the maximum value of the N-factor, depends on the turbulence rate following the Mack law:

$$N = -8.43 - 2.4 \ln Tu$$

The evolution of the N-factor limit is plotted in Fig. 1

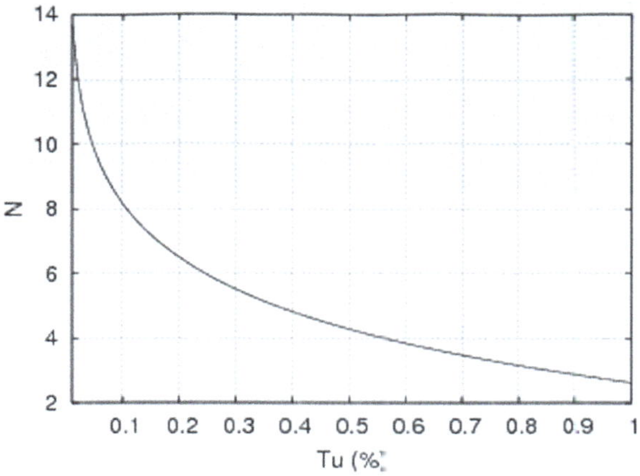

Fig. 1 Evolution of the limit N-factor with the turbulence rate

1.1.2 Aerodynamic Design

Geometric and Flow Parameters

Given the foreseen wind tunnels where the profile is to be experimented, a 250 mm chord has been chosen. Computations were carried out at Pi = 1 bar, Ti = 290 K, at 0.7 and 0.75 Mach numbers, so that the Reynolds numbers based on the chord length are in a 3.24e6–3.37e6 range. The angle-of-attack for which the laminar zone should go up to the shocks covers a 1°–7° range in order to assess the behavior of the laminar profile close to the buffet.

Description of the CFD Codes

The design of the laminar airfoil relies on the use of DAAV in-house Navier-Stokes solver Aether. It is a compressible, finite-element code based on the GLS formulation of the symmetric form of the Navier-Stokes equations using entropic variables. See Chalot et al. [1] for more details. Different turbulence models are available from 1-equation RANS model to DRSM or LES. All the presented results are obtained using the 2-layer k-ε model.

The laminar-to-turbulent transition prediction is based on the assessment of the amplification N-factor, computed with a parabola method (e^N stability data base), see Arnal [2] which is implemented in a 3D boundary layer compressible code. Such a method has been used by DAAV with success for many in flight laminarity tests (Courty, Bulgubure, and Arnal, 24th–27th May 1993 [3]).

Geometry

The final geometry delivered to the partners involved in WP5 is called V2C and is plotted in Fig. 2.

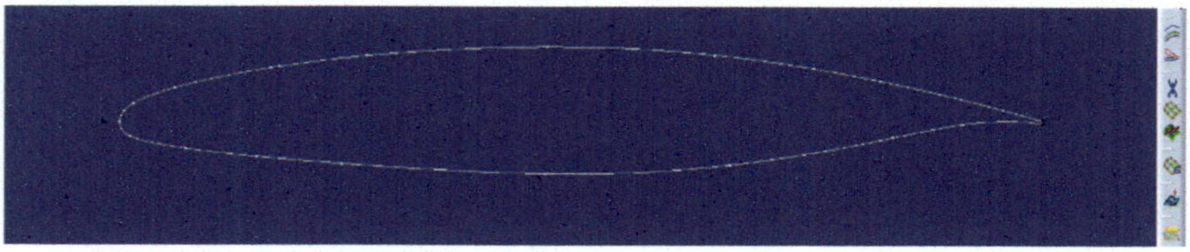

Fig. 2 Sketch of the laminar V2C profile

1.1.3 Computations

Mach Number 0.7

The pressure and friction coefficients for different angle-of-attack at M = 0.7 are shown in Fig. 3 for fully turbulent computations. A large angle-of-attack range is covered, from 1° to 7°. At $\alpha = 6°$ and $\alpha = 7°$, the Cp plateau is significant, and the flow separates. From the computed turbulent Navier-Stokes field, the transition is assessed by computing the instability amplification factor N. Figure 4 displays the evolution of the Mach number at the edge of the boundary layer and the N factor along the chord on the upper (solid lines) and the lower side (dashed lines). The amplification rate never exceeds the threshold value of 3 up to the shock ensuring the laminar region from the leading edge to the shock. At last, the Figs. 5 and 6 compare the fully turbulent field with the one computed with a fixed transition following the previous results, i.e. x/c = 30%.

Mach Number 0.75

The same methodology is used at M = 0.75. The pressure and friction coefficients for different angle-of-attack at M = 0.7 are shown in Fig. 7 for fully turbulent computations. The highest angles-of-attack show also a significant Cp plateau. Figure 8 shows the evolution of the Mach number at the edge of the boundary layer and the N factor along the chord on the upper (solid lines) and the lower side (dashed lines). The amplification rate never exceeds the threshold value of 2 up to the shock ensuring the laminar region from the leading edge to the shock. At last, the Figs. 9 and 10 compare the fully turbulent field with the one computed with a fixed transition.

1.1.4 Conclusions

The V2C profile is delivered to the partners of the Work package 5. The profile is designed to be laminar from the leading edge up to the shock at M = 0.7 and 0.75 from low lift coefficient up to the buffet onset. The maximum value of the N-factor never exceeds 3, so that it should ensure laminarity for "sufficiently" low turbulent wind-tunnel facilities. To have some margin, it is required that the freestream wind tunnel turbulence level is less than 0.6%. In any case preliminary experimental tests

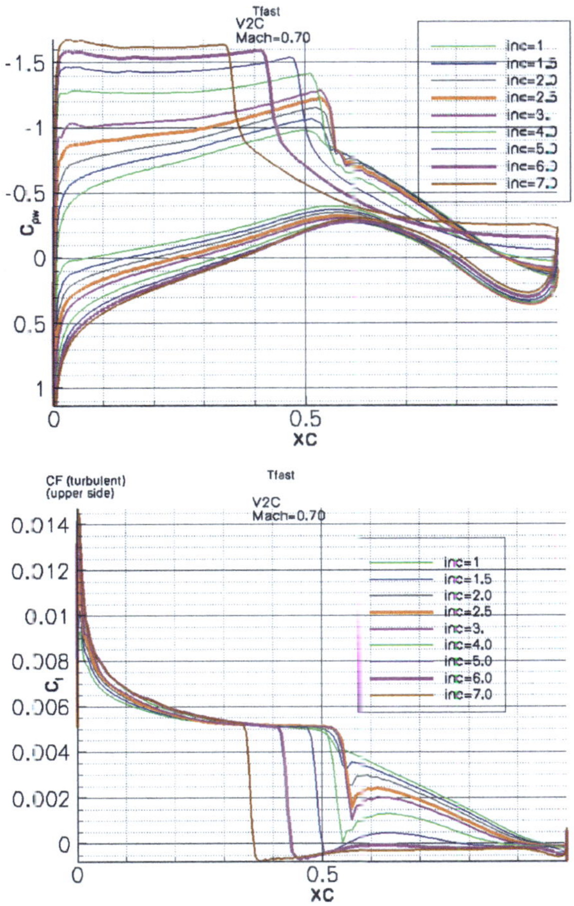

Fig. 3 Pressure and friction coefficients at M = 0.7

should be performed on a flat plate with a characterization of the transition location, to be compared to our computation in order to derive the appropriate N-factor limit.

1.2 UCAM

1.2.1 CUED No. 1 Supersonic Tunnel

Experiments were conducted in the CUED No. 1 Supersonic Tunnel (pictured in Fig. 11), an intermittent blow-down facility with run times up to 30 s, a working section 179 mm high and 114 mm wide and capable of operating at Mach numbers between 0.5 and 3.5. Free-stream Reynolds numbers are between 13 and 55 million

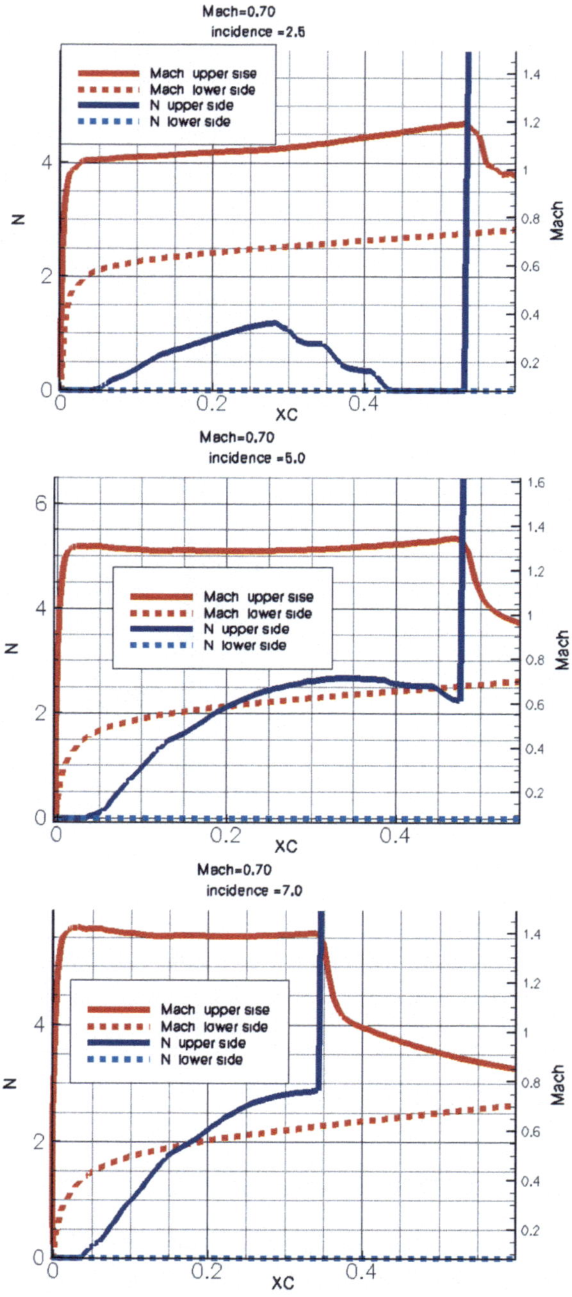

Fig. 4 Streamwise evolution of the Mach number and the N factor for $\alpha = 2.5$, 5.0, 7.0°, $M_\infty = 0.7$

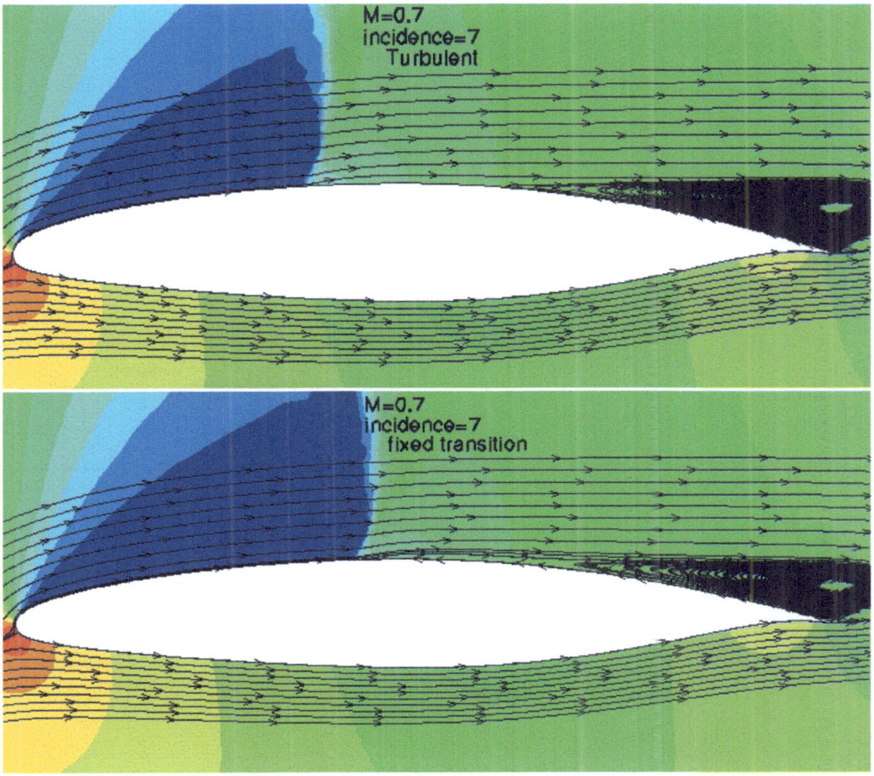

Fig. 5 Pressure contours and velocity streamlines for fully turbulent (*up*) and fixed transition at x/c = 30% (*down*) computation, at M = 0.7 and $\alpha = 7°$

per metre, with the stagnation pressure measured in the settling chamber and manually kept constant by trained technicians, with <0.1% fluctuation. The stagnation temperature (measured as the average of four T-type thermocouples in the settling chamber) increases during a typical run from around 294–300 K. The free-stream turbulence level of the tunnel in transonic mode was measured as 0.7% at Mach 0.7 (WP-5) and 0.5% at Mach 0.8 (WP-1 & 2) [4].

1.2.2 Experimental Methods

A 'Z-type' schlieren system with a horizontal knife-edge is used to visualise the flow, with high frame rate images captured by a Photron FASTCAM ultima APX high speed camera. Imperfections in the optical path are removed by subtracting a 'wind off' background image from each 'wind on' picture. Variation in intensity along a row of pixels for each frame is used to extract shock position versus time, and

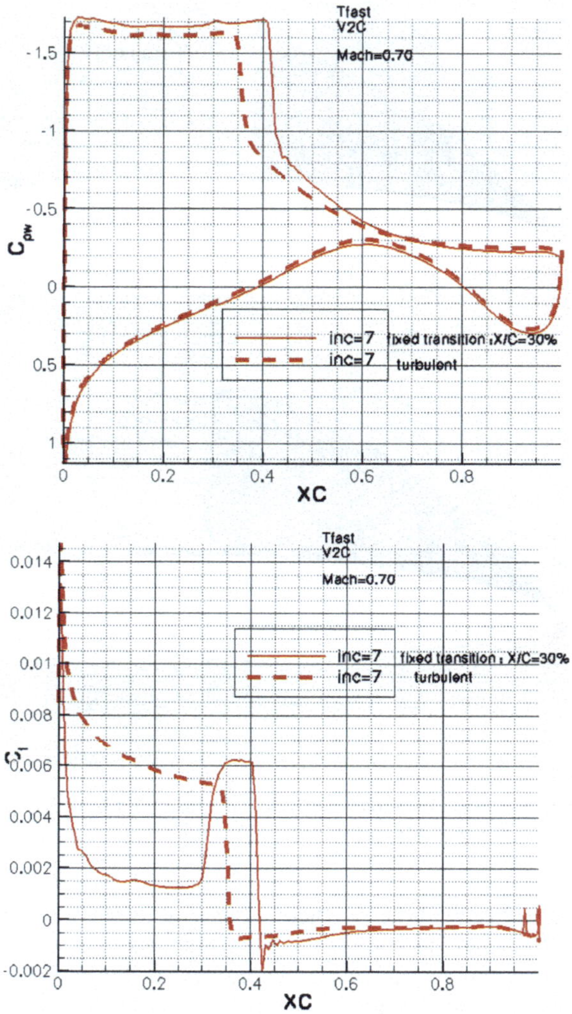

Fig. 6 Pressure and friction coefficients for fully turbulent (dashed) and fixed transition (solid) computation, at $M = 0.7$ and $\alpha = 7°$

the fluctuation throughout a run is analysed to measure any oscillation amplitudes and frequencies.

Pressures are recorded with small tappings in the walls and test models connected to a NetScanner 9116 pressure transducer system with a quoted accuracy of 0.05%. The total experimental error of measured pressures is then <0.2%, and as such the error in calculated Mach numbers is <0.5%. A greater contribution to inaccuracy in Mach number is slight fluctuations in free-stream conditions, which for the most unsteady flows can result in r.m.s. variations of ±0.03. Surface pressure data can be taken over a wide area by using pressure sensitive paint (PSP). When illuminated with UV, the paint luminesces with an intensity re-lated to surface pressure. An ISSI

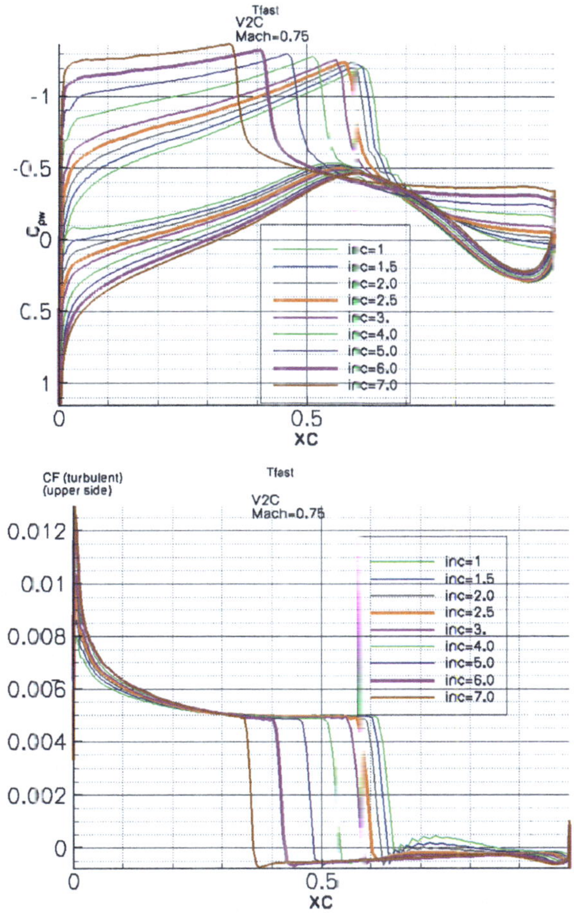

Fig. 7 Pressure and friction coefficients at $M = 0.75$

LM2x-400 serves as the light source, with images taken by an Apogee Alta U2000 cooled CCD camera. The intensity also varies with the thickness of the paint layer, so 'wind on' images are referenced to 'wind off' background images to remove this dependency, and additional images are taken without UV to subtract any stray light. The relative intensity is then calibrated with pressure taps. The paint coat is rougher (Ra ~ 0.8 μm) than the hand polished surface, and so causes the boundary layer to transition (evidenced by the lack of lambda structure seen in schlieren imaging).

A two-component laser Doppler velocimetry (LDV) system from TSI is used in forward-scatter mode to measure velocities. The current implementation uses a Coherent Innova 70C-5 Argon- Ion laser at 1.5 W, with beam wavelengths of 514.5 and 488 nm; an FBL-2 fiberlight box for beam manipulation, containing a 40 MHz Bragg cell; and a TR260 emitter with a TLN06-350 lens (focal length 363 mm), mounted so the measured vectors are at 45° to x & y and the lower beam pair is angled slightly below horizontal. This lens is chosen for its small measurement

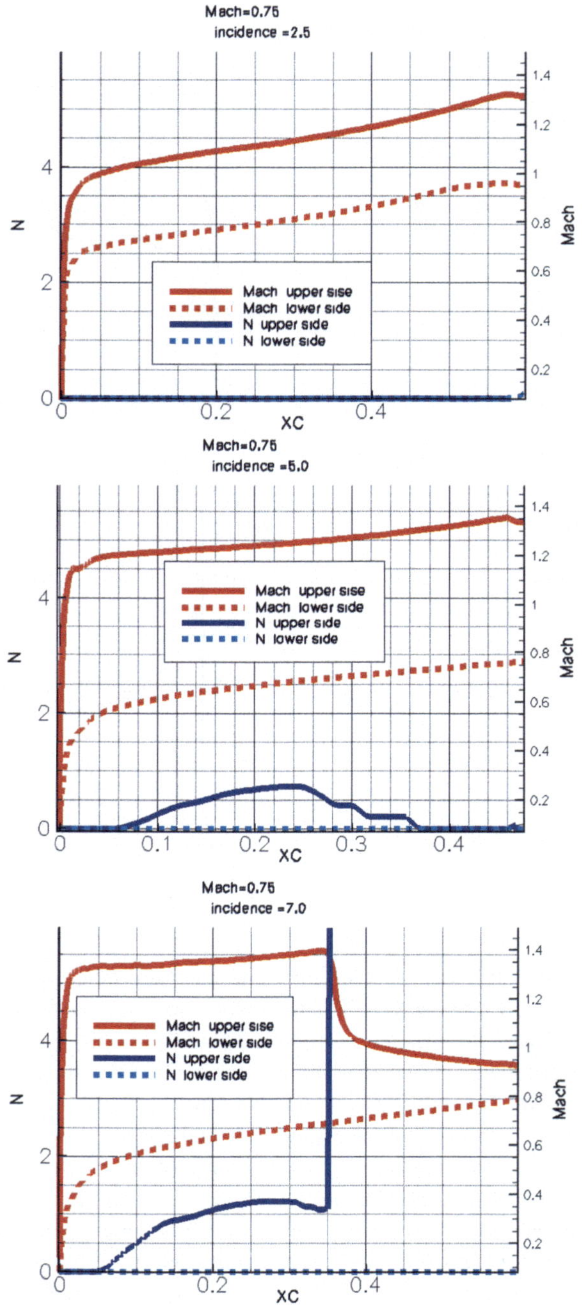

Fig. 8 Streamwise evolution of the Mach number and the N factor for $\alpha = 2.5, 5.0, 7.0°$, $M_\infty = 0.75$

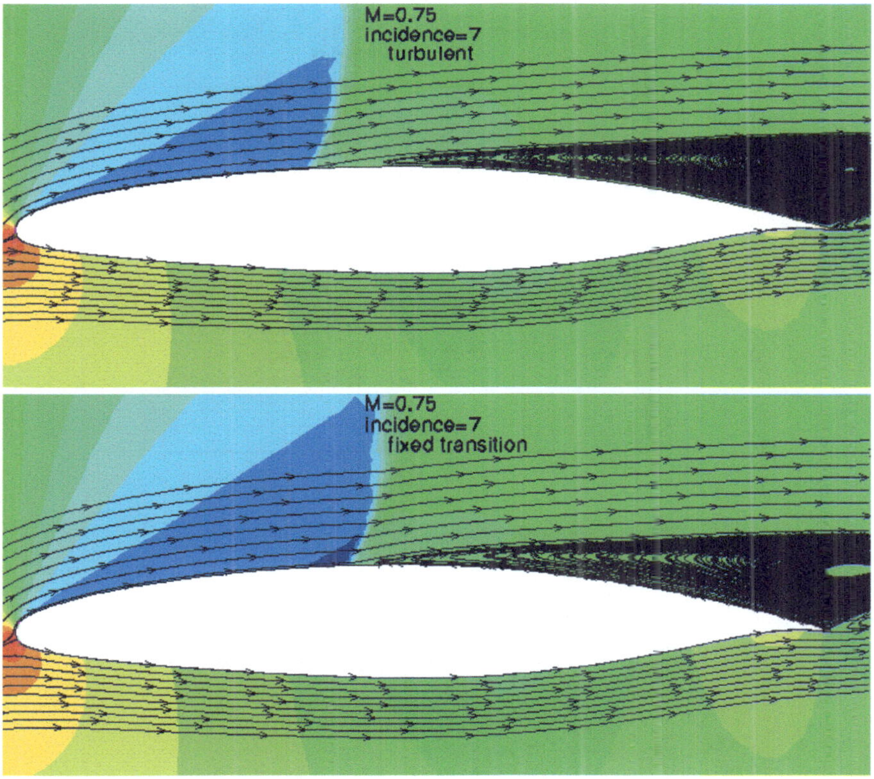

Fig. 9 Pressure contours and velocity streamlines for fully turbulent (*up*) and fixed transition at x/c = 30% (*down*) computation, at M = 0.75 and $\alpha = 7°$

volume, producing beam overlap waists of 90 and 85 μm for the two wavelengths respectively. The corresponding fringe spacings are 3.7 and 3.5 μm, and the beam overlap lengths are 1.3 and 1.2 mm (span-wise). A band pass filter of 40–120 MHz allows velocities from 0 m/s up to 300 and 284 m/s respectively to be measured. A TLN06-500 lens (focal length 512 mm) is used for the aerofoil wake traverses, where instantaneous velocities can frequently be negative, as at the maximum bandwidth setting (20–175 MHz) this allows velocities between −106/−100 and 712/675 m/s to be measured. This lens gives beam waists of 127 and 120 μm, with fringe spacings of 5.3 and 5.0 μm, and overlap lengths of 2.6 and 2.5 mm.

The receiving head is an RV70 with a 500 mm lens, mounted to point 11.25° down and 10° downstream which reduces reflections and decreases the effective length of the measurement volume. The receiver is connected to a PDM 1000-2 photo-detector box and an FSA 4000-2 frequency/burst spectrum analyser, with data recorded by TSI's FlowSizer software.

Droplets of paraffin, ~0.5 μm in diameter (as determined by a shock-lag method), are injected through a rake in the settling chamber and provide good seeding levels

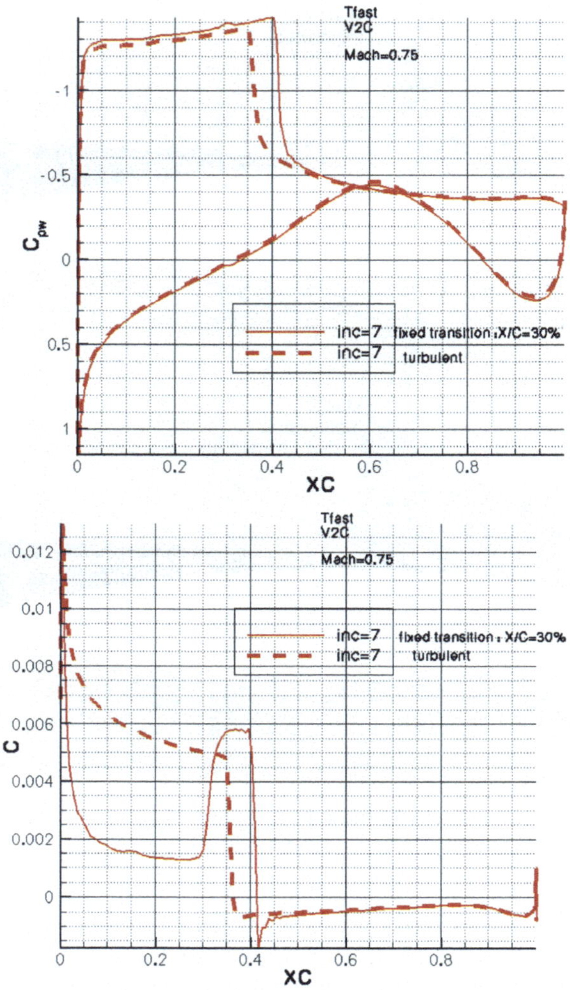

Fig. 10 Pressure and friction coefficients for fully turbulent (dashed) and fixed transition (solid) computation, at $M = 0.75$ and $\alpha = 7°$

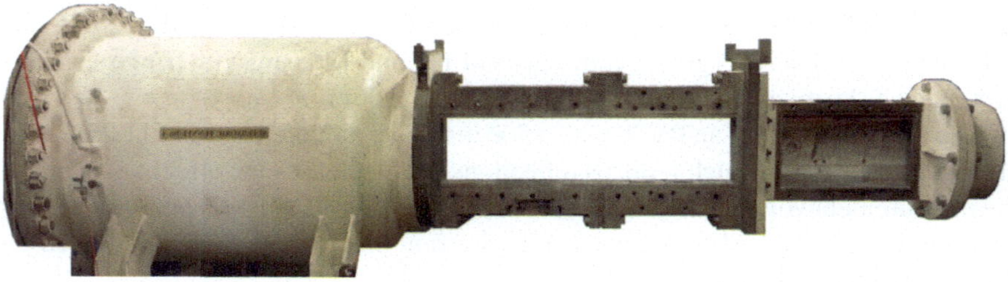

Fig. 11 CUED No. 1 SST, showing the settling chamber, working section and first diffuser

around the centre-span, following the flow well [5]. In the free stream, valid burst rates are in the range of 25–40 kHz, although this drops dramatically in the boundary layer. Averages are taken over 0.2 s at each point of the traverse, which typically ensures at least 5,000 samples in the bulk flow. However, very near the wall (within a few dozen microns) this can fall as low as 500 per data point. Samples outside of three standard deviations are rejected for the purposes of determining the mean and r.m.s. velocities.

The manufacturer's quoted accuracy for the optics and burst analysis is 0.1%, but larger errors are present due to the physical geometries of the emitting head (for example, aligning the emitter with the flow and determining the angle between the beams), which result in an estimated uncertainty of 1.5% for measured velocity components away from the wall [5]. Of greater impact are the difficulties in measuring the flow within a millimetre or so of the wall. Here, errant reflections contribute to a drop in signal-to-noise ratio; low seeding levels reduce particle counts and the accuracy of the measured mean and fluctuating velocities; velocity variation across the measurement volume can be significant in smaller boundary layers, giving increased scatter; and at the extreme end of the measurement range the volume can actually intersect with the floor, exacerbating all of these problems. As such, when the boundary layer thickness is much below 1 mm (as is typical upstream of the shock), confidence in the data is much reduced and the technique is only trusted to give an indication of δ, and not expected to accurately determine profiles and integral parameters.

Integral quantities (δ^*i, θi) are calculated with a straight line assumed between points (trapezium method) and a no-slip condition at the wall. This tends to underestimate the velocity between points, leading to overestimates in the thicknesses, and also simulates a less full profile, and so $Hi = \delta^*i/\theta i$ is likely to be overestimated as well. More complex methods, such as those discussed in Titchener et al. [6], are not employed here as they involve a priori assumptions about the state of the boundary layer which are unsuitable for transitional flows where this is subject to change. Incompressible quantities are used here as they are less dependent on Mach number and so a better universal parameter [7].

Two styles of trip were added to the models during the experimental campaigns, sketched in Fig. 12. The first is a two-dimensional step, created by a strip of gloss spray paint approximately 2 mm wide and 10 μm high. The second tripping method is three-dimensional and involves carefully applying drops of poly(vinyl acetate)

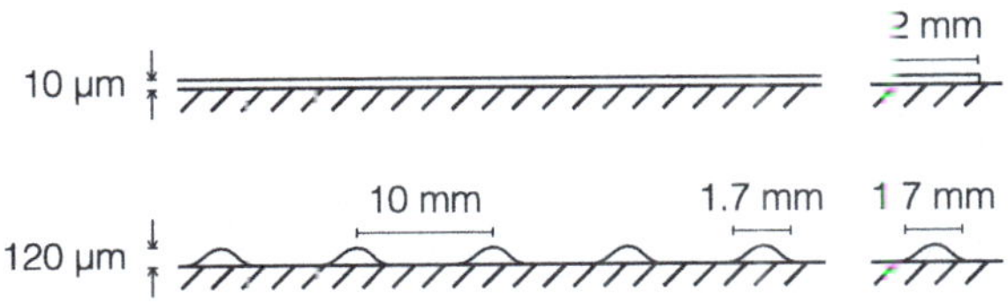

Fig. 12 Sketches of the two trip styles: (top) 2D trip—paint strip; (bottom) 3D trip—glue dots (mean values given); (left) stream-wise view; (right) span-wise view

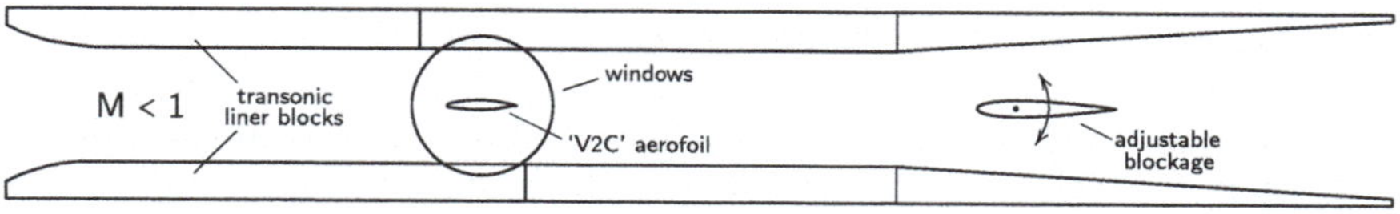

Fig. 13 The test section layout used for the transonic aerofoil investigation

glue at centre-span (z = 0) and then at 10 mm intervals along the span. Dot heights are between 90 and 180 μm with an average of 120 μm, and the diameters vary from 1.4–2.2 mm with a mean of 1.7 mm.

1.2.3 Work Package 5: 'V2C' Aerofoil Experiments

The tunnel configuration used for the transonic laminar aerofoil investigation is shown in Fig. 13. The free-stream Mach number is controlled by adjusting the blockage at the tunnel exit. The aerofoil has the 'V2C' profile designed by Vincent Levasseur of Dassault Aviation; a chord, c, of 100 mm; and was manufactured by wire-erosion and then hand-polished to a surface roughness of Ra < 0.5 μm. The model is held between the tunnel windows to avoid the flow disturbance of a stand or mount; though blockage effects are still significant due to the small size of the tunnel. Two styles of trips were added to the upper surface as detailed above in a paint strip or glue dots (Fig. 12).

1.2.4 Steady Flow Results

PSP measurements were performed using a tapped model to compare the designed and observed flows. Pressure distributions are plotted in Fig. 14 against calculations by DAAV which simulated an unbounded stream with a fully turbulent boundary layer. Due to tunnel confinement effects in the experiments, numerical comparisons are performed at a higher Mach number, but very similar trends can be seen in the data. Additionally, no lambda shock structure was visible in the schlieren images for these runs, suggesting that the paint on the surface caused a significant roughness increase which promoted transition and so the boundary layer was turbulent, as in the calculations. Both data sets show the shock moving forwards as angle of attack increases, with little variation in the peak suction level. Critically, the favourable pressure gradient is seen to extend all the way back to the shock, which helps the boundary layer stay laminar when the surface is clean.

A schlieren image of the clean model at zero degrees angle of attack is shown in Fig. 15a. The flow accelerates around the profile, with a supersonic region terminated by normal shockwaves on both the upper and lower surfaces. On the upper side, the flow appears to separate under the shock (the boundary layer turning is easily visible, but any separation bubble itself is too small to be seen) and a characteristic lambda

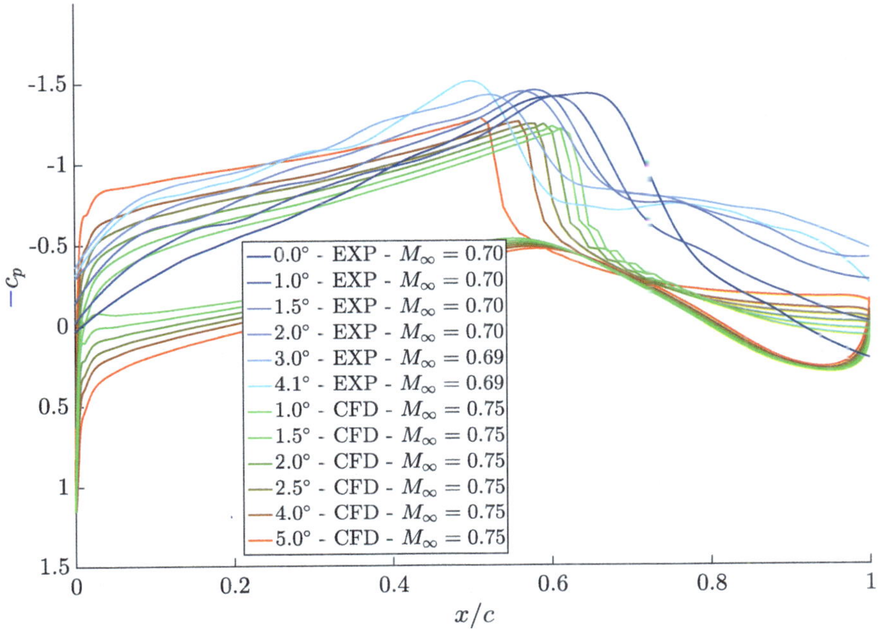

Fig. 14 A comparison at varying angles of incidence between measured surface pressures in the wind tunnel using PSP at Mach numbers around 0.7 and DAAV calculations for a fully turbulent unbounded flow at $M_\infty = 0.75$

shock structure is formed, much like in the clean flat plate interactions. (The small waves upstream of the interaction are around the screws that hold the two halves of the model together, but these are located close to the side-walls and so don't interfere with the bulk of the flow.)

Figure 15b shows the same flow but with a coating of matte black spray paint on the upper surface of the solid aerofoil model. This was applied to prevent reflections which were expected to be detrimental to LDV measurements. The schlieren image shows that the separation and lambda structure are now missing, and the upstream boundary layer is thicker than in the clean case. It is concluded that the slight increase in surface roughness (from Ra ~ 0.2 to 0.5 μm) is sufficient to promote transition at these high unit Reynolds numbers, as did the PSP. It is also seen that the upper edge of the wake is much more sharply defined. (Near the mid-chord, there are a pair of weak waves in the schlieren image which are caused by imperfections in the manufactured profile, but they don't appear to affect the flow significantly.)

In Fig. 15c, d, discrete transition devices are added to the model at x/c = 0.4, in the form of a paint stripe and trip dots respectively. The behaviour of the flow around the trips and at the interaction is very similar to the corresponding flat plate cases, including significant thickening of the boundary layer over the trip itself.

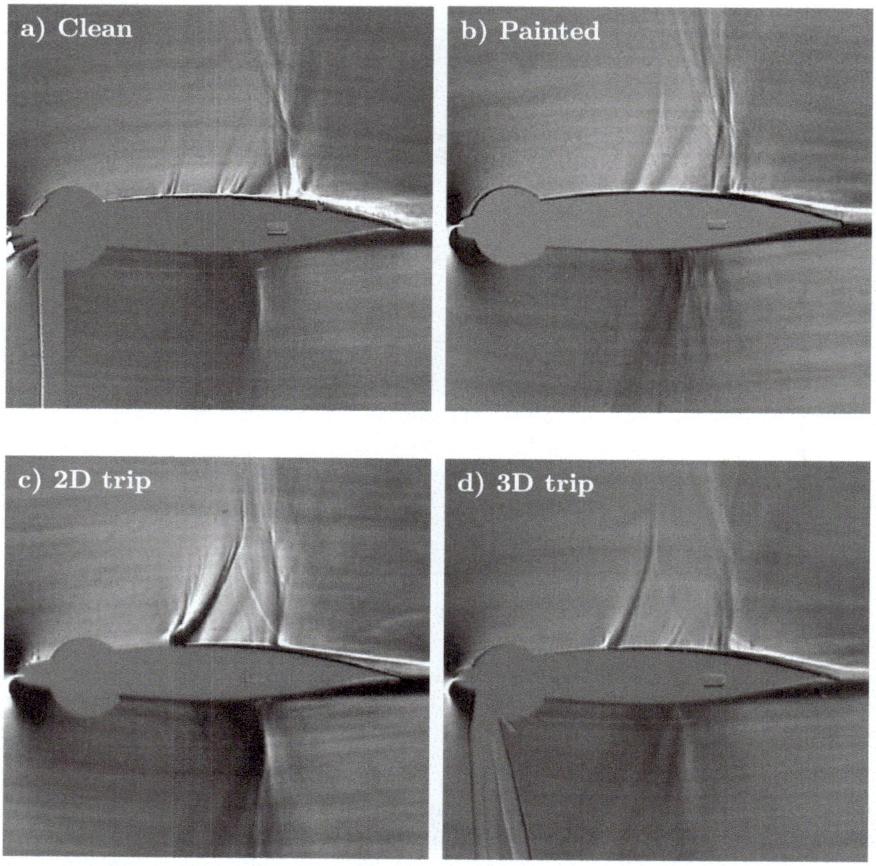

Fig. 15 Aerofoil schlieren images at $\alpha = 0°$, $M_\infty = 0.705$: **a** clean tapped V2C aerofoil, $Re_c = 1.82 \times 10^6$; **b** painted solid aerofoil, $Re_c = 1.84 \times 10^6$; **c** solid aerofoil model with paint stripe at $x = 0.4c$, $Re_c = 1.84 \times 10^6$; **d** tapped model with trip dots at $x = 0.4c$, $Re_c = 1.85 \times 10^6$

Additionally, as seen with the fully painted surface, the wake edge is better defined leaving the aerofoil.

A number of LDV traverses were performed through the wake for three of the aerofoil cases at different stream-wise positions, with examples plotted in Fig. 16a. The available range of data was limited by the location of the seeding within the tunnel and by the size of the observation windows, and so the measurements are restricted to the area near the aerofoil. However, taking inspiration from Hue et al. [8], the data is extended by fitting profiles to the experimental data by combining three skewed Gaussian profiles (one for the viscous wake and one for the shock wave on each surface) which asymptote to a far-field value, all matching the LDV data with R^2 values greater than 0.995. There is a clear pattern in the data of the tripped flow

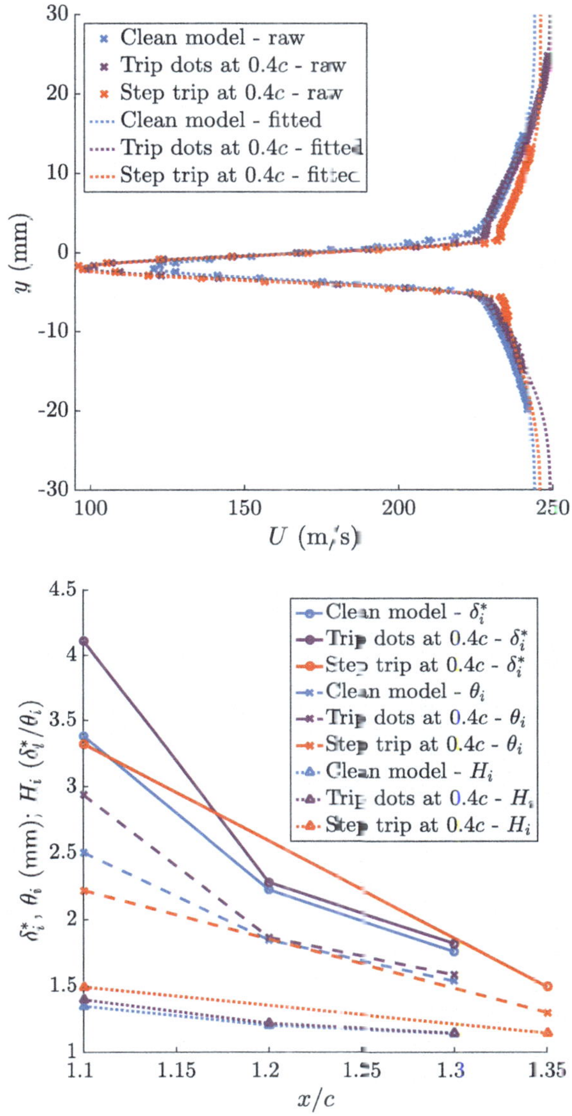

Fig. 16 **a** Wake traverse examples at $\alpha = 0°$, $M = 0.705$, $x/c = 1.1$, $z/c = 0$, with raw data and fitted profiles. **b** Wake integral properties calculated from the fitted profiles

wakes having a greater maximum velocity deficit and more sharply defined edges to the viscous wake (also seen in the schlieren images).

Figure 16b presents the integral properties calculated using the fitted wake profiles. Of course, the traverse locations are close to the aerofoil itself and so the flow clearly has not yet reached equilibrium pressure, but there is little distinction between the three cases. The step tripped case does show a slightly higher Hi nearer the shock, which correlates well with the flat plate results. Once again, the wake traverses were made along the centre-line (z = 0), and so behind one of the trip dots in the three-dimensional tripped case. However, LDV suggests that span-wise non-uniformities have smoothed out by the measured wake positions.

1.2.5 Buffet

At Mach numbers around 0.711 and above, buffet can be observed for all surface conditions at zero incidence.[1] The shock motion has been captured using high speed schlieren imaging and typical cycles are shown in Figs. 17 and 19, with the frequency and magnitude of the shock and wake oscillations calculated in Fig. 18. A range of experimental and model conditions have thus been investigated and reported in Table 1: typical frequencies vary from 290 to 390 Hz, corresponding to St = 0.13–0.16. Crucially, the buffet onset Mach number does not seem to depend on the surface condition or the boundary layer state. This is an excellent illustration of the main conclusion of this research, both on the aerofoil and the flat plate: although the shock

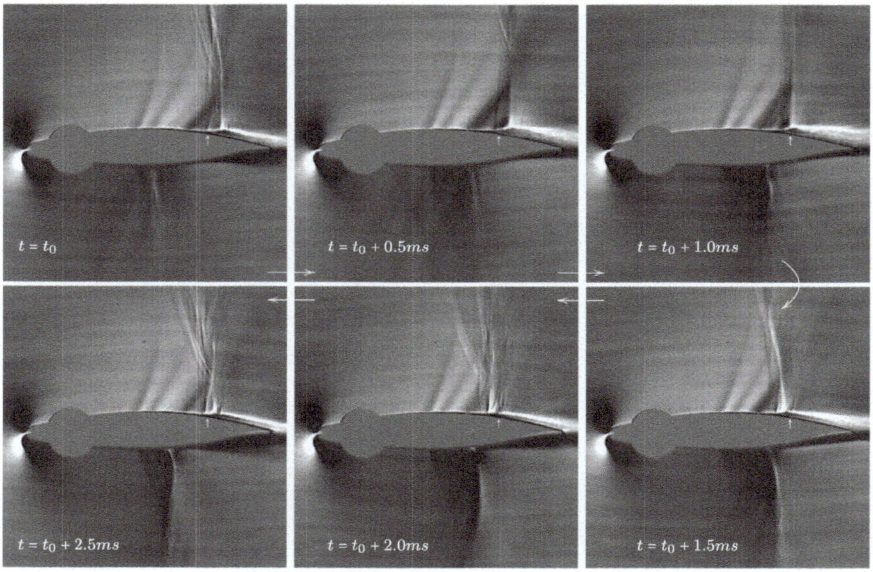

Fig. 17 Buffeting on the clean aerofoil model at $M_\infty = 0.711$, $Re_c = 1.83 \times 10^6$ (Arrows indicate mean shock position)

[1]Zero incidence buffet has been observed by previous investigators both experimentally and numerically, predominantly in research on thick biconvex aerofoils in small wind tunnels. Lee [28], Mabey [29] in particular came to the same conclusion as here that for thickness-to-chord ratios greater than 12% the buffet boundary at $Re_c = 0.6 \times 10^6$ (as detected by a wall-mounted pressure transducer) had little dependence on transition (though at very low Reynolds numbers the untripped flow began buffeting at a lower Mach number, with higher pressure fluctuations). No buffet was observed when the aerofoil was given an angle of attack, due to increased confinement effects.

structure looks substantially different in the schlieren image, the actual overall effect of a laminar versus turbulent SBLI on the flow is minimal. This seems to be due to the low pressure rise required to separate a laminar boundary layer, which in turn means any separation is actually very thin, and transition can then occur before an only slightly weakened rear shock and the rest of the flow is very similar to the turbulent interaction case.

1.2.6 Conclusions

Two distinct interactions have been observed on the flat plate and on the transonic aerofoil's upper surface at zero angle of attack: a lambda structure, indicating separation, when the model is in a clean configuration; but just a small amount of smearing at the shock foot when anything is added to trip the flow. However, downstream of the shock the various flow cases exhibit no significant differences. The lack of impact on the flow of the boundary layer state upstream of the shock is evidenced most clearly in the fact that the buffet onset Mach number remains the same in the aerofoil experiment. As such, it is concluded that it is not necessary to trip the flow ahead of the shock, as the clean interaction has no real disadvantages and it is safe to let the normal shock wave meet the laminar boundary layer.

1.3 IoA

1.3.1 Trisonioc Wind Tunnel N-3

IoA TFAST experimental tests were conducted in the trisonic N-3 wind tunnel, blow-down type with the partial re-circulation of the flow (see Figs. 20 and 21a). The square cross section dimensions are: 0.6 m × 0.6 m, length 1.58 m. The lower and upper walls of the N-3 wind tunnel test section could be mounted in option: perforated or solid and side walls are solid. In TFAST contribution wind tunnel test section was equipped with solid walls only in order to have the comparable tests condition as in CFD calculations.

The Mach number range of a subsonic and transonic flow is 0.3–1.15, and supersonic 1.5, 2.3. Possible changeability of Reynolds numbers in this Mach number range is presented in Fig. 21b (on the right).

It has been noticed that significant differences in pressure distribution are observable at high angle of incidence. The Cp distributions at the angle of incidence $\alpha i = 7°$, Mach number $M = 0.7$ are presented in plots (Figs. 28, 29, 30 and 31).

As it can be seen at the pressure distribution at incidence $\alpha i = 7°$ (Figs. 28 and 29), for models with smother upper surface (after polishing) the under pressure drop appears more rapidly and far downstream in the vicinity of shock wave. That results in higher lift coefficient. The polished cases also characterise far lower drag coefficient value.

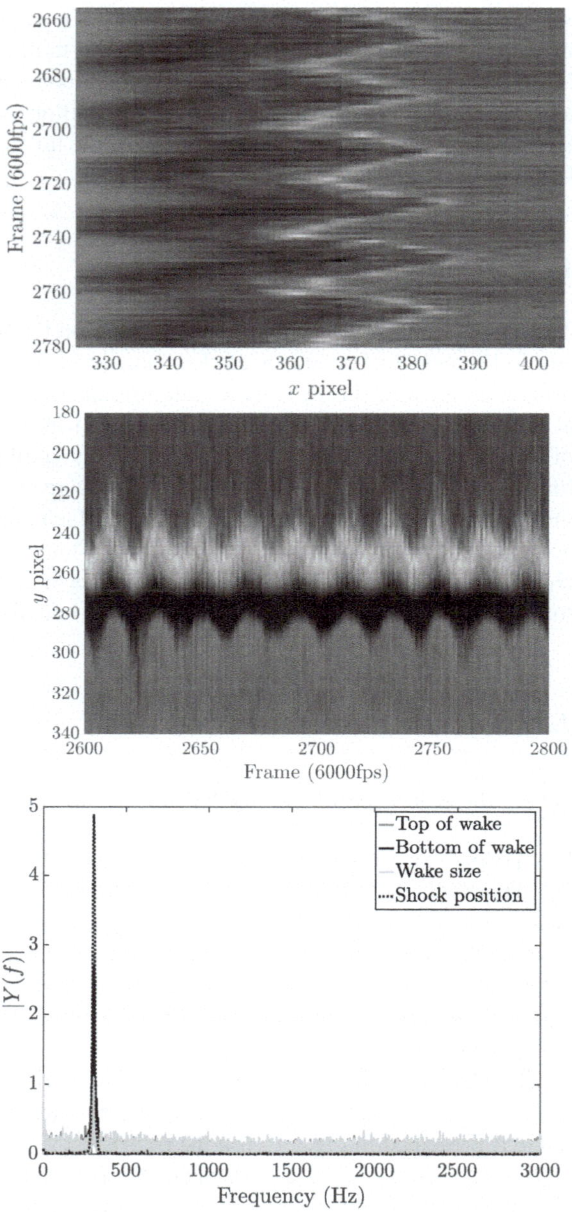

Fig. 18 Analysing the buffet of Fig. 17: **a** observing the shock oscillation by examining a pixel row for each frame of the high-speed schlieren video; **b** observing the wake fluctuations with a pixel column; **c** calculated frequency

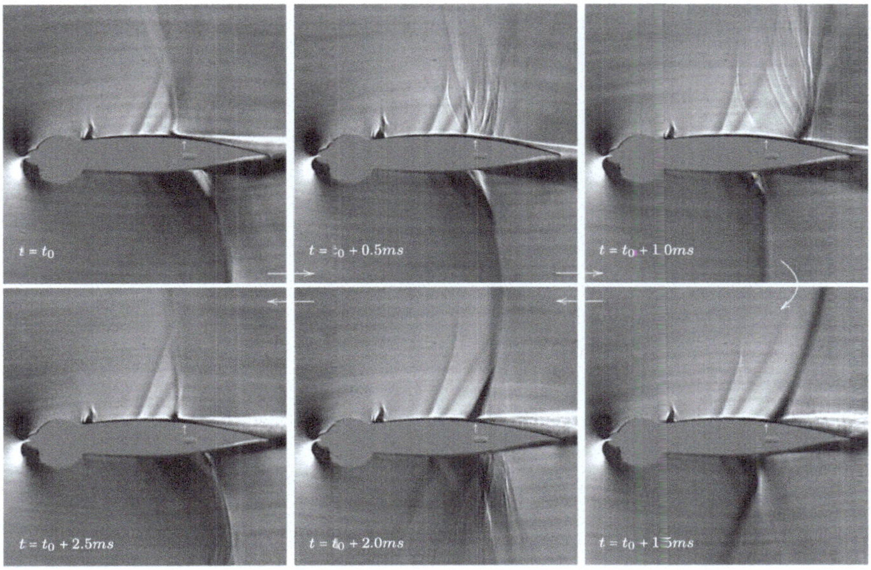

Fig. 19 Buffeting on the aerofoil model at $M_\infty = 0.712$, $Re_c = 1.82 \times 10^6$ with a paint strip at 0.25c (Arrows indicate mean shock position)

Description to (Fig. 21b):

- E1 ejector 1,
- E2 ejector 2,
- DF changeable subsonic and supersonic diffuser,
- KP test section with plenum chamber, changeable fixed nozzle section,
- KD fixed convergent nozzle in contraction chamber,
- KS setting chamber,
- S set of antiturbulence nets and honeycomb.

The contraction and test section of the N-3 wind tunnel is presented in Figs. 22 and 23. The upper and lower walls were deflected of 0.5° from each other. During the wind tunnel run, the Mach number was set on base local static and total pressure measurements.

The contraction and test section of the N-3 wind tunnel is presented in Figs. 22 and 23. The upper and lower walls were deflected of 0.5° from each other. During the wind tunnel run, the Mach number was set on base local static and total pressure measurements.

1.3.2 Tested Model

The V2C airfoil model, as mentioned in 6.1 DAAV, was laminar type. The airfoil model of chord 0.2 m and span 0.6 m had the relative thickness approx. 15% chord.

Table 1 Shock motions for different flows and model conditions

Condition	α (°)	M_∞	Re_c ($\infty 10^6$)	$X_{shock, r.m.s.}$ (pixels)	f_{buffet} (Hz)	St_{buffet}	Figure
Clean	0.0	0.700	1.84	7			
	0.0	0.702	1.83	9			
	0.0	0.704	1.82	5			
	0.0	0.705	1.82	4			13(a)
	0.0	0.706	1.81	9			
	0.0	0.709	1.81	8			
	0.0	0.710	1.84	8			
	0.0	0.711	1.83	11	308	0.132	16
	0.0	0.712	1.83	14	305	0.130	
	0.0	0.713	1.82	18	325	0.139	
	0.0	0.714	1.83	14	363	0.155	
	0.0	0.715	1.84	12	369	0.157	
	0.0	0.717	1.84	19	331	0.111	
Clean	2.0	0.698	1.84	5			
	2.0	0.707	1.85	5			
	2.0	0.713	1.86	6			
	4.2	0.705	1.82	5			
	4.6	0.704	1.85	4			
	4.6	0.700	2.17	4			
Painted	0.0	0.705	1.84	4			13(b)
	0.0	0.709	1.84	4			
	0.0	0.710	1.85	7			
	0.0	0.711	1.85	7			
	0.0	0.712	1.85	12	294	0.126	
2D trip (0.25c)	0.0	0.712	1.82	32	355	0.152	17
	0.0	0.716	2.63	36	387	0.165	
	0.0	0.718	2.64	3	–	–	
2D trip (0.4c)	0.0	0.701	1.84	5	–	–	
	0.0	0.705	1.84	6			13(c)
	0.0	0.706	1.84	6	–	–	
	0.0	0.710	1.85	29	343	0.147	
2D trip (0.4c)	2.0	0.709	1.84	4			
	4.6	0.703	1.84	3	–	–	
3D trip (0.4c)	0.0	0.705	1.85	8			13(d)
	0.0	0.711	1.85	15	322	0.138	

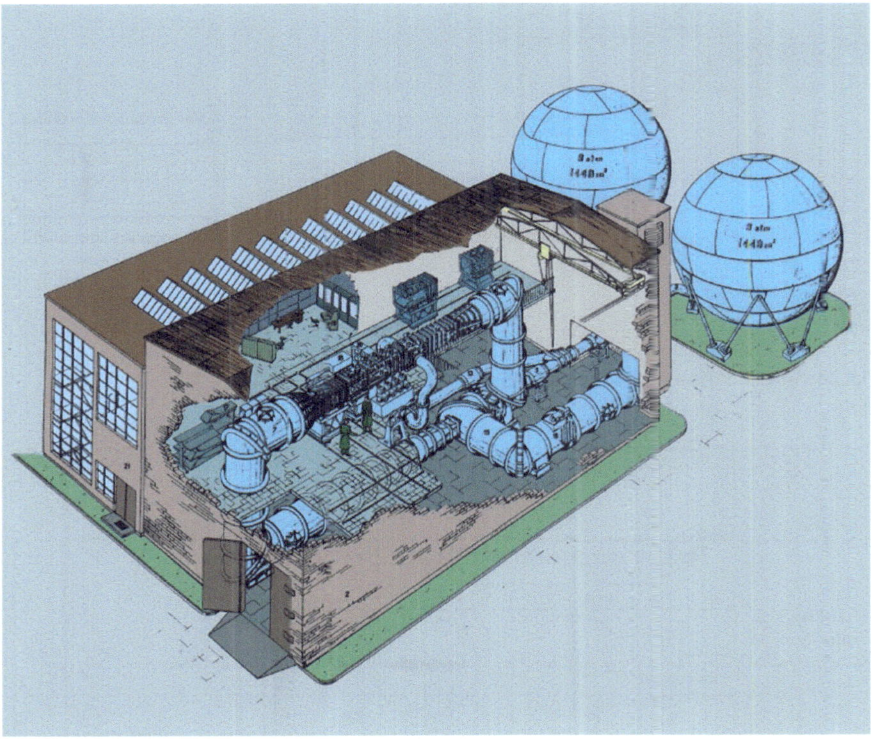

Fig. 20 IoA N-3 transonic wind tunnel view

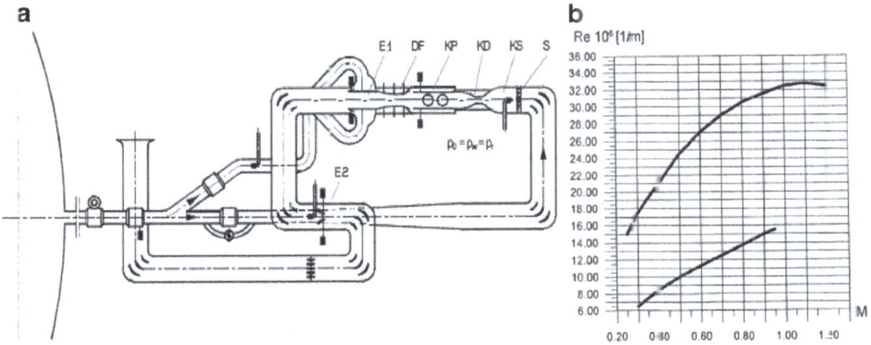

Fig. 21 **a** IoAN-3 trisonic wind tunnel, **b** the range of the free-stream Mach and Reynolds numbers: normal—(bottom curve) and pressurised—conditions per meter (upper curve)

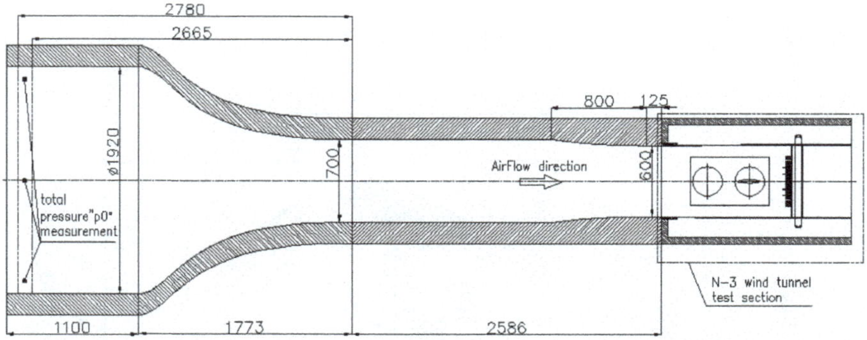

Fig. 22 IoA N-3 trisonic wind tunnel contraction chamber

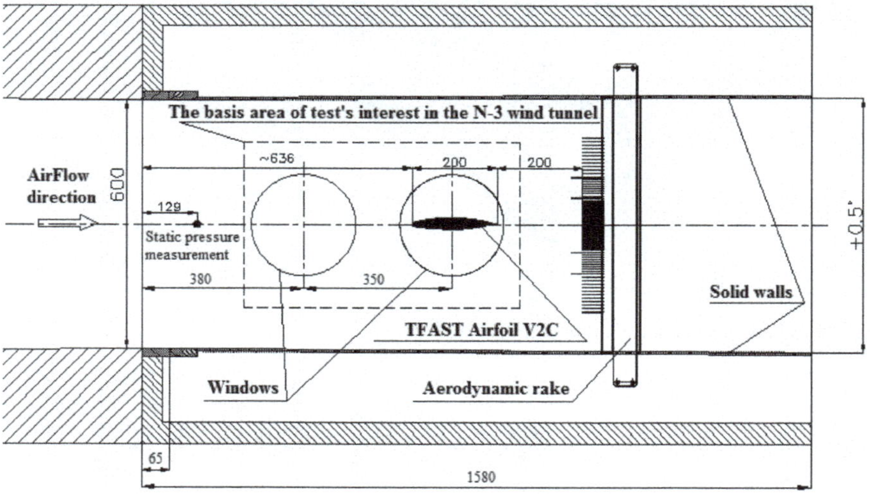

Fig. 23 IoA N-3 trisonic wind tunnel test section with TFAST "V2C" model mounted—scheme

Initially, model shape accuracy was equal ±0.06–0.08 mm and surface roughness accuracy ~1 μm. The model surface was darkened. During experimental investigation shape accuracy and roughness of the model had being changed. In further tests, V2C airfoil 2D model shape and roughness accuracy were improved. (up to ±0.03–0.05 mm and <0.5 μm, respectively).

1.3.3 Experimental Setup

Measurements of the airfoil pressure distribution along the airfoil chord (mean value of static pressure) were performed by two 32-channel electronic scanners ESP-32HD

a b

Fig. 24 **a** Model mounted in wind tunnel (on the left), **b** the pressure tubes inside the model (on the right)

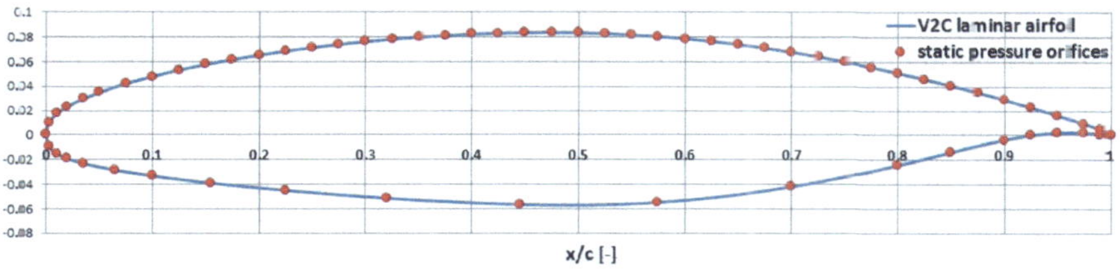

Fig. 25 Static pressure orifices location at the V2C airfoil

DTC INITIUM. The pressure system at maximum scan rate 650 Hz was connected by pressure tubes mounted inside the model (Fig. 24) to 64 pressure taps of 0.5 mm diameter on the airfoil (two rows: on the upper and bottom model surface). The side view of the orifices distribution on profile presented in Fig. 25.

The aerodynamic rake was installed behind the model profile at distance of 1 model chord (see Fig. 23). It was used for the measurement of the loss of momentum in trail, and on this base an aerodynamic drag was estimated. The aerodynamic rake contained 115 measured points. The aerodynamic rake could measure vertical cross section area 222 mm up and down from tunnel axis.

KULITE pressure transducers were adapted for unsteady pressure measurements in the vicinity of the shock location. Analysis of RMS fluctuations was performed for estimation of the buffet onset phenomena; similarly as it was in UFAST project.[2]

SENFLEX sensors were used for laminar–turbulent transition location measurement on a profile upper surface (Fig. 26). The 50-sensor array was mounted in the earlier prepared cavity on the top surface, filled by resin in order to avoid thermal influence during tests.

For data acquisition and processing homemade LabVIEW and MATLAB software was prepared.

[2]Unsteady Effects of Shock Wave Induced Separation: Specific Targeted Research Projects AST4-CT-2005-012226.

Fig. 26 The SENFLEX array mounted on the TFAST model in the N-3 wind tunnel

The following instruments were used in the measurements and described above:

- Kulite LQ-125-25D sensors
- SENFLEX SF1303 hot-film 50-sensor array
- ESAM Traveller 1 CF.

To investigate SCHLIEREN photography the IoA Schlieren optical system and the 60 Hz video camera was used. The record mode was on when wind tunnel test run. From recordings reduction the movies and pictures has been achieved.

The visualization and pressure tests have been made at the same time.

The PIV laser pulse laser impulse time was 8 ns and repetition of the system was 7 Hz. The light sheet thickness was about 2 mm and the laser light was provided with a use of periscope system downstream of the test section. This configuration enabled positioning the light sheet parallel to the incoming flow providing good particles illumination conditions in the test section form the trailing edge of the model to approximately $x/c = 0.2$, where $x = 0$ corresponds to the leading edge. The applied PIV flow visualisation method (Particle Image Velocimetry) was described by Stryczniewicz et al. [9].

A boundary-layer transition was initially triggered by usage of the CARBORUNDUM strip, fraction f180, of approx. 0.1 mm height and 4 mm width (Fig. 27). Carborundum grains were attached to a model surface using speeded varnish. The main advantage of this mean is that the test strip can be placed/removed from the surface of the model. The height of carborundum grain were based on a local Reynolds number derived in IoA CFD calculations and were related to NACA-TN-4363 report.

The position of fixing trigger on the upper model surface (model marked as: Pre-test model and Final model_v2) (in respect to airfoil chord) was: 50, 40, 30, 20, 10%c from the leading edge of the V2C model. The carborundum was fixed through the wingspan.

In further tests, the boundary-layer transition was triggered on the model upper surface by carborundum means (with the use of the model with higher shape accuracy: Final model and Final model_v2 polished). The three fractions f180, f100 and f80

Fig. 27 Transition trigger: strip of carborundum f180 (on the left)

(approx. 0.1, 0.15 and 0.26 mm height) of the 4 mm width carborundum strip were used. The positions of a fixing trigger on the upper model surface were as previously.

1.3.4 Results

Baseline Configuration

Four quality models at baseline configuration, with lower and higher quality of upper surface, have been compared at Mach number M = 0.7 (Pre-test, Final_v2 with poorer shape accuracy and Final, Final_v2 (polished) with greater shape accuracy) (Fig. 28). It was observed that models with more accurate shape accuracy characterise smaller aerodynamic wake and thus smaller drag coefficient (At Mach number M = 0.75 wake tendency was similar as at M = 0.7).

This is apparently because of the lower shock wave drag component due to lower oscillations amplitude of the shock for smother airfoil (observed from schlieren visualisation). The momentum loss comparison between Pre-test and Final model (polished) reveals, that wake behind the more accurate model is also narrower and so the measured value of drag is lower too (Fig. 29). The shape of the shock wave in the Schlieren picture for less accurate model might indicate, that at the angle of incidence $\alpha_i = 7^\circ$ and Mach number M = 0.7 there is no more laminar SWBLI according to Kim or/and laminar bubble occurrence. However, the shock wave position in both cases (less and more accurate model) remained almost the same.

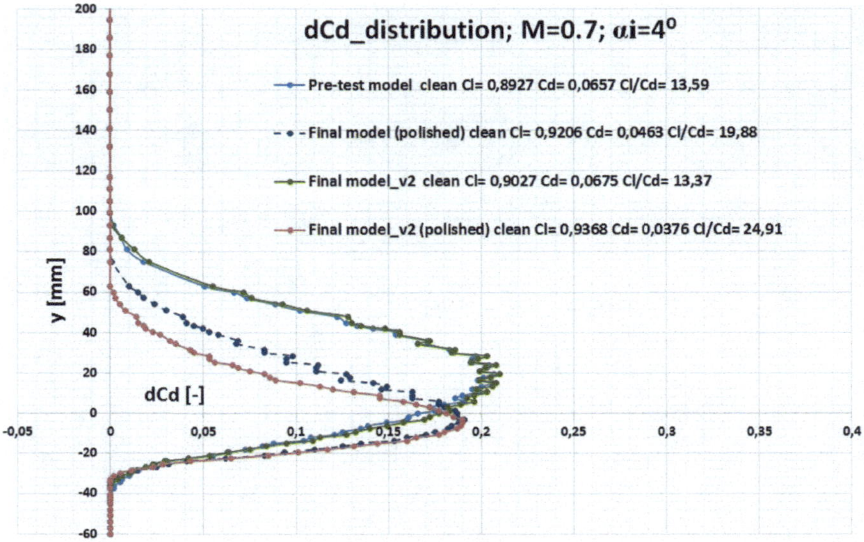

Fig. 28 Wake behind the baseline configuration model; M = 0.7; $\alpha_i = 4°$

The flow separation development process of the boundary layer up to fully separated flow at Mach number M = 0.7 was described by Stryczniewicz et al. [10]. Description is related to more accurate V2C model (Final model (polished)).

When focusing on Cp distribution over the airfoil, at Mach number M = 0.7 polished models stand out greater area with more distinct laminarity at shock, where characteristic plateu indicates on laminar type shock wave boundary interaction, according to Shapiro and Tijdeman (Fig. 30). The identification method of the boundary layer shock wave interaction type, based on the Cp distribution and other wind tunnel data, was described in paper by Placek and Stryczniewicz [11].

(At Mach number M = 0.75 airfoil Cp distribution differencies were less significant).

It has been noticed that significant differences in pressure distribution are observable at higher angle of incidence. The Cp distributions at the angle of incidence $\alpha_i = 7°$, Mach number M = 0.7 are presented in plots (Figs. 31, 32 and 33). As it can be seen at the pressure distribution at incidence $\alpha i = 7°$ for model with smother upper surface (after polishing), the pressure rise at the shock wave appears more rapidly and far downstream. That in turn, results in higher lift coefficient. The polished baseline configurations characterise also far lower drag coefficient value.

On plots (Figs. 31 and 32) can be observed the Cp plateu occurence versus incidence angle rise. Results indicates that for lower quality model a Cp plateu presence is up to incidence $\alpha i = 4°$, and for polished model up to incidence $\alpha i = 5°–6°$. At incidence $\alpha i = 7°$ there is no Cp plateu and thus probably no fully laminar SWBLI interaction for all tested models.

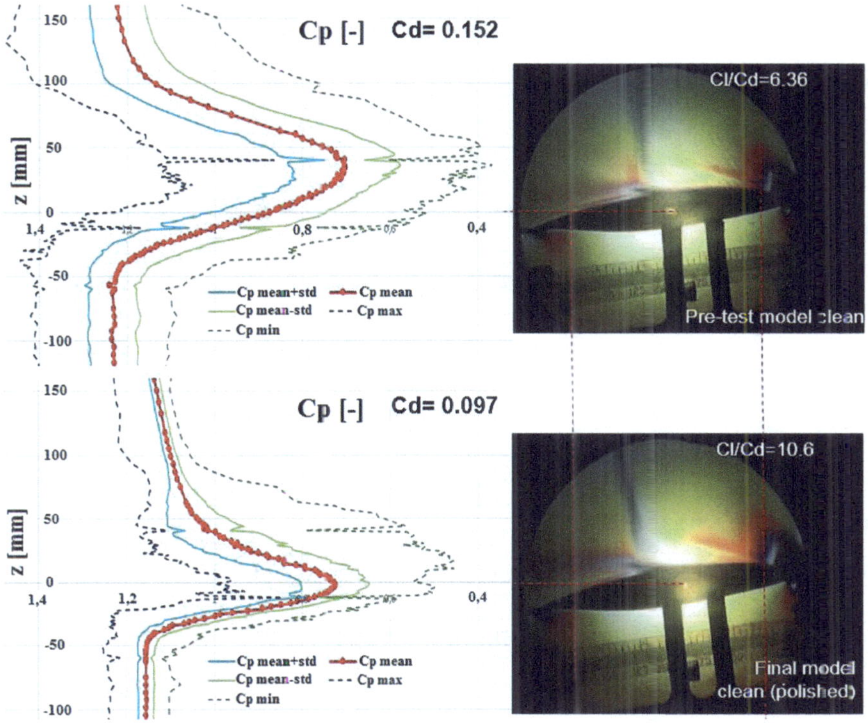

Fig. 29 Momentum loss and Schlieren pictures comparison of the baseline configuration; $M = 0.7$; $\alpha_i = 7°$: Pre-test model (top) and Final model (polished) (bottom)

The comparison of the pitching moment fluctuation was presented in Fig. 34. The pithing moment fluctuations amplitude is lower for the polished model. Taking into account that pitching moment fluctuations are related to shock wave movements, results confirm that shock oscillations amplitude along chord wise direction for polished model is lower too.

When increasing the incidence, the RMS value of pressure fluctuations in the vicinity of the front foot of the shock wave also in crises (Figs. 35 and 36). The rise of it begins from $\alpha_i \approx 3°$ for model with lower quality upper surface airfoil and from $\alpha_i \approx 4°$ for polished model. Following the pressure fluctuation rise is much greater for the first model.

Roughness Configuration

The roughness influence, was described by Placek et al. [12]. Paper contains aerodynamic characteristic comparison and unsteady measurement results of the less accurate V2C model (Pre-test model) for baseline configuration and chosen turbulent carborundum strip cases.

The roughness influence on pressure the Cp distribution and on aerodynamic coefficients for polished model (Final_v2 (polished)), was presented on plots (Figs. 37

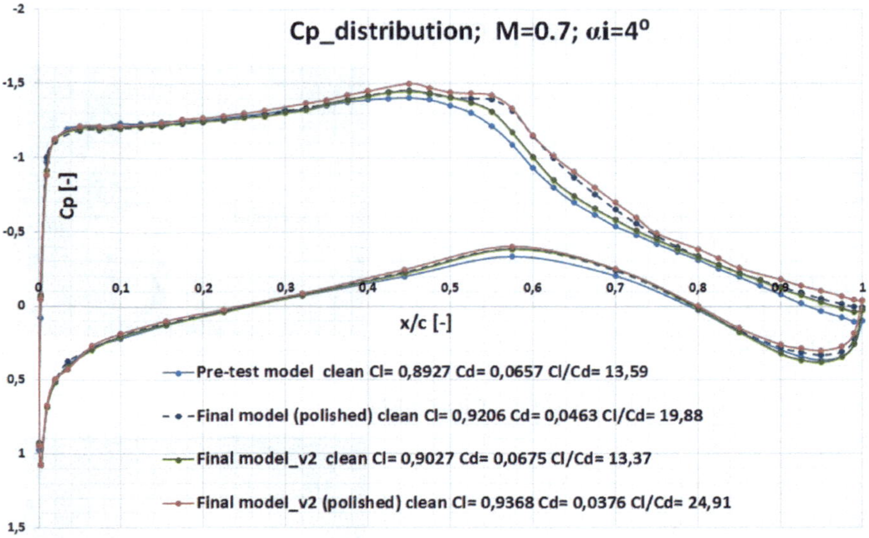

Fig. 30 Cp distribution comparison of baseline configuration; M = 0.7; $\alpha_i = 4°$

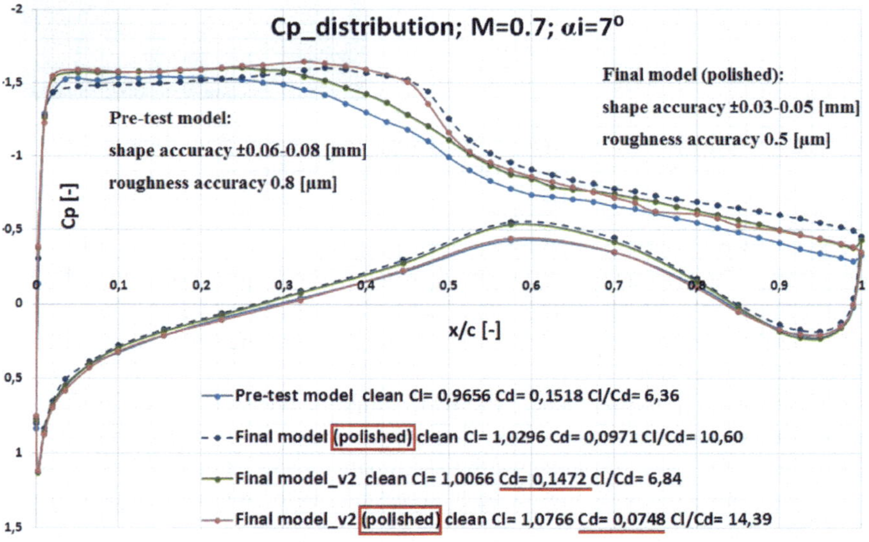

Fig. 31 Cp distribution comparison of baseline configuration; M = 0.7; $\alpha_i = 7°$

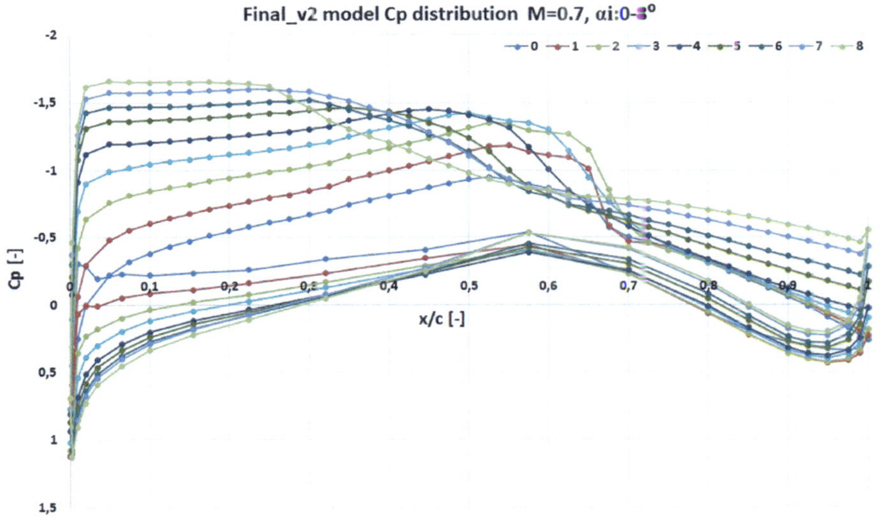

Fig. 32 Cp distribution of baseline configuration of the Final model_v2; M = 0.7; $\alpha_i = 0°–8°$

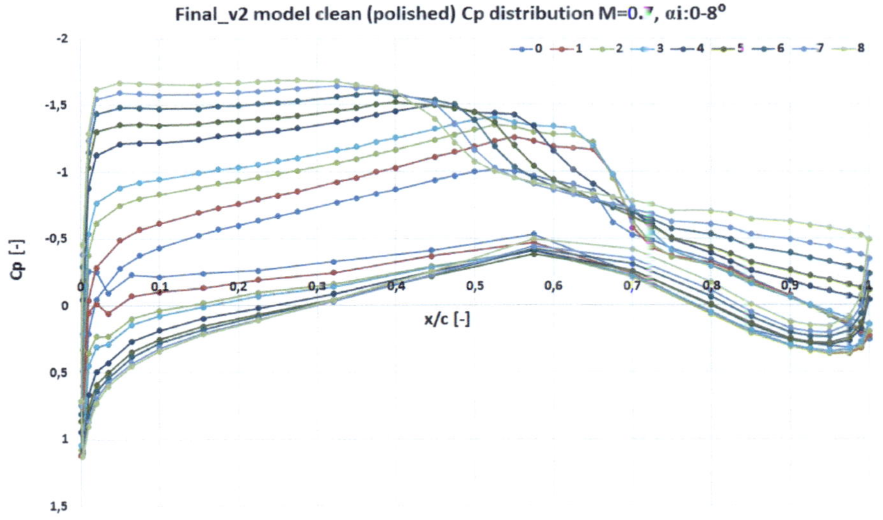

Fig. 33 Cp distribution of baseline configuration of the Final model_v2 (polished); M = 0.7; $\alpha_i = 0°–8°$

and 38). Carborundum strips (h = 0.1 mm) change Cp plateau of the baseline configuration into step pressure rise. Upstream positions of carborundum configurations cause pressure rise occurence closer to the leading edge occurrence. Simultaneously, lower negative pressure values up to the shock wave occurrence take place. The closer to the leading edge turbulizers positions also characterise the lower lift coefficients

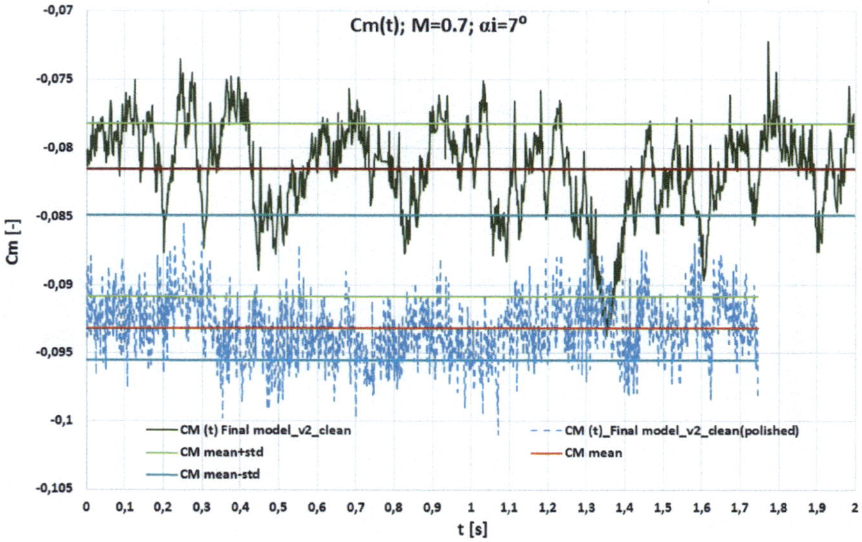

Fig. 34 Baseline configuration momentum versus time plot comparison; M = 0.7; $\alpha_i = 7°$

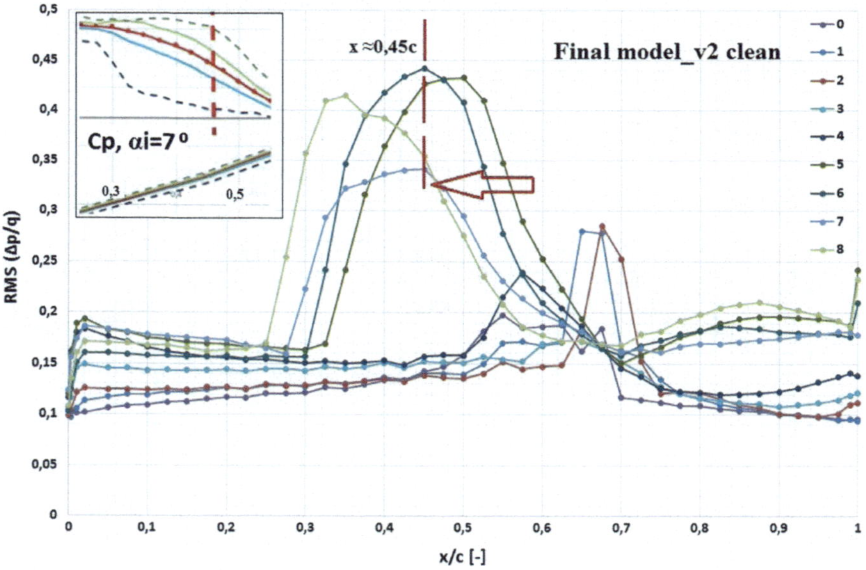

Fig. 35 Baseline configuration pressure fluctuation RMS of the Final model_v2; M = 0.7; $\alpha_i = 0°–8°$

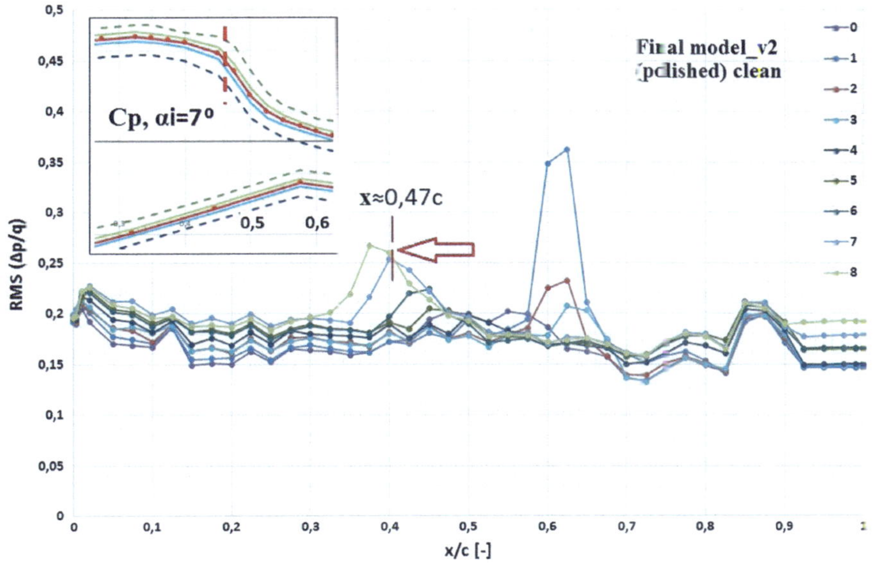

Fig. 36 Baseline configuration pressure fluctuation RMS of the Final model_v2 (polished); M = 0.7; $\alpha_i = 0°–8°$

values. However, the drag value is often, but not always, being reduced (especially at moderate angle of incidence).

On Figs. 39, 40 and 41 the influence of carborundum strip grain size for x/c = 0.4 location was presented. The higher height of turbulizing strip caused stronger change of pressure fluctuation at turbulizer location. Also for bigger carborumdum grains configurations, the lift coefficient was getting smaller. The most optimal carborumdum configuration in terms of drag and CL/CD ratio was for the lowest from tested strip size (h = 0.1 mm). However, at different x/c strip locations and depending on the angle of incidence it wasn't the rule (Fig. 42).

1.3.5 Conclusions

- The model shape and mainly (upper) surface quality have significant impact on boundary layer development and its further interaction with shock wave.
- Carborundum presence change the boundary layer shock wave interaction character into turbulent one. The pressure rise (seen in pressure distribution Cp plot) is then more rapid than for baseline configuration.
- At the low angle of incidence, before the model vibration onset appears (buffet onset indication was described further, in Sect. 2.2.) SW drag component apparently is less significant than friction drag. (thus a weak separation is seen

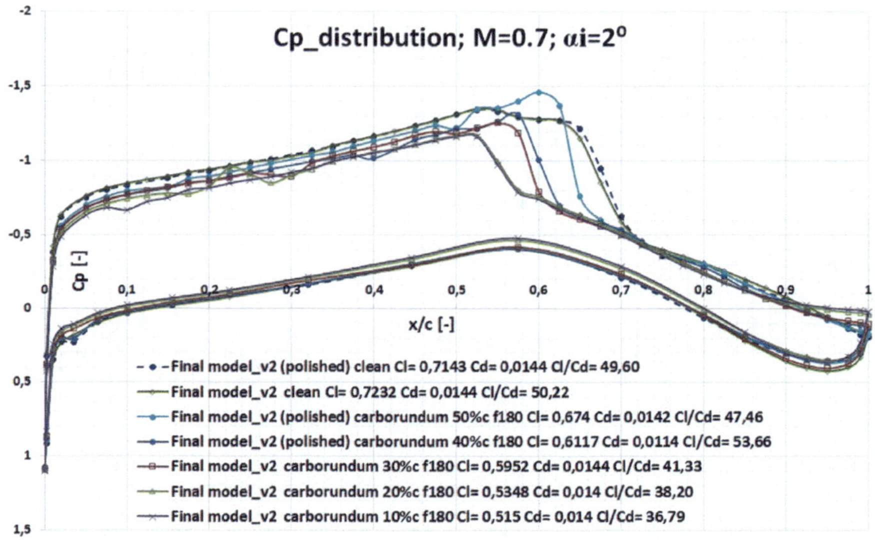

Fig. 37 Cp distribution comparison of baseline and with roughness configuration (h = 0.1 mm); M = 0.7; $\alpha_i = 2°$

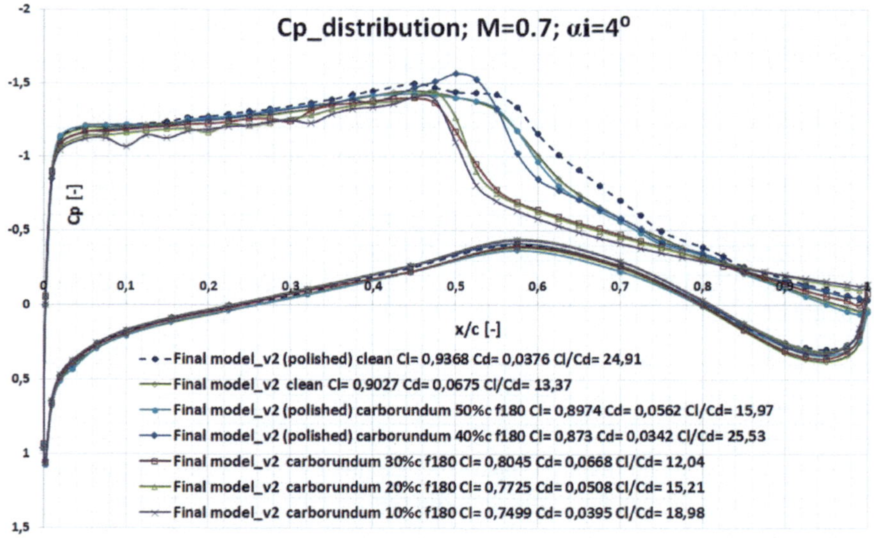

Fig. 38 Cp distribution comparison: baseline and with roughness configuration (h = 0.1 mm); M = 0.7; $\alpha_i = 4°$

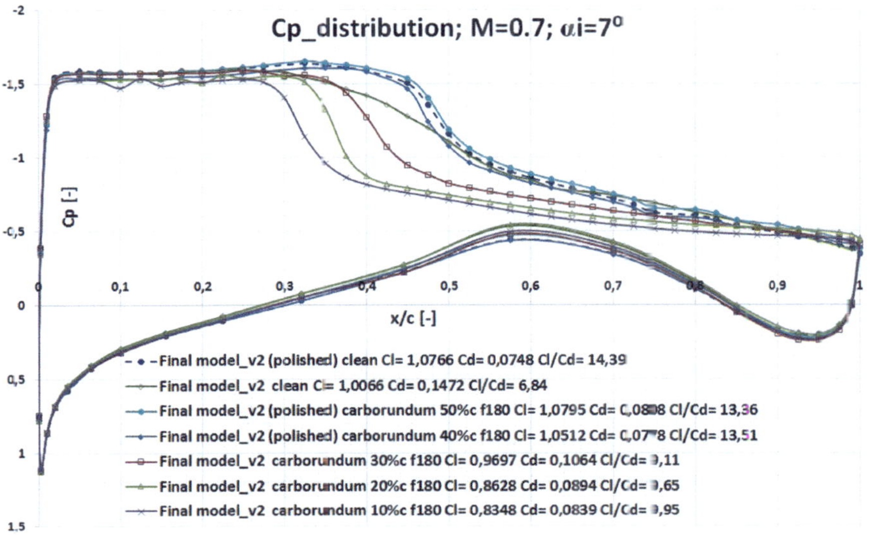

Fig. 39 Cp distribution comparison: baseline and with roughness configuration (h = 0.1 mm); M = 0.7; $\alpha_i = 7°$

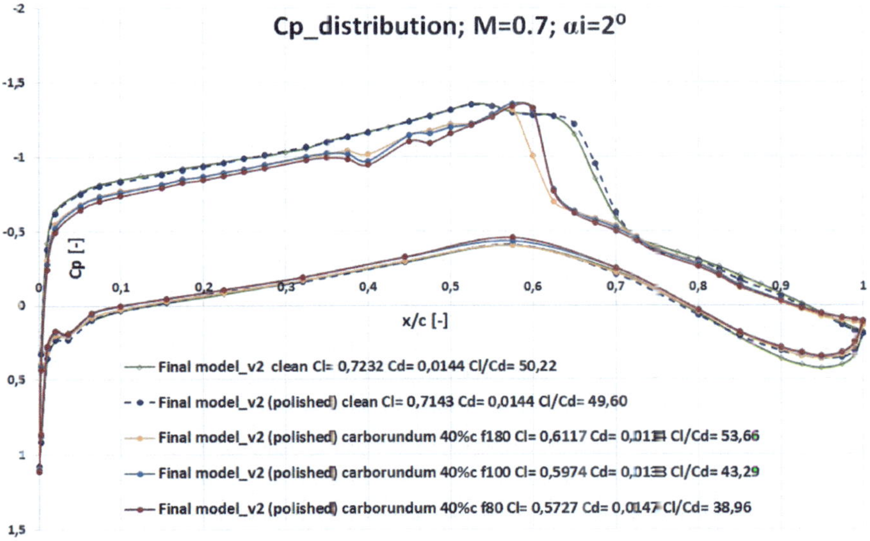

Fig. 40 Cp distribution comparison: baseline and with roughness configuration (h = 0.1; 0.15; 0.26 mm); M = 0.7; $\alpha_i = 2°$

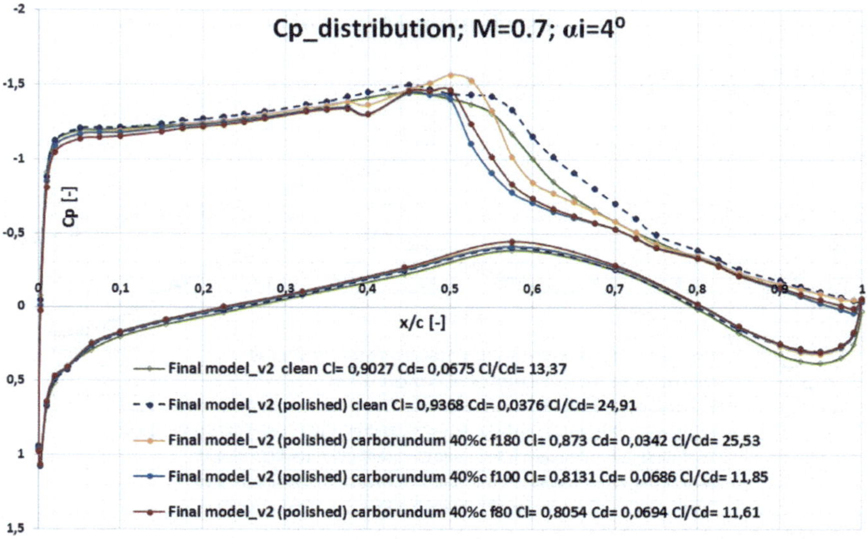

Fig. 41 Cp distribution comparison: baseline and with roughness configuration (h = 0.1; 0.15; 0.26 mm); M = 0.7; $\alpha_i = 4°$

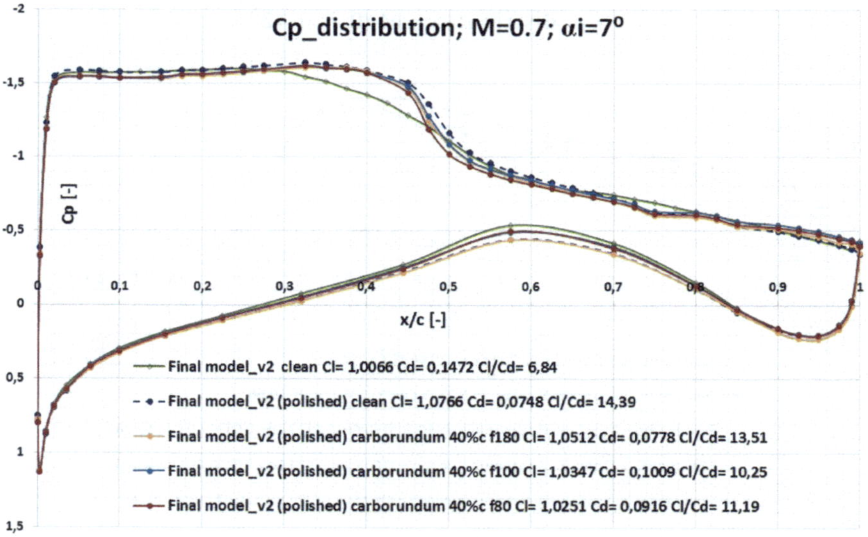

Fig. 42 Cp distribution comparison: baseline and with roughness configuration (h = 0.1; 0.15; 0.26 mm); M = 0.7; $\alpha_i = 7°$

closer to the TE). The baseline configurations without turbulent triggers, especially with quality upper surface of airfoil model, indicates on better aerodynamic performance.

- The beneficial usage of turbulent triggers was observed at moderate angle of incidence. For almost all test cases, the most optimum carborundum position was located just ahead of the front foot of the SW (at x/c 0.4–0.5 of the model chord). Exception was the vibration onset angle (shown in Sect. 2.2), where at the most front position of turbulent trigger was the most beneficial in terms of CL/CD value.
- At high angle of incidence the turbulent triggers don't affect on shock wave boundary effectively in order to cause less severe shock induced separation. In the almost all case tested the quality baseline configuration had the higher performance
- Changing the carborundum strip position into the leading edge direction, independently on the angle of incidence, results in earlier pressure rise (on Cp plot), causing the SW moves upstream. Simultaneously, the lift coefficient value gets smaller.

2 Buffet Limit for Laminar Profile

2.1 DAAV

2.1.1 Introduction

WP5 is devoted to the study of transition location effect on separation size, shock structure and unsteadiness. In particular it should address how close to the shock location transition has to occur to avoid detrimental effects, if any, associated with laminar shock-induced separation, and what is the best way to trigger transition. This task, WP5.2, focuses on the experimental and numerical study of laminar, transitional and turbulent shock boundary layer interaction (SBLI) on the V2C laminar profile, designed by DAAV in WP5.1. In the scope of the WP5.2 task of the TFAST project, the contribution of DAAV consisted in:

- The adaptation of an automatic transition location predicting algorithm to flows with laminar separation. So far the algorithm was only developed for attached boundary layer, and has been adapted to cope with separating flows during the TFAST project. This has required combining iterative algorithm for the detection of transition and imposition of an intermittency field taken from the literature to correctly predict the laminar separation bubble. The method has been applied on the low Mach number SD7003 configuration, and the study also covered effects of turbulence modeling employed along with the intermittency field. The next step would be the prediction of transitional shock-induced separation, such as the one observed on the present V2C configuration.

- Numerical simulations of the buffeting on the V2C profile using the Zonal Detached Eddy Simulation (ZDES) in fully turbulent and transitional regimes. Due to the considerable efforts of WP5 partners, IMFT, URMLS and LIV, devoted to unsteady scale-resolving simulations of buffeting on V2C in WP5.2, and availability of UCAM and IoA experimental data, as well as more recent availability of URMLS DNS data of laminar and turbulent SBLI at buffet, DAAV has undertaken additional work using its in-house code, for the simulation of turbulent and transitional buffet, thus providing additional elements for the cross-comparison survey amongst numerical and experimental studies, in terms of buffeting frequency, aerodynamic coefficient, surface pressure coefficient and RMS pressure.

2.1.2 Description of the Tools

The in-house Navier-Stokes solver Aether is a compressible, finite-element code based on the GLS formulation of the symmetric form of the Navier-Stokes equations using entropy variables. See Chalot et al. [1] for more details. Different turbulence models are available from 1-equation RANS models to DRSM or ZDES and LES. The laminar-to-turbulent transition prediction is based on the assessment of the amplification N-factor, computed with a parabola method (e^N stability data base), see Arnal [2] that is implemented in a 3D boundary layer compressible code. Such a method has been used by DAAV with success for many in flight laminarity tests (Courty, Bulgubure, and Arnal, 24th–27th May 1993 [3]).

2.1.3 Automatic Prediction of Transitional Separation Bubble

The standard methodology for automatic prediction of transition location relies on the coupling between a Navier-Stokes (NS) solver and a Boundary Layer (BL) code which integrates a robust method for determination of transition criteria, the parabola method (Perraud et al. [13]). However this method only works for attached boundary layer and for cases with flow separation, such as the presence of a laminar separation bubble, the alternative consists in performing full linear stability analysis directly on the NS velocity profiles. The conceptual implementation of both methods is depicted in Fig. 43.

Both methods provide an evolution of the N-factor, N(x), x the stream-wise coordinate, and transition occurs when N-factor overcomes a threshold provided by the Mack law:

$$N = -8.43 - 2.4 \ln Tu$$

It was sought in the course of the TFAST project to address respective shortcomings of the two above options:

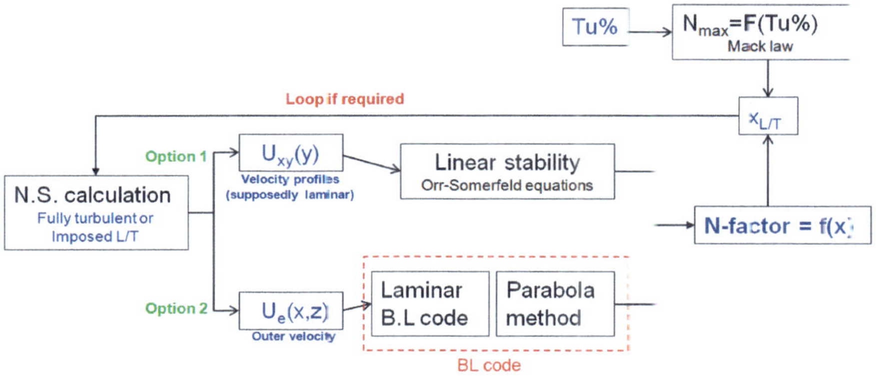

Fig. 43 Methodology of the transition prediction tool chain

Fig. 44 Reference mesh of the SD7003 profile

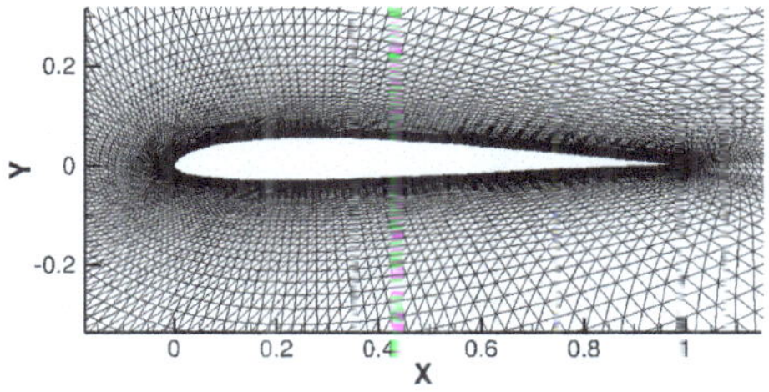

- For option 1: The linear stability solver assumes laminar profiles as input. When a NS calculation, whether fully turbulent or with imposed transition too early upstream, is performed to provide input to the linear stability code, the latter will classically under-predict the N-factor growth and in practice the Mack-law threshold is never reached.
- For option 2: The BL code is not suitable for separated flows and the N-factor calculation stops at separation point.

Adaptation of both Option 1 and 2 to flows with laminar separation bubble were therefore proposed. The case study is the SD7003 profile: it is a low Mach number configuration which depicts a long bubble, around 15–20% of the chord length. The geometry and the mesh can be seen in Fig. 44.

2.1.4 Improving Option 1

Option 2: N factor (thick black line) provided by the BL code
Figure 45 compares the output of both options. The linear stability analysis is first performed over a range of frequencies to determine which is the most dominant. The simpler parabola method of option 2 provides a N-factor up to the shock but the calculation stops further downstream (Fig. 46).

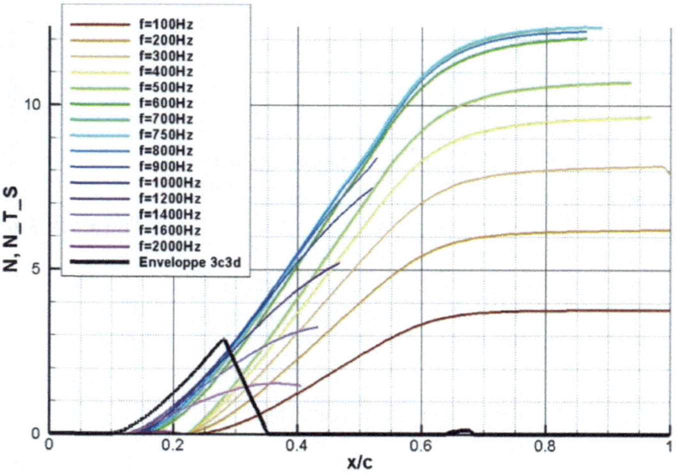

Fig. 45 SD7003 profile. Option 1: frequency sweep provided by the linear-stability solver

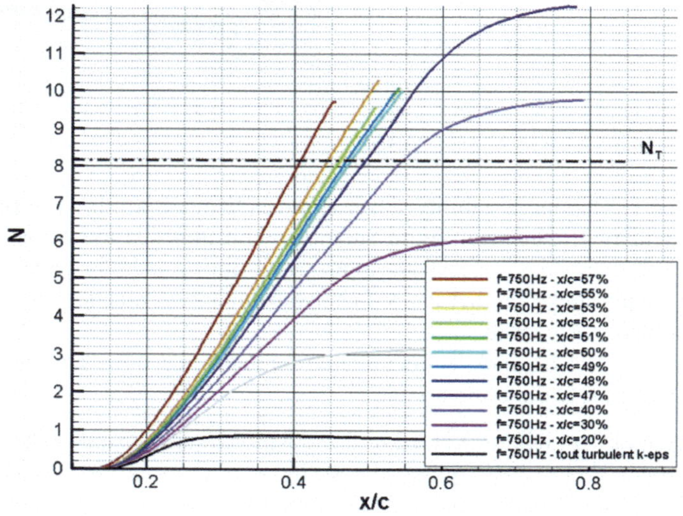

Fig. 46 SD7003 profile. Option 1: N-factor calculated by the linear stability code for the dominant frequency (f = 750 Hz)

Following option 1, after determining the dominant frequency the predictive tool chain can be run iteratively until convergence of the transition location, although it was noticed a large number of iterations is required in practice. Moreover a range of acceptable values for the transition location should be known beforehand: indeed if the transition location is imposed too early the linear stability input are fully turbulent profiles and the calculated N-factor never reaches the threshold. If the transition location is imposed too late the laminar separation bubble remains in the laminar region and the flow may separate, yielding unsteady calculation.

To understand how the number of iterations can be reduced, a series of NS calculations were performed with imposed transition at various locations, and the N-factor was determined for the dominant frequency for each NS calculation with the linear stability code (Fig. 47). The transition location predicted by the latter occurs when the N factor overcomes the Mack law threshold (indicated by dashed black line on Fig. 47). The number of iterations required for convergence of the transition location will be all the more important as the transition location in the input NS simulation and that subsequently estimated by the linear stability code differ. Those are reported in Table 2. It may be noticed however that the average of both transition locations is nearly constant.

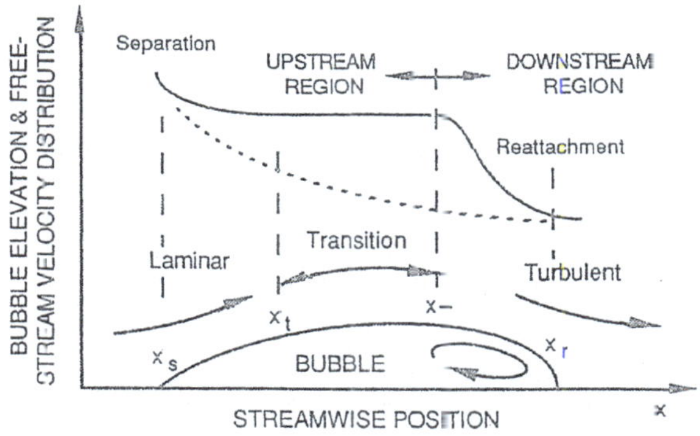

Fig. 47 Schematic of a laminar bubble

Table 2 SD7003 profile. Option 1: Fig. 45 results. Transition location of the input NS flow and that calculated by the linear stability code

Imposed transition in the NS code (%)	Transition location predicted by the linear stability code (%)	Averaged of both location (%)
57	40.7	48.9
55	44.3	49.7
53	45.8	49.4
52	46	49
51	47.1	49.1
50	47.6	48.8
49	47	48
48	49.6	48.8
47	49.5	48.3
40	54.5	47.3

To use Option 1 in an automatic and robust manner the following algorithm is thus proposed:

- An initial NS calculation is run with transition imposed at $x_{NS} = 50\%$ chord. If the calculation does not converge (the laminar region reaches the laminar separation bubble), a new simulation is performed with imposed x_{NS} further upstream, repeatedly until a converged simulation is obtained.
- Stability analysis is performed on the last NS solution. If the N-factor does not reach the Mack law threshold, meaning the transition location is too early upstream, a new NS simulation is performed with imposed x_{NS} further downstream, repeatedly until N-factor reaches the Mack law threshold. This transition location is noted x_{LS}.

The final transition location is defined as $(x_{NS} + x_{LS})/2$.

2.1.5 Improving Option 2

The above algorithm, "improved option 1", allows determining the transition location in case of a laminar separation bubble with a minimum number of NS simulations required. However in such flows, imposing transition at the right location does not prove sufficient to secure correct prediction of the flow topology, in particular the length of the laminar separation bubble. For instance, a high sensitivity to the turbulence model was observed. This motivates more accurate transition modeling such as using an intermittency field.

Figure 47 shows a schematic of a laminar bubble. It was numerically assessed that prescribing a gradually increasing field of intermittency γ between x_t and x_T in the calculation (in turbulent NS simulations, γ generally factors turbulent viscosity wherever used) noticeably improves the prediction of the flow topology (bubble size, most visible on Cp plots plateau) and independently on the turbulence model.

The model of Mayle [14] is used: $(\mathrm{Re}_x)_{st} = \mathrm{Re}_{xt} - \mathrm{Re}_{xs} = A_{st}$ and $(\mathrm{Re}_x)_{sT} = \mathrm{Re}_{xT} - \mathrm{Re}_{xs} = A_{sT}\mathrm{Re}_{\theta s}^{0.7}$. Coefficients A_{st} and A_{sT} are flow dependant and have been tuned for the SD7003 flow. This is used in conjunction with the model of Dhawan and Narasimha [15] to provide the intermittency field.

$$\gamma = 1 - \exp[-\hat{n}\sigma(\mathrm{Re}_x - \mathrm{Re}_{xt})^2]$$

with $\hat{n}\sigma = 2.28 \times 10^{-5}/\mathrm{Re}_{\theta s}^{1.4}$ for a separated flow.

An improved Option 2 algorithm is proposed.

- NS simulation followed by a BL code simulation for determination of the separation point x_s, and the momentum thickness Re_{θ_s}.
- Definition of an intermittency field using the Narasimha formula.
- NS simulation with imposed transition at x_t and the intermittency field.

The improved Option 2 algorithm is applied to the same SD7003 case.

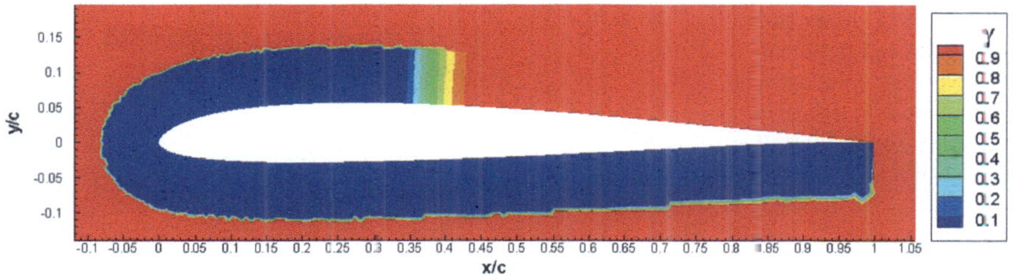

Fig. 48 Intermittency map on the SD7003 case provided by the Narasimha relation

The NS results with the above intermittency map, Fig. 48, are shown on Fig. 49. Results are compared to a fully turbulent simulation, as well as to that from an advanced Reynolds Stress Model, performed by Probst et al. [16]. Current "Option 2" results using a generic k-ε model are in very good agreement with the RSM results,

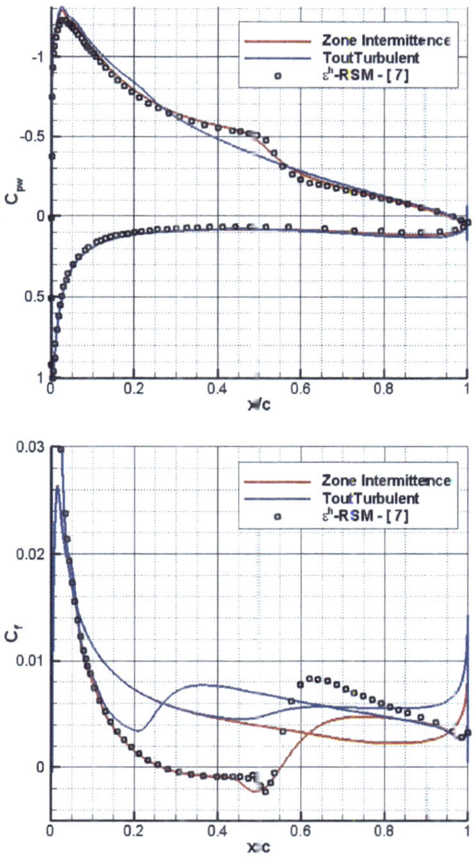

Fig. 49 Pressure and skin-friction coefficient. Comparison between Option 2, fully turbulent and RSM results from Probst et al. [16]

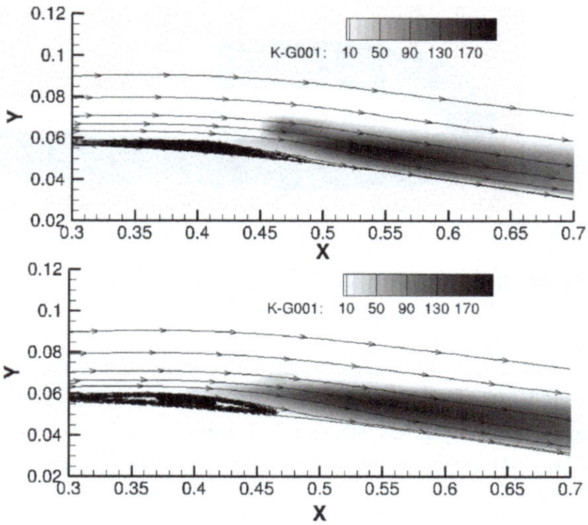

Fig. 50 RSM computation: streamlines and turbulent kinetic energy contours with k-ε computation. Effect of the tripping intensity (left and right)

in particular in the laminar bubble region, and the use of an intermittency field allows the correct Cp plateau to be returned.

2.1.6 Further Investigation on Turbulence Modeling Sensitivity

Part of the good agreement observed on Fig. 49 is due to the use of the Narasimha intermittency field, Eq. 1. Further turbulence modeling sensitivity study was performed on the same SD7003 case, with this time, a sharp transition imposed in the calculation, at $x_T = 47\%$ of chord length (i.e. $\gamma = 0$ for $x < x_T$ and $\gamma > 1$ for $x > x_T$), with the same k-ε model as well as the RSM of Gerolymos and Vallet [17].

The DRSM computations show an effect of the tripping intensity set for prescribing the transition, Fig. 50. Compared with the k-ε computation (Fig. 51), the DRSM induces an earlier reattachment point and thus a shorter bubble. This is confirmed looking at the Cp and Cf distribution, Fig. 52. The RSM offers a better agreement with the reference RSM data [16] (Fig. 52) and also with reference Experiment [18] and Implicit LES [19] data in terms of lift and drag coefficient (Table 3). Comparison between the k-ε with Narasimha intermittency field (Fig. 49) and RSM with sharp transition (Fig. 52) shows superiority of transition modeling over turbulence modeling in this case.

2.1.7 Scale Resolving Simulation of Buffeting

Simulations were performed on the V2C profile for buffeting conditions ($M = 0.7$, $\alpha = 7°$) using Aether. The ZDES method [20] was used based on two turbulence models: the 2-layer k-ε and the Spalart-Allmaras model. The mesh used is shown on

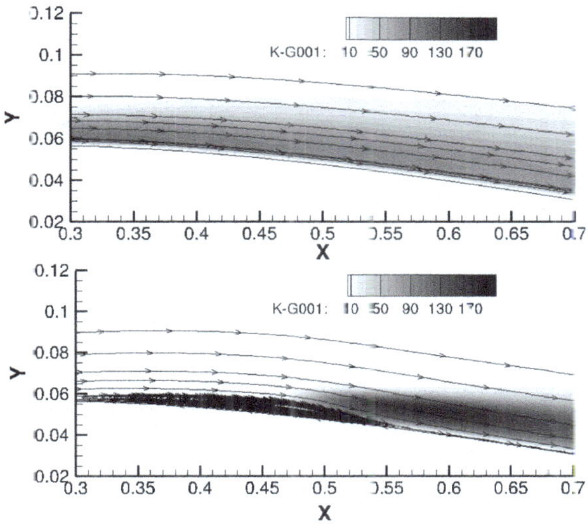

Fig. 51 k-ε computation: streamlines and turbulent kinetic energy contours with. fully turbulent (right) and fixed transition

Fig. 53, and details are summarized on Table 4. A time step of 1×10^{-6} was used. Mesh was extruded over 33% chord on the spanwise direction with 80 grid points in span.

Three calculations were performed: 2 simulations with a turbulent boundary layer from the leading edge (termed "fully turbulent"), using the ZDES-SA and the ZDES-k-ε models respectively, and one simulation with the ZDES-k-ε with a boundary layer tripped at 24% chord. This tripping position was imposed at what was observed to be the most upstream shock position in the fully turbulent ZDES-k-ε case (thus preventing the SBLI from occurring in the laminar region, which is a situation that cannot be handled satisfactorily by turbulence modeling with the present state of the art). The tripping applied in the present case is sharp: i.e. Heaviside intermittency.

2.1.8 Lift, Drag and Surface Pressure Coefficient

Lift, drag and surface pressure coefficients are presented in Figs. 55, 56 and 57 respectively, with results from the DNS of URMLS (boundary layer tripped at 10% chord) added for comparison. The SA-URANS and SA-DDES results from URMLS are also added for comparison, as well as the k-ε OES results from IMFT for Figs 54 and 55.

The many DDES run in the course of the TFAST project for turbulent SBLI at buffeting have all reported an exaggeratedly large shock oscillation amplitude, with a shock travelling all the way to the leading edge, reflected by large CL and CD oscillations, with a sawtooth shape, as clearly illustrated by the URMLS SA-DDES results shown here. This was shown to be typical of DDES simulations and this behavior has not been reported in URANS or OES simulations, as, again, shown here. The above has motivated the present ZDES study performed by DAAV: it

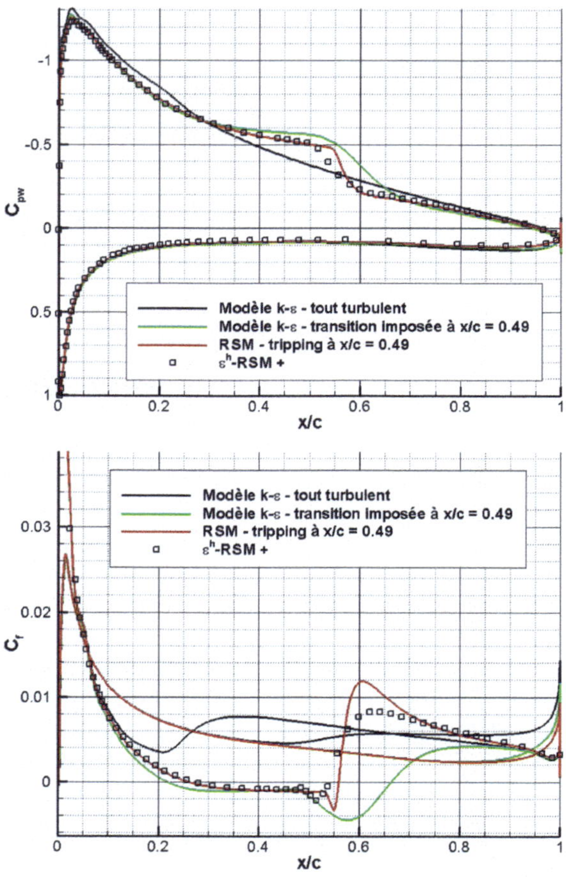

Fig. 52 Prediction of the wall-pressure coefficient (left) and the skin-friction coefficient (right) for the k-ε and the RSM simulation, compared to the reference data of [16]

Table 3 Lift and drag coefficients for reference data compared to the present CFD results

	References	Modeles				
	Exp.	εh-RSM	ILES	k-ε fully turbulent	k-ε imposed transition 49%c	RSM imposed transition 49%c
C_L	0.55	0.5809	0.59	0.604	0.615	0.588
C_D	–	0.0203	0.022	0.0214	0.0207	0.0205

was sought to determine whether the reported behavior of previous DDES studies was to be attributed to the DDES method itself or to other parameters (such as the grid resolution). As shown in Figs. 54, 55 and 56, the shock oscillation amplitude is smaller for the present three ZDES, with a rather sinusoidal evolution of CL(t) and CD(t). The comparison with reference DNS results is satisfactory although not consistent: for the drag coefficient, the agreement with DNS data is best for the fully

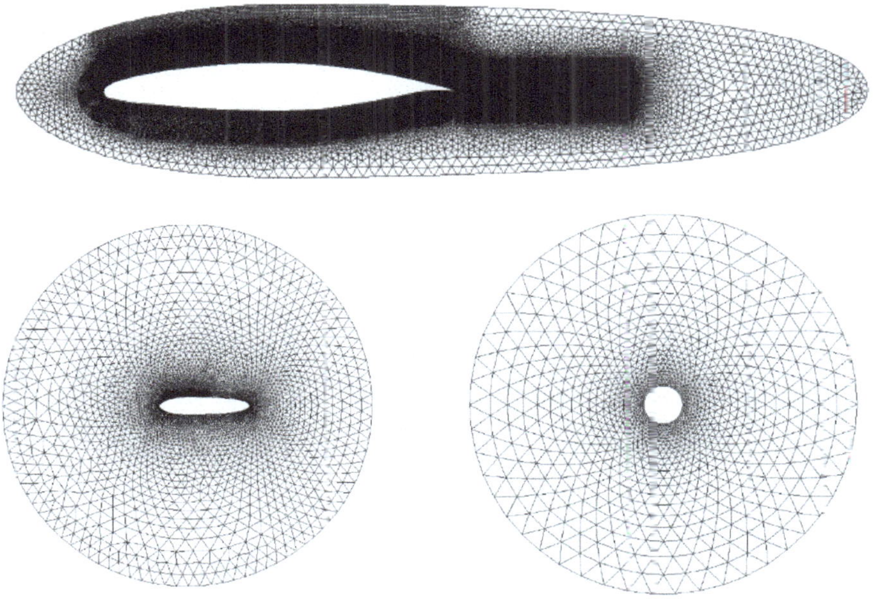

Fig. 53 Mesh used for the ZDES simulation: the mesh in the profile region is embedded into a ellipsoidal sub-domain (top), embedded itself into a circle (bottom left), itself embedded into the circular domain boundary (bottom right)

Table 4 Characteristics of the mesh (node number, mesh-step size in natural and wall-unit)

Nodes around profile	2000		Near-wall region	LES region
Total number of nodes	13 millions	$\Delta x+$	150	150
First off-wall node distance to the wall	0.5 μm	$\Delta y+$	0.25	150
Δx upper side	0.3 mm	$\Delta z+$	410	410
Δx lower side	0.3 mm			

turbulent ZDES-SA results, whereas the Cp distribution coming from the ZDES-k-ε (with boundary layer tripped at 24% chord) is in best agreement with the DNS results. The overall improvement offered by the present ZDES results may well be attributed to the unstructured meshing strategy allowed by DAAV in-house code, with 2000 computational points clustered in the streamwise direction around the profile, for a total points count of less than 13 millions. For comparison the structured grid used by URMLS for their DDES calculation has 400 points around the profile, and DAAV ZDES streamwise resolution around the profile is closer to URMLS DNS resolution.

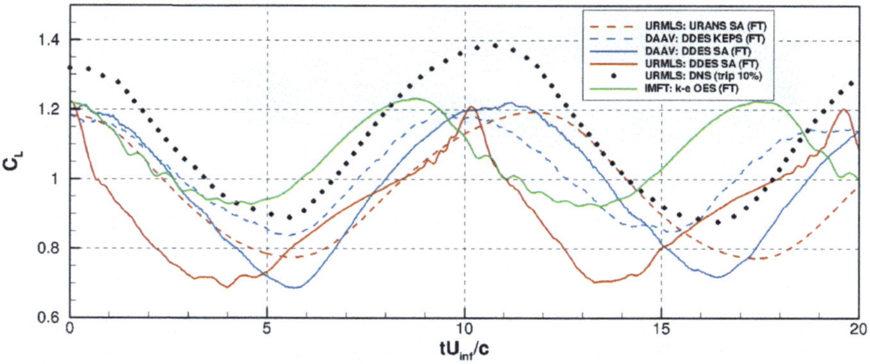

Fig. 54 Evolution of the lift coefficient over time (convective units) for DAAV DDES-SA and DDES-k-ε versus URMLS DDES-SA and URANS-SA. Reference data are from URMLS DNS tripped at 10% chord length

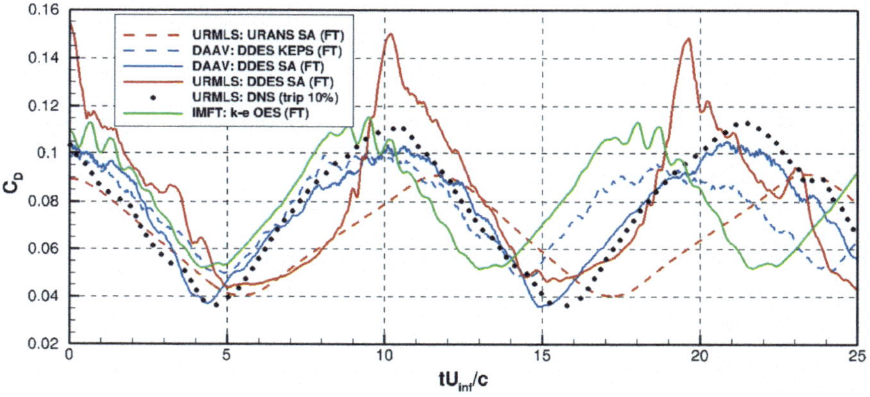

Fig. 55 Evolution of the drag coefficient over time (convective units) for DAAV DDES-k-ε versus URMLS DDES-SA and URANS-SA. Reference data are from URMLS DNS tripped at 10% chord length

2.1.9 Shock Structure and Oscillation Amplitude

Figure 57 shows the instantaneous pressure field over the V2C profiles for the three present ZDES calculations. The presence of possible spanwise oscillation of the shock structure could not be experimentally characterized by IoA, due to absence of PIV planes in the span direction, although URMLS DNS have reported such oscillation for the laminar interaction DNS. The present fully turbulent ZDES evidence a straight shock structure, but in the 24% chord tripped case, a single wave structure appears: it is hard to determine whether this could be a boundary layer thickness effect (i.e. in the tripped case, the boundary layer is thinner at shock-interaction) or

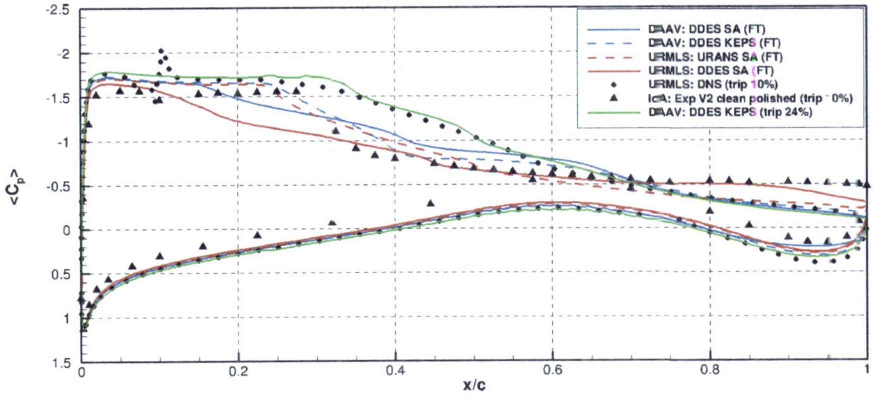

Fig. 56 Evolution of the surface pressure coefficient for DAAV DDES-SA and DDES-k-ε versus URMLS DDES-SA and URANS-SA. Reference data are from URMLS DNS as well as IoA Exp data (V2 polished) both tripped at 10% chord length

simply numerical. Indeed only 80 points were used in the spanwise direction with a 1/3 chord domain length in span.

In Fig. 58, plots are provided for the most upstream and downstream shock position respectively. It can be clearly seen that the shock oscillation in the stream-wise direction depends on the turbulence models. It oscillates between 15 and 42% chord length for the ZDES-SA (that is a 27% chord length amplitude), and between 25 and 40% chord length for the ZDES-k-ε (that is a 15% chord length amplitude). The tripping only shifts the averaged shock position further downstream but does not change the oscillation amplitude. Such variability with respect to turbulence modeling is also reported with other WP5.2 partners, as illustrated on the Cp plots, Fig. 56.

2.1.10 Flow Pattern and Buffeting Mechanism

Figure 58 shows instantaneous Schlieren visualization of the three calculations, Fig. 59 presents surface pressure spectrum map, for the continuous range of locations along the profile upper surface. Figure 60 provides PSD spectra (i.e. "cuts" of Fig. 59 plots) for specific locations along the profile and in the wake. Schlieren pictures allows characterizing Kelvin-Helmholtz (KH) instabilities in the separated shear layer downstream of the shock-induced separation, and the Von Kármán pattern (VK) in the wake, all present in the three calculations to different extent. The buffeting frequency is 106 Hz and 119 Hz, respectively for the ZDES-SA and the ZDES-k-ε (corresponding Strouhal number St = 0.093 and St = 0.104). Generally, KH are more noticeable for the ZDES-k-ε model than for the ZDES-SA with characteristic frequency of about 10 kHz, whereas the VK are stronger for the ZDES-SA results. VK frequency is 1970 Hz and 2700 Hz for the ZDES-SA and the ZDES-k-ε

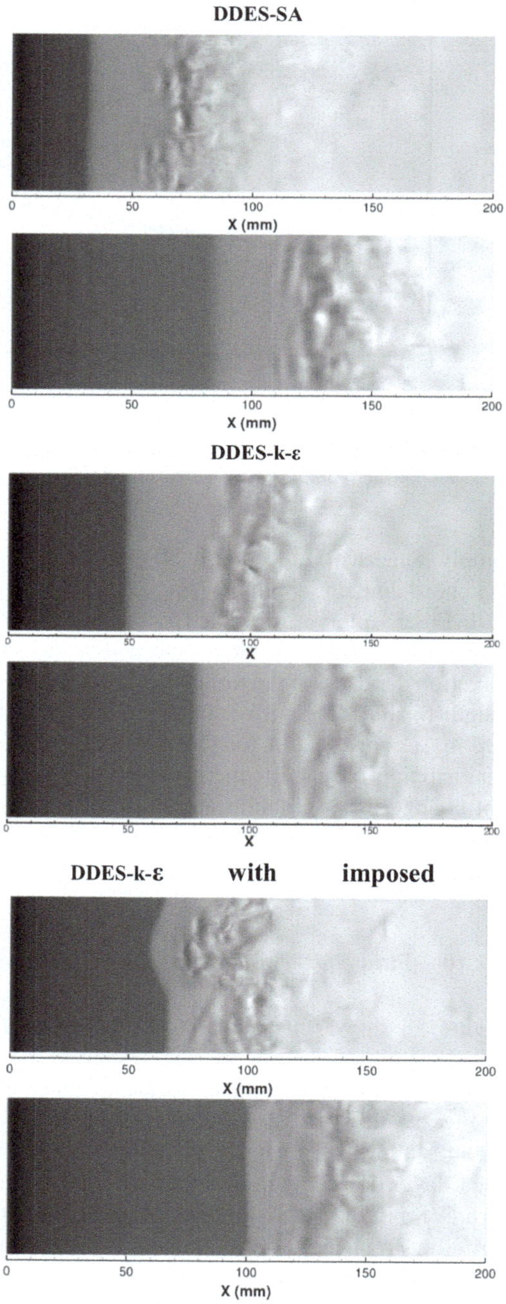

Fig. 57 Instantaneous pressure field over the upper surface of the V2C profile: snapshots for the two extremes shockwave position

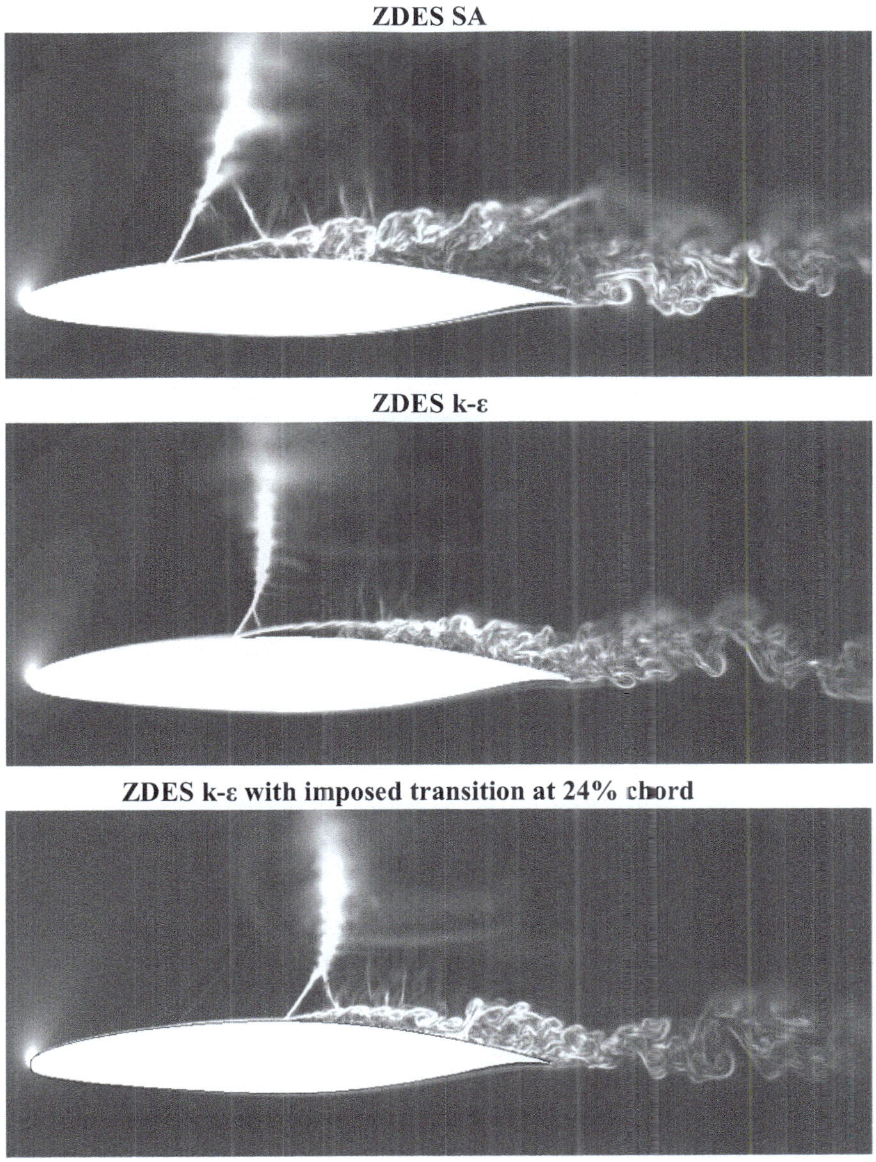

Fig. 58 Instantaneous Schlieren visualization of the three simulations

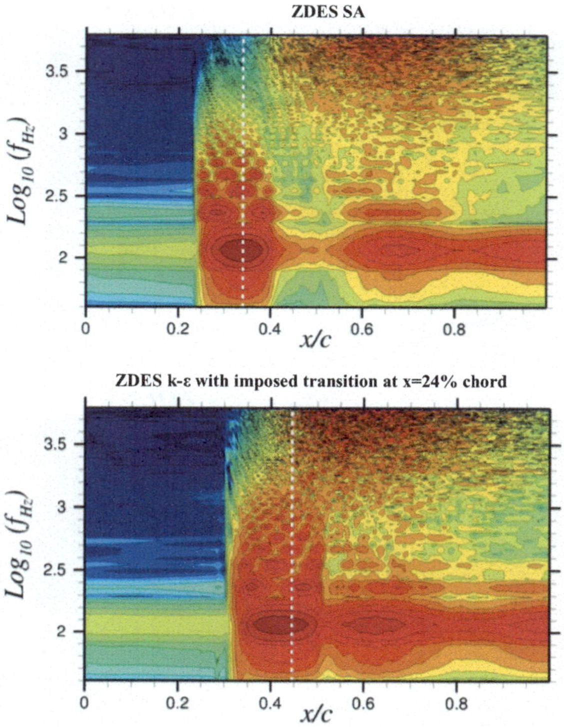

Fig. 59 Surface pressure PSD for the three simulations

respectively. Spectral map evidence the sustenance of instabilities at buffeting frequency throughout the separated shear layer, although the analysis at this stage could not ascertain coupling between the buffeting, KH and VK mechanism. The effect of tripping at 24% chord (creating a weak oblique shock at the tripping location, as seen on the Schlieren picture) is not reflected on the spectral information: the buffeting frequency is unchanged and VK and KH are not noticeably affected (Fig. 61).

2.1.11 Surface Pressure Fluctuations

Normalized RMS surface pressure fluctuations are plotted on Fig. 62, along with DNS data from URMLS, both for the turbulent interaction (BL tripped at 10% chord) and for the laminar interaction (clean configuration). Although differences are visible amongst models (ZDES-SA returns smaller P_{RMS} at buffet than the ZDES k-ε but larger values in the separated flow region) the most crucial observation is that for the tripped ZDES k-ε, P_{RMS} maximum values in the shock region are not affected by tripping but P_{RMS} is substantially increased further downstream, compared to the fully turbulent ZDES k-ε.

Fig. 60 ZDES-SA: power density spectra for four probes, at several locations over the upper surface and in the wake

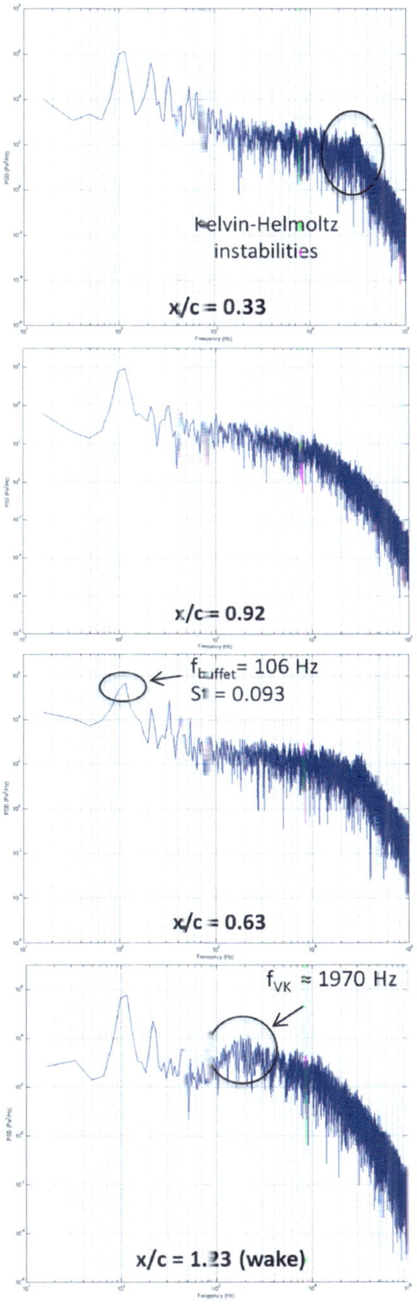

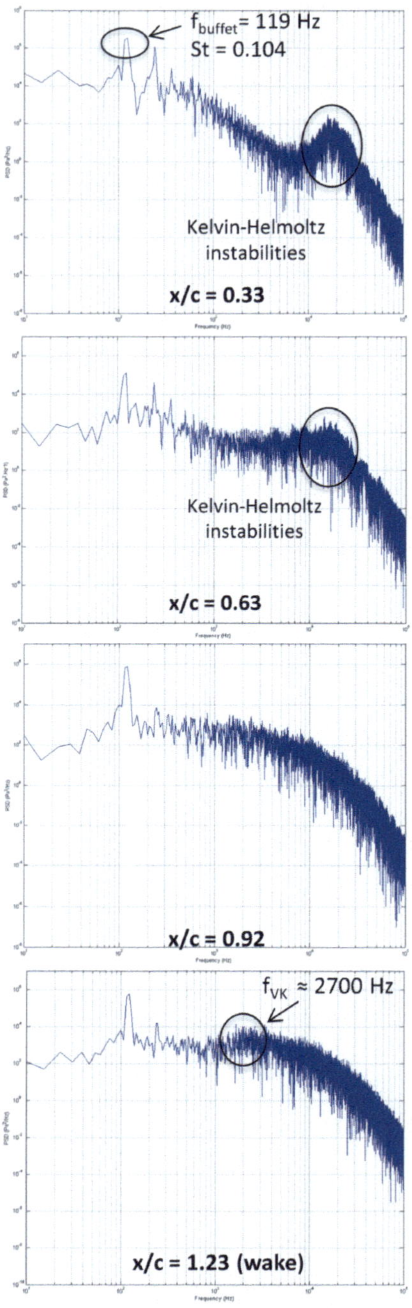

Fig. 61 ZDES-k-ε: power density spectra for four probes, at several locations over the upper surface and in the wake

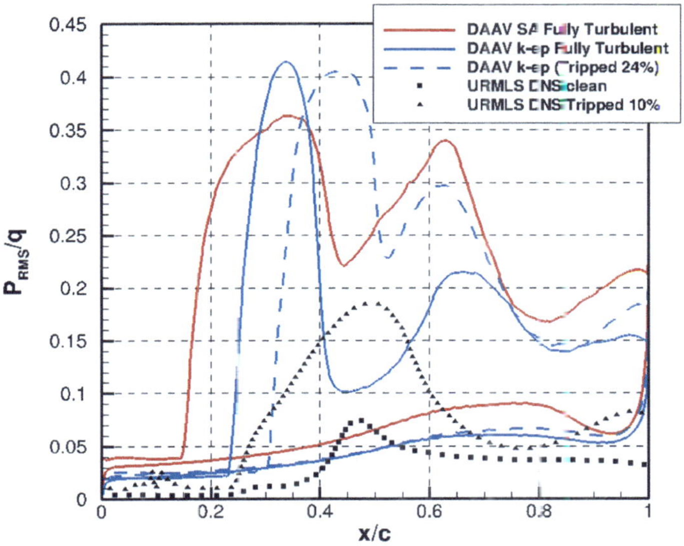

Fig. 62 Surface pressure RMS for the three DAAV simulations compared to URMLS reference DNS data

This is in contrast with the DNS results, for which clean airfoil P_{RMS} results are considerably smaller than for the tripped case: this would tend to demonstrate that the laminar interaction behavior is not the limiting case of tripped turbulent SBLI when the tripping position tending to the shock position with strong differences between a laminar SBLI and a turbulent SBLI (however thin the boundary layer may be).

2.1.12 Conclusions

Automatic prediction of transitional separation bubble: Extending the applicability range of the current in-house transition prediction tool has allowed the correct determination of the transition location and flow topology of the laminar bubble on the SD7003 profile. Two options were explored:

- Coupling a NS solver with a linear stability code
- Coupling a NS solver with a BL code and parabola method.

CFD results show that the knowledge of the separation point (provided by the boundary layer code) combined with an algebraic expression of an intermittency field provides the best predictions of the transitional separation bubble. The study has shown a stronger influence of transition modeling (through intermittency field) compared to turbulence modeling.

Scale resolving simulation of buffeting: Three zonal DES of buffeting were performed on an unstructured mesh with sensitivity study to underlying RANS model and tripping position. The results analysis reveals a strong influence on the RANS model used, in terms of buffeting frequency, amplitude and pressure fluctuation, as well as instability mechanisms. The tripping effect mostly affects the pressure fluctuations downstream of the shock. The unstructured mesh used herein has a resolution, in terms of nodes around the profile, close to the DNS of URMLS, which yields considerably improved predictions compared to other DES studies of the same case with structured meshes.

2.2 *IoA*

2.2.1 Experiment

The determination of buffet onset boundaries for Pre-test model and Final_v2 (less accurate models) were done through the measurement of RMS of the bending tensions, registered by a strain gauge bridge (mounted at 42% of model chord near wall).

In a graphs below for Mach number M = 0.7 (Figs. 63 and 64) it is seen that the jump of tension begins at the angle of incidence $\alpha_i = 3°–4°$. There is a tendency that with the shifting of a carborundum strip on the model upper model surface in the leading edge direction, the jump of tension vibration point is shifted of about $\Delta\alpha_i = 1°$. Then the level of model vibrations increased.

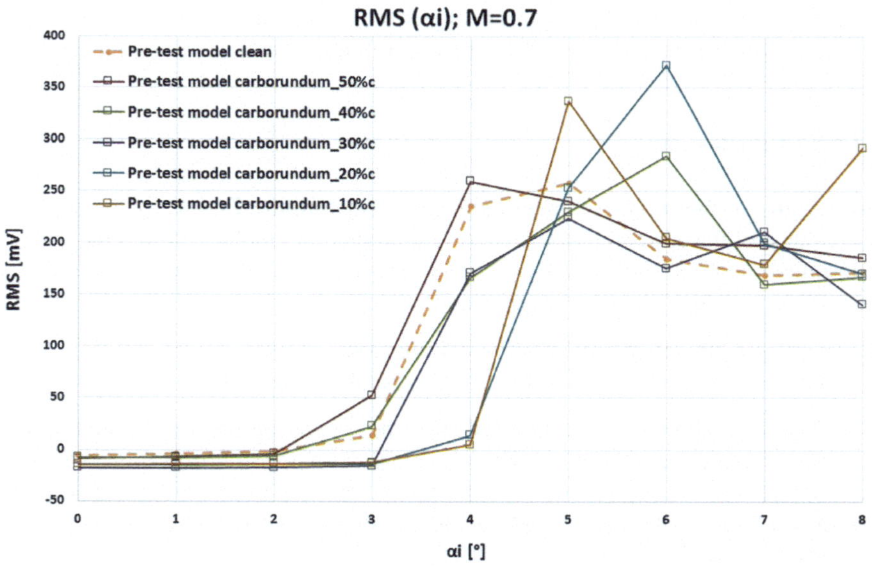

Fig. 63 Graph presentation of RMS for Pre-test model at Mach number M = 0.7

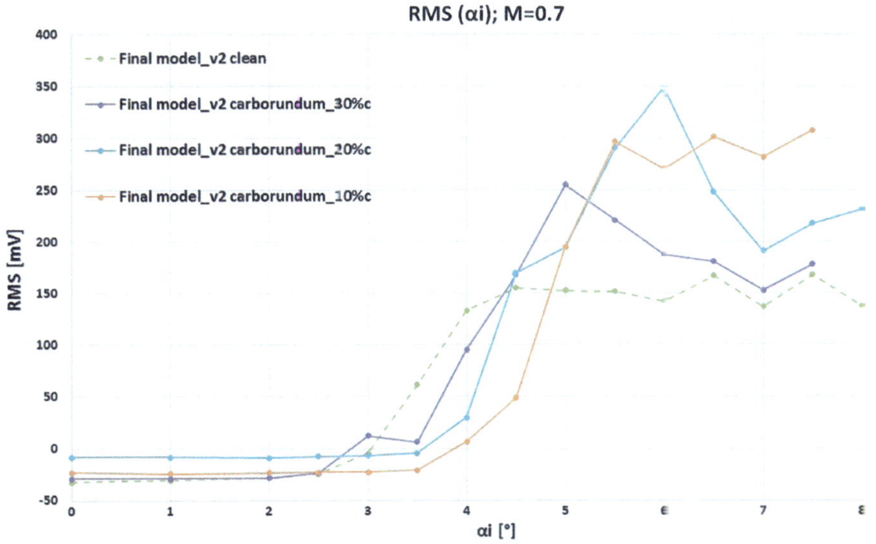

Fig. 64 Graph presentation of RMS for Final model_v2 at Mach number $M = 0.7$

At Mach number $M = 0.75$ the tension level initially increased from the angle of incidence 0° up to $\alpha_i = 2°–3°$. Then the level of each case was about equal.

The shifting a carborundum strip on the model upper model surface in leading edge direction, caused the jump of tension vibration less rapid. At higher angles of incidence for forward carborundum positions the RMS level of model vibrations was the lowest (Fig. 65).

Additionally the frequency analysis of model vibrations from strain gauge balance measurement at incidence $\alpha_i = 4°$ (applying Fast Fourier Transform) was presented for clean cases (for Pre-test and Final model) (Fig. 66). The strain gauge balance was fixed to model upper surface at 42% of chord position. Characteristic frequencies at the incidence close to buffet onset $\alpha_i = 4°$ were about 85.4 Hz for poor quality model and 136.7 Hz for polished model.

Additionally tests with unsteady KULITE pressure sensors were computed. In Fig. 67, the result of Fast Fourier Transform (FFT) analysis for Final_v2 model is presented. Pressure vibrations starts at about 3°–3.5° angle of incidence.

Characteristic frequencies for lower quality model at incidence $\alpha_i = 4$ were estimated using FFT analysis from pressure KULITES sensors. Characteristic frequencies were found: 80.6 Hz and 136.7 Hz (at 50%c), 80.6 Hz (at 50%c) and 80.6 Hz (at 70%c) of chord position. On the higher incidence $\alpha_i = 7°$ the vibration signal from sensor was much more weaker and similar frequencies haven't been found (dominating frequencies had low values) (Fig. 68).

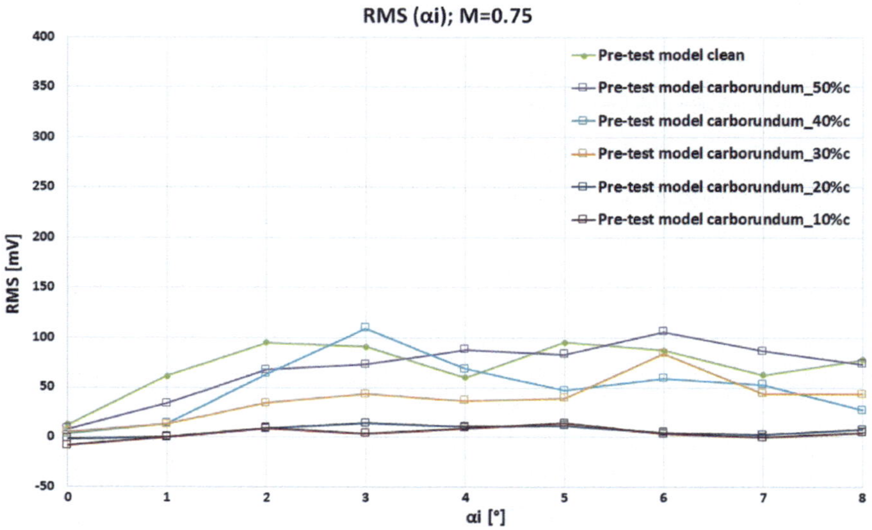

Fig. 65 Graph presentation of RMS for Mach number M = 0.75

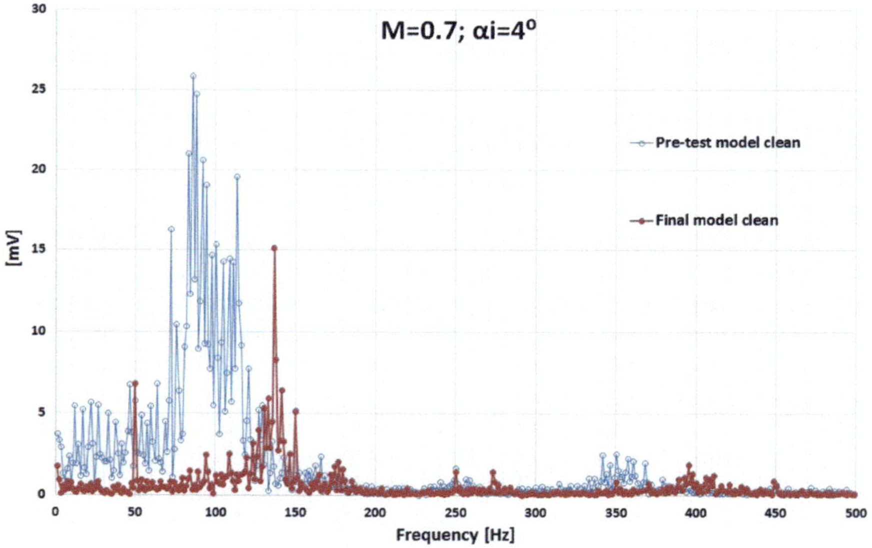

Fig. 66 The frequency analysis (FFT) from strain gauge balance measurement for Pre-test and Final model (polished). M = 0.7, $\alpha_i = 4°$

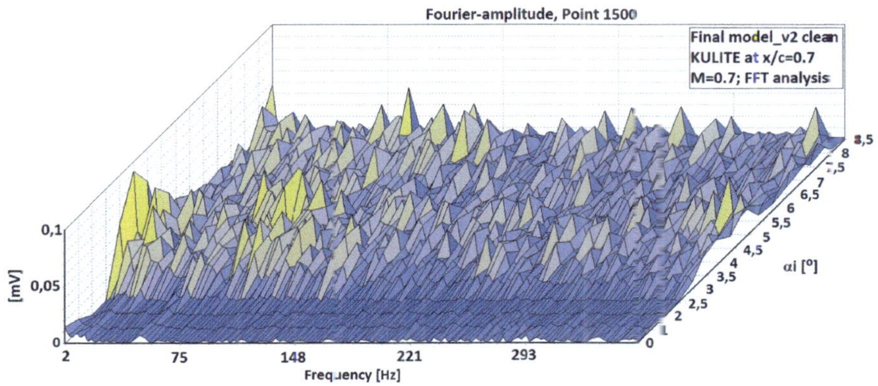

Fig. 67 Graph presentation of FFT analysis for Final model_v2 clean configuration

Conclusion

- The application of carborundum strips caused the delay of the buffet onset appearance. The shifting of carborundum strips to the leading-edge position (applied in the front of the SW) caused that buffet onset delay was more noticeable (both for the pre-test and final_v2 model).
- The highest amplitudes of SW oscillations were noticed after buffet onset phenomena occurrence and were correlated not only with the SW position. The tendency of the SW amplitude level modification for the carborundum cases was also in dependence of the Mach number.
- For all tested configurations at low angles of incidence there were no SW oscillations noticed. From the angle of buffet onset appearance the characteristic frequencies and amplitude level were indicated by sensors (mainly by these closest to the SW position).

2.2.2 Numerical Simulations

Calibration of the Turbulence Intensity Level at the Inlet Boundary Condition Surface

Motivation for numerical investigation of effects on turbulence intensity on transition onset was concern for differences in results of wind-tunnel tests conducted in wind-tunnels of different turbulence intensity. The numerical investigations were conducted using ANSYS FLUENT code and its capabilities of varying turbulence intensity in the computational domain by change of settings of boundary conditions on the boundaries of the computational domain.

First computations were conducted for domain representing empty test section of IoA trisonic wind-tunnel of 0.6 m. by 0.6 m. width and height in order to find

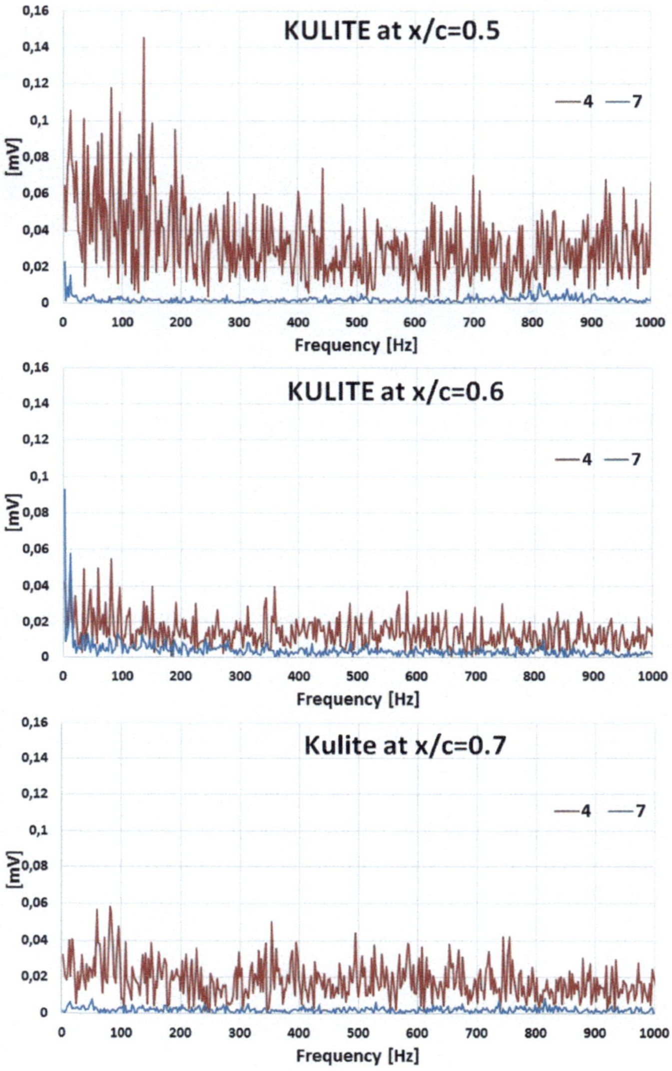

Fig. 68 The frequency analysis from KULITE pressure sensor measurement for Final_v2 model. $M = 0.7$, $\alpha_i = 4°$ and $7°$

correct boundary condition settings for assumed turbulence intensity in location of investigated airfoil. This was necessary due to strong dissipation of turbulence along the computational mesh, which effects presents Fig. 69 Zero on the X axis corresponds to the location of airfoil leading edge when placed in the test section.

The computations, results of which are presented in Fig. 69 were conducted for turbulence intensity of 10% and turbulent viscosity ratio of 1, set on the inlet boundary

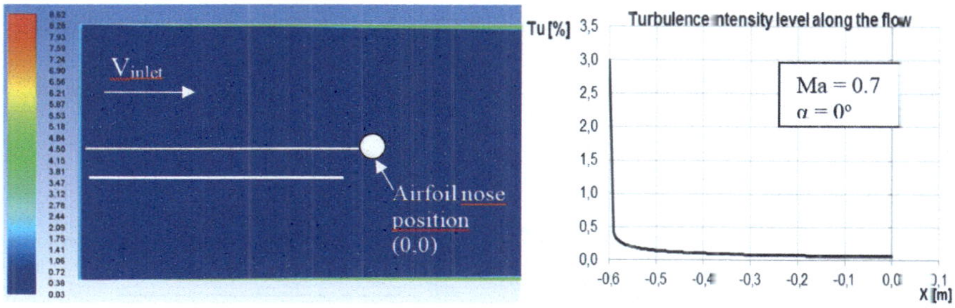

Fig. 69 Location noise probe and turbulent intensity distribution

condition surface in front of the position of airfoil leading edge. Value of turbulent intensity dropped from the level of 10% at the inlet to 0.055% at the airfoil leading edge. Further calculations were performed for constant value of turbulence intensity level of 10% and a various turbulent viscosity ratio level. The results are collected in the Table 5.

The higher the value of turbulent viscosity ratio the higher is the turbulence intensity level at the airfoil's leading edge location Fig. 70. The level of turbulence intensity at the airfoil nose location can be controlled by changing the value of turbulent viscosity ratio at the inlet. However, setting the value bigger than 100 is not recommended because the skin friction can deviate significantly from the laminar value.

Effects of Turbulence Intensity on Transition Onset

Flow conditions of the most interest in the TFAST project were defined by Mach number of 0.70 and a range of angles of attack between 1° and 7°. In the lower part of this range solution of URANS equations with 4-equation TRANSITION SST turbulence model has revealed oscillating shockwave with Strouhal number of 0.08, characteristics for buffet phenomenon. On the other hand, at an angle of attack of 7° shockwave was in steady position and these conditions were chosen to investigate effects of turbulence intensity on transition onset. With turbulence intensity on the inlet boundary condition set to 10% and by varying turbulent viscosity ratio, the values of turbulence intensity at airfoil leading edge presented in Table 5 have been imposed in the computations of flow or the investigated airfoil

First analysis was conducted for turbulence intensity at airfoil leading edge of 0.055% (first position in Table 5). In Fig. 71 is shown distribution of pressure coefficient with characteristic zones: the laminar flow region with increasing flow velocity,

Table 5 Dependence of turbulence level at airfoil nose on turbulent viscosity ratio level at inlet to computational domain

Turbulent viscosity ratio (Inlet)	1	2	5	8	10	20	50	100
Tu_{NOSE} (%)	0.055	0.083	0.139	0.181	0.204	0.299	0.496	0.724

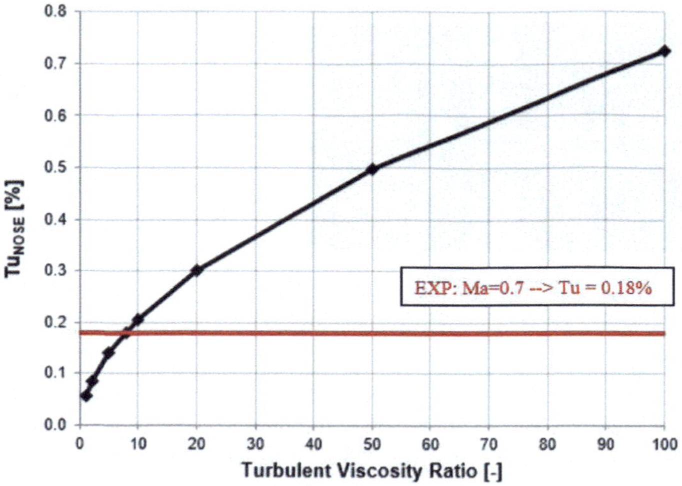

Fig. 70 Calibration of the turbulence intensity level at the inlet for required turbulence intensity at airfoil location

pressure plateau in front of the shockwave where laminar-turbulent transition occurs, region of shockwave, and region of turbulent, subsonic flow towards the trailing edge.

In the investigated flow conditions the onset of laminar-turbulent transition is most conveniently detected by inspection of contours and plots of intermittency (probability of boundary layer being turbulent) read at close distance from surface and of tangential stress on the airfoil surface where rise of tangential stress indicates turbulisation of boundary layer.

In order to analyse laminar-turbulent transition on the upper surface of airfoil, plots of intermittency along the control line at a height of 0.3 mm above the surface and of surface tangential stress in X-direction were prepared. The choice of distance of 0.3 mm from airfoil surface for the control line of intermittency can be explained by start of rise of intermittency at this height, shown in contour in Fig. 73. Comparison of plots of intermittency and of tangential stress in Fig. 74 shows that first rise of intermittency corresponds to separation of laminar boundary layer at the foot of shockwave. After the first rise intermittency stabilizes at the value of approximately 0.3 at the end of supersonic region. Full turbulisation with intermittency reaching unity occurs at the beginning of region of strong flow recirculation behind the shockwave, visible as large increase of negative tangential stress at 0.45% chord in the right plot of Fig. 72.

By varying turbulent viscosity ratio on inlet boundary condition surface, applying values presented in Table 5 new flow solutions were obtained for increasing values of turbulence intensity at the position of airfoil leading edge. Positions of first rise of intermittency along the control line at the height of 0.3 mm and positions of its final rise to unity are presented in Fig. 74. It can be seen that position of the first rise of intermittency moves upstream and the position of final rise of intermittency moves

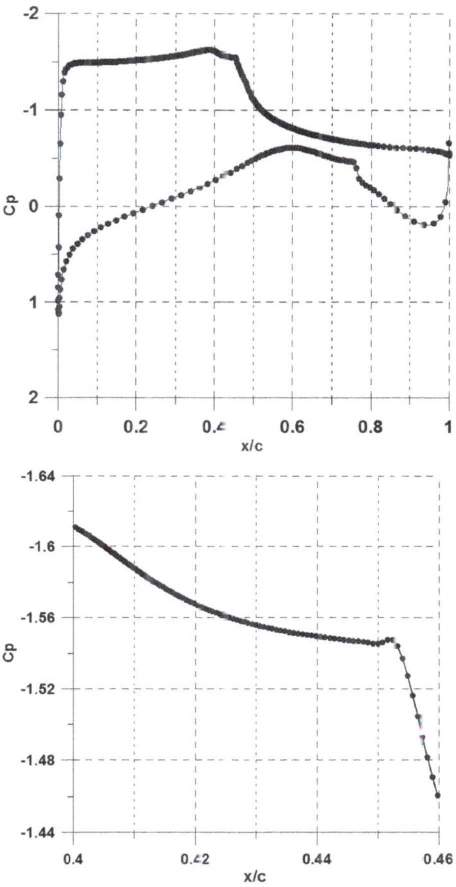

Fig. 71 Plots of pressure coefficient with enlarged region of "pressure plateau" where natural laminar-turbulent transition takes place

downstream, enlarging the transition region. The changes of positions of these points are moderate and almost linear with the rise of turbulence intensity.

Effects of Location of Laminar-Turbulent Transition on Structure of Shockwave-Boundary Layer Interactions and on Aerodynamic Characteristics of the V2C Laminar Airfoil

Two concepts of turbulators were studied in CFD analyses. The first one involved array of delta-shaped micro vortex generators (VGs) producing chordwise-oriented vortices, shown in Fig. 75. The VG height was chosen such, that it was fully submerged in the boundary layer. Inclination of mVG with respect to flow direction was 14°. The other concept of turbulators consisted in rectangular micro vanes (mVs) submerged in the boundary layer, perpendicular to airfoil surface and to flow, shown in Fig. 76, producing vortices with axes parallel to wing span.

First run of analyses was conducted for angle of attack of $\alpha = 4°$, Ma = 0.70, Re = 2.67 mln. Two heights of delta-shaped mVGs were tested: 0.10 and 0.15 mm, and

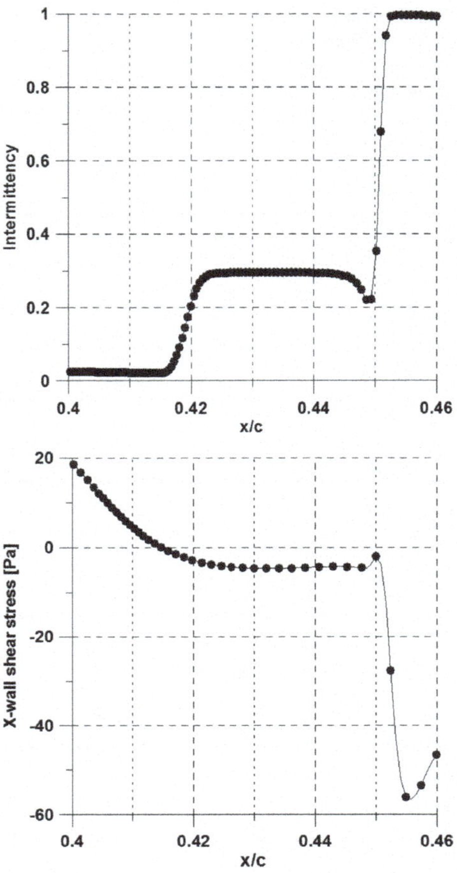

Fig. 72 Plots of intermittency along the control line 0.3 mm above airfoil surface and of surface tangential stress in the region of natural laminar-turbulent transition

they were placed in 20, 30, 40 and 50% of airfoil chord. Other dimensions of the VG were kept constant. Values of relative height, measured with respect to boundary layer thickness for these combinations of VG height and chordwise position are presented in Table 6. By the application of symmetry boundary conditions on the lateral sides of the wing strip shown in Fig. 75 an array of counter-rotating vortex generators was modelled. The distances between mVG vertices and lateral sides of the wing strip were chosen such that there is a gap of 10% mVG chord between the grid-resolved mVG and its mirror-image modelled with the symmetry b.c.

For both heights of vortex generators changes in intermittency at the distance of 0.15 mm were read from the flow solution, as well as changes of the tangential stress on airfoil surface. Comparison of these changes is shown in Figs. 77, 78, 79 and 80.

In plots of changes of intermittency and surface tangential stress presented in Figs. 77 and 78 it can be seen that the lower VGs were generally more effective at tripping laminar boundary layer, producing higher rise of intermittency and surface tangential stress then the higher ones. An exception was the case of rapid growth

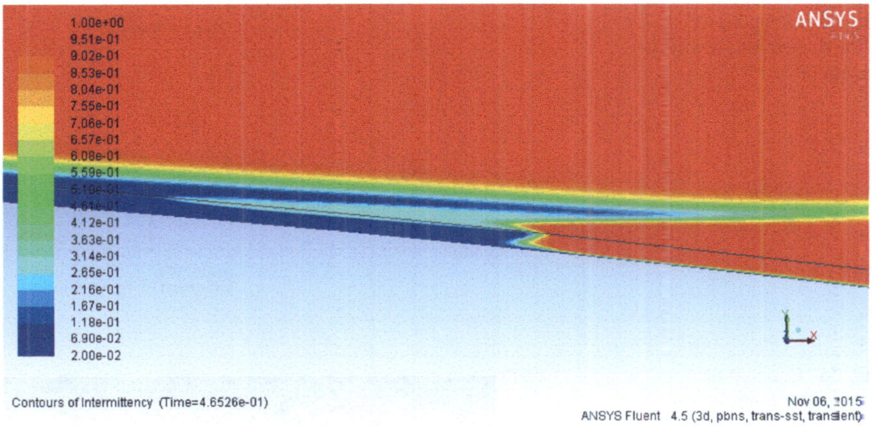

Fig. 73 Contour of intermittency in the boundary layer in the region of natural laminar-turbulent transition with marked control line at the distance of 0.3 mm from airfoil surface

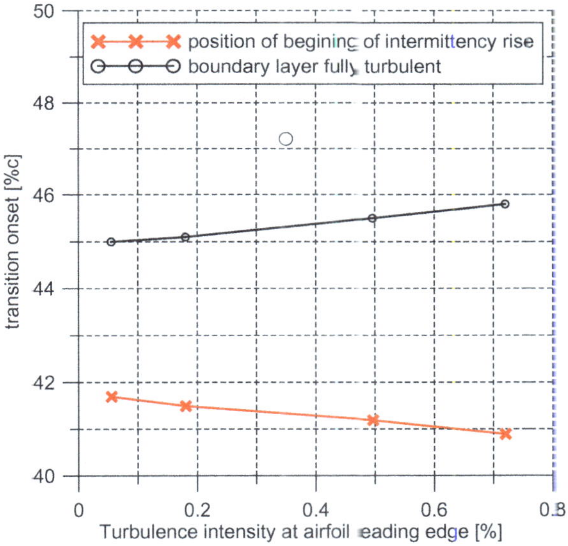

Fig. 74 Change of position of first rise of intermittency on the control line and of location of fully-turbulized boundary layer in conditions of natural laminar-turbulent transition

of b.l. thickness near the shockwave where effectiveness of lower VGs significantly decreased. It also can be seen, that higher rise of intermittency corresponds to higher rise of surface tangential stress, and, in spite of not obtaining fully turbulent b.l. behind the VGs, the rise of tangential stress is similar in the investigated cases. Other important observation is elimination of the laminar separation region at the foot of the shockwave, visible in Figs. 79 and 80 as a short, slightly negative fragment of the tangential stress curve at 45–50% chord.

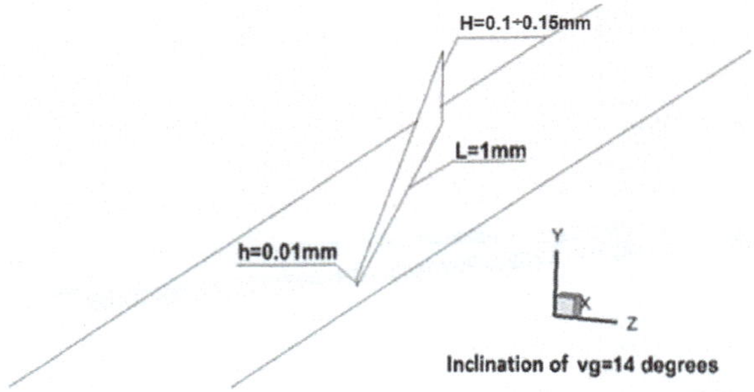

Fig. 75 Geometry of delta-shaped micro vortex generator

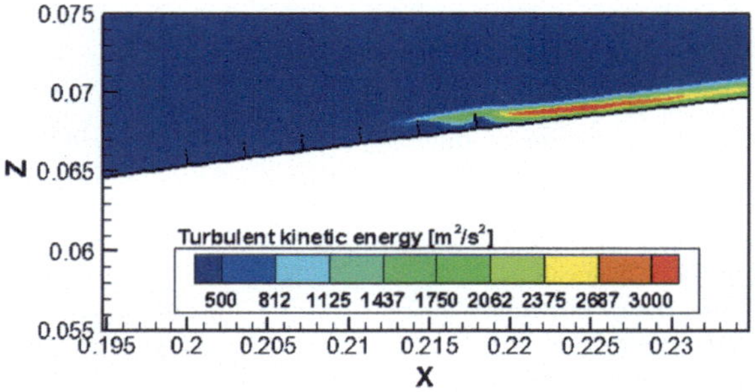

Fig. 76 Concept of rectangular micro vanes perpendicular to airfoil surface

Table 6 Relative height of delta-shaped mVGs, related to local thickness of boundary layer

x/c	0.2	0.3	0.4	0.5
δ	0.30	0.24	0.21	0.05

Changes in lift and drag coefficients caused by the VGs are presented in Figs. 83 and 84. The most important effect of tripping of laminar b.l. is elimination of buffet oscillations of shockwave, which is shown as unsteady solution of URANS equations converging to steady-state values of aerodynamic coefficients. Depending on the position of VGs either small decrease or small increase of lift coefficient can be observed with respect to its averaged value for clean configuration and higher VGs produced larger reduction of lift in frontal position and larger increase in rear position than the lower ones.

Changes of drag coefficient with respect to averaged value for clean airfoil case are positive for all investigated combinations of mVGs height and position. The reason for this is increase of tangential stress as well as pressure drag on the VGs. It can be seen that the more effective is vortex generator at tripping of laminar boundary layer, the higher is drag increase due to this action. Particularly high increase of drag for VGs placed in 50% chord is due to increased suction in the part of airfoil with negative slope, behind maximum thickness, as a result of chordwise vortices produces by VGs. This effect is shown in Fig. 85 where plots of pressure coefficient in region of "pressure plateau" of clean airfoil are shown. It can be seen that the

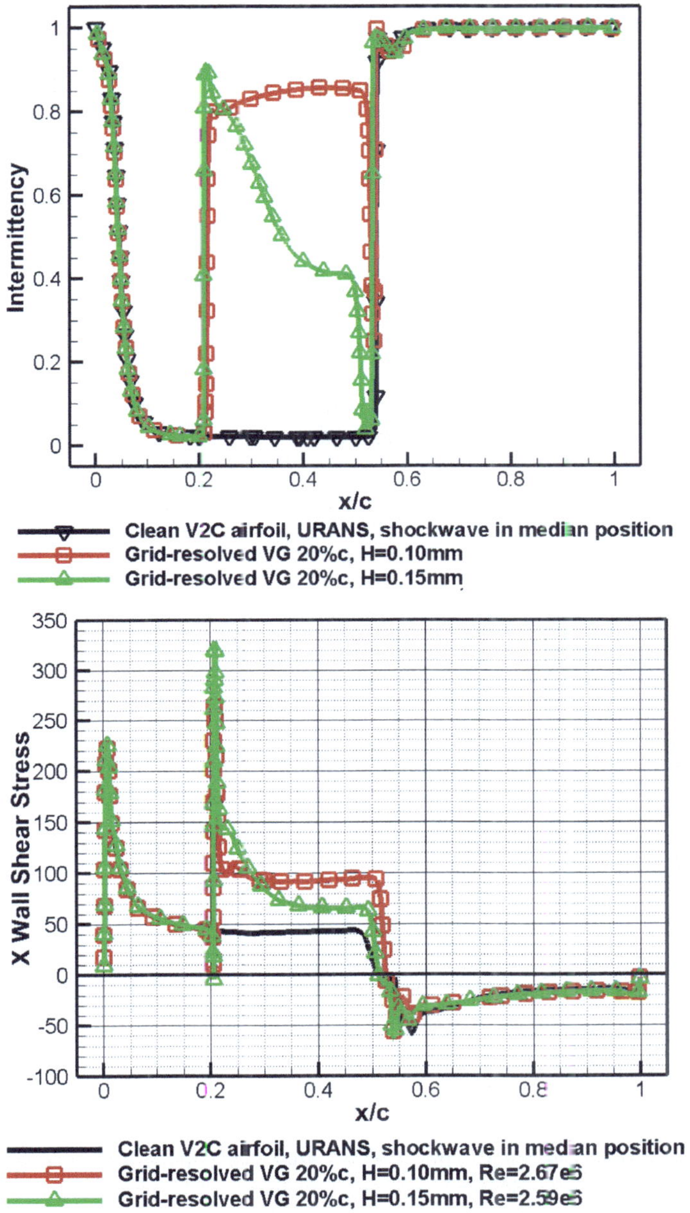

Fig. 77 Plots of intermittency at control line of 0.15 mm behind vortex generators and of surface tangential stress in X direction for VGs located in 20% chord

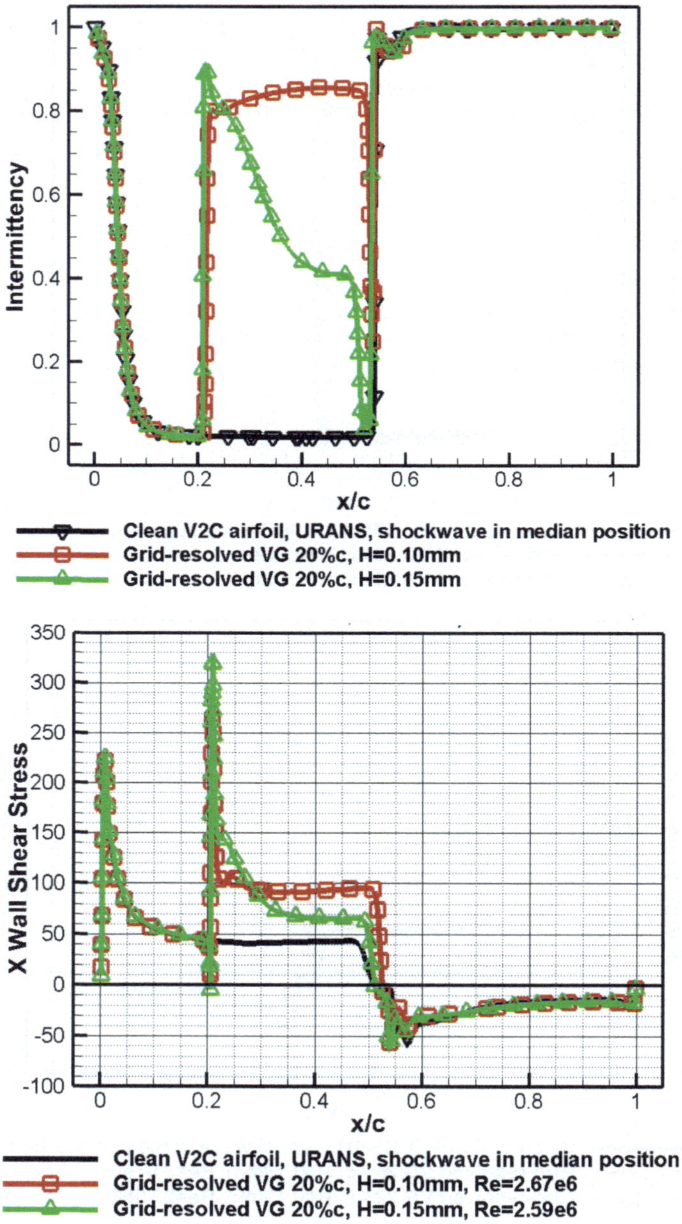

Fig. 78 Plots of intermittency at control line of 0.15 mm behind vortex generators and of surface tangential stress in X direction for VGs located in 30% chord

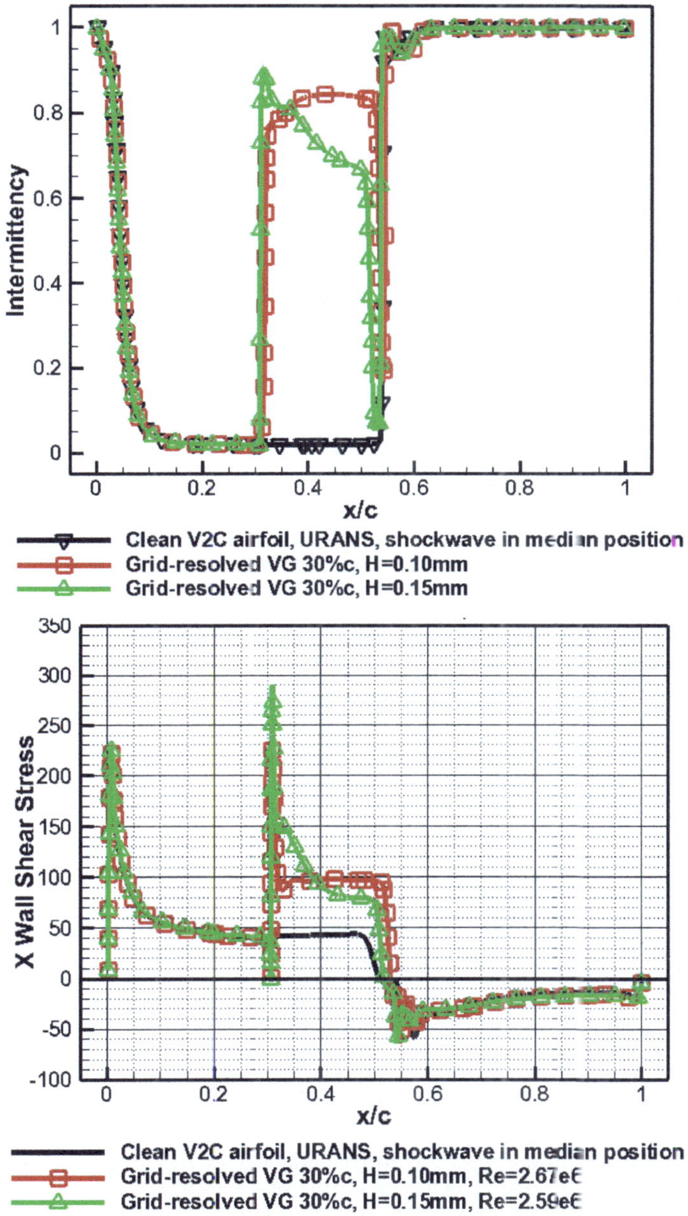

Fig. 79 Plots of intermittency at control line of 0.15 mm behind vortex generators and of surface tangential stress in X direction for VGs located in 40% chord

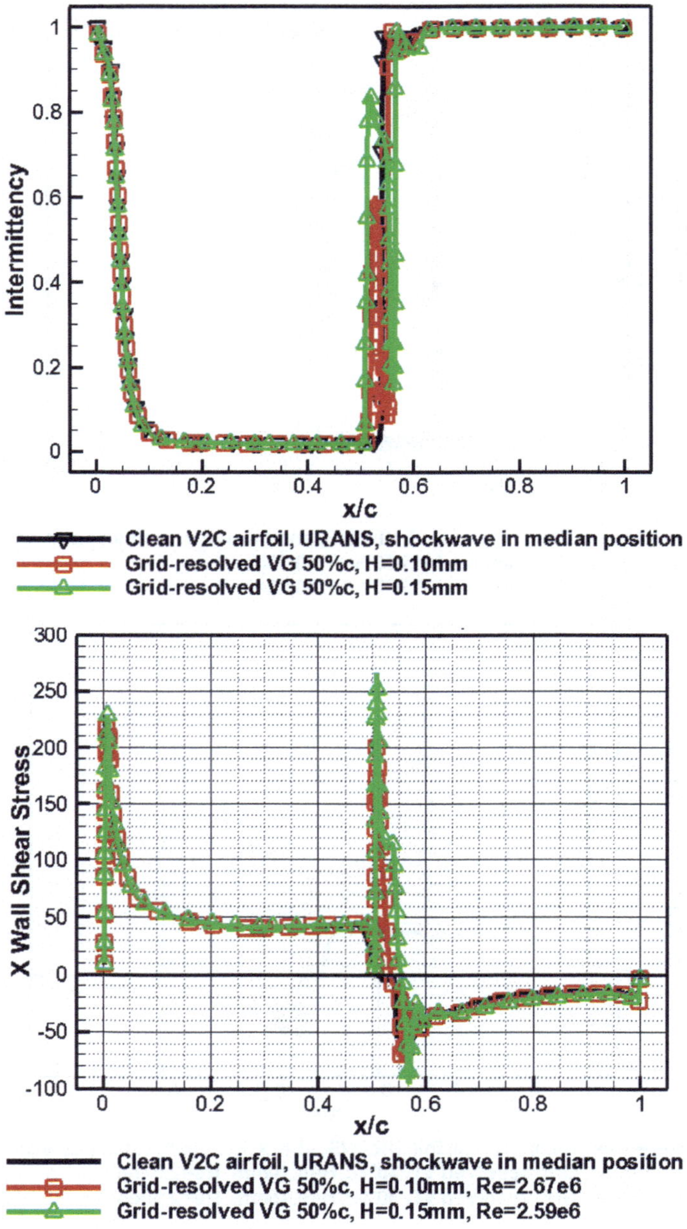

Fig. 80 Plots of intermittency at control line of 0.15 mm behind vortex generators and of surface tangential stress in X direction for VGs located in 50% chord

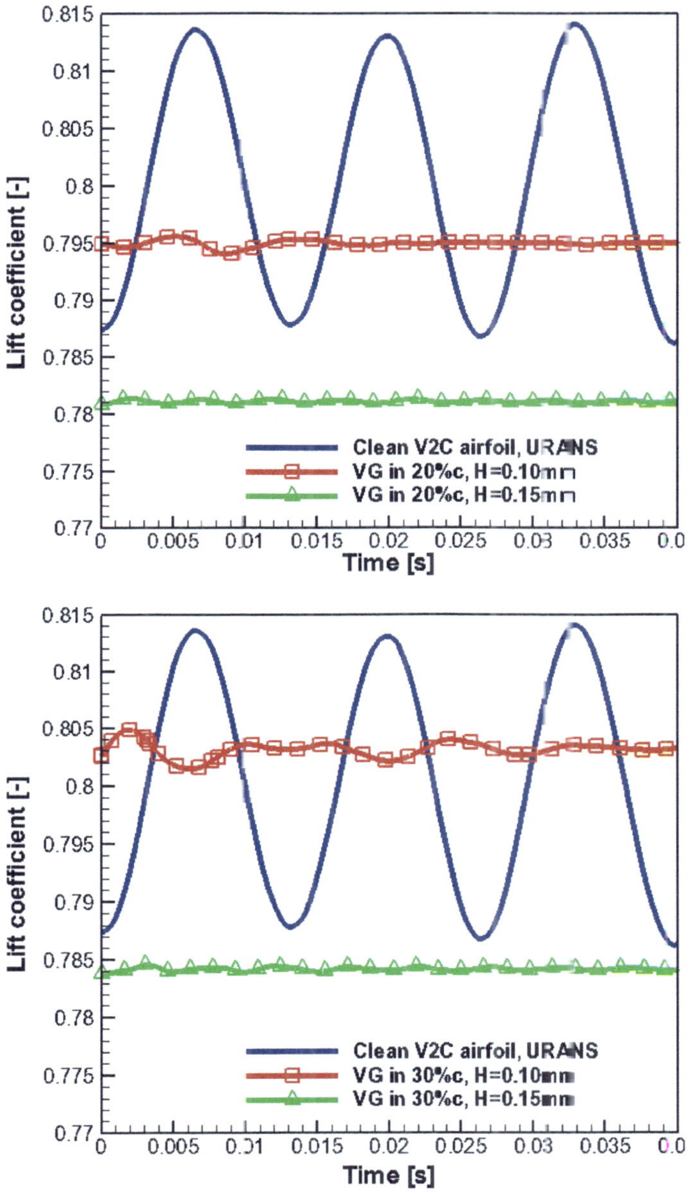

Fig. 81 Comparison of lift coefficient for clean-airfoil case and for cases with VGs located in 20% chord (left) and in 30% chord (right)

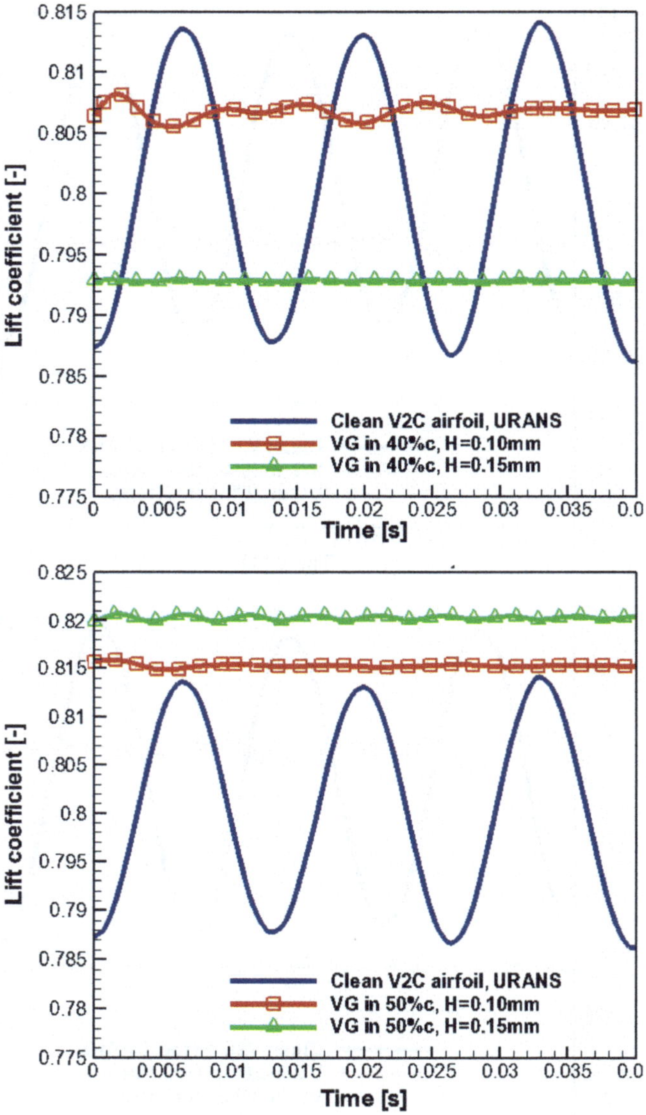

Fig. 82 Comparison of lift coefficient for clean-airfoil case and for cases with VGs located in 40% chord (left) and in 50% chord (right)

chordwise-oriented vortices enlarge area of accelerating flow and move shockwave slightly streamwise. Thus it may be concluded that VGs producing chordwise vortices are not recommended as turbulizers for chordwise locations close to 50% chord.

Changes of lift and drag coefficient due to tripping of laminar boundary layer at an angle of attack of 4° lead to decrease of lift-to-drag ratio, shown in Fig. 86. However,

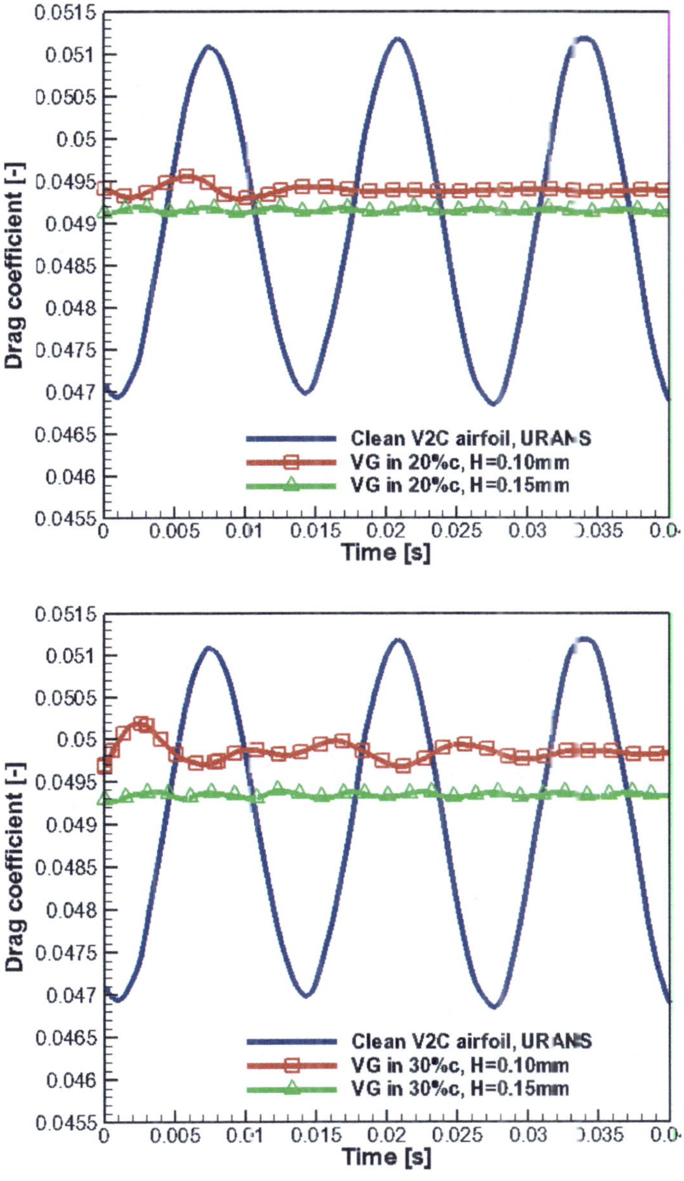

Fig. 83 Comparison of drag coefficient for clean-airfoil case and for cases with VGs located in 20% chord (left) and in 30% chord (right)

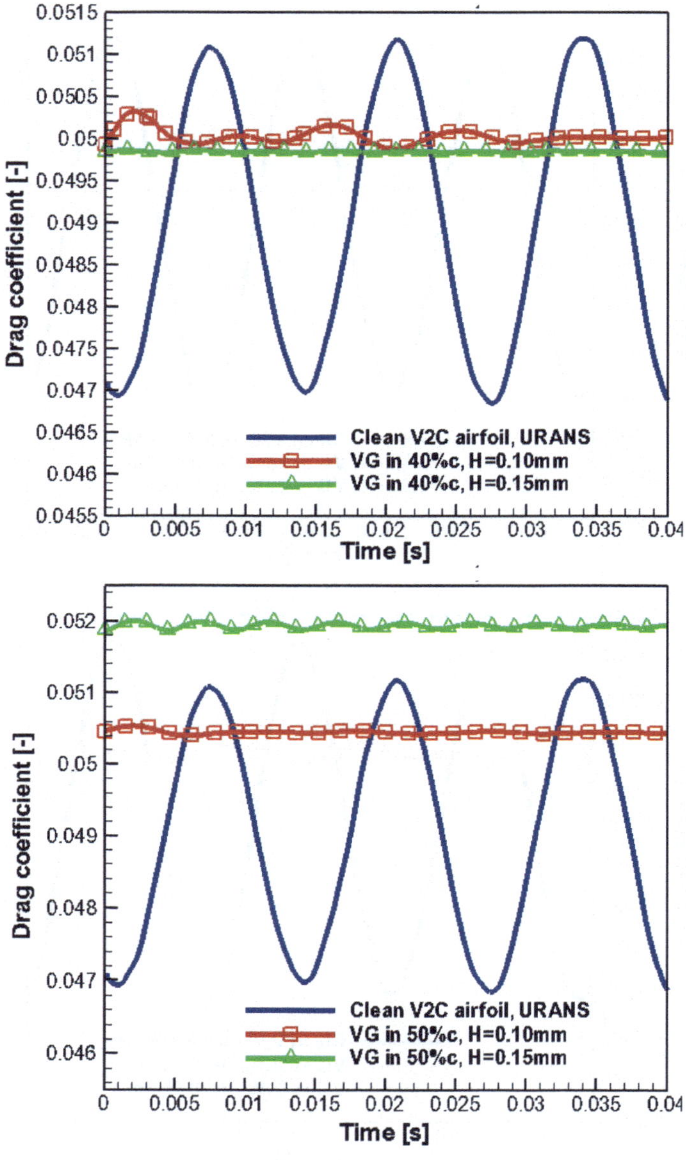

Fig. 84 Comparison of drag coefficient for clean-airfoil case and for cases with VGs located in 40% chord (left) and in 50% chord (right)

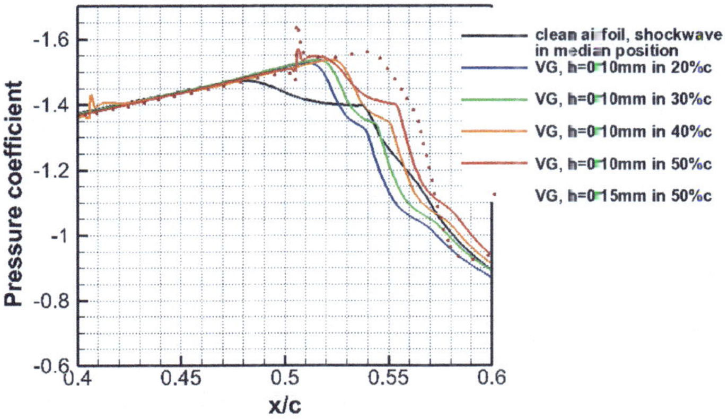

Fig. 85 Comparison of distributions of pressure coefficient in the pressure plateau region for different locations and height of VGs

Fig. 86 Changes of lift-to-drag ratio as effect of tripping laminar boundary layer with delta-shaped VGs at angle of attack of 4°

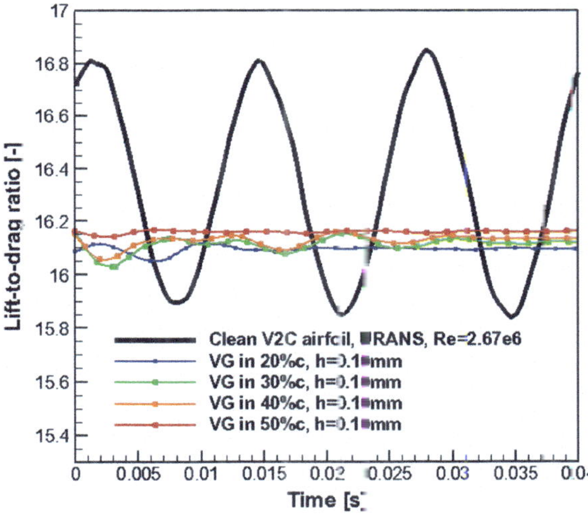

when angle of attack is increased to 7° the shockwave and "pressure plateau" region move forward, to location between 40 and 45% of airfoil chord.

In this position, due to less negative slope of the upper surface it was possible to obtain positive change of lift-to-drag ratio, shown in Fig. 87, in spite of drag increase due to presence of vortex generators. This shows, that improvement of aerodynamic characteristics due to elimination of laminar separation bubble is more likely to be achieved as increase of lift-to-drag ratio, than as drag reduction, particularly when using mechanical turbulators.

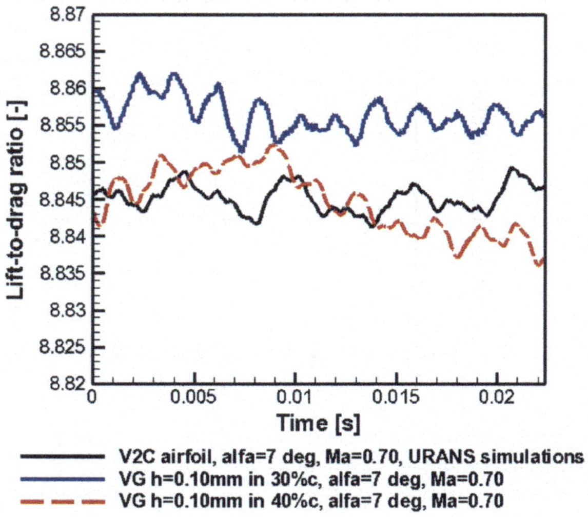

Fig. 87 Changes of lift-to-drag ratio as effect of tripping laminar boundary layer with delta-shaped VGs at angle of attack of 7°

Table 7 Relative height of micro-vanes, related to local thickness of boundary layer

x/c	0.2	0.3	0.4
δ	0.45	0.36	0.30
δ	0.61	0.48	0.40
δ	0.75	0.60	0.50

Another investigated concept of tripping laminar boundary layer was array of micro-vanes (mVs) producing vorticity between plates perpendicular to airfoil surface, shown in Fig. 76. Three values of vane height were investigated: 0.15, 0.20 and 0.25 mm. The relative vane heights, with respect to b.l. thickness are presented in Table 7.

In Fig. 88 is shown comparison of surface pressure coefficients and tangential stress for vanes of different height, located in 20% chord. The flow conditions were Ma = 0.68 and $\alpha = 6°$ with steady flow, just below buffet limit. It can be seen, that for vane height higher than 0.15 mm, for which turbulisation of laminar b.l. was achieved, the shockwave has tendency of shifting towards leading edge, without enlarging the flow acceleration area as was the case with VGs producing chordwise vortices. For these cases flow turbulisation is recognisable by increase of surface tangential stress behind vanes and by elimination of laminar separation at the foot of the shockwave, which can be seen in the right plot of Fig. 88.

Turbulators consisting of the micro-vanes were applied for the investigations of the effects of induced turbulisation of laminar b.l. on the buffet phenomenon. Buffet phenomenon, consisting of oscillations of shockwave with Strouhal number of approximately 0.1 limits operational range of angle of attack and Mach number of transport aircraft and may be particularly harmful for laminar-flow wing because shockwave in laminar flow is usually more intensive than in turbulent flow. The

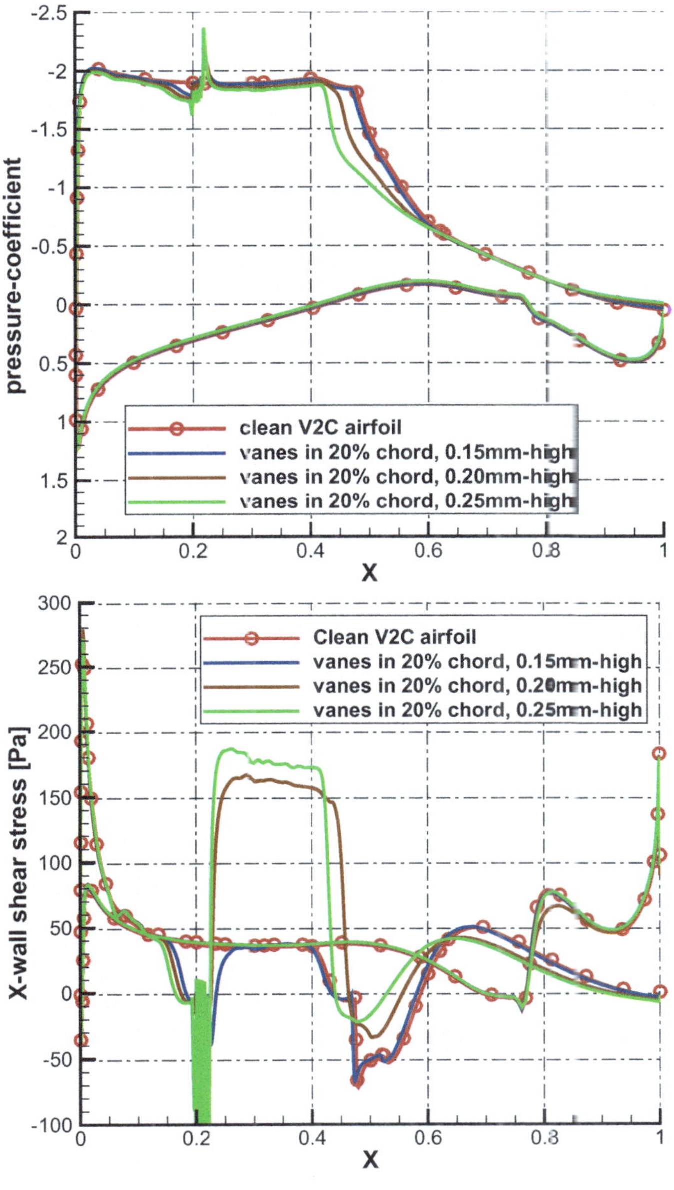

Fig. 88 Comparison of distribution of pressure coefficient (left) and of surface tangential stress (right) for clean-airfoil case and for cases with rectangular micro-vane turbulators of different height, placed in 20% of airfoil chord

analysis was conducted by simulation of flow with continuously increasing free-stream Mach number, set as a User-Defined Function of ANSYS FLUENT solver on far-field boundary condition. Flow simulations started with low Mach number of steady flow, without oscillating shockwave. Open-space boundary conditions were used with pressure far-field and pressure outlet boundaries placed at the distance of 12 chords from the airfoil. The model wing was a segment of 20 cm chord, 1 cm-width with boundary conditions of periodicity on lateral sides of computational domain. Computations were conducted at constant angle of attack of 6°. In Fig. 89 is presented comparison of pitching moment coefficient of clean airfoil and of configurations with vane height of 0.25 mm, placed in 20, 30 and 40% airfoil chord. It can be seen that buffet phenomenon is eliminated for configurations with vane turbulators in the investigated range of Mach numbers. In Fig. 90 is presented comparison of lift coefficient versus Mach number for the previously mentioned cases. The comparison shows, that the more frontal is position of the turbulators, the larger is reduction of lift coefficient. This effect suggests application of the investigated turbulator as active flow-control device with anti-buffet and gust-load alleviation function because entering buffet conditions may take place in gust suddenly increasing aircraft angle of attack.

In Fig. 91 is presented change of lift-to drag ratio as a result of activation of the micro-vane turbulators at constant angle of attack of 6°. The comparison of results for different locations of turbulators shows that at Mach numbers close to conditions of

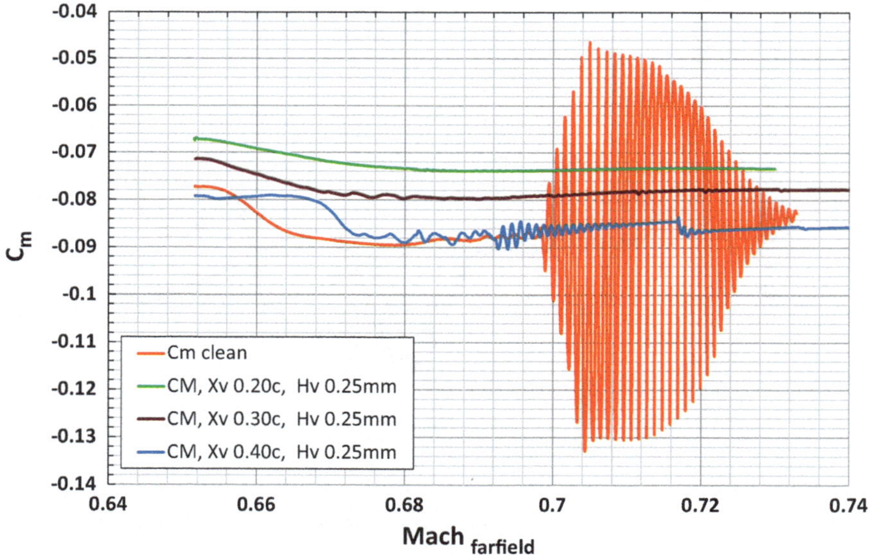

Fig. 89 Comparison of pitching moment coefficient as a function of Mach number for clean-airfoil case and for cases with 0.25 mm-high rectangular micro-vane turbulator placed in different chord positions

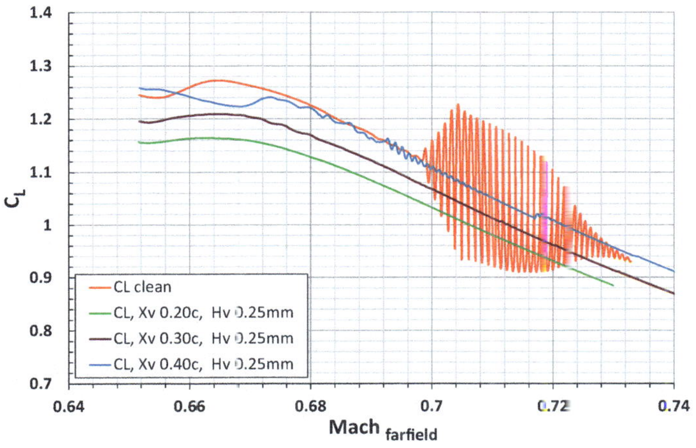

Fig. 90 Comparison of lift coefficient as a function of Mach number for clean-airfoil case and for cases with 0.25 mm-high rectangular micro-vane turbulator placed in different chord positions

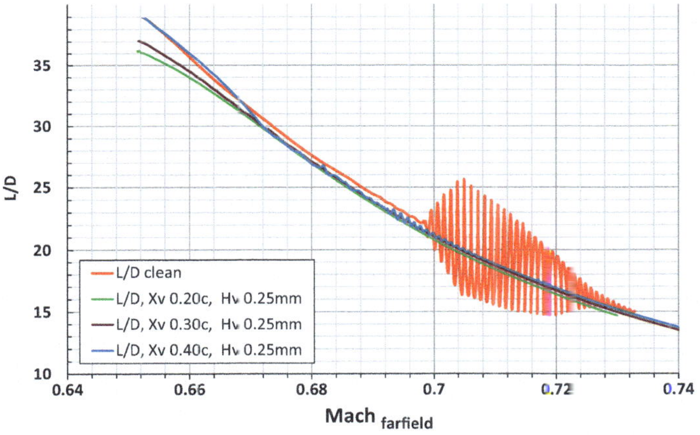

Fig. 91 Comparison of lift-to-drag ratio as a function of Mach number for clean-airfoil case and for cases with 0.25 mm-high rectangular micro-vane turbulator placed in different chord positions

buffet the reduction of lift-to drag ratio with respect to clean-airfoil values is almost the same for all investigated positions of turbulators.

2.2.3 Conclusions

The investigated turbulators produced additional effects in the flow, apart from their main function of tripping laminar boundary layer. The vortices produced by delta-shaped vortex generator tended to enlarge the velocity acceleration zone into the region of negative slope of airfoil upper surface, bringing about pressure drag increase. The rectangular micro-vanes perpendicular to flow and airfoil surface tended to move shockwave forward, reducing the velocity acceleration region on the laminar airfoil.

Turbulent boundary layer obtained as an effect of tripping of laminar flow stabilises the shockwave, eliminating buffet phenomenon in the investigated Mach range. The turbulator consisting of rectangular micro-vanes has revealed potential for applications as an active flow-control device with anti-buffet and gust-load alleviation function due to its lift-reducing capability.

Improvement of airfoil characteristics due to elimination of laminar separation bubble is more likely to occur as increase of lift-to-drag ratio than as drag reduction. This is due to increased friction in turbulised boundary layer and due to increased drag as an effect of introducing turbulators.

2.3 URMLS

2.3.1 Results

A preliminary grid convergence study was performed using the Spalart-Allmaras model in unsteady mode (URANS) for the following flow conditions:

- Mach number $M_\infty = 0.7$
- Reynolds number (based on the chord length) $Re = 3.2 \times 10^6$
- AoA $\alpha = 5°$.

An example of the (C-type) mesh used for the computations is reported in Fig. 92. Three different grids were tested, whose main parameters are reported in Table 8.

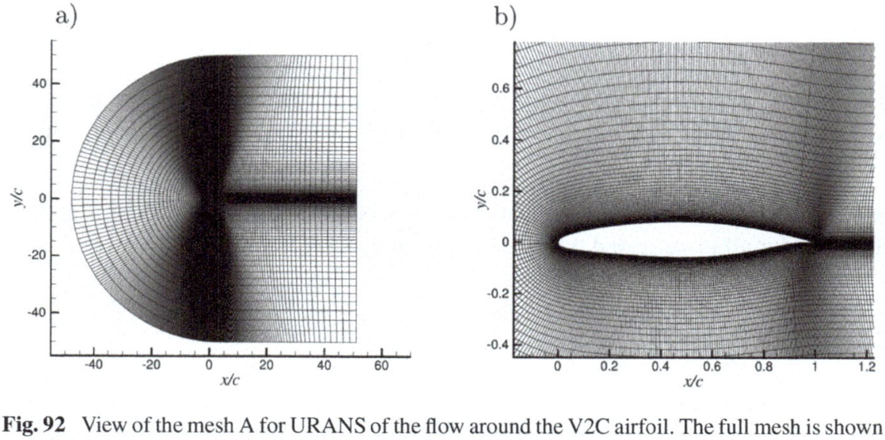

Fig. 92 View of the mesh A for URANS of the flow around the V2C airfoil. The full mesh is shown in panel (**a**) and a zoom in the airfoil region is reported in panel (**b**)

Table 8 Number of cells of the various meshes used for the grid convergence study

Grid	Upper surface	Lower surface	Wake	Normal direction
A	96	96	96	96
B	192	192	128	128
C	384	384	128	128

All the meshes have been obtained starting from the V2C geometry provided by DAAV using a hyperbolic mesh generator, placing the external boundary of the computational domain at a distance of 50 chords from the airfoil surface.

The distribution of the lift coefficient obtained on the three meshes is reported in Fig. 93 as a function of time. All the curves predict the absence of a buffet regime, with C_l attaining a limiting (constant in time) value after an initial transient period. Apparently, mesh B and C converge to the same value of the lift coefficient. Grid convergence is also achieved in the distribution of the mean pressure coefficient on the airfoil surface, reported for the three meshes in Fig. 94 and compared with the data obtained by URANS performed by IMFT with the SST model. Again, the curves

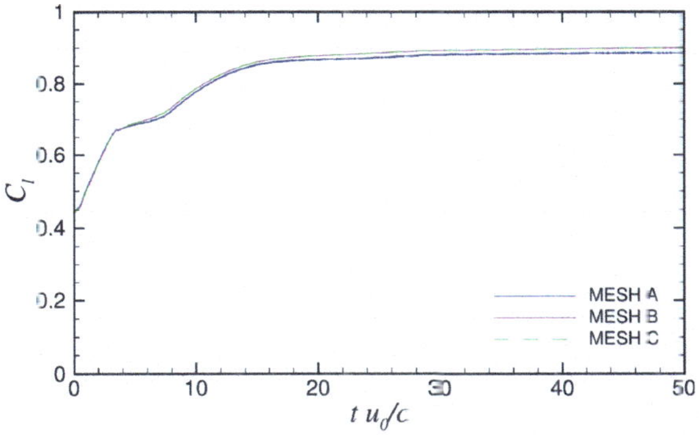

Fig. 93 Timeevolution of the lift coefficient for the three meshes reported in Table 1

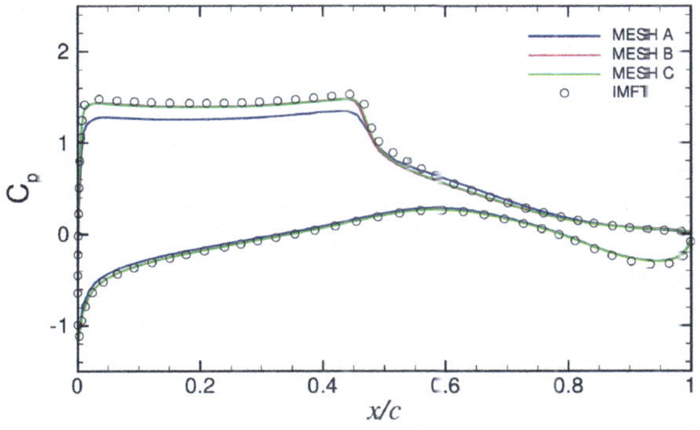

Fig. 94 Distribution of the pressure coefficient on the airfoil surface

for mesh B and C perfectly collapse and they are also in very good agreement with the data obtained by IMFT.

On the basis of these results grid B was selected to perform a series of URANS at the same flow Mach number and Reynolds number, but for different values of the airfoil incidence, from $\alpha = 0°$ to $\alpha = 7°$. Contours of the instantaneous pressure coefficient are shown In Fig. 95. The figure well highlights the changes in the pressure distribution occurring when the angle of attack increases, and in particular the formation of a lambda-shock structure on the suction side of the airfoil, which is apparent starting from $\alpha = 3°$ and whose strength increases with the incidence. This shock is not always steady and the onset of buffet occurs at $\alpha = 6°$ as can be appreciated by Fig. 96 where the evolution of the lift coefficient with time is reported. While until $\alpha = 5°$ the curves tend to an asymptotic (steady) value, at $\alpha = 6°$ and $\alpha = 7°$ periodic oscillations are observed, which are the evident signature of the shock movement. These oscillations determine also the behavior of the mean pressure coefficient on the suction side, reported in Fig. 97, which shows a smoothed pressure rise, whose spatial extent increases with the angle of attack.

As a further step, a series of URANS were run to find the onset of buffet in the $\alpha - M_\infty$ plane, which is identified in the map of Fig. 98, where green circles indicate unsteady configurations characterized by shock oscillations and red circles correspond to steady configurations. The behavior of the flow as steady or unsteady was decided looking at the evolution of the lift coefficient. The black line reported in the figure is an attempt to draw the stability boundary for the V2C airfoil. Interestingly, such boundary is made of two different branches, a lower and an upper branch, a result similar to that obtained by Crouch et al. [21] for a NACA0012 airfoil using numerical simulations and global stability analysis.

The Spalart-Allmaras model here implemented was selected because it is also the basis of the DDES approach, which allows to perform three-dimensional, unsteady computations at a reasonable cost. One of the goal of this task is indeed that of obtaining the buffet limit for the V2C airfoil by means of unsteady 3D computations, which are expected to provide more accurate results than that obtained with URANS. As a first step, a series of DDES have been performed, by fixing the Mach number ($M_\infty = 0.7$) and changing the AoA in the range $4°-8°$. These simulations were run using the two-dimensional mesh indicated as C in Table 8, extruded in the spanwise direction for a width $L_z = 0.5c$, and discretized with $N_z = 96$ points. Concerning the buffet onset the simulations provided the same response of URANS, with the appearance of buffet between $\alpha = 5°$ and $6°$. However, the quantitative predictions of the two approaches were found significantly different, especially for flow cases characterized by large shock oscillations.

A three-dimensional visualization of the flow at $\alpha = 7°$ is reported in Fig. 99, where vortical structures are visualized through an iso-surface of the swirling strength criterion and the shock location is identified through contours of the pressure field in a x–y slice. The figure well reflects the main features of the DDES approach, with the boundary layer turbulence fully modeled and the large eddies of the separated region in the aft part of the airfoil directly resolved on the computational mesh.

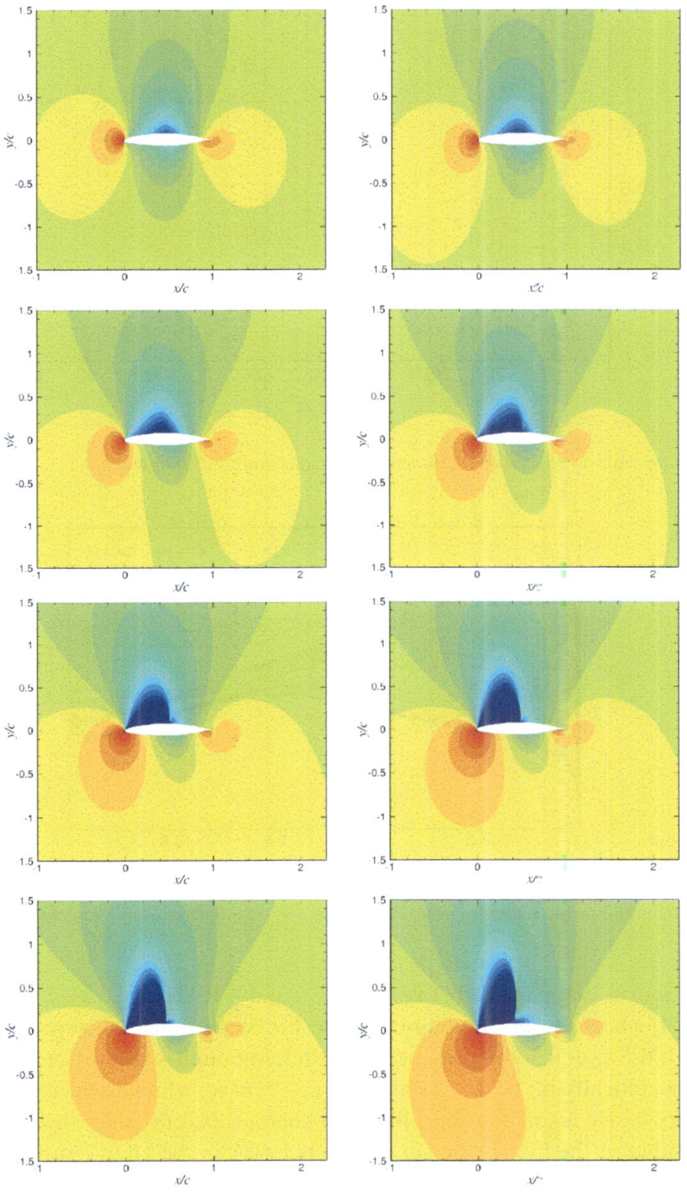

Fig. 95 Contours of pressure coefficient for different values of the incidence ($\alpha = 0°–7°$). Mach number $M_\infty = 0.7$, Reynolds number $Re = 3 \times 10^6$. The incidence increases from top to bottom, left to right. Thirty-six contours from -1.5 to 1.1

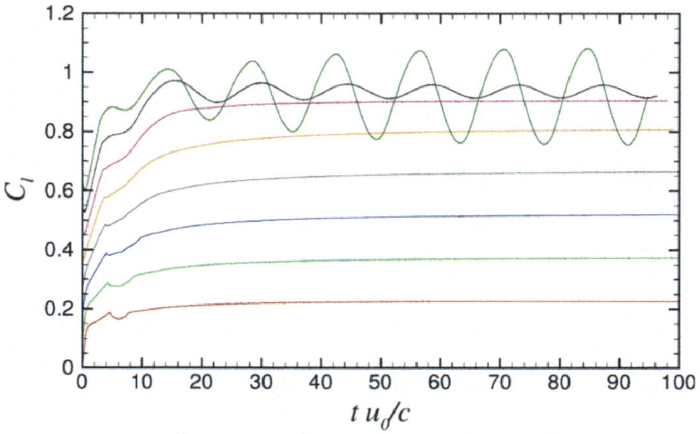

Fig. 96 Time evolution of the lift coefficient for various angles of incidence

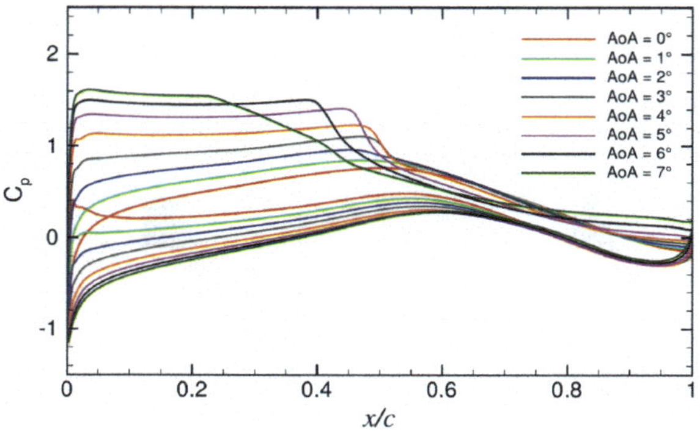

Fig. 97 Distribution of the mean pressure coefficient for various angles of incidence

A quantitative comparison between DDES and URANS at $\alpha = 7°$ is reported in Fig. 100a, where the time evolution of the lift coefficient is shown. Similarly to URANS, DDES predicts the presence of buffet, although some differences can be observed, specifically in: (i) the shape of the $c_l - t$ curve, which in the case of DDES is characterized by a saw tooth profile with sharper extrema than in URANS; (ii) the mean value of the lift coefficient, which is lower in the DDES simulation; (iii) a different spectral content between the two curves, as shown by the corresponding frequency spectra reported in Fig. 100b. Indeed, while in URANS the associated spectrum is characterized by a single tone located at $St = fc/u_\infty = 0.09$, the DDES spectrum shows a broader range of excited frequencies.

A comparison of the mean pressure coefficient between URANS and DDES is reported in Fig. 101 for two cases, one corresponding to a steady configuration ($\alpha = 4°$) and one to a case with buffet ($\alpha = 7°$). The experimental data provided by IoA are

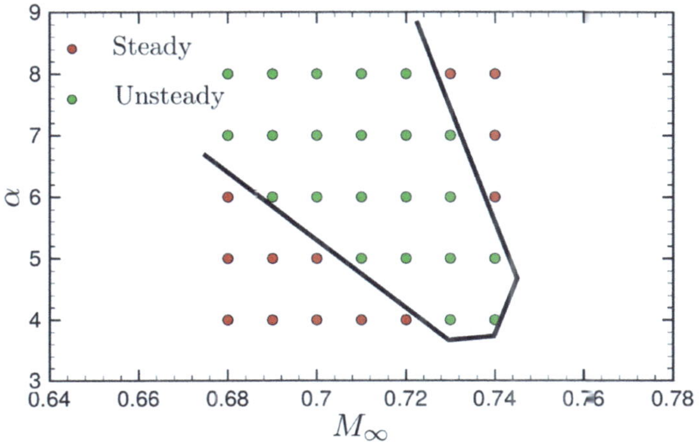

Fig. 98 Buffet limit for the V2C airfoil obtained with URANS

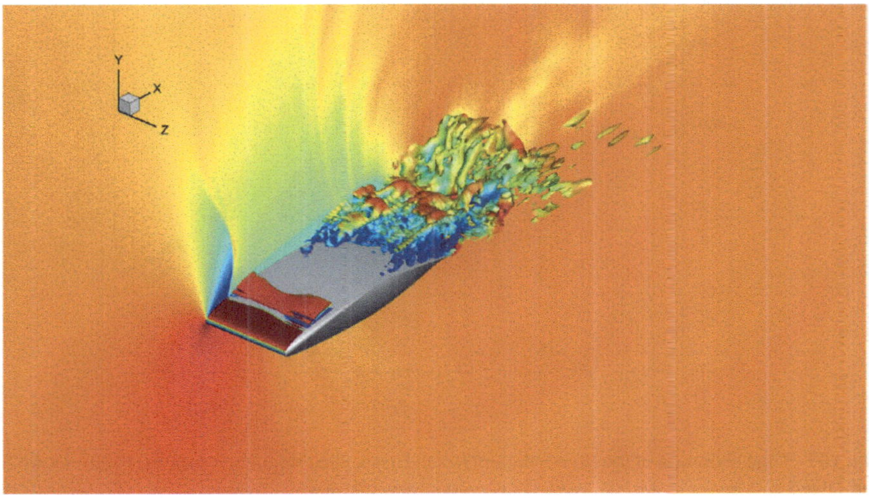

Fig. 99 DDES at $M_\infty = 0.7$, $Re = 3 \times 10^6$, $\alpha = 7°$. Vortical structures are shown through an iso-surface of the swirling strength

also reported in the plot for comparison. At $\alpha = 4°$ the numerical results are relatively close, except for the mean position of the shock, which is better predicted by DDES, in very good agreement with the experimental data. At $\alpha = 7°$ the reverse trend is observed and the URANS results are closer to the experimental measurements than DDES. Moreover, the smoothed shape of the DDES pressure coefficient on the suction side indicates the presence of large excursions of the shock, almost reaching the leading edge (and confirmed by movies available on the TFAST website). This

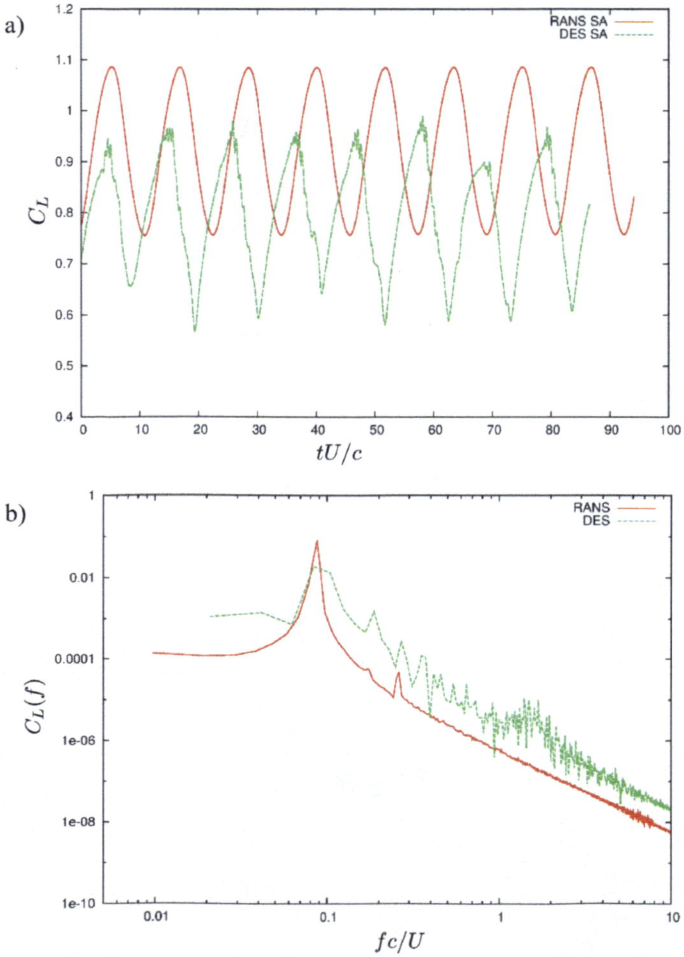

Fig. 100 Time evolution of the lift coefficient (**a**) and associated frequency spectra (**b**) from RANS and DDES

strange behavior is not found in the experiment but is consistent with the DDES predictions of IMFT (see Fig. 102) and it may indicate a failure of DDES.

We also tried to check the possible influence of the spanwise size of the domain on the buffet behavior by performing two additional simulations at $\alpha = 7°$, by using $L_z = c$ and $L_z = 3c$ as shown in Fig. 103 where a three-dimensional visualization of the vortical structures is reported. The results are shown in Fig. 104 in terms of mean pressure coefficient and highlight a slight influence of the spanwise width of the computational domain.

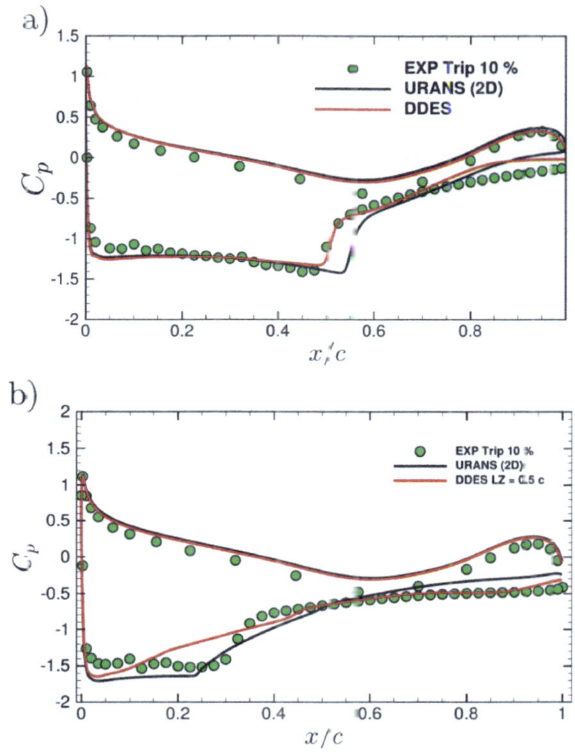

Fig. 101 Distribution of the mean pressure coefficient at **a** $\alpha = 4^\circ$ and **b** $\alpha = 7^\circ$

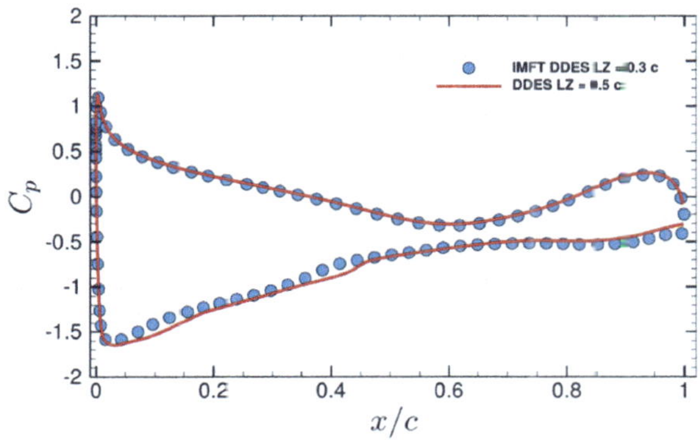

Fig. 102 Distribution of the mean pressure coefficient at $\alpha = 7^\circ$. Comparison with IMFT results

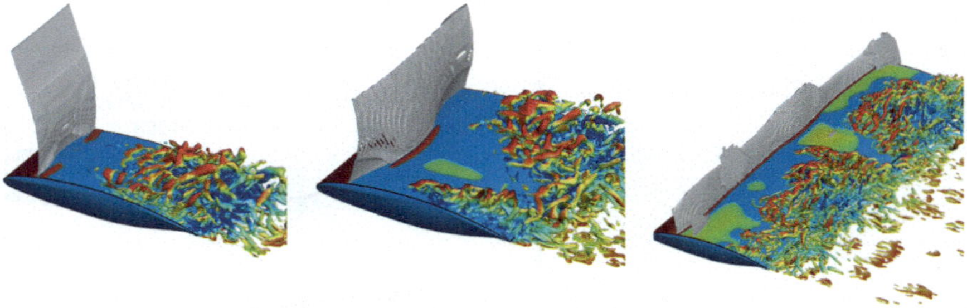

Fig. 103 DDES of V2C airfoil at $\alpha = 7°$. Effect of spanwise width of the computational domain. Vortical structures are shown through an iso-surface of the swirling strength

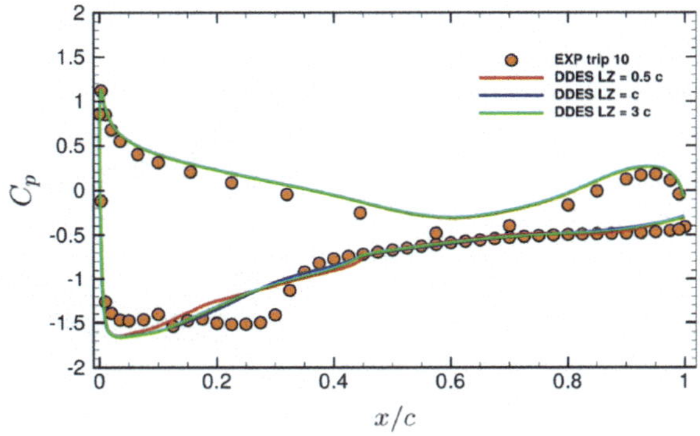

Fig. 104 Distribution of the mean pressure coefficient at $\alpha = 7°$. Effect of spanwise width of the computational domain

2.3.2 Results for the Laminar Interaction

In this section we report results obtained from numerical simulations of laminar interaction on the V2C airfoil. As usual in TFAST, the term laminar only refers to the state of the incoming boundary layer, which can be tripped in the proximity of the shock foot as a consequence of the associated strong adverse pressure gradient, similarly to what has been observed in the numerical simulations of the laminar interaction performed in WP 1 for the flat plate.

This kind of numerical simulation is extremely complex in the context of the DDES methodology applied for the turbulent interaction, because the exact location of the boundary layer transition (if any) is not known a priori. Furthermore, turbulence modeling for transitional flows has not reached a sufficient level of maturity to be considered reliable. Therefore, in our approach, we decided to switch off the turbulence model to perform numerical simulations of a laminar interaction at $M_\infty = 0.7$, $\alpha = 7°$ and $Re_c = 3 \times 10^6$. Two different grids were considered, the first one (referred to as coarse mesh) identical to that used for DDES of the fully turbulent interactions analysed in the previous section, whereas the second mesh (denoted as fine mesh) consists of $N_x \cdot N_y \cdot N_z = 2560 \cdot 512 \cdot 768$ grid points. The latter mesh

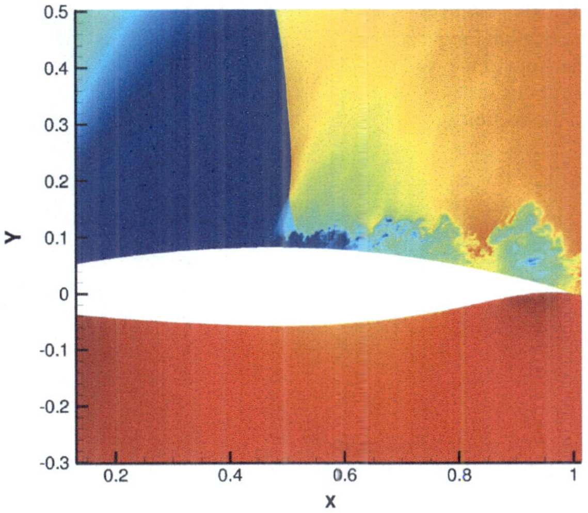

Fig. 105 Laminar interaction. Contours of the instantaneous density field in x–y plane (fine mesh case)

was selected on the basis of a trade-off between the computational time available and the need of running for a time span sufficiently long to capture several buffet cycles. The simulation was run on the IBM super computer Fermi by using 16,384 cores and a total number of 7 M hours.

Contours of the instantaneous density field are reported in a two-dimensional x–y plane for the computation performed on the fine mesh in Fig. 105. The figure well highlights the ability of the simulation to capture the salient features of the interaction: the very thin incoming laminar boundary layer which is tripped in the proximity of the shock foot and the lambda shape structure of the main shock.

A movie of the density field reported in Fig. 105 is available on the TFAST website. It allows to appreciate the presence of a global motion of the shock characterized by a limited spatial excursion with respect to the case predicted by URANS and DDES for the fully turbulent interaction and more similar to the experimental findings of IoA. This low-amplitude buffet is well visible in the spectral map reported in Fig. 106 showing the frequency spectrum of the wall pressure at all streamwise stations of the suction side. Two main peaks can be identified: (i) a low frequency peak located at $f \approx 100$ Hz, $x/c = 0.47$, associated with the low-amplitude buffet mentioned above and ii) a second peak at higher frequencies ($f \approx 1000$ Hz, $x/c = 0.47$), associated with the boundary layer separation and transition.

The evolution in time of the lift and drag coefficients is reported in Fig. 107, where a comparison is shown between the simulations performed on the two grids. For the purpose of comparison, the figure also includes data obtained from DDES of the fully turbulent interaction at the same flow conditions.

A good agreement is found between the two simulations of the laminar interaction, which provides support to the reliability of the obtained results. Interestingly, the lift coefficient associated with the laminar interaction is higher than that predicted by the DDES for the fully turbulent interaction whereas the average value of the drag

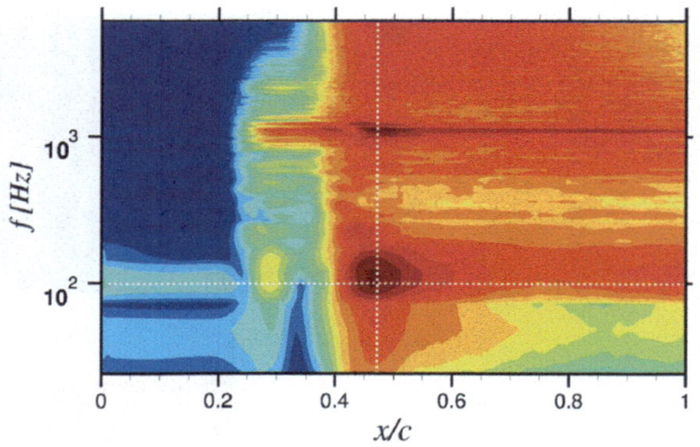

Fig. 106 Laminar interaction. Frequency spectrum of the wall pressure at various streamwise station on the suction side

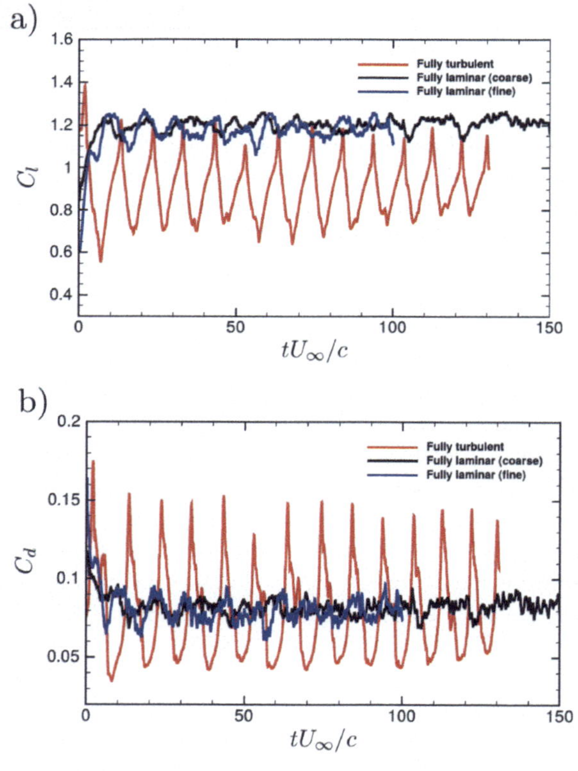

Fig. 107 Evolution of **a** lift and **b** drag coefficients

coefficient is found to be approximately the same. This suggests a higher efficiency of the laminar configuration.

2.3.3 Results for 10% Tripped Interaction

As a final step for WP 5.2 a direct numerical simulation of the V2C airfoil with tripping of the boundary layer on the suction side has been performed. Due to the high computational cost required to perform this kind of simulation, a single run was performed. The flow conditions are identical to the laminar interaction case analysed in the previous section, i.e. $M_{\infty} = 0.7$, $\alpha = 7°$ and $Re_c = 3 \times 10^6$. Boundary layer tripping was obtained by means of a two-dimensional roughness element along the spanwise direction, which is directly resolved by the simulation thanks to the immersed boundary approach, whose implementation has been already detailed in the deliverable D1.2.6. The position of the tripping device is set at $x_t = 10\%c$, a value selected after a preliminary discussion with the project partners. The results are presented in the following and compared with both the DNS of the laminar interaction and the URANS/DDES of the fully turbulent case at the same flow conditions.

An overall perception of the flow behavior can be gained looking at Fig. 108, where contours of the instantaneous density field are reported using the same style as in Fig. 105. Besides the different location of the main shock on the suction side, which depends on the particular selected time instant, the figure appears qualitatively similar to that of the laminar interaction, the most salient features being again the thickening of the boundary layer induced by the adverse pressure gradient and the formation of the lambda-shape structure at the shock foot. The main difference is the status of the incoming boundary layer, that in the tripped interaction becomes turbulent well ahead of the shock, although this aspect is not visible in the figure due to the small thickness of the upstream boundary layer. On the other hand, the presence of the tripping device is revealed by the formation of an oblique shock at $x \approx 0.1c$ and the boundary layer transition can be monitored looking at the skin

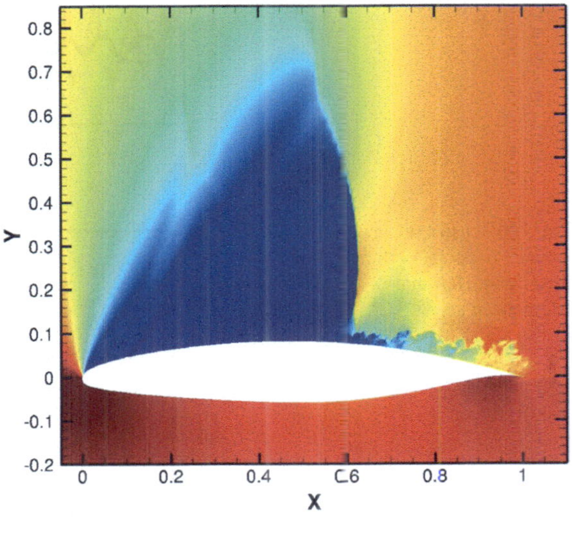

Fig. 108 10% tripped interaction. Contours of the instantaneous density field in x–y plane (fine mesh case)

friction coefficient on the suction side, which is reported in Fig. 110. The main effect of tripping the boundary layer can be appreciated by looking at Fig. 109, where the evolution of the lift and drag coefficients is shown.

Note that the curves of the tripped case are appended to that of the laminar interaction because the simulation has been performed using the last laminar solution for restart, and artificially imposing the tripping device at the selected location. The figure clearly highlights that tripping the boundary layer has a major effect on the flow behavior and in particular on buffet. In particular large oscillations of the lift and

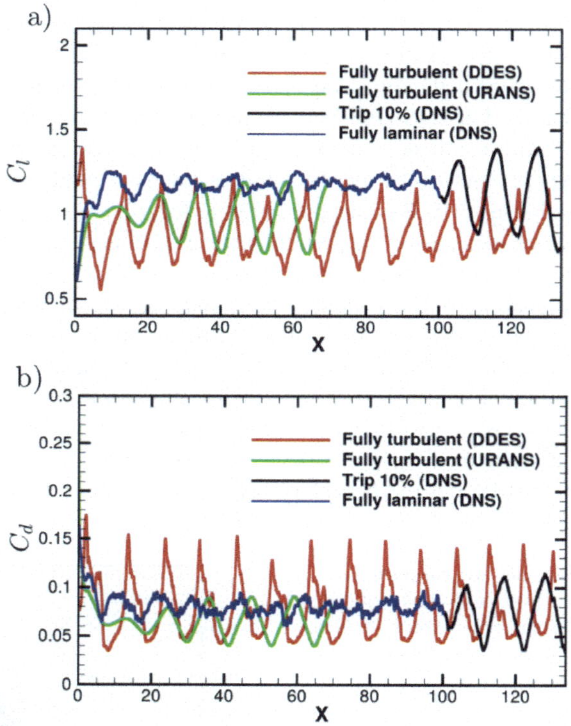

Fig. 109 Evolution of **a** lift and **b** drag coefficients

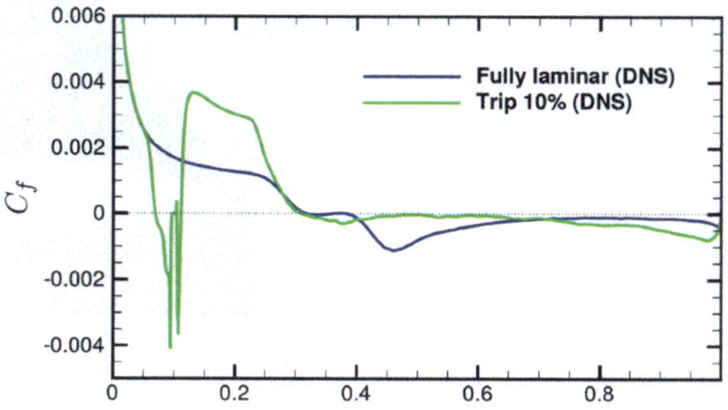

Fig. 110 Distribution of average friction coefficient. Lines: blue, laminar interaction; green, 10% tripped

drag coefficients appear, very similar to that predicted by URANS of the fully turbulent case. Significant differences are found with the DDES predictions of the fully turbulent case, which is a further confirmation of the failure of the DDES approach for this flow configuration.

The large oscillations observed in the force coefficients reflect a global motion of the shock which is also visible in a movie of the density field, which is available on the TFAST website. This movie shows that the shock motion extends between 0.3 < x/c < 0.6, as also can be deduced by the mean pressure coefficient, reported in Fig. 111. Note that in the case of tripping the comparison with the experiment is rather poor, and in particular the experimental C_p distribution suggests the absence of a true shock motion. The reasons for such differences are still not clarified and require further analysis (Fig. 112).

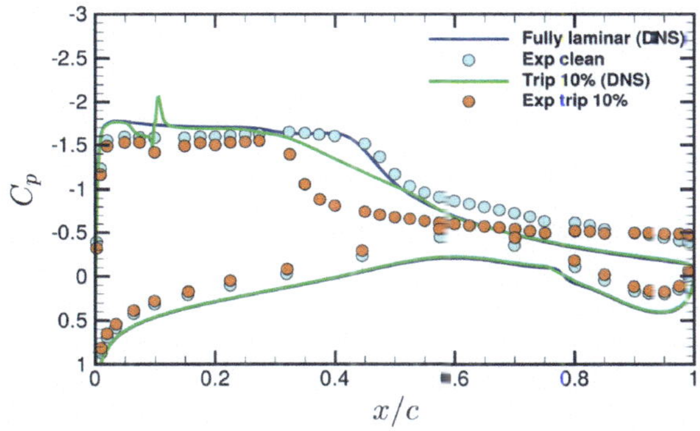

Fig. 111 Distribution of mean pressure coefficient. Lines: blue, laminar interaction; green, 10% tripped. Solid circles denote experiments of IoA

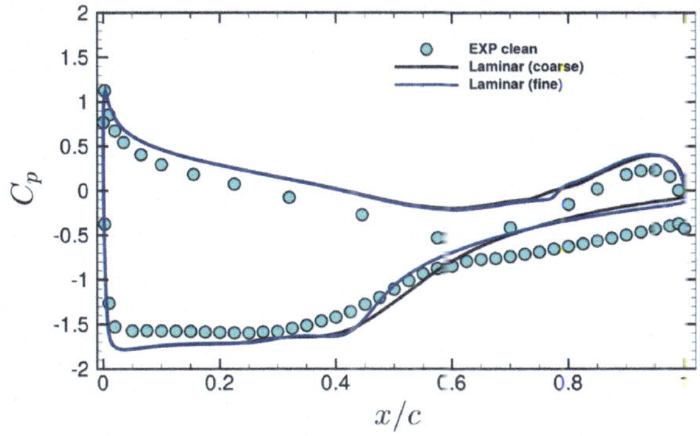

Fig. 112 Laminar interaction. Distribution of mean pressure coefficient

2.4 ULIV

2.4.1 UCAM Configuration

Mesh Generation

The employed mesh generation strategy allowed for easy introduction of vortex generators and roughness elements using the same basic multi-block topology. Partial views of the grid are shown in Fig. 113 for the UCAM case.

A typical C–H type topology was employed around the wing surface. All walls were considered as solids and at all cases a 10^{-5} c (where c is the aerofoil chord) distance was selected for the first cell. The initial mesh consisted of 659,802 grid points distributed in 510 blocks. The number of grid points around the upper and lower surface, the trailing edge and at the spanwise direction of the V2C aerofoil in the University of Cambridge wind tunnel can be summarised in the Table 9.

The inlet parameters were defined through imposed profiles while the pressure at the end of the computational domain was adjusted to provide the correct Mach number ahead of the aerofoil.

CFD Results for UCAM Configuration

For the UCAM case, results were obtained using the k-ω SST [22] and k-ω-γ $Re_{\theta t}$ (LCTM) [23] models at 3° and 5° of pitch angle. At the angle of 3°, extra computations were conducted with the Scale Adaptive Simulations (SAS) [24] and the SST-DDES [25] turbulence models. These initial computations were performed using free-stream conditions and did not account for the effects of the wind tunnel walls.

The streamlines in Fig. 114a demonstrate that the flow was separated at the 60% chord downstream of the leading edge. The pressure distribution of the flow on the upper surface (Fig. 114b) shows symmetry in the span-wise direction with respect to the mid-span. For this reason, two sections were selected, one at 10% of the span and another one at the mid-span.

Figure 115 presents the surface pressure and the skin friction distributions for the aforementioned stations on the section. The surface distribution at the mid-span shows some broad agreement with the experimental data received of UCAM for a pitch angle of 3°. The skin friction and surface pressure coefficient suggest a shock location at 55% of the chord with flow separation downstream.

Figure 116a, b suggest that there was substantial blockage at these angles and the shock reached the upper wall of the tunnel. A numerical Schlieren image at the mid-span of the aerofoil can be seen in Fig. 117.

At the same conditions but for a pitch angle of 5°, the blockage continued, and the shock was stronger. This can be seen in Fig. 118a, b where the Mach number field is shown at two span-wise stations. The flow blockage remained the same, and the shock wave reached the upper wall of the wind tunnel.

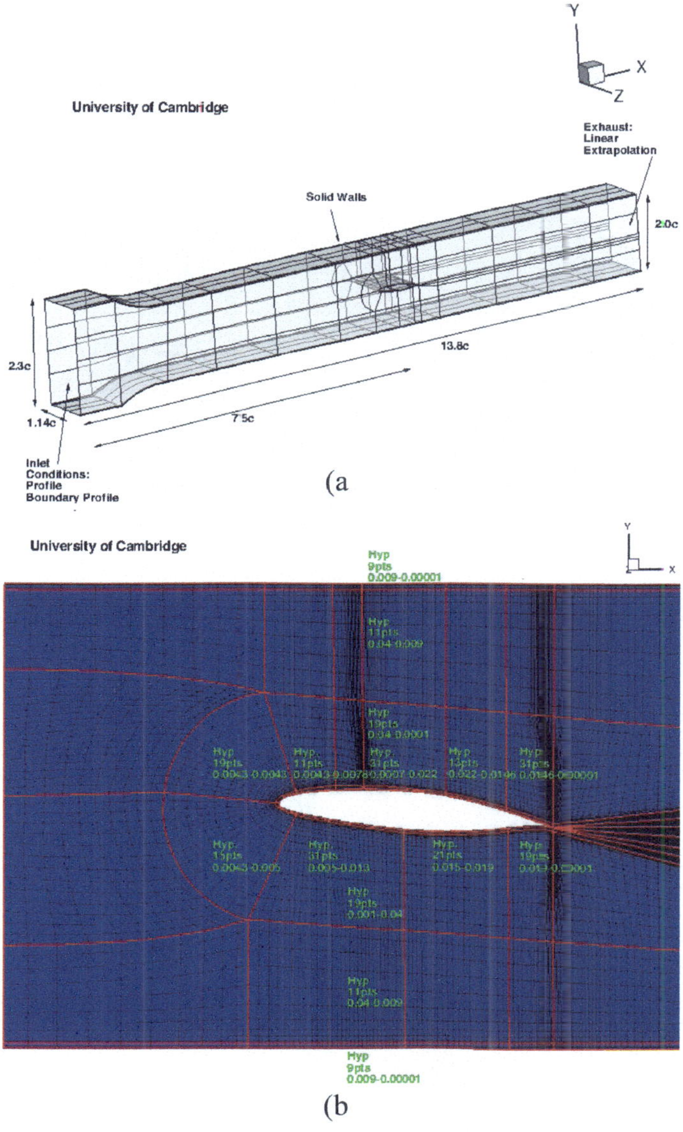

Fig. 113 Dimensions (**a**) and a detailed view of the mesh (**b**) around the UCAM configuration

Table 9 Number of points

	Upper surface	Lower surface	Trailing edge	Spanwise
Mesh points	98	86	18	39

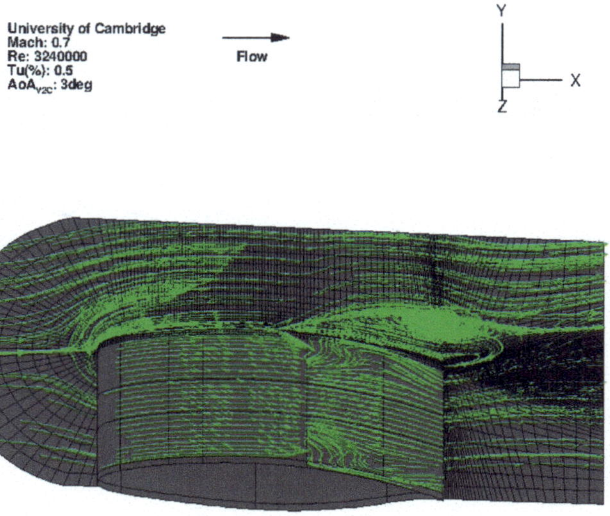

(a

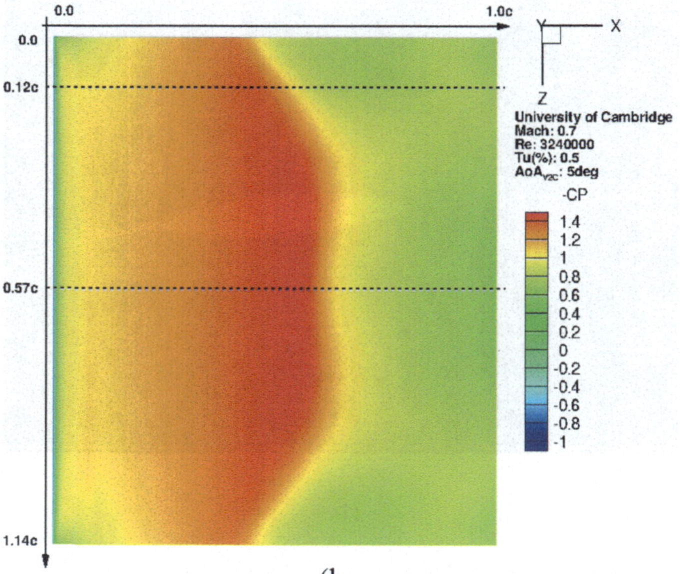

(b

Fig. 114 Surface flow visualisation (**a**) and surface pressure distribution (**b**) of the flow on the upper surface of the V2C aerofoil. Calculations made with k-ω SST turbulence model (Mach number 0.7, Re 3.24×10^6 and 3° of pitch)

Fig. 115 Surface pressure (**a**) and skin friction (**b**) coefficients at two span-wise stations. Calculations made with k-ω SST turbulence model (Mach number 0.7, Re 3.24 × 10^6 and 3° of pitch)

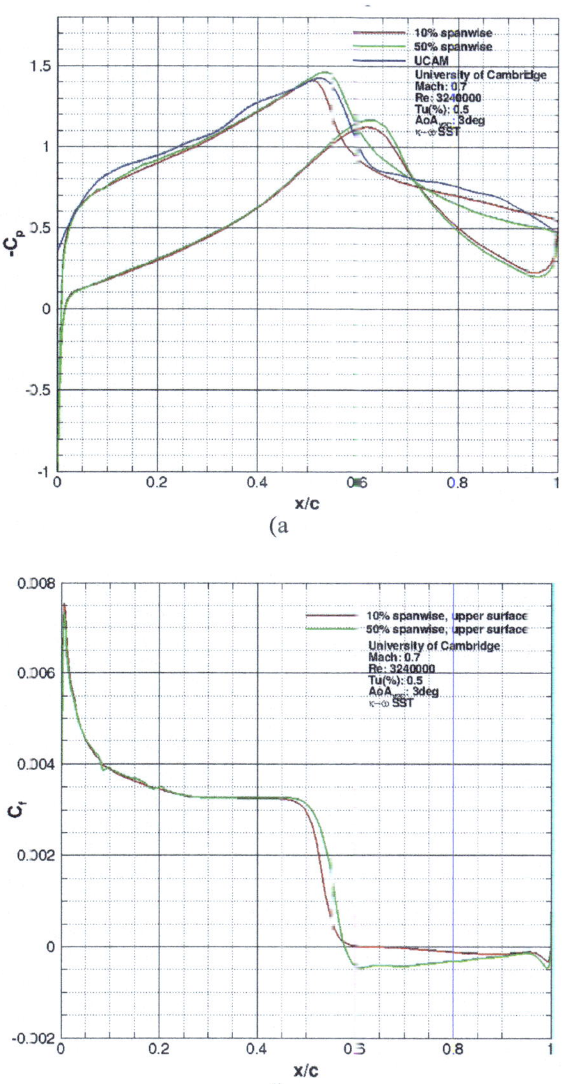

(a)

(b)

There was good agreement for the surface pressure distribution at this pitch angle with the experimental data (4.6°). The shock moved slightly downstream in comparison to the 3° case. The surface pressure and skin friction on the upper surface of the aerofoil distributions can be seen in Fig. 119a, b.

The computations were repeated using the k-ω-γ $Re_{\theta t}$ (LCTM) transition model at the same conditions. The flow remained laminar for about 15% of the chord as can be seen in Fig. 120a, b. The surface pressure distributions show higher peaks with

Fig. 116 Mach number visualisation at two span-wise stations of the UCAM wind tunnel, 10% (**a**) and 50% (**b**). Calculations made with k-ω SST turbulence model (Mach number 0.7, Re 3.24 × 10^6 and 3° of pitch)

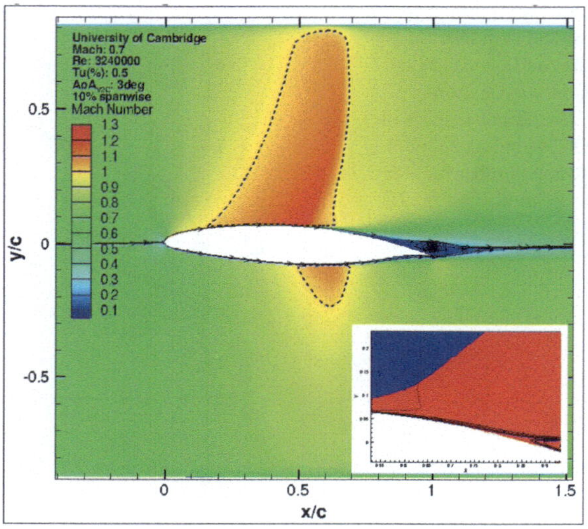

(a

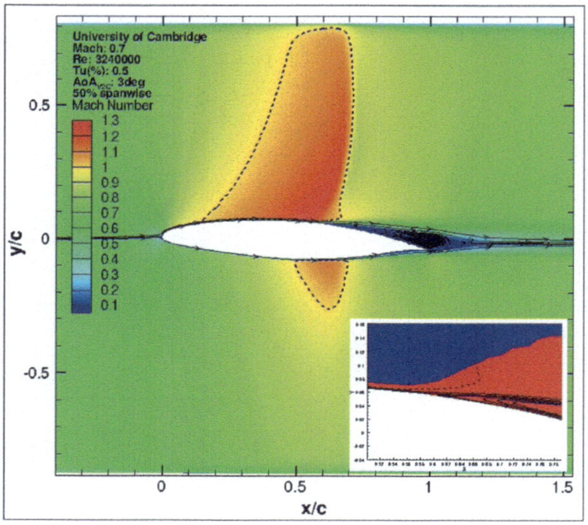

(b

the LCTM is used and there was larger flow separation downstream the shock in comparison to the results obtained with the fully turbulent SST model. Figure 120b shows no separation for the LCTM transition model.

Figure 121a, b show the Mach number and the intermittency fields of the LCTM computations. Both figures further corroborate the previous remarks on the LCTM performance, and its early transition onset (intermittency higher than 0.1).

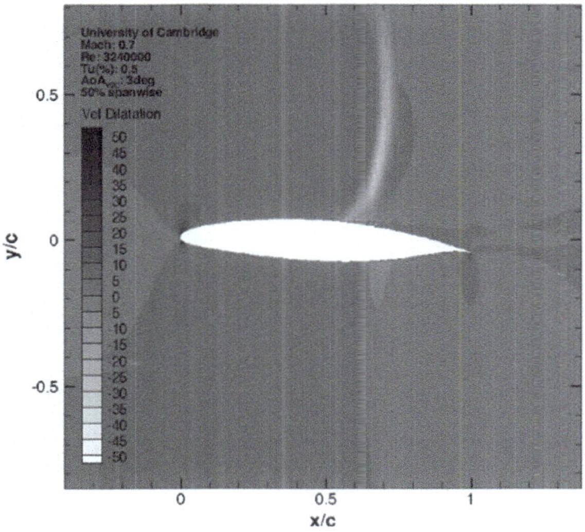

Fig. 117 Schlieren image at the mid-span of the section in the UCAM wind tunnel. Calculations made with k-ω SST turbulence model (Mach number 0.7, Re 3.24 × 10^6 and 3° of pitch)

Similar calculations were conducted for the case of 5° pitch. The skin friction coefficient distribution suggests that the flow remained fully turbulent during the calculations (Fig. 122).

The substantial blockage observed at the previous cases was present here too with the LCTM transition model (see Fig. 123a, b). The shock at both stations reached the upper wall of the tunnel.

Following from the rather disappointing results of the LCTM, computations were undertaken with fixed transition at 30%c.

The streamlines of Fig. 124a suggest that the flow separates after 65% of the section. The intense separation shown with the streamlines can also be observed with the Mach number field where the strength of the shock has been increased creating a blockage at both sides of the wind tunnel based on the results presented in Fig. 124b at the mid-span of the blade.

The surface pressure and the skin friction distributions on the upper surface of the wing can be seen in Fig. 125a, b, respectively. The results are for two span-wise locations based on the symmetry observed during the computations. The shock moved downstream in comparison to the SST fully turbulent computations while larger separation is observed.

Although most computations for the UCAM case resulted in steady flows, unsteady computations were performed. Indicative results with the SST turbulence model are now presented. For these calculations, a dimensionless time step of $t^* = tU/c = 0.0023$ was obtained. Probes were added on the wing surface, and 20 of these were placed at each span-wise station. The distribution of the each probe can be seen in Fig. 126, with the majority of the probes on the upper surface and every 5%c after the 40%c. The four sections are located at 10.5, 25, 50 and 75% of the span.

Fig. 118 Mach number visualisation on two different span locations of the section of the CAM wind tunnel, 10% (**a**) and 50% (**b**). Calculations made with k-ω SST turbulence model (Mach number 0.7, Re 3.24 × 10^6 and 5° of pitch)

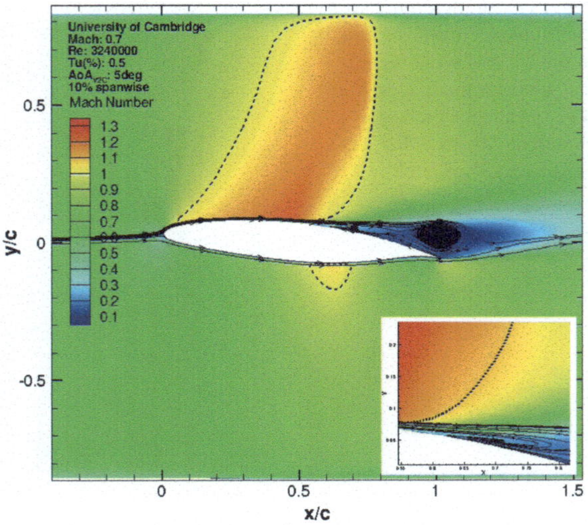

(a

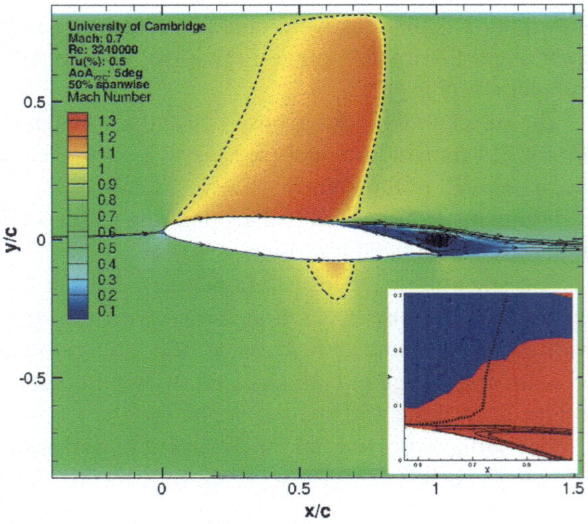

(b

The streaklines at t* = 4.76 can be seen in Fig. 127a while the pressure distribution on the upper surface of the aerofoil can be seen in Fig. 127b.

The obtained Mach number field is given in Fig. 128a, b at a non-dimensional time scale t* = tU/c = 4.76. The results are for two stations, one close to the sidewalls of the wind tunnel and one at the mid-span of the tunnel. The shock has moved downstream and large flow separation can be seen in Fig. 129b.

More calculations were conducted using the Scale Adaptive Simulation (SAS) model. These were unsteady computations are now discussed. Similar to the URANS

Fig. 119 Surface pressure (a) and skin friction (b) coefficients on two different span locations of the section, 10% (a) and 50% (b). Calculations made with k-ω SST turbulence model (Mach number 0.7, Re 3.24×10^6 and 5° of pitch)

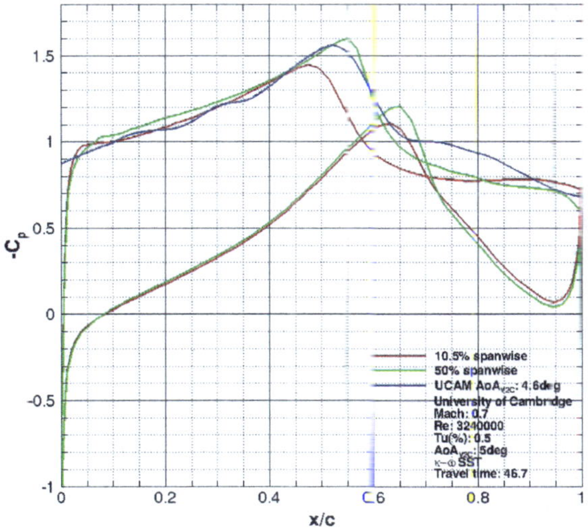

(a

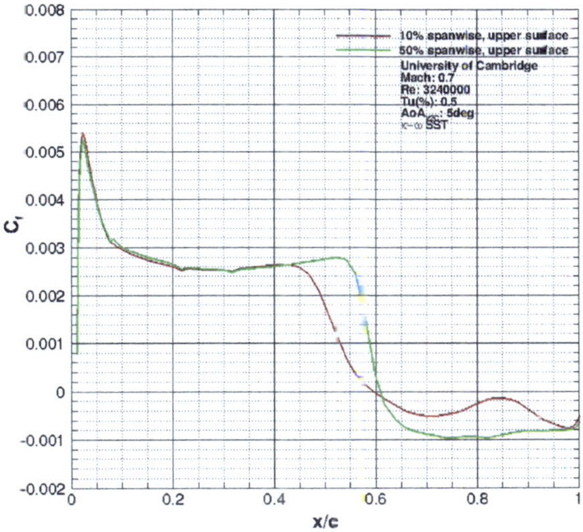

(b

computations, the SAS computations were performed with a dimensionless time step of 0.0023. Probes were added also on the wing.

The streamlines at $t^* = 4.76$ can be seen in Fig. 130a while the pressure distribution on the upper surface of the aerofoil can be seen in Fig. 130b. Both figures indicate separation close to the walls.

The obtained Mach number field is given in Fig. 131a, b at a non-dimensional time scale $t^* = tU/c = 4.76$. The results are for two stations, one close to the sidewalls

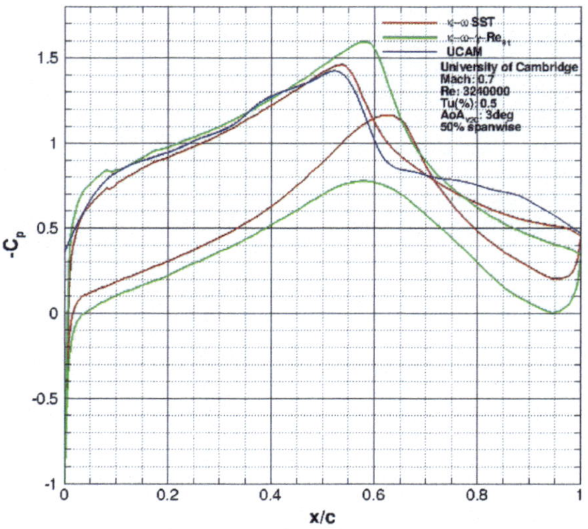

(a

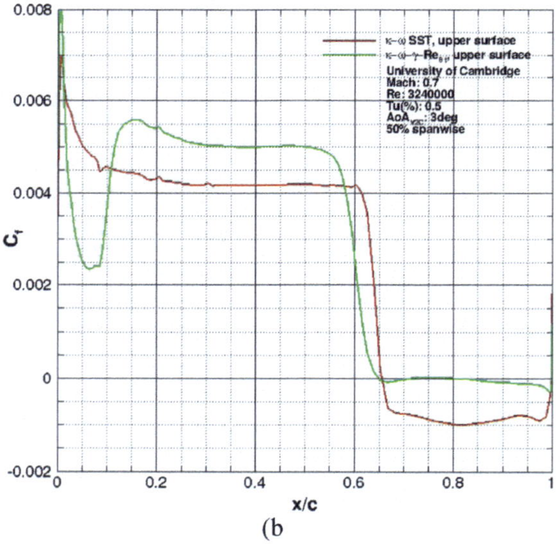

(b

Fig. 120 Surface pressure (**a**) and skin friction (**b**) coefficients at two span-wise stations. Calculations made with k-ω-γ $Re_{\theta t}$ transition model (Mach number 0.7, Re 3.24 × 10^6, 3° of pitch)

of the wind tunnel and one at the mid-span of the tunnel. The shock has moved downstream and large flow separation can be seen in Fig. 132b.

2.4.2 IoA Configuration

The investigations continued with numerical simulations of the investigations performed in the Institute of Aviation (IoA) N-3 wind tunnel. The N-3 wind tunnel of the

Fig. 121 Mach number (**a**) and intermittency (**b**) fields at the mid-plane of the wind tunnel. Calculations made with k-ω-γ $Re_{\theta t}$ transition model (Mach number 0.7, Re 3.24×10^6, 3° of pitch)

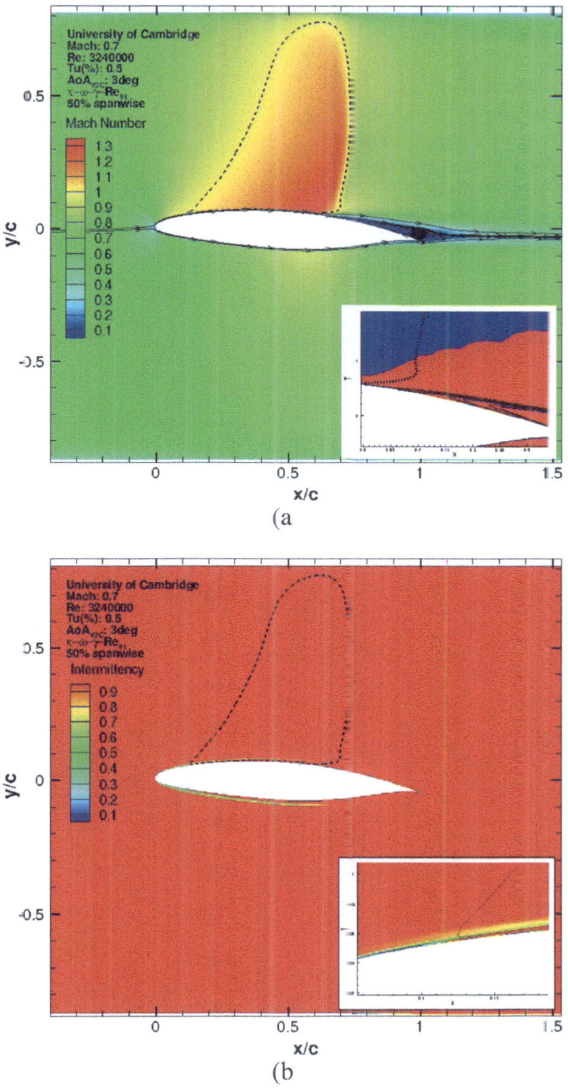

IoA is a transonic blown down wind tunnel and Mach number range of subsonic and transonic flow is 0.3 and 1.15. The transonic test section has a square cross section of 0.6 m × 0.6 m and a length 1.58 m.

The model of the aerofoil initially was thought to have a chord of 100 mm and its span was equal to the width of the test section. Roughness, conventional vortex

Fig. 122 Surface pressure (**a**) and skin friction (**b**) coefficients at two span-wise stations. Calculations made with k-ω-γ $Re_{\theta t}$ transition model (Mach number 0.7, Re 3.24 × 10^6, 5° of pitch)

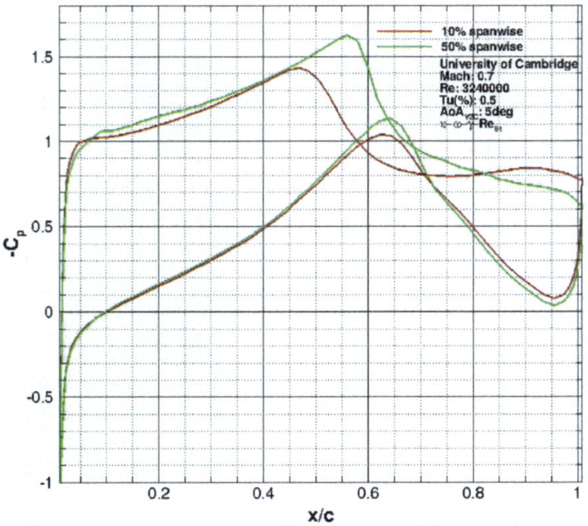

(a

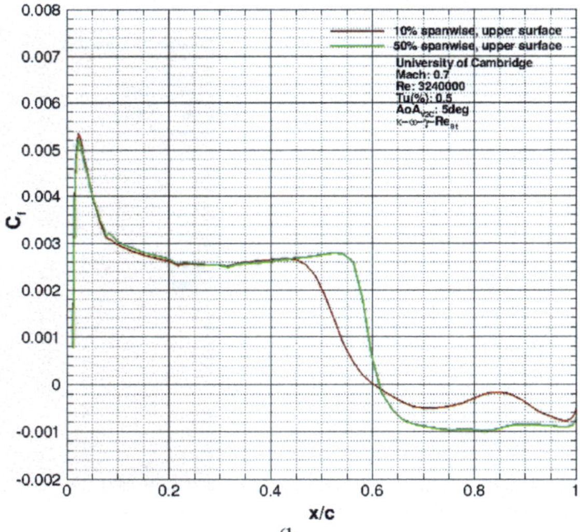

(b

generators and air-jet vortex generators were initially selected to trip the laminar-turbulent transition.

Mesh Generation

Following the same strategy as for the UCAM case, a C-H type topology was employed around the V2C aerofoil for the IoA case. All walls were considered as solids and at all cases a 10^{-5} c distance was selected for the first cell. The properties of the CFD mesh are summarised in the Table 10. The grid for this case consists of

Fig. 123 Mach number visualisation at two span-wise locations of the UCAM wind tunnel, 10% (**a**) and 50% (**b**). Calculations made with k-ω-γ $Re_{\theta t}$ transition model (Mach number 0.7, Re 3.24 × 10^6, 5° of pitch)

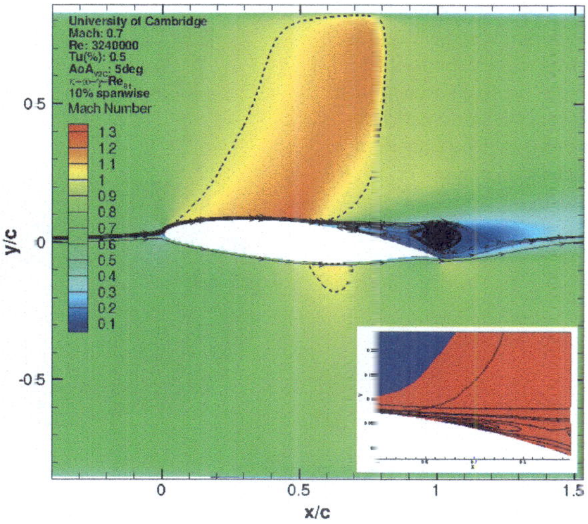

(a

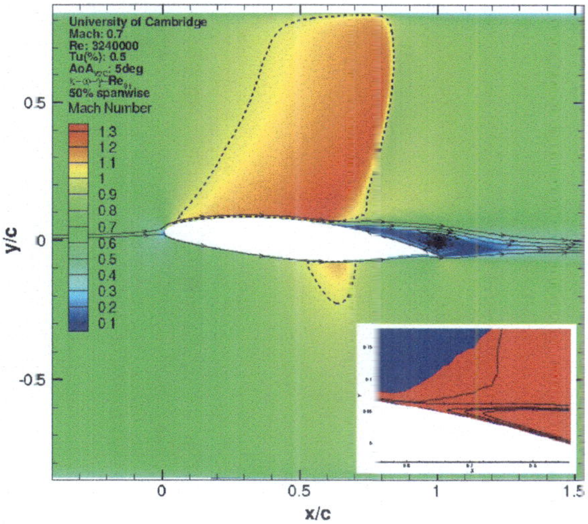

(b

3,924,648 grid points and 560 blocks. Partial views of the initial grid for the N-3 Wind tunnel can be seen in Figs. 133.

CFD Results for IoA Configuration

For this case, the computations were performed using URANS at an incidence angle of 5°. The streamlines of Fig. 134a show that the flow is separated after 45% of the section. The pressure distribution of the flow on the upper surface, Fig. 134b shows

Fig. 124 Streamlines and Mach number visualisation at mid-span stations. The flow is for fixed transition at 30%c. Calculations made with k-ω-γ $Re_{\theta t}$ transition model (Mach number 0.7, Re 3.24×10^6 and 3° of pitch)

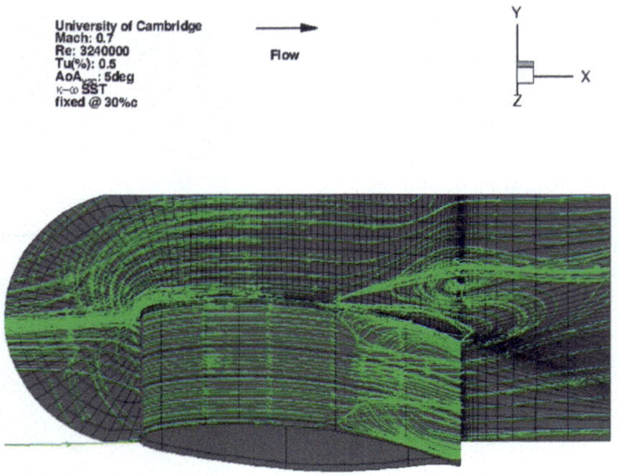

(a

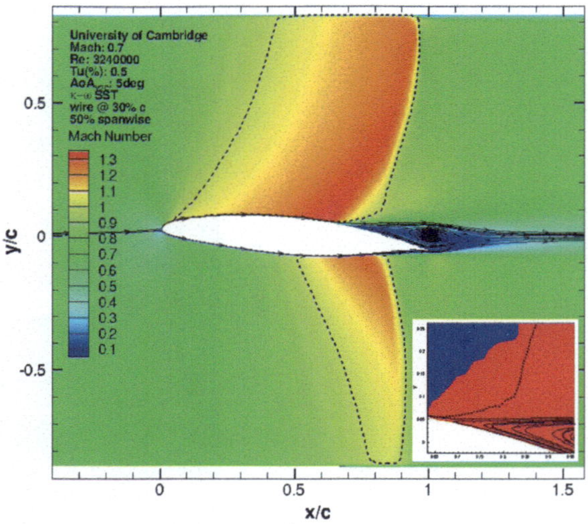

(b

the symmetry observed in the span-wise direction. For this reason two stations were selected, one at 15% of the span and another one at the mid-span.

Figure 135 presents the surface pressure and the skin friction distributions for the aforementioned span locations of the section. The shock is located at about 50% of the chord in the mid-span station while the separation at that location is more intense. These results can be seen in Fig. 136a, b where Mach number visualization is presented.

Fig. 125 Surface pressure (**a**) and skin friction (**b**) coefficient at two span-wise stations. Calculations made with k-ω SST turbulence model with fixed transition at 30%c (Mach number 0.7, Re 3.24 × 10^6 and 3° of pitch)

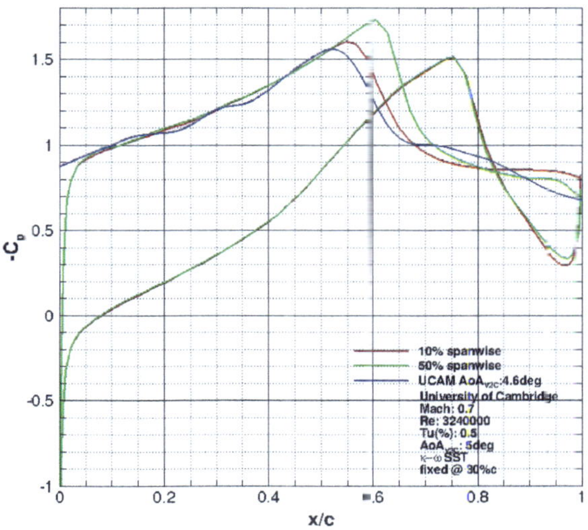

(a

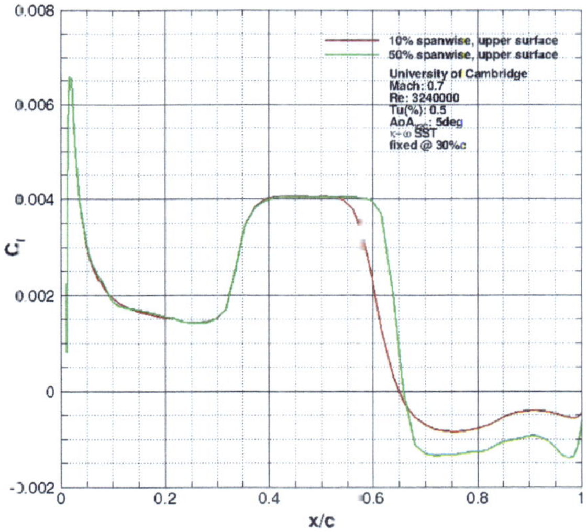

(b

Vortex Generators

Mechanical vortex generators were also tested to control the shock interaction with the separated boundary layer. For the case of IoA, co-rotating convectional VGs (CoC1 VGs) were considered.

The co-rotating conventional VGs employed were triangular elements fixed at 18° pitch with respect to the main flow direction. Five CoC1 VGs are located in the span-wise direction of the wing, upstream of the shock region A view of the CoC1 vortex generators is shown in Fig. 137.

Fig. 126 Distribution of probes on the V2C wing

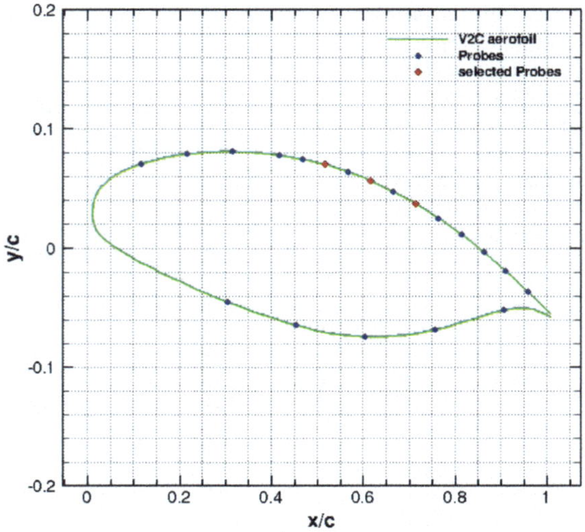

(a

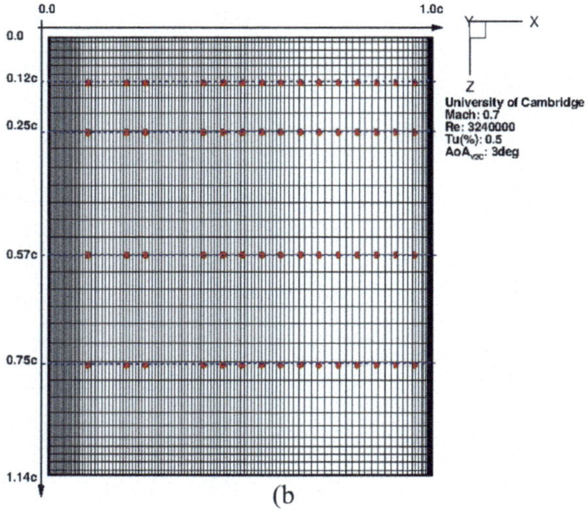

(b

The calculations were conducted for a pitch angle of 5° and the results can be seen in Fig. 138 where the pressure distribution on the upper surface of the wing is presented. Sections of the obtained Mach field are shown in Fig. 139 (Figs. 140 and 141).

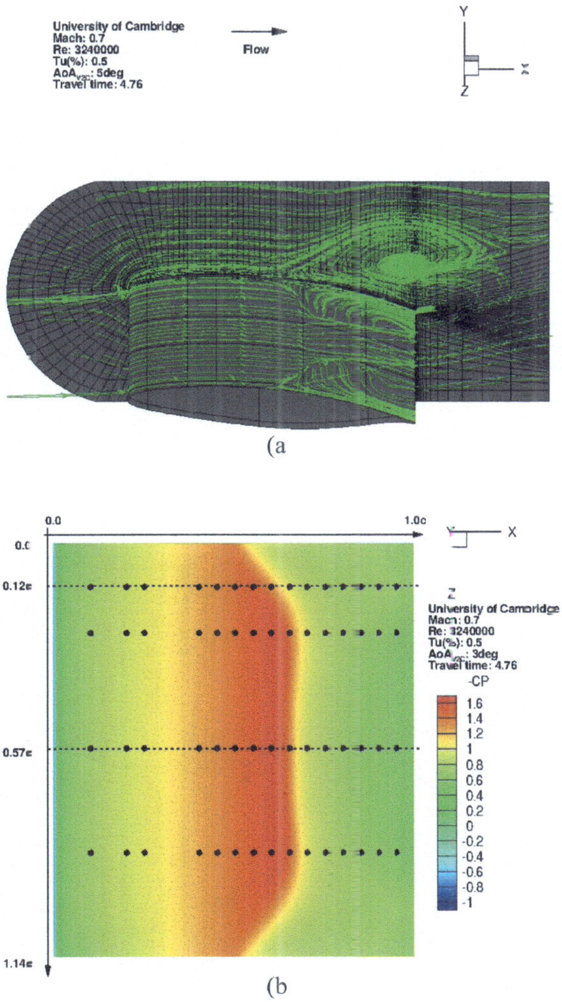

Fig. 127 Surface flow visualisation (**a**) and surface pressure distribution (**b**) of the flow on the upper surface of the V2C aerofoil. Calculations made with k-ω SST turbulence model (Mach number 0.7, Re 3.24 × 10^6 and 3° of pitch)

2.4.3 Conclusions

This work was mainly conducted with URANS methods to include the effect of the wind tunnel walls on the buffet boundary. Computations using k-ω SST turbulence model, k-ω-γ $Re_{\theta t}$ transition model and Scale Adaptive Simulation were conducted. The results show at least qualitative agreement with the measurements though further computations are certainly needed for detailed comparisons to be obtained against the experimental data of TFAST.

Fig. 128 Mach number visualisation at two span-wise locations of the UCAM wind tunnel, 10.5% (**a**) and 50% (**b**). Calculations made with k-ω SST turbulence model (Mach number 0.7, Re 3.24 $\times 10^6$ and 3° of pitch)

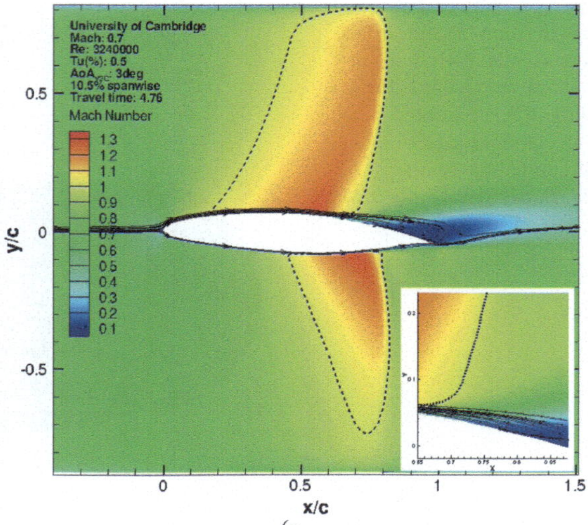

(a

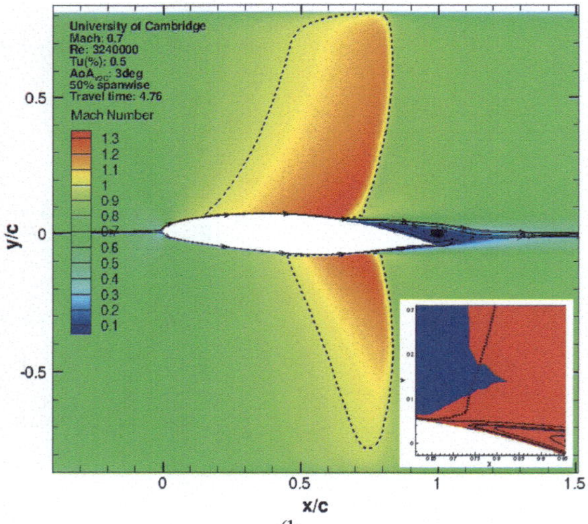

(b

Fig. 129 Surface pressure (**a**) and skin friction (**b**) coefficients at two span-wise stations. Calculations made with k-ω SST turbulence model (Mach number 0.7, Re 3.24 × 10^6 and 3° of pitch)

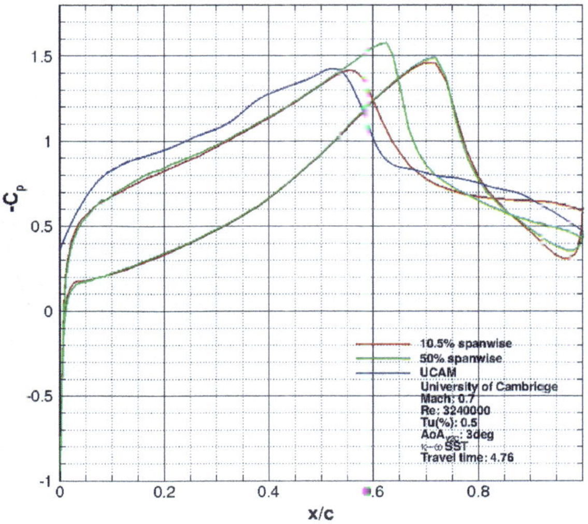

(a

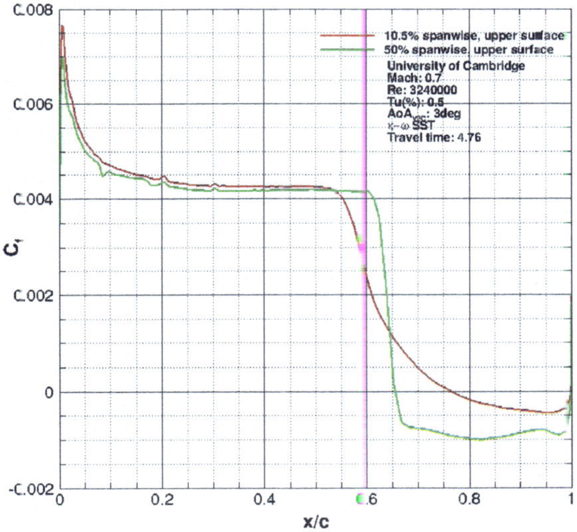

(b

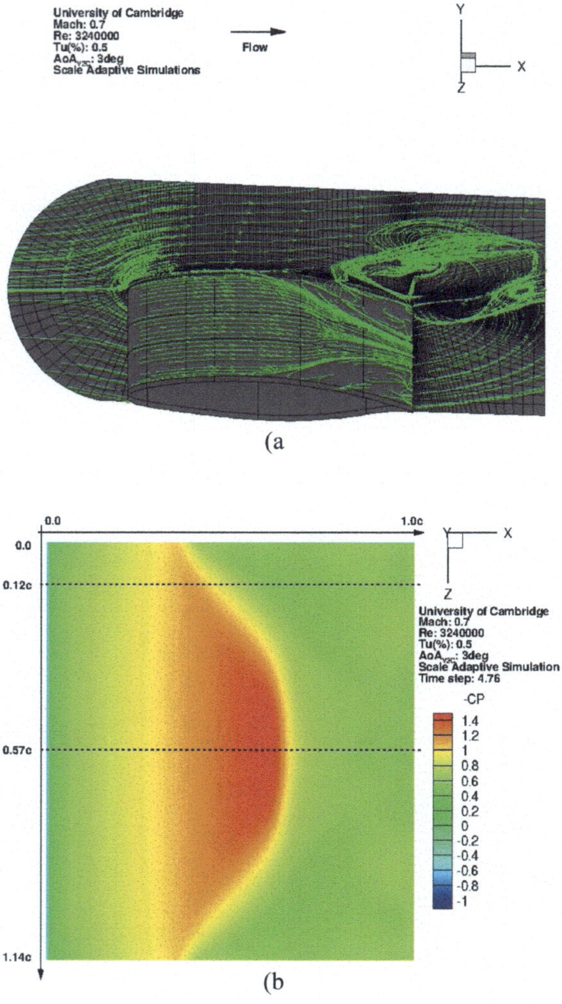

Fig. 130 Surface pressure visualisation (**a**) and surface pressure distribution (**b**) of the flow on the upper surface of the V2C aerofoil. Calculations made with SAS model (Mach number 0.7, Re 3.24×10^6 and 3° of pitch)

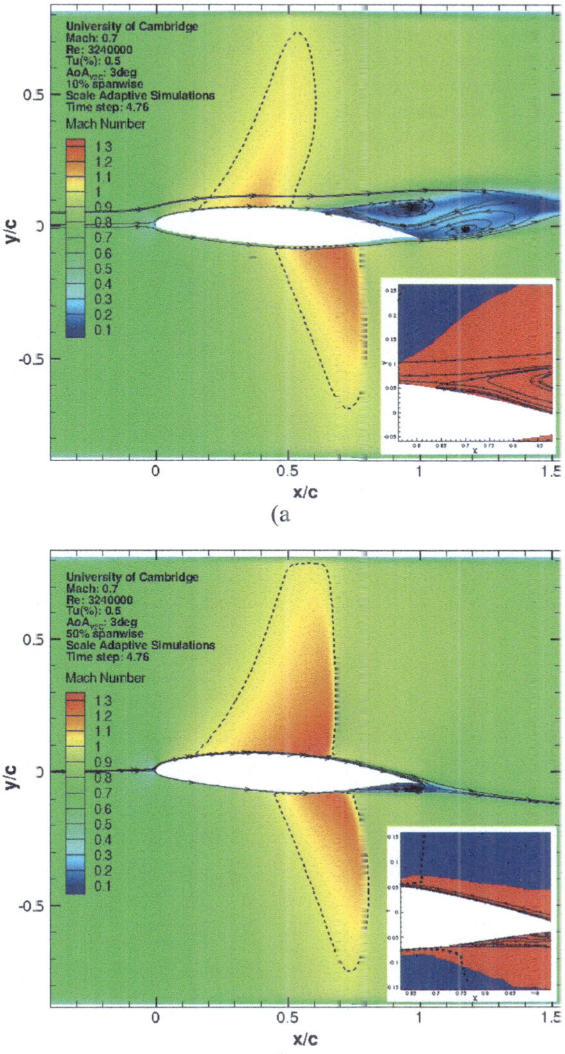

Fig. 131 Mach number visualisation at two span-wise locations of the UCAM wind tunnel. 10.5% (**a**) and 50% (**b**). Calculations made with SAS model (Mach number 0.7, Re 3.24×10^6 and 3° of pitch)

Fig. 132 Surface pressure (a) and skin friction (b) coefficients at two span-wise stations of the section. Calculations made with SAS model (Mach number 0.7, Re 3.24×10^6 and 3° of pitch)

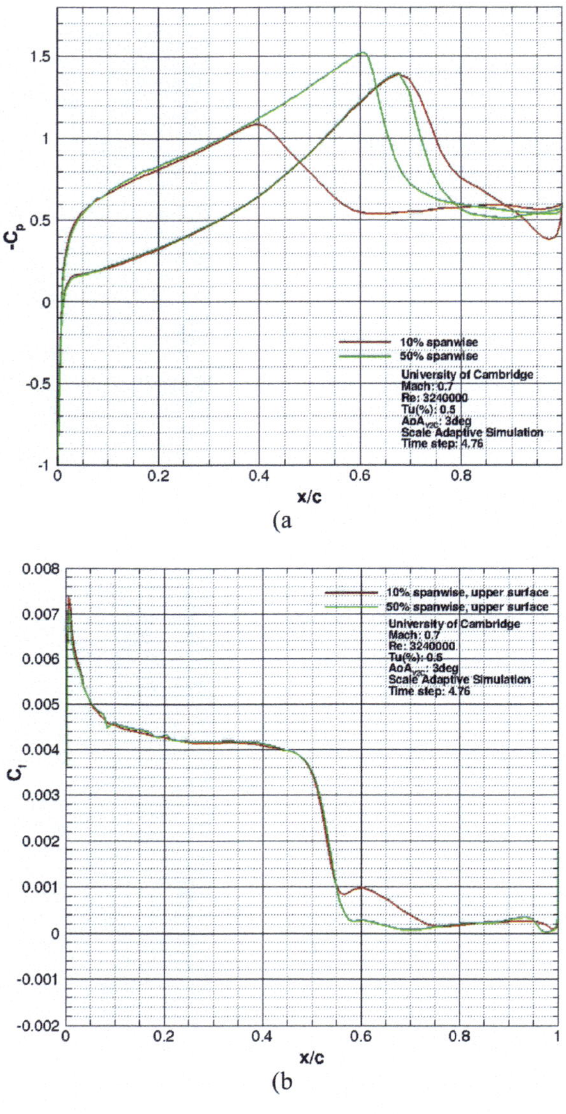

Table 10 Number of points

	Upper surface	Lower surface	Trailing edge	Spanwise
Grid points	140	91	34	91

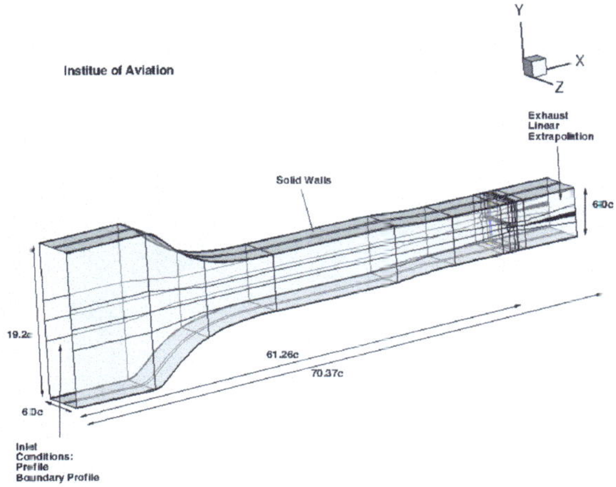

(a

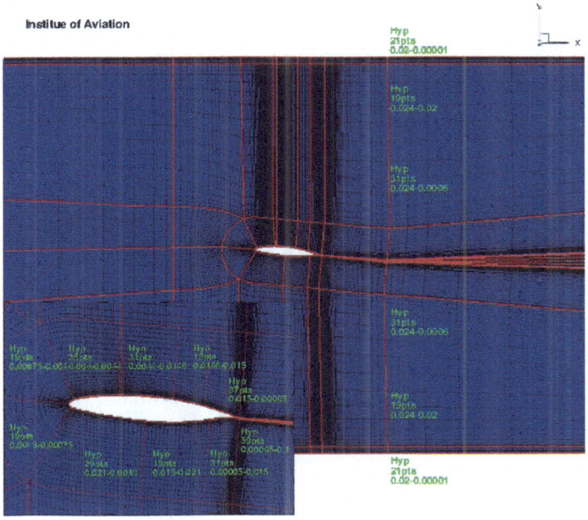

(b

Fig. 133 Dimensions (**a**) and a detailed view of the mesh (**b**) around the IoA configuration

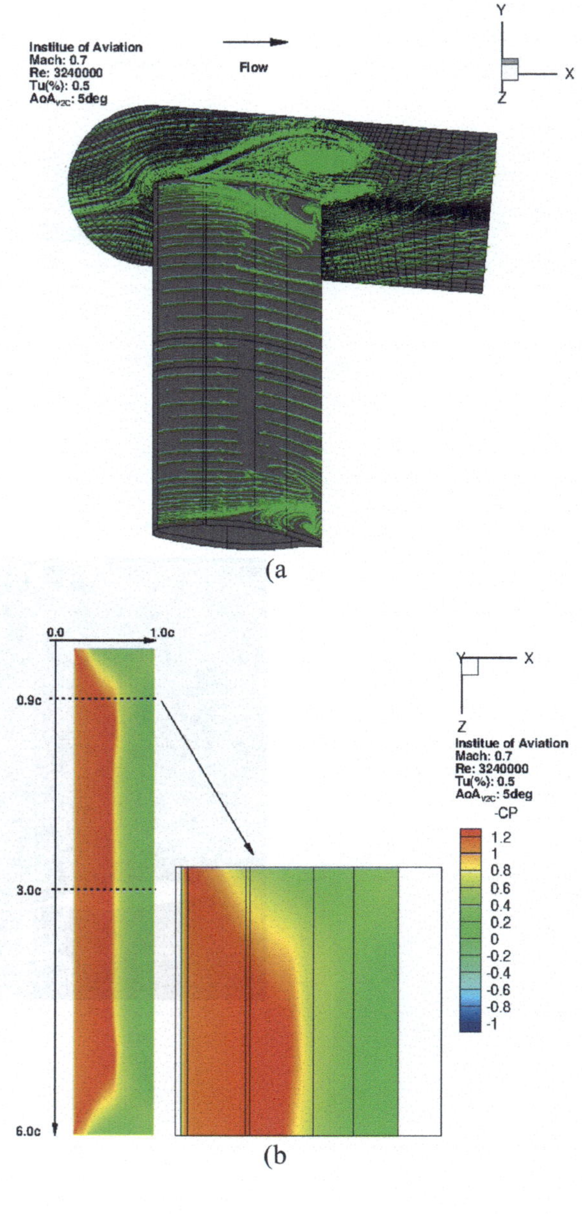

Fig. 134 Streamlines (**a**) and pressure distribution (**b**) of the flow on the upper surface of the V2C aerofoil. Calculations made with k-ω SST turbulence model (Mach number 0.7, Re 3.24 $\times$ 10^6 and 5° of pitch)

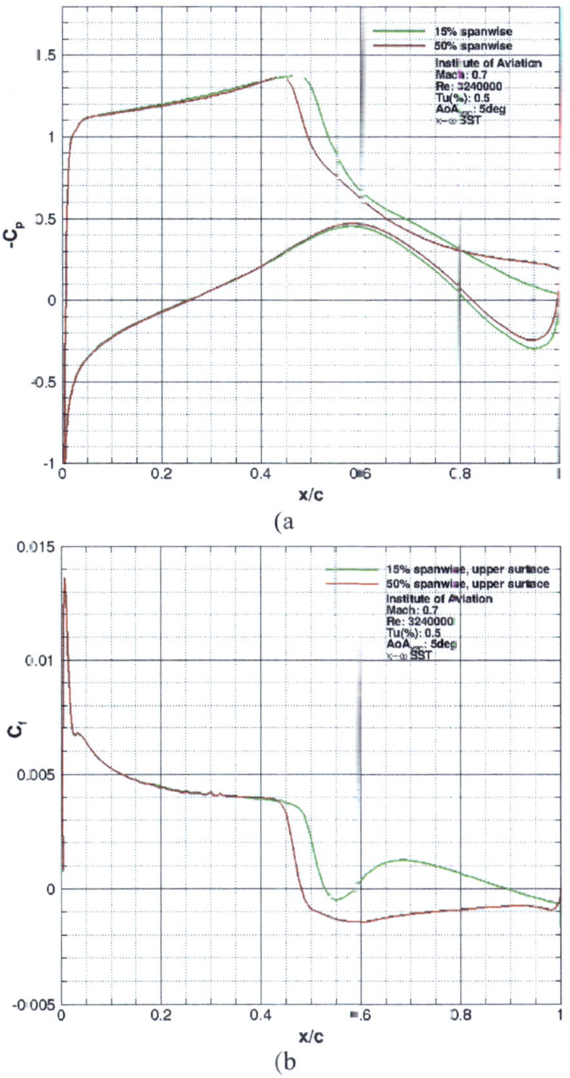

Fig. 135 Surface pressure (**a**) and skin friction (**b**) coefficients at two span-wise stations. Calculations made with k-ω-γ $Re_{\theta t}$ turbulence model (Mach number 0.7, Re 3.24 × 10^6 and 5° of pitch)

Fig. 136 Mach number visualisation at two span-wise stations of the IoA wind tunnel. 15% (**a**) and 50% (**b**). Calculations employed the k-ω SST turbulence model (Mach number 0.7, Re 3.24×10^6 and 5° of pitch)

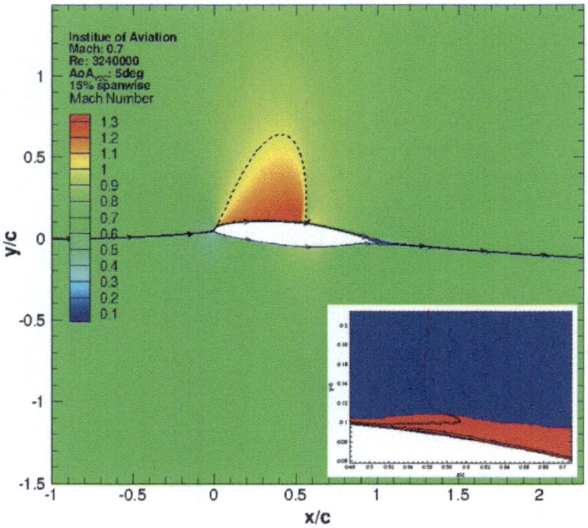

(a

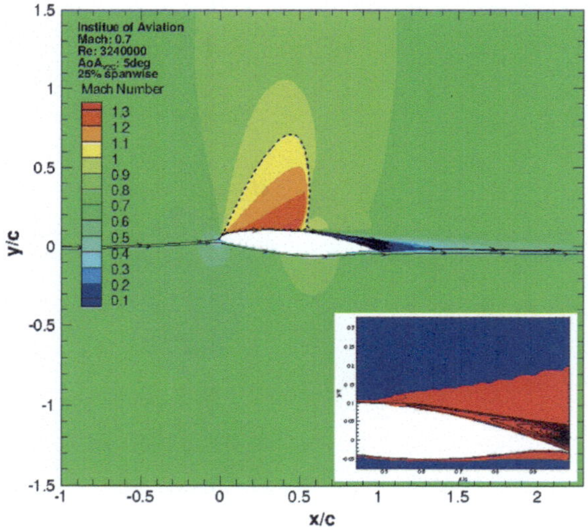

(b

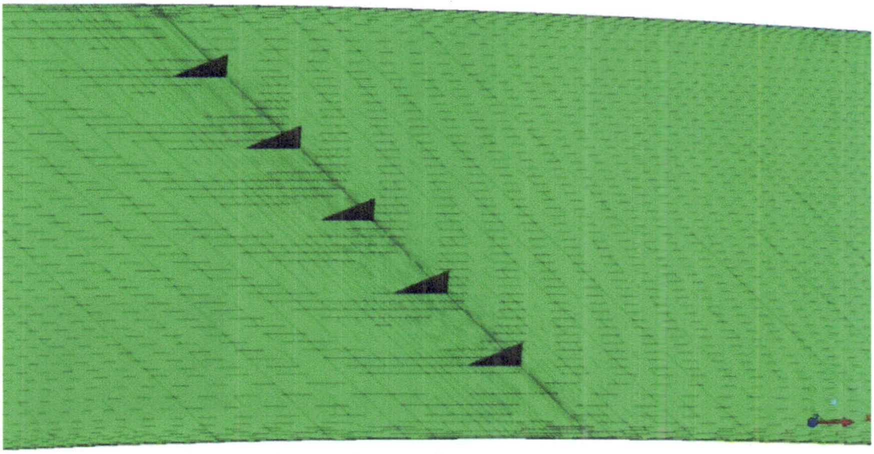

Fig. 137 Location and the details of the grid around the vortex generators

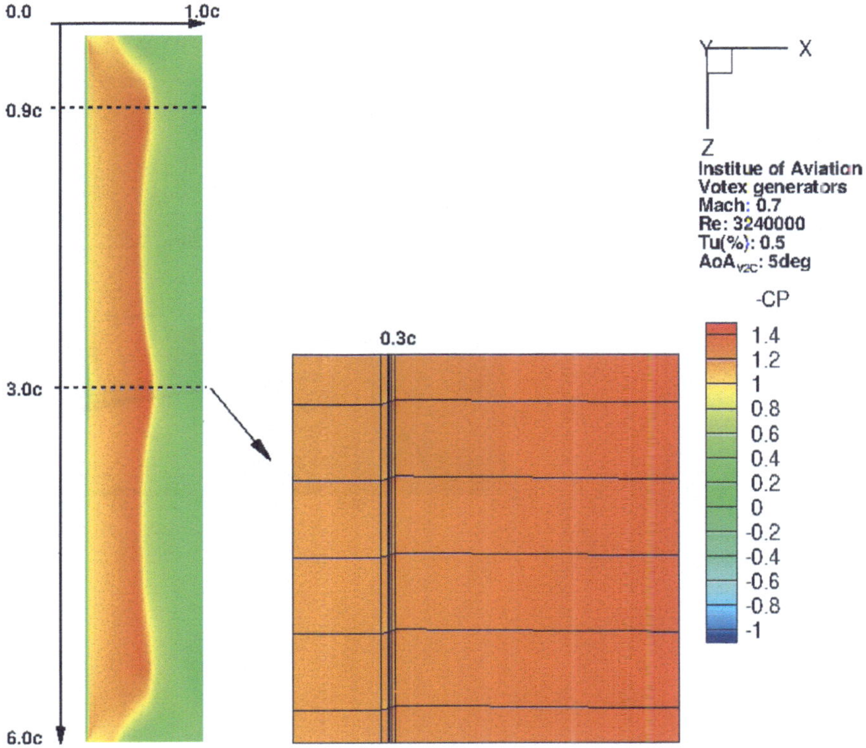

Fig. 138 Pressure distribution of the flow on the upper surface of the V2C aerofoil. Calculations made with k-ω SST turbulence model (Mach number 0.7, Re 3.24×10^6 and 5° of pitch)

Fig. 139 Mach number visualisation on two different span locations of IoA wind tunnel. 15% (left) and 50% (right). Calculations made with k-ω SST turbulence model (Mach number 0.7, Re: 3.24×10^6 and 5° of pitch)

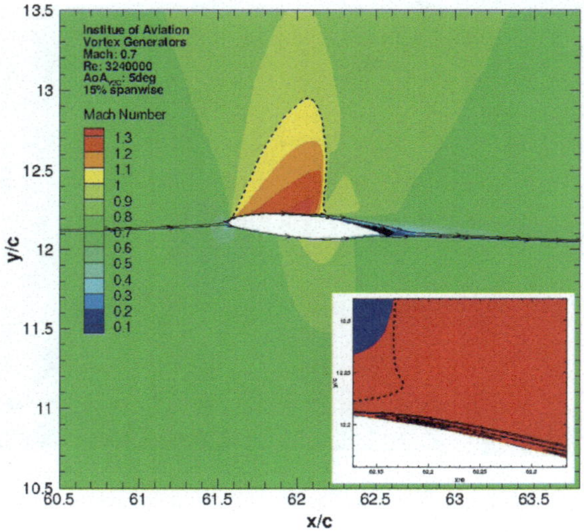

(a

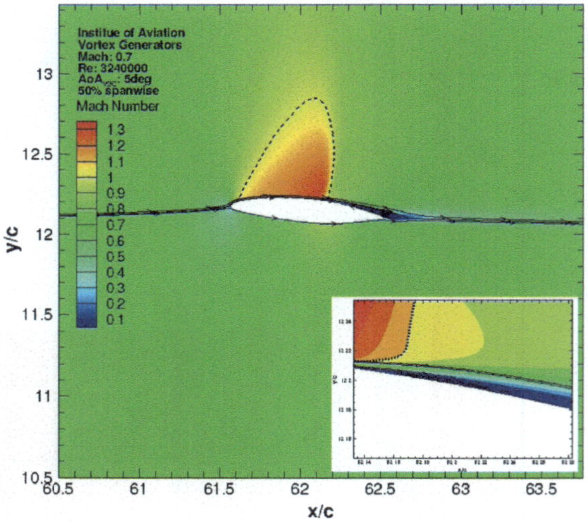

(b

Fig. 140 Visualisation of vortices downstream the generators for the IoA case. Calculations made with k-ω SST turbulence model (Mach number 0.7, Re: 3.24×10^6 and 5° of pitch)

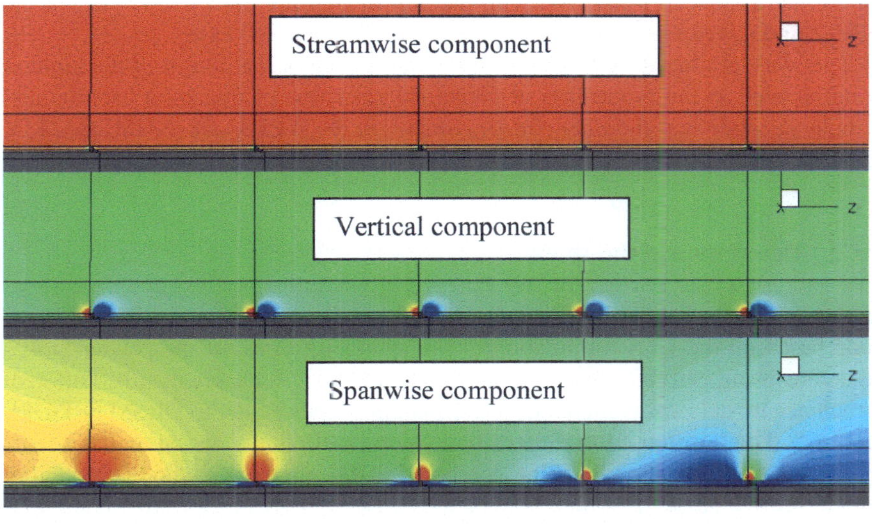

Fig. 141 Velocity fields behind the vortex generators for the IoA case. Streamwise (upper), vertical (middle) and spanwise (lower) components. Calculations made with k-ω SST turbulence model (Mach number 0.7, Re: 3.24×10^6 and 5° of pitch)

3 Numerical Simulation of 3-D Wing

3.1 DAAV

3.1.1 Introduction

WP5 is devoted to the study of transition location effect on separation size, shock structure and unsteadiness. In particular it should address how close to the shock location transition has to occur to avoid detrimental effects, if any, associated with laminar shock-induced separation, and what is the best way to trigger transition. The WP5.2 tasks was dedicated to experimentally and numerically understand the differences between laminar, transitional and turbulent SBLI in terms of surface pressure distribution, flow topology (presence of laminar separation bubble at shock foot) and, in the buffeting limit, in terms of shock oscillation amplitude, frequency and pressure fluctuations downstream of the shock. Then the WP5.2 covered the study of tripping devices effects on these flow characteristics, and their dependence on their location upstream of the shock, and produce models for CFD reproducing these effects. The WP5.3 is a purely numerical task (so called "blind test"): the methods validated in WP5.2 were applied on a laminar wing, based on the V2C profile and produced by DAAV, with the aim of figuring out to which extent the WP5.2 findings can be transposed to this rather different 3-D case.

The work produced by DAAV in WP5.3 consisted in the design of the laminar wing, as well as the production of guidelines (in terms of transition location and turbulence modeling) for other partners involve in the WP5.3 task: NUMECA, LIV and IMFT.

3.1.2 Description of the Tools

Aether The in-house Navier-Stokes solver Aether is a compressible, finite-element code based on the GLS formulation of the symmetric form of the Navier-Stokes equations using entropic variables. See Chalot et al. [1] for more details. Different turbulence models are available from 1-equation RANS model to DRSM or DDES and LES. **The laminar-to-turbulent transition prediction** is based on the assessment of the amplification N-factor, computed with a parabola method (e^N stability data base), see Arnal [2] which is implemented in a 3D boundary layer compressible code. Such a method has been used by DAAV with success for many in flight laminarity tests (Courty, Bulgubure, and Arnal, 24th–27th May 1993 [3]).

3.1.3 Design of the Laminar Wing

To initiate WP5.3 work, DAAV has designed geometry of a 3-D laminar wing (Fig. 142). As for the V2C profile (WP5.1) the design is such that the flow should

Fig. 142 DAAV laminar wing geometry

Table 11 Geometrical parameter of the DAAV laminar wing

Aspect ratio	8
Taper ratio	1/3
Leading edge sweep	20°
Twist angle	2° (linear)
Wing root	3.75 m
Wing tip fairing	1.25 m
Span	10 m
Sref	25 m^2

remain laminar from the leading edge to the shock location. The details of the geometry are given on Table 11. It is based on the V2C profile with a leading edge sweep angle and a twist angle.

The geometry was delivered to partners as an "igs" file late March 2014. In order to assess laminarity of the flow, preliminary calculations for $M = 0.725$ and $\alpha = 2°$ are performed. The tool chain, as presented before, is used to assess laminarity criteria: a Navier-Stokes (NS) calculation is first performed and the velocity field at the edge of the boundary layer is extracted as input to the laminar boundary layer code. The latter includes a robust method for determination of transition criteria: the parabola method.

The maximum instability angle Ψ and N-factor are represented on the leeward (extrados) and windward (intrados) sides, Fig. 143. On the windward side the evolution of Ψ indicates an alternating sequence of cross flow (CF) and Tollmien-Schlichting (TS) instabilities, which are not detrimental for the purpose of the present workshop. On the leeward side, the flow prior to shock is mainly dominated by CF instability ($\Psi = 90°$) for which the threshold $N = 14$ is never reached before the shock. The flow laminarity up to the shock is therefore demonstrated.

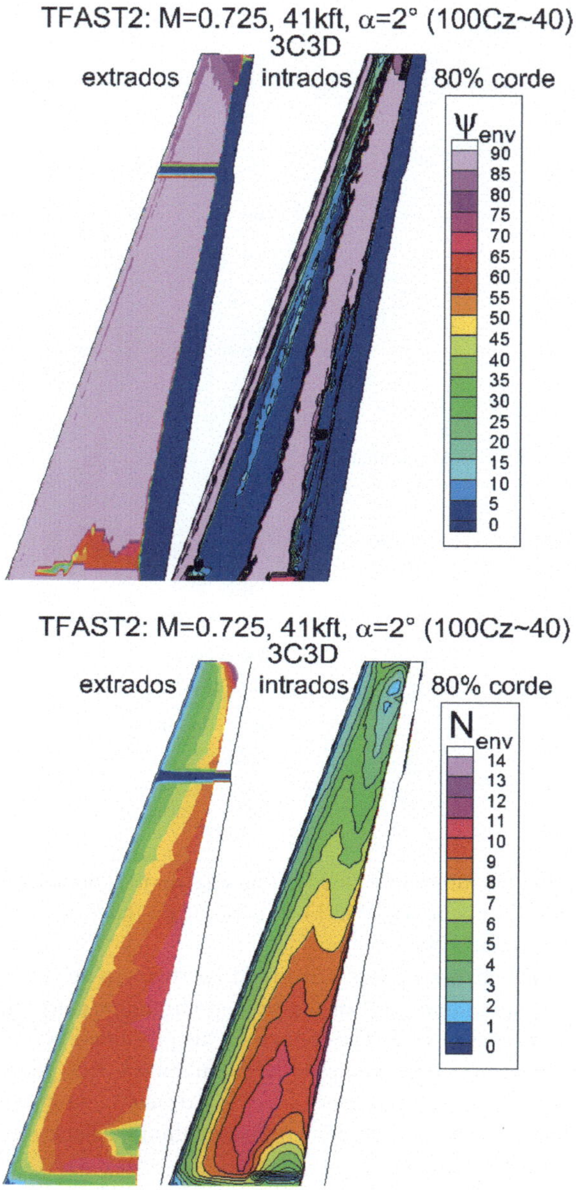

Fig. 143 Transition criteria provided by the parabola method

It was also investigated whether possible flow contamination from the wing root could occur. Figure 144 shows the Rbar contamination parameter. Contamination is possible for Rbar > 250 which is never the case here.

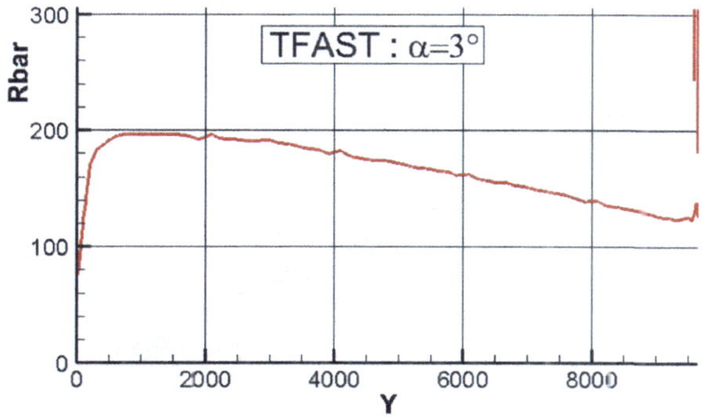

Fig. 144 Contamination parameter over a line extending along the span-wise direction (Y) over the wing

After confirmation of the flow laminarity up to the shock the next task was to determine three flow regimes for the WP5.3 study: steady shock, buffet onset and established buffet.

Several RANS simulations for varying the Mach number and the incidence angle enabled to characterize three regimes, based on the analysis of the shape factor (Fig. 145) in the area between the shock and the trailing edge:

- Below buffet: the area of possible flow separation (shape factor H > 2.5) never extend from the shock to the trailing edge at a given span Y position.
- Buffet onset: there is at least one position in span (Y) where the shape factor H overcomes 2.5 entirely between the shock and the trailing edge.
- Established buffet: there is a large and continuous range of span-wise locations where the shape factor H overcomes 2.5 entirely between the shock and the trailing edge.

This analysis should have been done, in theory, with a laminar flow up to the shock. However it is not numerically possible, with current state of the art, to obtain a converged solution when the flow is laminar at the shock. The analysis was thus performed while imposing a transition line at 55% chord (as close as possible to the shock). In the region upstream of this line, the turbulent viscosity is simply set to 0.

The three flow regimes recommended to WP5.3 partners are summarized on Table 12. The calculations were performed under condition of 41,000 feet and the equivalent Reynolds number is also provided.

3.1.4 Specifications of Guidelines for Numerical Simulations

The three flow regimes summarized in Table 12 were assessed based on shape factor analysis of steady RANS 2-layer k-ε results, with laminar/transition tripping at 55% chord. Partners subsequently followed these guidelines but with various turbulence models, scale-resolving level, meshes discretization, numerical schemes. Most note

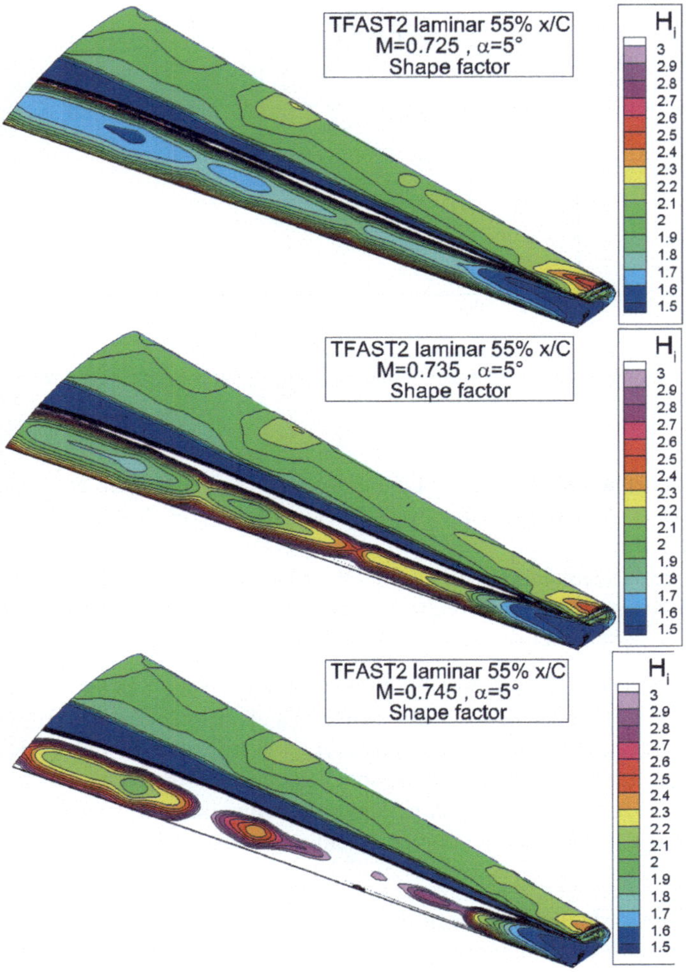

Fig. 145 Shape factor map for the three selected flow regimes

Table 12 Parameters for simulation of the three flow regimes

Regime	Mach	Reynolds/m/10^6	Incidence	Transition
Below buffet	0.725	4.34	5°	55% chord
Buffet onset	0.735	4.40	5°	55% chord
Established buffet	0.745	4.46	5°	55% chord

Table 13 Calculations performed (green: new, grey: existing, for reference)

	Fully turbulent	Tripped at 55% chord
2-layer k-ε (reference)	green	grey
k-ω SST	green	green
Spalart Allmaras	green	green

worthily, some favored "fully turbulent" calculations as their preliminary run. The first feedback was that buffet was difficult to obtain even for the highest recommended value for Mach number. It was therefore here sought to investigate whether the above three Mach number recommended values were still relevant when varying the turbulence model and/or the tripping location. Table 13 tabulates the new simulations performed.

Figure 146 compares shape factor H map of 2-layer k-ε results, tripped at 55% chord, for reference and fully turbulent. The shape factor is noticeably lower for fully turbulent calculations, revealing that H-based criteria to define "buffet onset" and "full buffet" regimes cannot be met for fully turbulent calculation. Subsequently DAAV circulated the first recommendations for the partners to avoid fully turbulent calculations, by either using a tripping (ideally 55% chord) or a transition-predictive turbulence model. Buffet cannot be predicted by fully turbulent calculations at these recommended Mach number values as they do not allow separation at the shock.

Figure 147 compares shape factor provided by simulations using 2-layer k-ε, k-ω SST and Spalart-Allmaras models, for laminar/transition (L/T) location imposed at 55% chord. The shape factor returned by the 2-layer k-ε model is noticeably higher than that for the two other turbulence models. Amongst models generally used in the DDES framework, the k-ω SST model predicts slightly higher values for the shape factor, indicating conditions closer to buffet. But again the H-factor based criteria defining "buffet onset" and "full buffet" do not seem to be met for these two models.

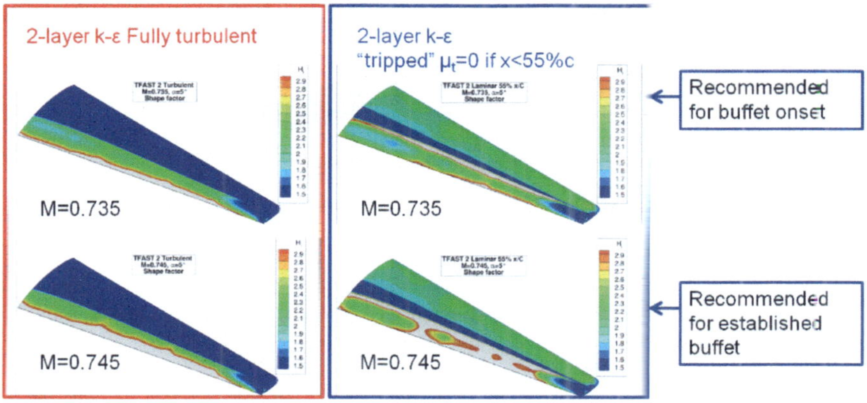

Fig. 146 Shape factor map: fully turbulent/ tripped at 55%

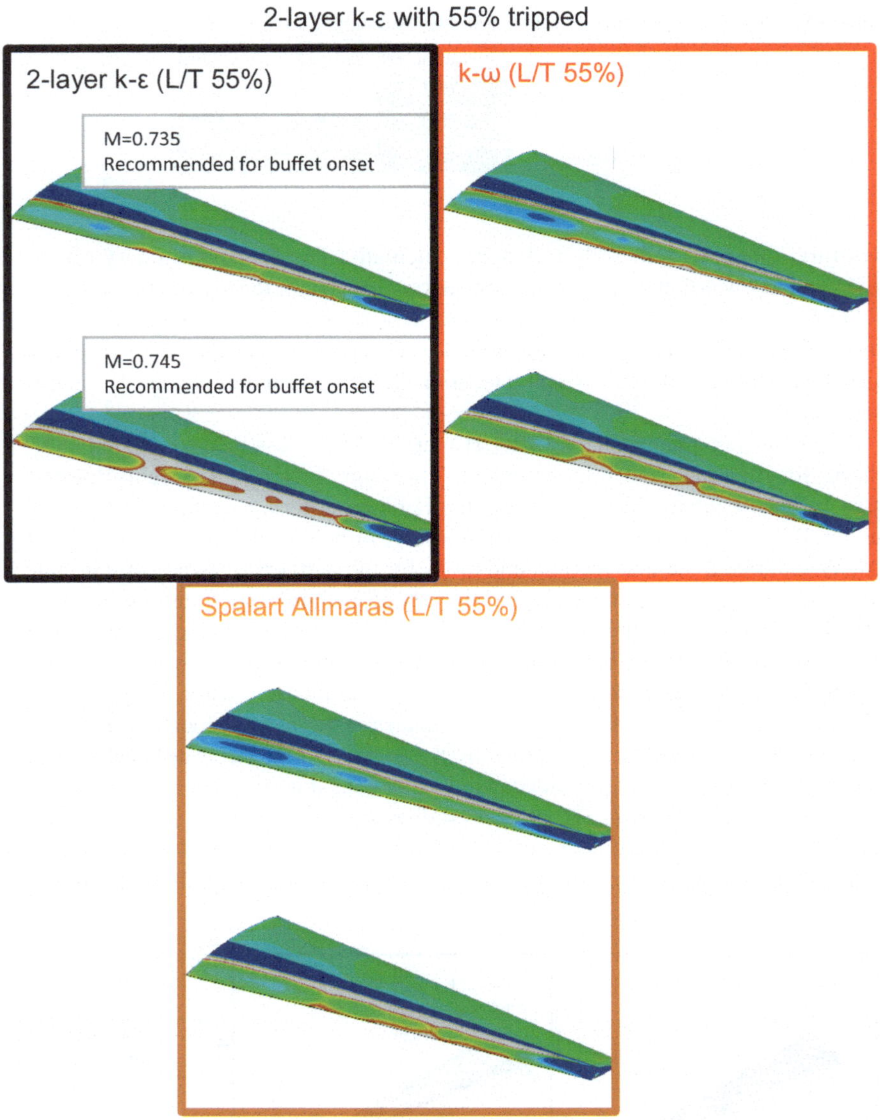

Fig. 147 Shape factor map: M = 0.735 (first row), M = 0.745 (second row), for turbulence model sensitivity

For those, higher values for the Mach number and/or the angle of attack may be preferred to obtain buffet.

3.1.5 Conclusions

As the main purpose of task WP5.3 is the numerical representation of buffet (in laminar, turbulent or transitional SBLI) partners investigating this case, first attempted DAAV provided guidelines in terms of angle of attach and Mach number, with persistent difficulties to obtain buffet (i.e. unsteady shock). At light of the tripping location/turbulence model sensitivity study presented above, DAAV has formulated further recommendations:

- For a mesh resolution and numerical scheme similar to that used in DAAV: "buffet onset" and "full buffet" can be obtained with unsteady simulation using the 2-layer k-ε model (or DDES based on this model) with laminar/turbulent tripping at 55% chord.
- In order to obtain buffet for "fully turbulent" simulations (URANS or DDES) it may be necessary to increase some of the following: mesh resolution, numerical scheme order, Mach or angle of attack.
- In order to obtain buffet for calculation tripped at 55% using alternative models to the 2-layer k-ε (either URANS or with DDES) it may be necessary to increase some of the following: mesh resolution, numerical scheme order, Mach or angle of attack, **but to a lesser extent**.

3.2 NUMECA

3.2.1 Introduction

The effect of laminar to turbulent transition on the ability of the DAAV 3D wing to become unstable at various free-stream Mach numbers is investigated. In a first step the Explicit-Algebraic Reynolds Stress (EARSM) turbulence model is used. It has been coupled to a transport model for laminar-to-turbulent intermittency as originally described in Langtry [30]. Other correlations available in the scientific literature are also investigated. Furthermore, and in order to confirm the role of a laminar interaction or the ability to produce buffeting, final unsteady tests are performed with a prescribe location for laminar-to-turbulent transition.

3.2.2 Test Configuration

The 3D laminar wing that has originally been computed by DAAV. It does not correspond to any existing wing. Wing characteristics are the followings: the wingspan is **10** m and the root chord is **3.8** m. The wing has a rounded tip and a blunt trailing edge.

Flow conditions correspond to three different free-stream Mach number—**0.725**, **0.735**, and **0.745**. The former one should corresponds to bellow buffet onset condition, the second one to buffet onset, and the third one to above buffet onset condition. The incidence is set to **5°** and the unit Reynolds number is equal to **4.35** $\times$ **10⁶** m^{-1}. According to DAAV inputs, the boundary layer along the wing should remain laminar over a large part of the chord—at least up to 55% of the chord length.

3.2.3 Numerical Model

The commercial package Fine^{TM}/Turbo is used for this study. This package consists of tools, covering all related parts of Computational Fluid Dynamics (CFD) simulations: a grid generator, a flow solver and a visualization system. All components are adapted to turbomachinery applications. The flow solver is a three-dimensional, density-based, structured, multi-block Navier-Stokes code using a finite volume method. Central-difference space discretization is employed for the spatial discretization with Jameson type artificial dissipation. A four-stage explicit Runge-Kutta scheme is applied for the temporal discretization. Multi-grid method, local time-stepping and implicit residual smoothing are used in order to speed-up the convergence.

State-of-the-art turbulence models are available in the Fine^{TM}/Turbo flow solver and we will not describe in the present report. More emphasis will be put on new models that have been developed during the course of the TFAST project.

3.2.4 Transition Model

The laminar to turbulent transition could be characterised by the intermittency, i.e. the fraction of time the flow is turbulent or laminar. The intermittency range from zero to unity: from laminar to fully turbulent flow.

Among the various ways to treat laminar-to-turbulent transition in CFD codes, one can discriminate models based on global correlations from models that only rely on local variables. Though the great interest of the first class of model for simple configuration—such as the one of Abu-Ghannam and Shaw [31]—, such a model cannot be applied in complex configuration as can encounter by the industry. Therefore, a new class of model has been developed and are based on a transport equation of the intermittency.

This model has been originally proposed by Langtry [30] and is based on the resolution of two additional transport equations: one for the intermittency and one for the Reynolds number based on the momentum thickness. The major advantage of such a model is that it only relies on local variables. Furthermore, it requires no additional user input and is not restricted to simple geometries. The description of the two transport equations and associated source terms can be found in Langtry [30]. However, it should be stressed that this model still relies on empirical correlations which link the transitional Reynolds number to the local turbulent intensity

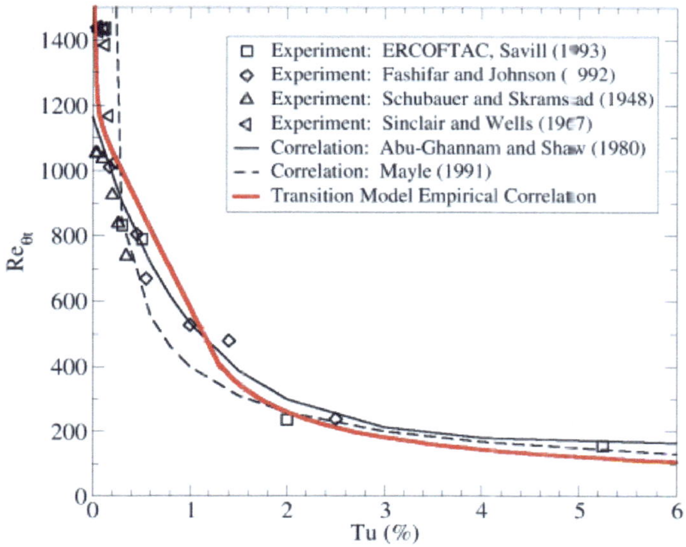

Fig. 148 Transition onset momentum thickness Reynolds number ($Re_{\theta t}$) predicted by the new correlation as a function of turbulence intensity (Tu) for a flat plate with zero pressure gradient (from [30])

(Fig. 148). The model will be denoted LCTM for Local Correlation Transition Model in the rest of this report. The coupling of the intermittency with the classical k-ω equations is done via the production and dissipation terms in the k equation and a slight modification of the blending function F_1. Is should be stressed that this method has not been extensively tested for shock-induced laminar to turbulent transition.

Furthermore, since the work of Langtry [30] a series of articles have proposed various correlation that are supposed to better represent the laminar to turbulent transition process. A few of those variants have been implemented in Fine^{TM}/Turbo: the correlation from Malan et al. [32], Sorensen [33], Kelterer et al. [35], and Toyoda et al. [36]. Comparison between those variants on the 3D wing is presented afterwards.

Both global and local methods are available in Fine^{TM}/Turbo flow solver and both are used in the present context.

3.2.5 Explicit-Algebraic Reynolds Stress Model

Beyond the classical approach based on linear eddy-viscosity models (LEVM) that are currently used in the industry, it is proposed to use hereafter a more advanced turbulence model. Indeed, Explicit-Algebraic Reynolds Stress Models (EARSM) have proved to be able to better represent secondary flows with strong adverse pressure gradient at a reduced additional CPU cost when compared to LEVM. Furthermore such models are also more robust than Reynolds Stress Models. This rather good compromise between accuracy, efficiency and robustness makes such EARSM a good candidate for using such a model in an industrial context. Therefore, it should be of interest to investigate its capabilities in the configuration studied in the scope of the TFAST project.

Several variants of EARSM have already been published in the literature. It has been decided here to use the one proposed in Menter et al. [37]. This model is a simplified version of the EARSM based on the "baseline" k-ω model of Menter [38]. This S-BSL-EARSM (for Simplified Baseline EARSM) has already been successfully tested in a turbomachinary context by Mehdizadeh et al. [39]. The detailed description of the model could be found in the above mentioned references together with validations on different configurations.

Furthermore, such a turbulence model has been coupled to the local correlation transition model of Langtry [30]. Correlations functions have been adapted to be compatible with EARSM.

Both options (fully turbulent or coupled to LCTM) could be accessible in Fine™/Turbo flow solver and are used in the present study.

3.2.6 Numerical Mesh and Settings

The geometry of the blade has been provided by DAAV and the computational mesh has been generated thanks to IGG™. It is a multi-block multi-grid mesh with 15 $\times$ 10^6 cells. The limits of the computational domain are 100 m away from the blade in the stream-wise and wing-normal directions, which corresponds roughly to twenty chords. The domain extension is 30 m in the span-wise direction. The mesh clustering has been controlled in order to reach a y^+ of the order to unity in the first inner cell close to the solid body. However, based on best practices, the wall clustering has been relaxed along the blunt trailing edge. About 200 points are used in the span-wise direction and 384 around the wing. The mesh is represented on Fig. 149.

Air assumed as a perfect gas is used for all those simulations.

For the boundary conditions, free-stream velocities, static pressure, and temperature along all the external boundary conditions are imposed. Free-stream turbulence is fixed so as to have a turbulent to laminar viscosity ratio equal to unity. A mirror boundary condition is used at the root of the wing. Simulations are performed at the three different free stream conditions corresponding to impose free-stream Mach numbers and with an incidence of $5°$.

3.2.7 Effect of Laminar-to-Turbulent Transition

A first series of simulation have been conducted using the EARSM turbulence with and without the LCTM model for the three different configurations. Therefore, six computations have been performed.

Fig. 149 Computational mesh along the solid part of the 3D wing from DAAV (one point over two in all the directions have been skipped)

The static pressure distribution along the wing suction-side for all those simulations is displayed on Fig. 150. It appears that LCTM module does not affect the pressure distribution along the blade. The shock remains almost at the same place with and without transitional model. This can be further identified by looking at the isentropic Mach number distribution 5 m away from the mirror plane (Fig. 151). No difference can be identified between the simulations with and without transition module on the wing pressure side. One the suction side, shock location is slightly

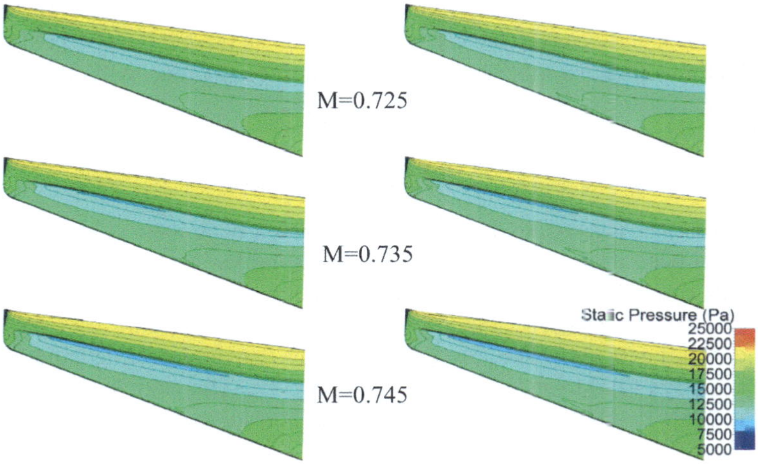

Fig. 150 Computed static pressure distribution along the wing suction side at three different freestream Mach numbers for fully turbulent calculation (left) and transitional calculations (right)

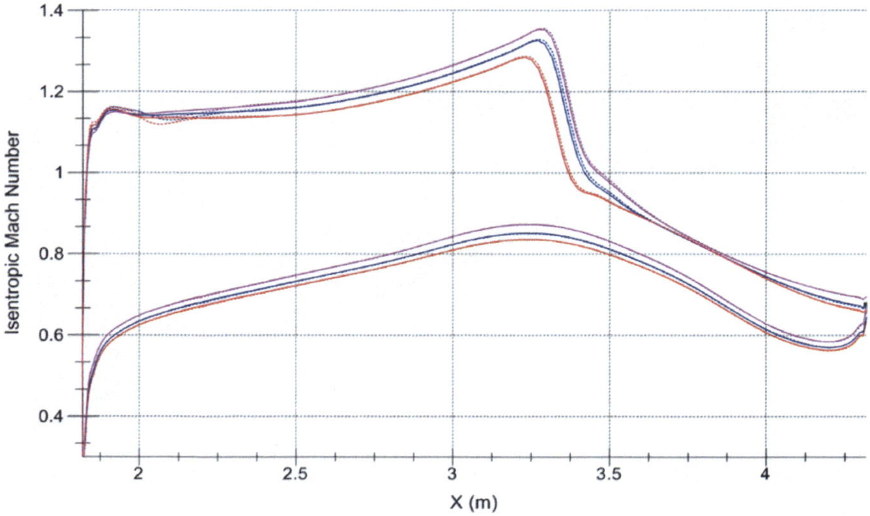

Fig. 151 Computed isentropic Mach number distribution at y = 5 m with (dotted lines) and without (continuous lines) transition modelling for the three different free stream Mach numbers: 0.725 (red), 0.734 (blue), and 0.745 (pink)

moved downstream as the free-stream Mach number increases and is not significantly affected when using transition model. Indeed, the LCTM module produces the laminar to turbulent transition to appear within the first ten percent of the wing suction side. This can be identified by looking at the friction coefficient along the wing suction side as displayed on Fig. 152. Region of lower skin friction do appears with transition module only close to the wing leading edge. In the interaction region, skin friction is not affected by the use of the transitional model. Location of laminar to turbulent transition is further independent of the free stream conditions.

3.2.8 Sensitivity to Transition Model

In order to further investigate the cause of such an early laminar-to-turbulent transition, it has been decided to test a series of existing variants of the LCTM with correlations coming from a series of authors. All those simulations have been conducted with the SST turbulence model on a coarse mesh and at the lowest free-stream Mach number, i.e. 0.7245. The various correlations that are tested hereafter are the ones of Langtry [30], Malan et al. [32], Sorensen [33], Kelterer et al. [35], and Toyoda et al. [36]. It should be mentioned that the latter correlation is known to be better adapted to wing configuration. Therefore, it has been specifically included in the Fine^{TM}/Turbo flow solver in order to perform such a sensitivity study. From Fig. 153 it appears that all the results are coherent with the already reported finding. All the correlations produce an early laminar-to-turbulent transition which remains

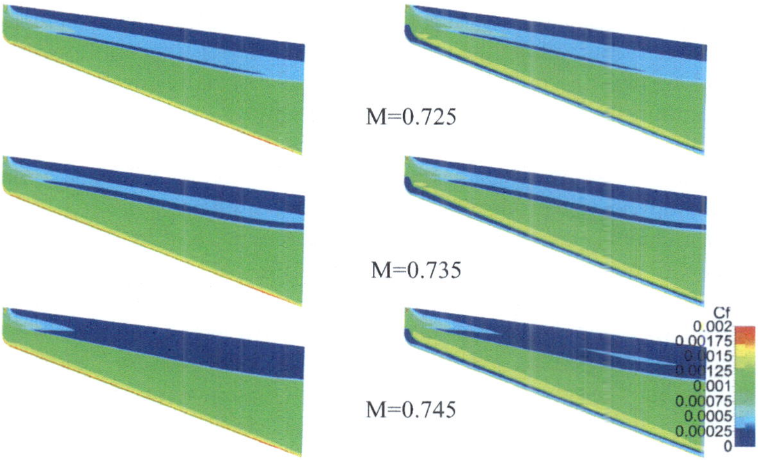

Fig. 152 Computed skin friction coefficient distribution along the wing suction side at three different free-stream Mach numbers for fully turbulent calculation (left) and transitional calculations (right)

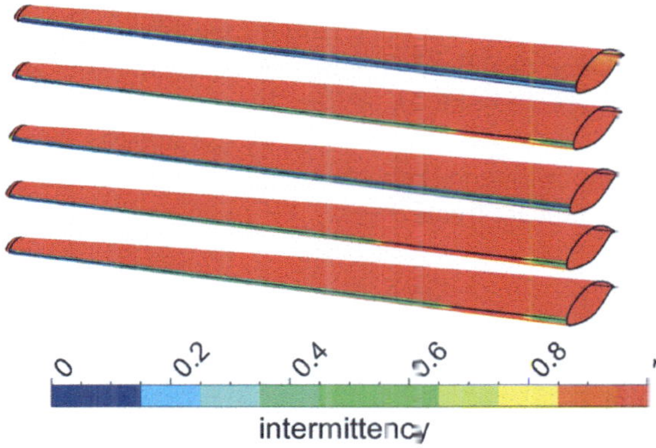

Fig. 153 Computed intermittency distribution on coarse mesh with SST turbulence model with five differents correlations coming, from bottom to top, from Langtry [30], Malan et al. [32], Sorensen [33], Kelterer et al. [35], and Toyoda et al.[36]

very close to the wing leading edge, i.e. far upstream from the shock-boundary layer interaction. Some small differences can be identified between all those models but their impact remains minor. Therefore, promoting transition by including some control devices cannot impact the state of the boundary layer close to the interaction region.

3.2.9 Effect of Delayed Transition

Even if the LCTM model produces the laminar-to-turbulent transition much too early, one can test the impact of delaying the transition by fixing the location of transition onset. This has been done by imposing the laminar-to-turbulent transition to occur at 55% of the chord length along the wing suction side and maintain a fully turbulent flow layer along the wing pressure side (as has been originally done by DAAV). By fixing such a transition location, upstream of the transition intermittency is equal to zero and is equal to unity downstream (following a step-wise function). Such an intermittency distribution has been coupled to the Spalart-Allmaras turbulence model by multiplying the production term by the intermittency.

Steady simulations have been produced on the fine mesh of the DAAV 3D wing for the three different flow conditions—corresponding to different free-stream Mach numbers. It appears that such a steady flow computations could be a good indicator of potential unsteady behaviour. Indeed, at the two lowest Mach number, both computations properly converge towards a steady solution. However, with the same numerical setting, the flow solver produces an oscillating solution (see Fig. 154). This confirms that at the highest Mach number unsteady buffeting is likely to occur. It should be mentioned that such an unsteady behaver was not observed for previous simulation with a too upstream laminar-to-turbulent transition location.

In order to further confirm such a finding it was decided to run an unsteady simulation with the same fixed transition at the highest Mach number. This simulation was performed again with Spalart-Allmaras turbulence model, a transition fixed at 55% chord length on the wing suction side, and with a time step of 2×10^{-3} s, which corresponds to one tenth of the convective time step. For such an unsteady simulation dual time-stepping has been used with 200 inner iterations per physical time-step. 200 time-steps have been performed and the computed time variations of lift can be observed on Fig. 155. It appears that the simulation exhibits strong temporal variations but not enough iteration has been conducted to be able to extract any significant frequency.

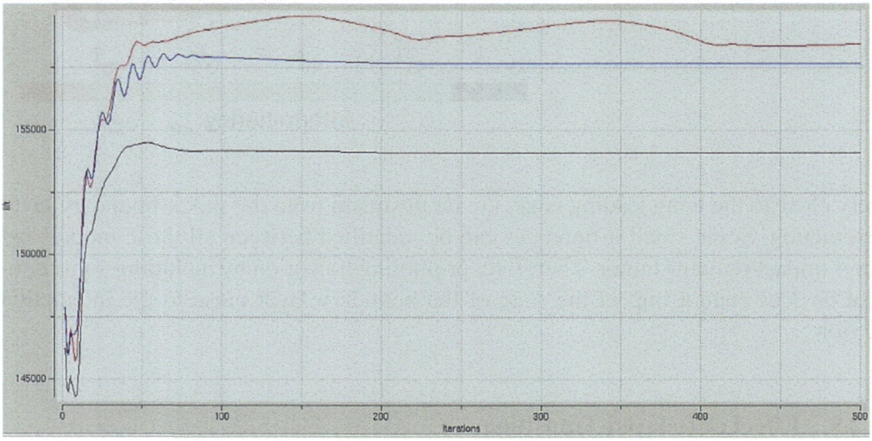

Fig. 154 Convergence history of lift [N] for steady simulations with fixed transition at Mach number equal to 0.725 (black), 0.735 (blue), and 0.745 (red)

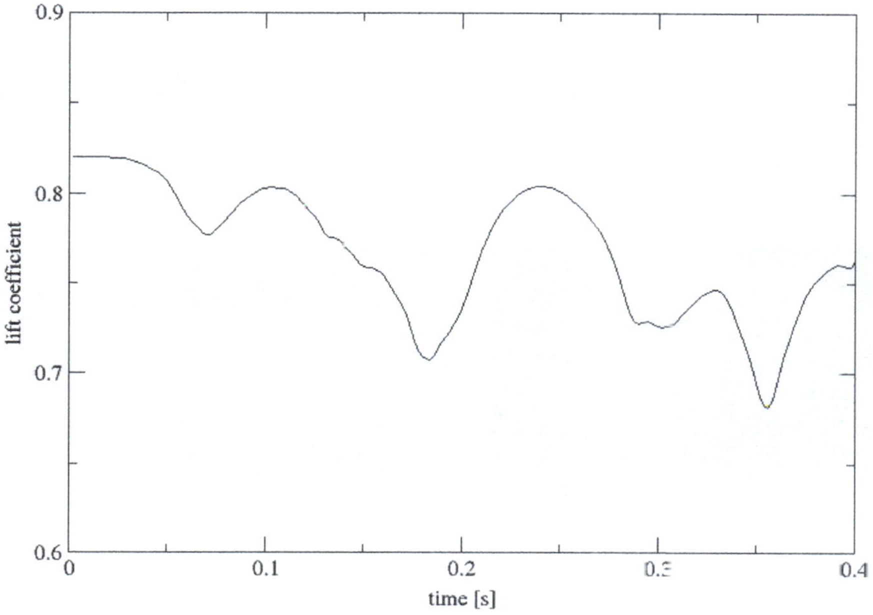

Fig. 155 Time evolution of computed lift coefficient at Mach number 0.745

3.2.10 Conclusions

Simulations with FINE^{TM}/Turbo flow solver have been performed on the 3D DAAV wing laminar-to-turbulent transition model. Contrary to original expectations, whatever correlation is used, the flow solver produces a much to early transition onset. Therefore, the shock-boundary layer interaction remains fully turbulent. Therefore, buffeting is not generated for the input Mach numbers and flow angle.

In order to avoid such a shortcoming, simulations have been produced with a prescribed location of transition onset. Unsteady results confirm that the wing is likely to generated buffeting at the highest Mach number as far as the boundary layer upstream of the shock remains almost laminar.

3.3 ULIV

3.3.1 Natural Transition of Flow Around a 3-D Wing

Grid Generation

For this case was developed of the CFD calculations. A typical C-H type topology is employed around the wing. The wing was considered at the middle of the test area.

Table 14 Mech parameters

	Grid points	Number of cells	Number of blocks	Wall distance
	2,076,219	3,337,488	108	10^{-5}
	Upper surface	Lower surface	Trailing edge	Spanwise
Grid points	113	113	31	55

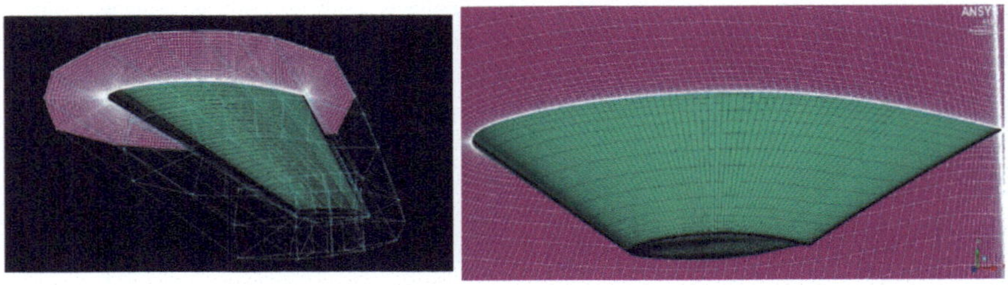

(a) (b)

Fig. 156 Blocking (**a**) and detailed view of grid (**b**) around the 3-D wing

All walls were considered as solids and at all cases a 10^{-5} c distance was selected for the first cell. The properties of the CFD mesh are summarized in the Table 14. The blocking and the computational grid of the 3-D wing can be seen in Fig. 156.

Steady Calculations

Initially, steady calculations were performed using the k-w SST model. For this case, Mach number was considered at 0.725 and the Reynolds number at 4.35×10^6. The Table 15 summarise the flow conditions.

The results were compared with the calculations provided by the DAAV. Figure 157 shows the surface pressure distribution for different spanwise locations and for two different incidence angles, at 2° and 3° respectively. The computations show to agree well with the results provided by DAAV for both incidence angles and especially close to the root of the wing (z/B = 0.1). For more outboard sections, there is a small disagreement at the comparison between the calculations close to the shock position.

For both cases of angle of attack, the transition location on the upper surface of the 3-D wing was estimated using the empirical correlations. The Michel and Cebeci-Smith models were applied and results from them are shown in Fig. 158. This was a two-step process where the initial flow field was generated and used to track the

Table 15 Flow conditions

Temperature	−56.5 °C
Pressure	5.278 mmHg
Density	0.1764
Speed of sound	573 kt

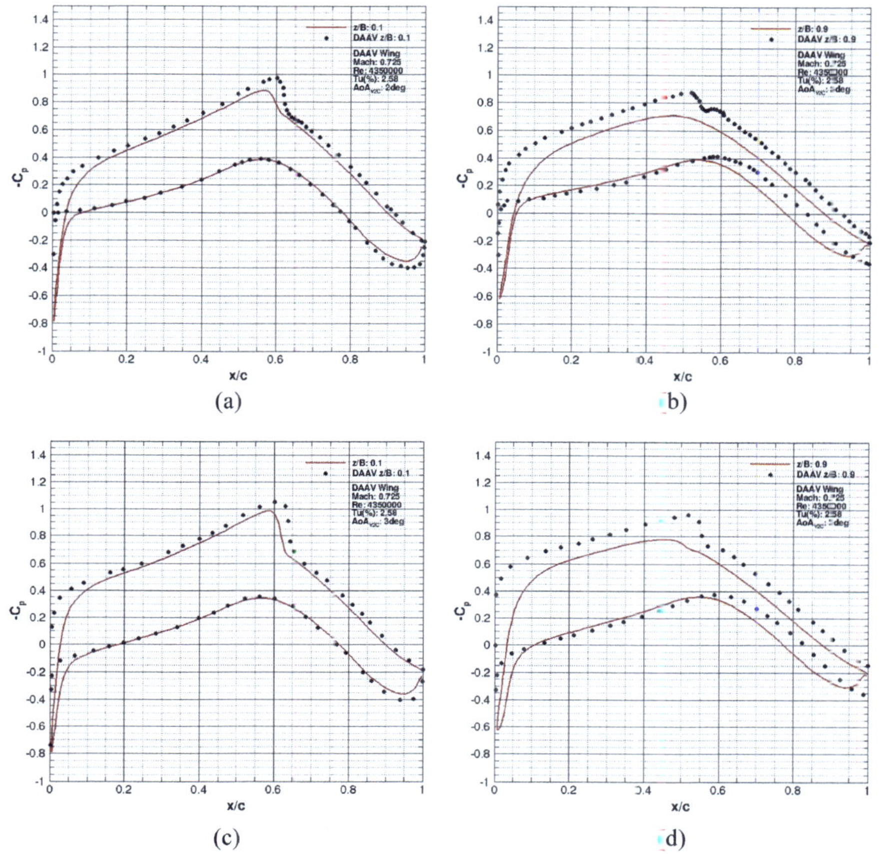

Fig. 157 Surface pressure distribution using k-ω SST model at two spanwise locations of the DAAV wing, 0.1z/B (**a**, **c**) and 0.9z/B (**b**, **d**) and for two incidence angles, 2 (**a**, **b**) and 3° (**c**, **d**) respectively (Mach number 0.725, Re: 4.35×10^6 and k-ω SST model)

flow around streamlines to identify an approximate transition point. The onset of transition is estimated around 60% of the chord, at the location of the shock. This estimates that the shock drives the transition location for the flow.

The computations were repeated using the LCTM transition model at the same flow conditions. The results show better agreement with the calculations from DAAV for the root of the 3-D wing. For the tip of the wing, the model shows to under predict the location and the strength of the shock. The results can be seen in Fig. 159.

Unsteady Calculations

First unsteady computations were performed using the SST model. For this case, the Mach number was at 0.735 and the Reynolds number at 4.4×10^6 following the instructions from DAAV. As can be seen from the time history of the lift coefficient in

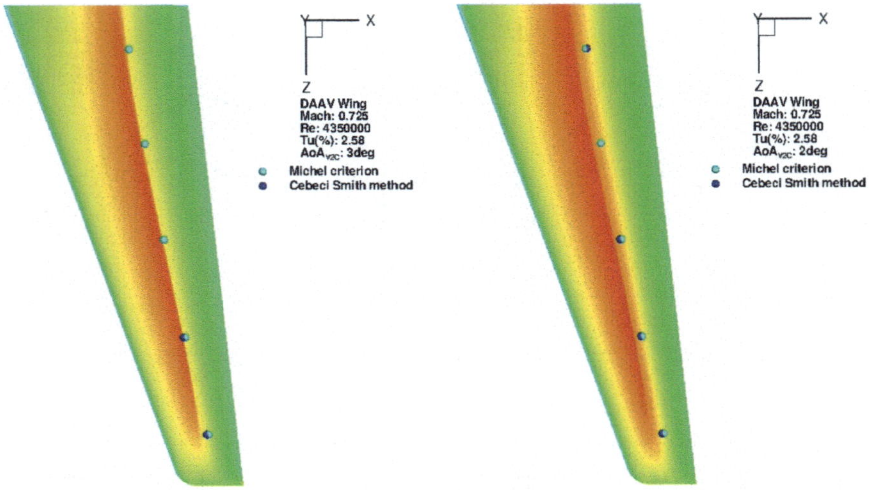

(a) (b)

Fig. 158 Transition location using the Michel criterion and Cebeci-Smith model for incidence angle of 2° (**a**) and 3° (**b**) (Mach number 0.725, Re: 4.35×10^6 and k-ω SST model)

Fig. 160, at 5° of incidence angle, the flow remains steady and no buffet is observed (Fig. 161).

The computations were repeated using the LCTM model at the same conditions. The flow remains steady even if some oscillations are observed in time history of the lift coefficient, Fig. 162. The separation at the foot of the shock and at the trailing edge is not large enough in order to have any interaction between them and provide with buffet. The Mach number visualisation at mid-span of the wing at two different time steps can be seen in Fig. 163.

Conclusions

The work was done with RANS and URANS methods to calculate the flow around the 3-D wind and estimate the buffet boundary. At this stage, computations using k-ω SST turbulence model and k-ω-γ $Re_{\theta t}$ transition model were done. For the computations with the k-ω SST turbulence model, the empirical transition models of Michel and Cebeci-Smith were used. The discrepancy observed at the steady calculations indicate the need for grid refinement.

3.3.2 Forced Transition of Flow Over a 3-D Wing

Flow Control Strategies

The shock oscillations were only predicted for the case with natural transition modelled using a 4-equation transition model and for a very narrow range of Mach

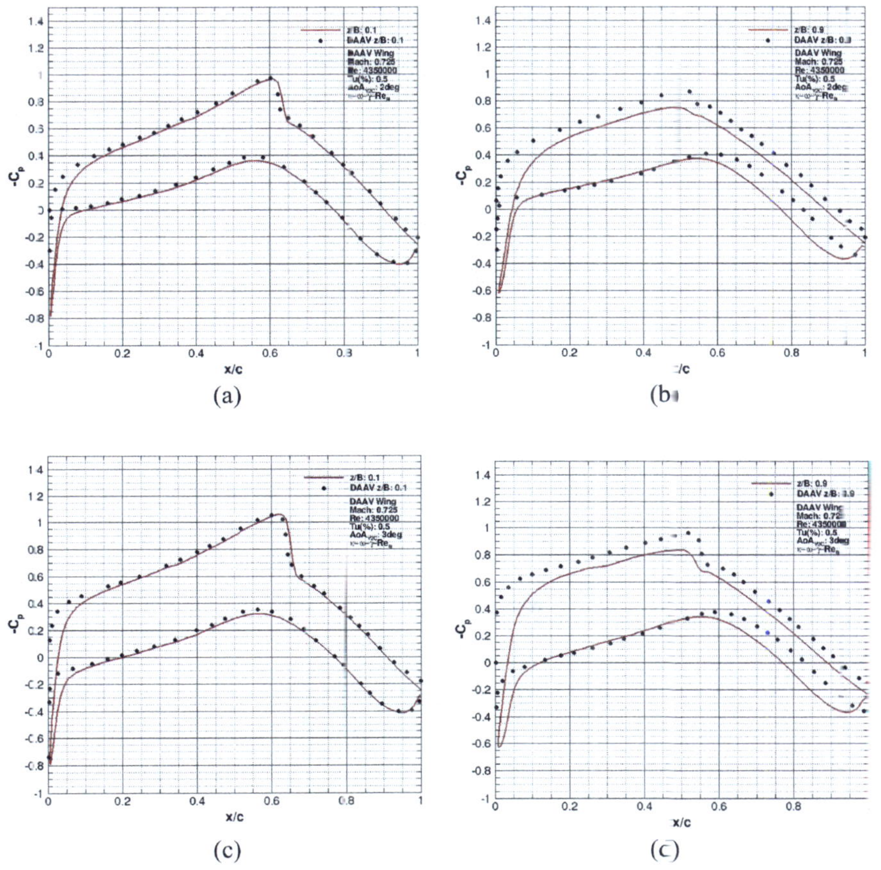

Fig. 159 Surface pressure distribution using k-ω-γ $Re_{\theta t}$ transition model at two spanwise locations of the DAAV wing, 0.1z/B (**a**, **c**) and 0.9z/B (**b**, **d**) and for two incidence angles, 2 (**a**, **b**) and 3° (**c**, **d**) respectively (Mach number 0.725, Re: 4.35×10^6 and k-ω-γ $Re_{\theta t}$ model)

numbers and angles of attack. This was in contrast to the 2D and quasi-3D case where buffet was more pronounced. This is the reason that flow control for this case was rather easy.

As pointed out in the analysis of the 2D and quasi 3D data, the flow is characterised by two separation regions that are located one at the foot of the shock and the second near the trailing edge of the section. These two separation regions vary in size during the buffet cycle and sine the flow direction between the two is co-rotating, they may also merge if their extent is long enough to cover the region of the aerofoil section aft the shock.

This dynamic behaviour of the two regions is key to the attempts to control buffet and two strategies were put forward in this task. The first strategy was concerned with altering the flow separation region at the food of the shock. This required tripping

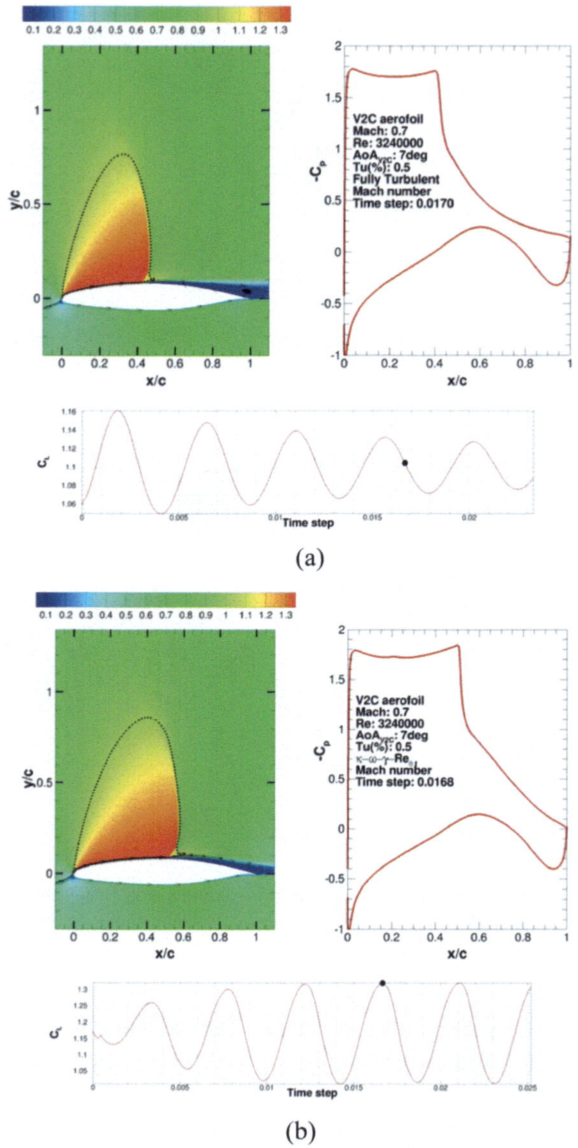

Fig. 160 Lift coefficient versus time for the DAAV wing using the k-w SST model (Mach number 0.735, Re: 4.3×10^6 and 5° pitch)

of the boundary layer ahead of the shock making sure that the properties of the boundary layer allow for a very small separation past the interaction with the shock. This limited separation at the foot of the shock breaks the flow mechanism of the buffet cycle and should be enough to alleviate the buffet. The challenge with this method is to ensure that tripping is just enough to allow for a reduced separation but

Fig. 161 Mach number visualisation mid-span of the 3-D wing at two time steps, t* = 0.3042 (**a**) and t* = 0.3347 (**b**) (Mach number 0.735, Re: 4.4×10^6, 5° of pitch and k-ω SST model)

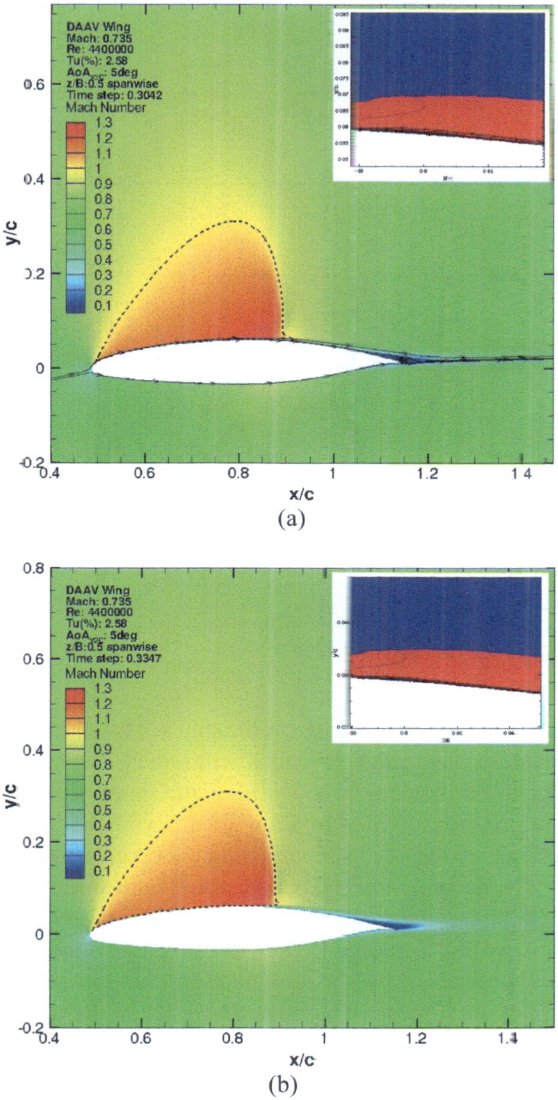

not at the expense of the laminar flow region ahead of the wing that is detrimental to the drag properties and be benefits achieved with laminar flow.

It has to be noted that most of the models considered for the natural flow transition for this case resulted in fully turbulent flow or with appropriate correlations, resulted in flow transitioning just ahead of the shock wave. In this later case, the flow separation aft the shock was large enough to cause buffet. Possible ways to trip the flow for this strategy include vortex generators as well as air-jet vortex generators placed in an

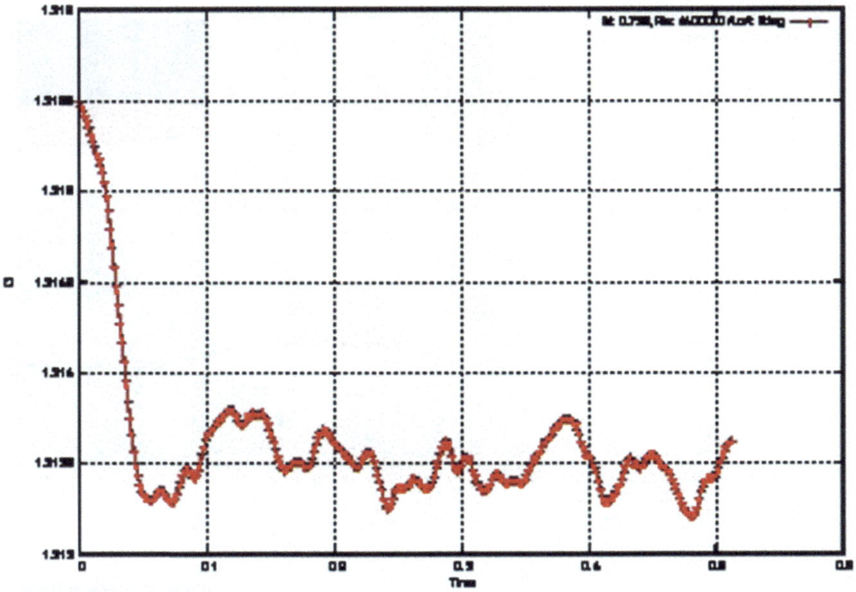

Fig. 162 Lift coefficient versus time for the DAAV wing (Mach number 0.735, Re: 4.3×10^6, 5° pitch and k-ω-γ $Re_{\theta t}$ transition model)

appropriate distance ahead of the shock. This distance is crucial since placing the trip devices far ahead of the shock will reduce the laminar flow region available, while placing the devices far too close to the shock will make them ineffective since there is a distance required to allow for the vortices generated by the device to become fully developed and start affecting the boundary layer.

The second strategy is concerned with direct manipulation of the flow region near the trailing edge of the section where buffet occurs along the span. This was a strategy also explored in the UFAST project and was effective in at least reducing the amplitude of the shock oscillations. For this strategy, the key idea is to reduce the size of the separation at the trailing edge of the wing via a flap that can be deflected in an appropriate manner. This strategy also required careful consideration since deflecting a flap by a negative angle (TE moves up) results in a loss of sectional lift, therefore reducing the overall lift of the wing. This in itself will call for a correction in the pitch angle of the wing so that the total wing lift is maintained. The trailing edge manipulation is perhaps the easier method to implement and for this reason it is put forward here too.

To support the above, Fig. 164 shows the arrangements introduced in the Liverpool CFD solver to allow for flexible trailing edge without distorting the CFD mesh and affecting in a negative way its qualities. The method is based in the Trans-Finite Interpolation (TFI) and allows for very smooth local deformations to be applied

Fig. 163 Mach number visualisation mid-span of the 3-D wing at two time steps, t* = 0.3042 (**a**) and t* = 0.3347 (**b**) (Mach number 0.735, Re: 4.4 × 10^6, 5° of pitch and k-ω-γ $Re_{\theta t}$ transition model)

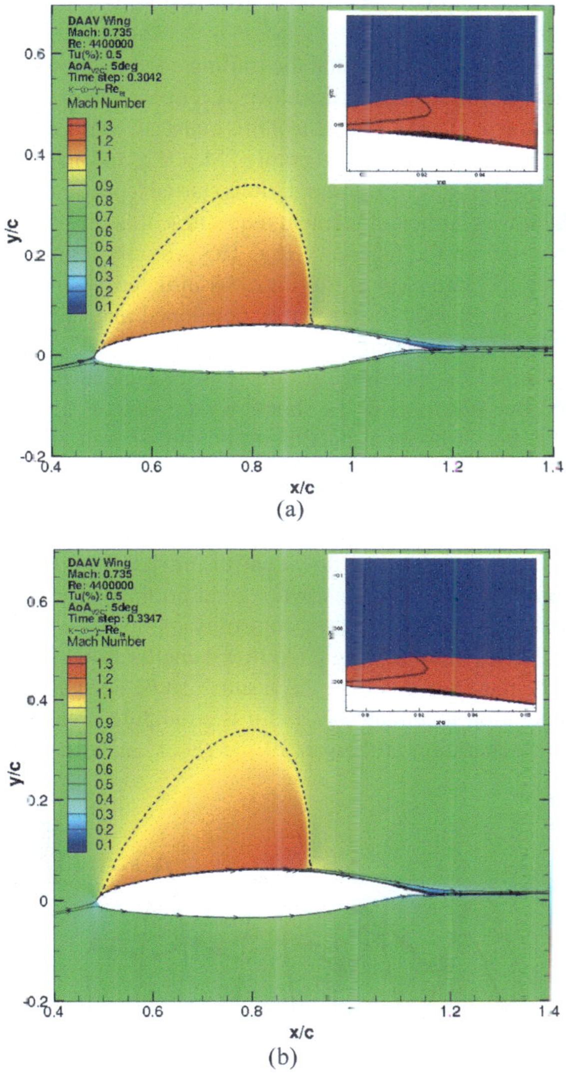

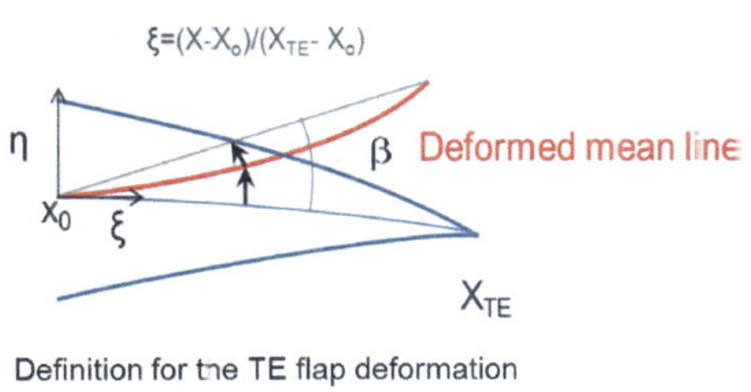

Fig. 164 Deformation method for trailing edge flow control

even on complex multi-block grids with fine spacing near the wall as necessary for turbulent flows.

Figure 165 shows that the trailing edge method is effective in reducing and eliminating buffet. The case shown is a quasi-3D example developed around the V2C section of DAAV. The span of this quasi-3D was 1 chord and the results were obtained near the well-established point of $M = 0.7$, and at Re based on the chord of the section of 3.24 million. The computations started from 5° of pitch but as the flap was deployed there was further adjustment required so that the overall lift is maintained. The angles reported on the legend of Fig. 165 are only nominal but correspond to similar lift once the actuation of the flap was initiated. As can be seen in Fig. 165, there is a well-established buffet that is developed and the action of the flap can effectively reduce this to zero.

Based on the above results, it was decided that in addition to the methods for tripping upstream of the shock, we are also to use the trailing edge method.

It has to be mentioned that the trailing edge flow can also be affected using the upstream devices as can be seen in the following figure where a clean section, trailing edge flow and Vortex Generators (VGs) are used. On the Fig. 166 the flow is clearly separated at the trailing edge and the onset of the separation is marked with a red line. The idea is to use the flow control for reducing the separation. The results with the vortex generators are showing great promise. The trailing edge is seen to be clearer in the area where the flow control is applied.

Equally, the trailing edge flow benefits from the deflection of a blended trailing edge flap. For this case the results can be seen as a blue line. The reduction of the separation is remarkable and it is enough to break the communication between the flow separation at the foot of the shock and that at the trailing edge. The effect of

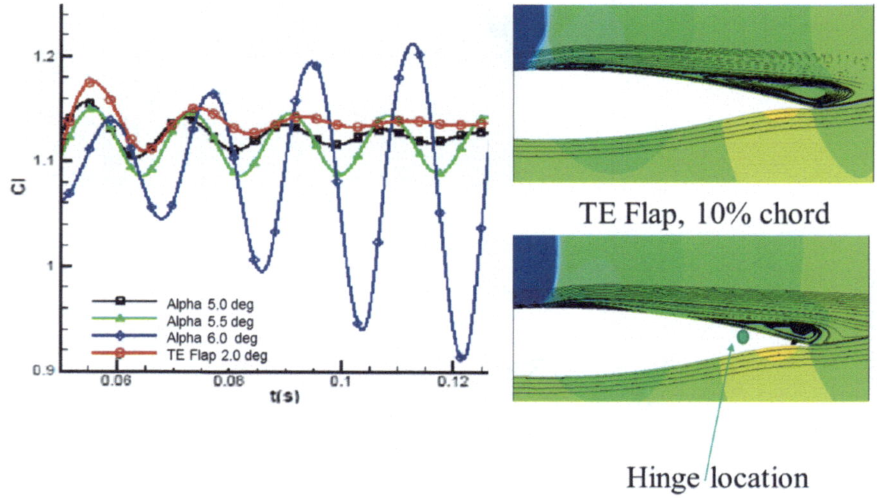

Fig. 165 Lift variation for the trailing edge of the section in a quasi-3D problem

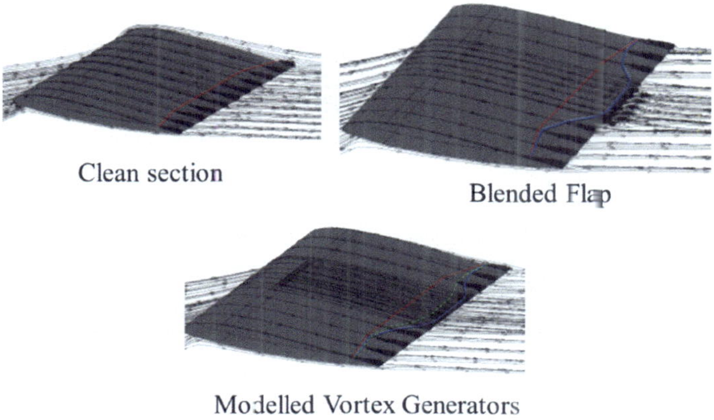

Fig. 166 Flow separation and control near the trailing edge of a section with various devices

this flow control can be seen in Fig. 167 where the performance of all assessed flow control devices is compared.

The results show that modelling the mesh deformation or using thin plates as vortex generators is highly effective.

With the currently employed method, devices that are thin with respect to the size of the wing (like vortex generators) are modelled using a resolved approach that is not relying on empirical coefficients like what is needed with methods based on momentum sources. Figure 168 shows the principle of the method that is based on representing thin structures as flow discontinuities. The Figure shows that in addition to the flow control devices, it is important to refine the mesh near these. This can be

Fig. 167 Flow separation and control near the trailing edge of a section with various devices

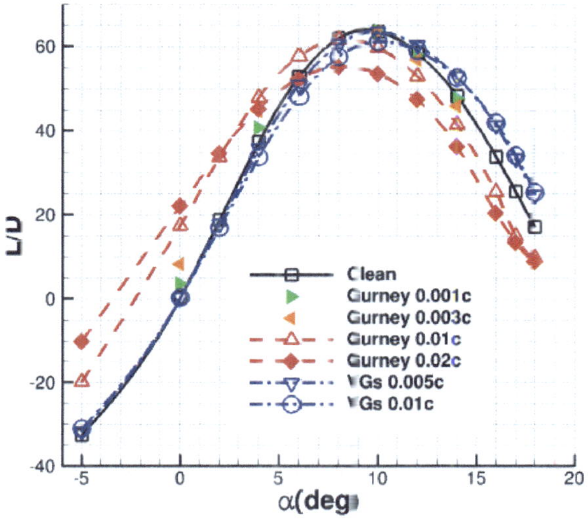

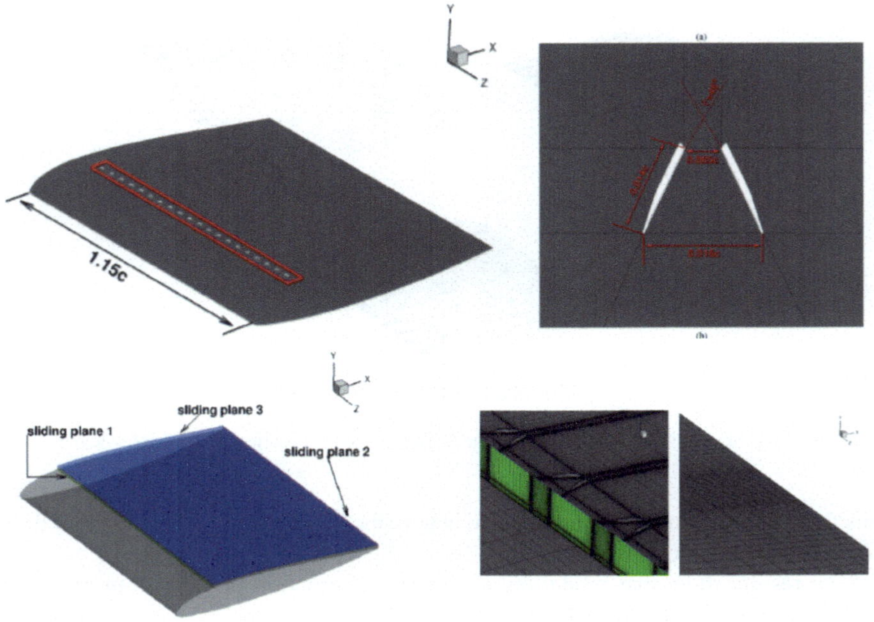

Fig. 168 Modelling thin flow control devices using a resolved approach

achieved using adaptive mesh refinement on structured or unstructured grids, or, as the case is here, by adding a small block of high density mesh over the flow region downstream of the flow control devices.

Air Jet Vortex Generators

For this case the flow control was applied using air-jet vortex generators that were operating based on the stagnation pressure. The plenum inside the wing or the inflow orifices of the AJVGs were not modelled. Instead, the total pressure condition was applied at the end of a short tube meshed inside wing as can be seen in the following Fig. 169.

The AJVGs were placed in an array including 10 AJVGs aligned with the area of the main shock oscillations centred around the 50 station along the span of the wing. These are relatively weak vortex generators and their role is simply to reduce the recirculation downstream the shock wave without affecting the flow and the lift of the wing.

First unsteady computations were performed using the SST model. For this case, the Mach number was at 0.735 and the Reynolds number at 4.4×10^6 following the instructions from DAAV. As can be seen from the time history of the lift coefficient in Fig. 170, at 5° of incidence angle, the flow remains steady and no buffet is observed.

There is a slightly increased Mach number for this case but this was necessary due to the sweep of the wing. In contrast to the quasi-3D cases analysed, this case showed no buffet without the control unless the Mach number was increased. The

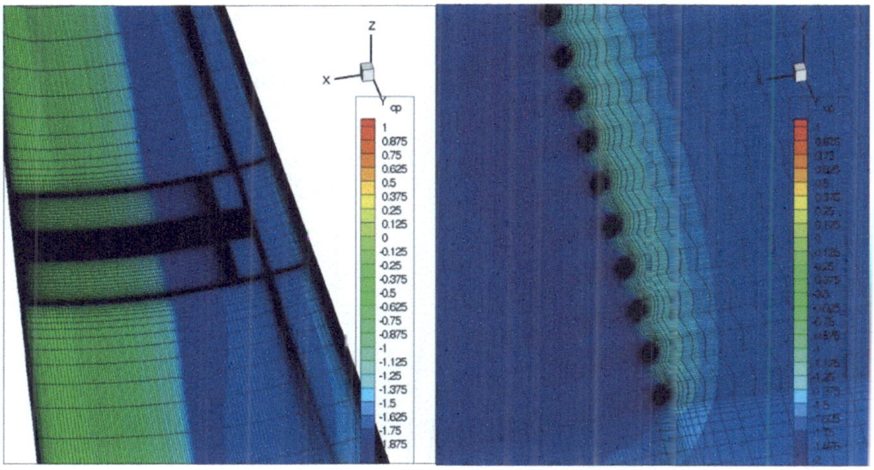

Fig. 169 AJVGs on the DAAV wing

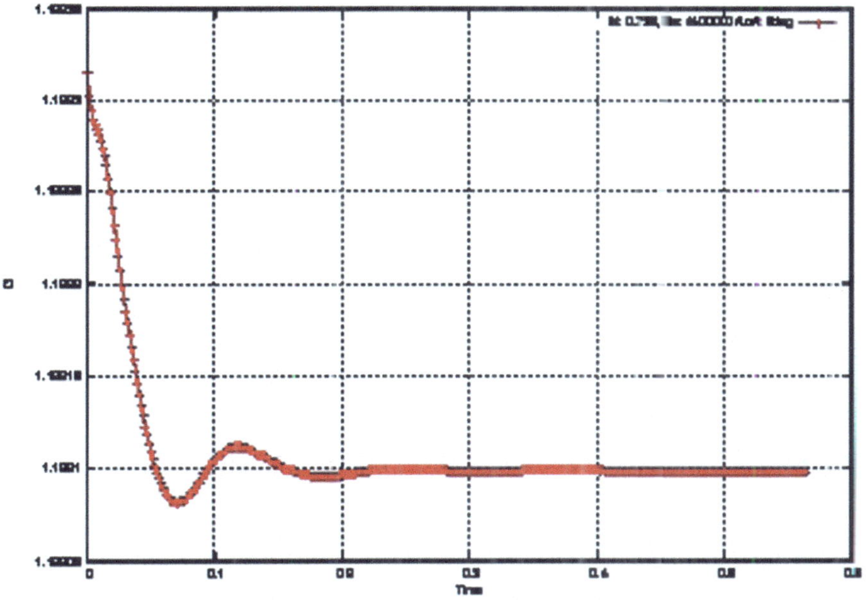

Fig. 170 Lift coefficient versus time for the DAAV wing using AJVGs (Mach number 0.735, Re: 4.3×10^6 and $5°$ pitch)

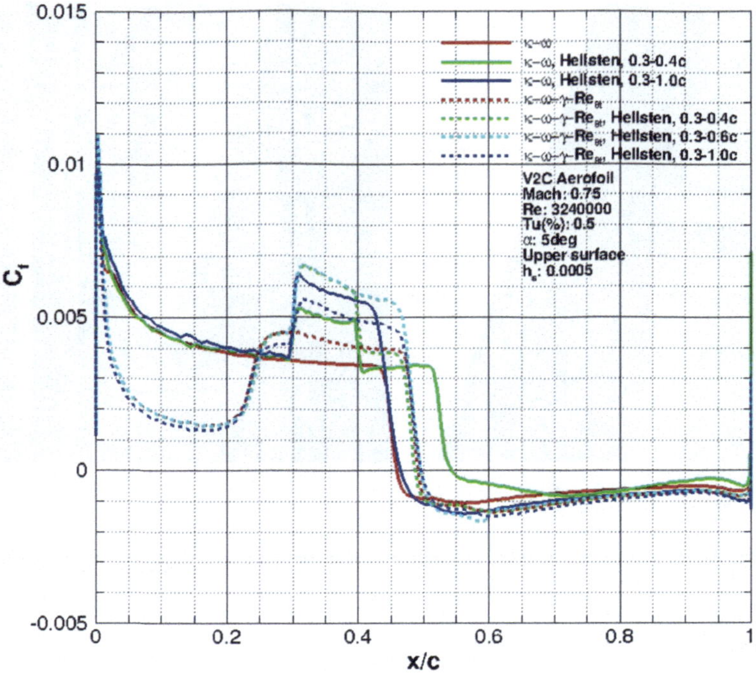

Fig. 171 Skin friction for the controlled DAAV wing case with AJVGs

results suggest that tripping is effective but it also has an effect on the predicted skin friction. This can be seen in Fig. 171 that includes results from a variety of models with different effects. The trip can be seen around 0.3 x/c and the results, especially for the green line reduce separation and remove the buffet effect.

3.4 IMFT

A first group of computations have been performed for the 2D wing of constant section along the span. The flow parameters were: a chord length of 0.25 m, a freestream Mach number of 0.70, Reynolds number of 3.245×10^6. The flow was found to separate between angles of incidence $\alpha = 6°$ and $7°$. The simulations have been performed by means of the Navier-Stokes Multi-Block (NSMB) solver [40, 41]. NSMB solves the compressible Navier–Stokes equations using a finite volume formulation on multi-block structured grids. For the present study, the upwind scheme of 3rd of accuracy by Roe [42] and the MUSCL flux limiter of van Leer [43] have been employed. 2nd-order central differencing has been employed for the diffusion terms. The time discretization has been accomplished by a dual-time stepping scheme, of 2nd order of accuracy. The time step was of 5 μs for 2D simulations and of 10^{-7}

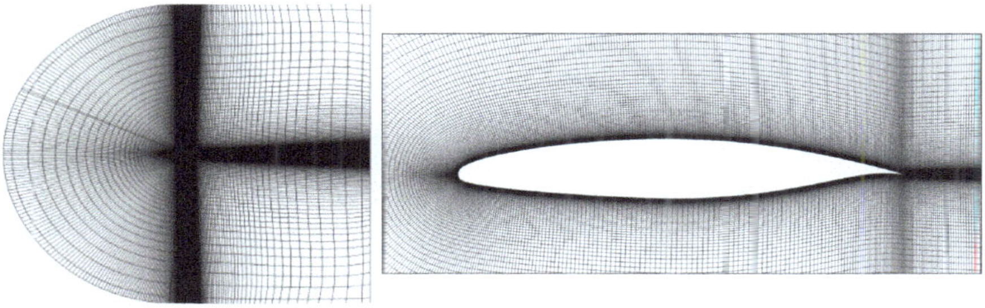

Fig. 172 Grid and zoom around the section for the case of constant section V2C wing

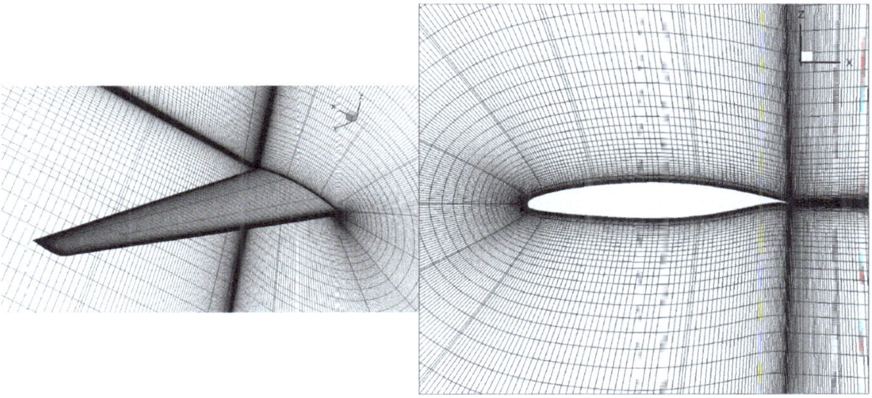

Fig. 173 Grid for the swept V2C wing: overview and zoom around a wing's section

s for the 3D simulations. The grid was of C–H type and of size 163,584 cells. The downstream distance of the computational domain is of 80 chords from the obstacle. A grid refinement study had been done during the TFAST project within the Ph.D. theses of Grossi [44] and Szubert [45]. A second group of computations has been done for the swept V2C wing for which the grid was of 9 Million cells corresponding to y+ values less than 0.03. The Mach number was 0.745 and the Reynolds number per meter 4.47×10^6. The time-step for the swept wing was 10^{-5}. These grids are shown in Figs. 172, 173 and 174.

3.4.1 Boundary and Initial Conditions

On the solid wall, impermeability and no-slip conditions were used. The far field boundary conditions are the characteristic variables extrapolated in time. The upstream free-stream boundary conditions are as follows: the total pressure ($P_\infty = 10^5$ Pa) and total temperature ($T_\infty = 290$ K). The upstream Reynolds number was

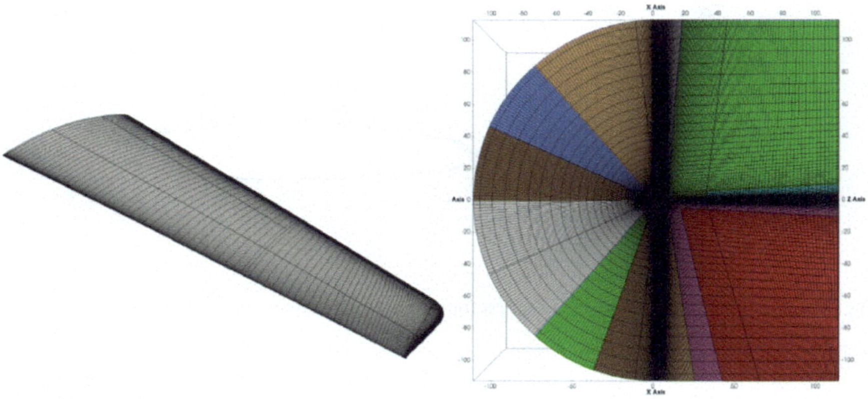

Fig. 174 Spanwise grid's surface on the wall for the swept V2C wing (left) and view of a plane section indicating the multi-blocks structure in the context of the NSMB code

3.245 Million and the upstream Mach number 0.70. The upstream turbulence intensity was Tu = 0.08%. Concerning the swept wing, the static pressure (free stream) was 17,900 Pa, the total pressure 25,900 Pa and the temperature 241 K. The inlet turbulence intensity Tu was 0.1% in this case. The initial conditions for both cases were those of a steady-state field, prior calculated on this purpose.

3.4.2 Turbulence Modelling

The following models have been used in URANS and Hybrid RANS-LES context respectively: the k-ω SST model of Menter [38] as well as the OES-k-ε [46, 51] as well as the DDES [47]: k-ω SST have been used with turbulence sustaining ambient terms [48] to prevent a decay of the turbulence generation, due to the dissipation transport equation.

3.4.3 Results for the Constant Spanwise Section V2C Wing

In Fig. 175, the buffet map is shown concerning the constant wing section case. The simulations carried out at 7° of incidence and free-stream Mach number correspond to buffet conditions. Indeed, in Fig. 176, the buffet phenomenon is clearly illustrated, as well as its interaction with the shear layer and von Kármán eddies. The 3D eddies structure strongly modified according to the buffet is clearly shown. The coherent vortices in the separated region (shear layer eddies and von Kármán ones) display a spanwise undulation according to well distinct wavelengths, due to amplification of secondary instability, in the same way as analysed by DNS in Persillon et al. [49]. Moreover, vortex dislocation patterns are illustrated as in DNS by Braza et al. [50]. The wake's width and the separation region, as well as the complexity of the 3D

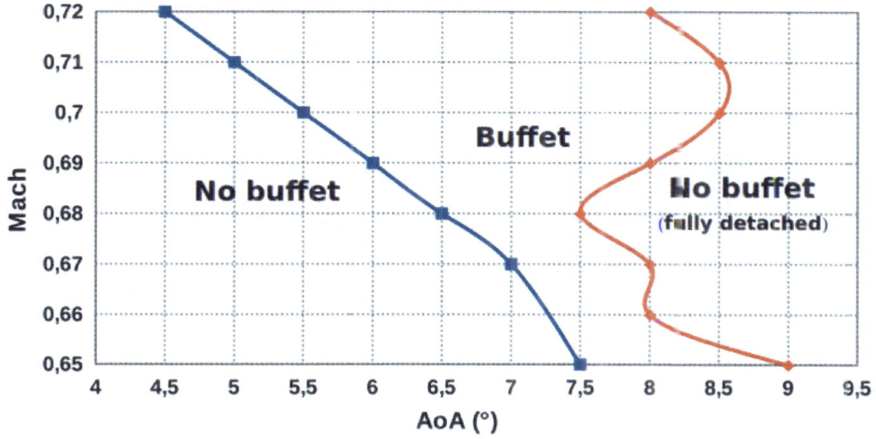

Fig. 175 Buffet map

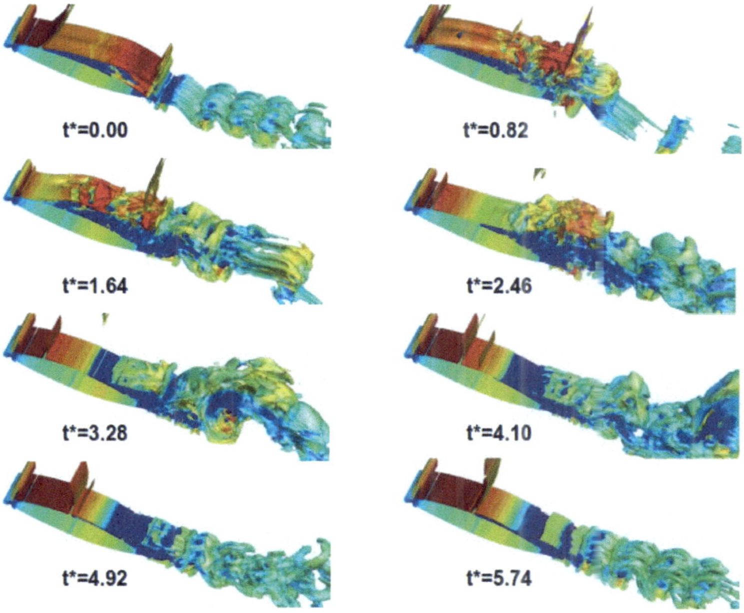

Fig. 176 Instantaneous Q-criterion iso-surfaces for $Q(c/U)^2 = 75$. DDES-SST model

vortex patterns increase as the shock wave is moving upstream, due to the buffet dynamics. Figure 177 shows the lift coefficient versus time, according to different turbulence models. The DDES-SST model provides a strong suction and creates a 'picky' behaviour of the lift coefficient oscillations. Table 16 shows comparison of the mean values with experimental data measured by the Institute of Aviation

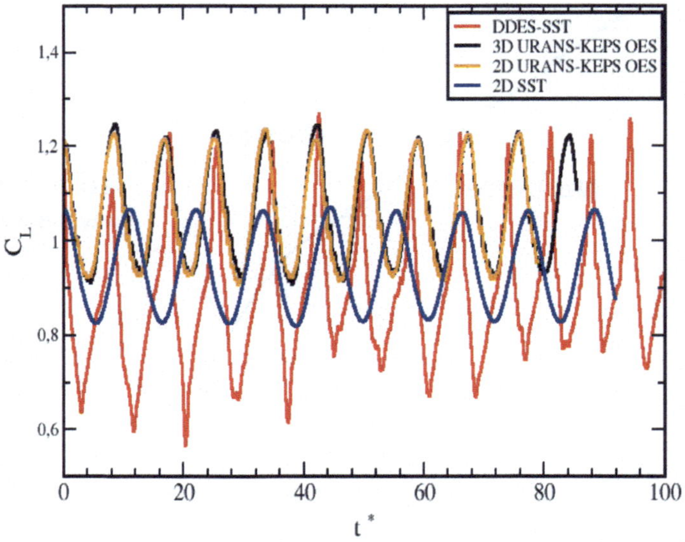

Fig. 177 Lift coefficient oscillations for different turbulence models used

Table 16 Constant section V2C wing: grid and zoom around the section

	EXPT	3D OES	2D OES	2D SST	DDES-SST
CL_avg	1.063	1.062	1.58	0.942	0.875

(IoA)—TFAST co-partner. The strong suction effect of the DDES is seen on these values, where the OES modelling provides a good agreement. Figure 178 shows the lift coefficient spectra where the DDES-SST displays a quite spread spectrum bump around the buffet frequency. The OES approach provides a more pronounced spectral bump. The dimensionless buffet frequency (Strouhal number) is of order 0.1, slightly varying from one model to the other (see the exact values displayed in the figure). Figure 179 shows iso-contours of the velocity vector divergence according to the DDES-SST and OES modelling.

It is recalled the OES approach has also a hybrid character regarding the coherent/non-coherent distinction of the turbulence structures to be resolved/modeled and is classed in-between URANS and LES approaches, needing more economic grids than the LES one.

In order to analyse the buffet mode interaction with the shear layer dynamics and the near wake structures, 3D Proper orthogonal Decomposition (POD), Berkooz et al. [26] has been carried out in case of the DDES. Figure 180 shows the energy distribution of the POD modes which displays quite significant values up to the first decade of modes.

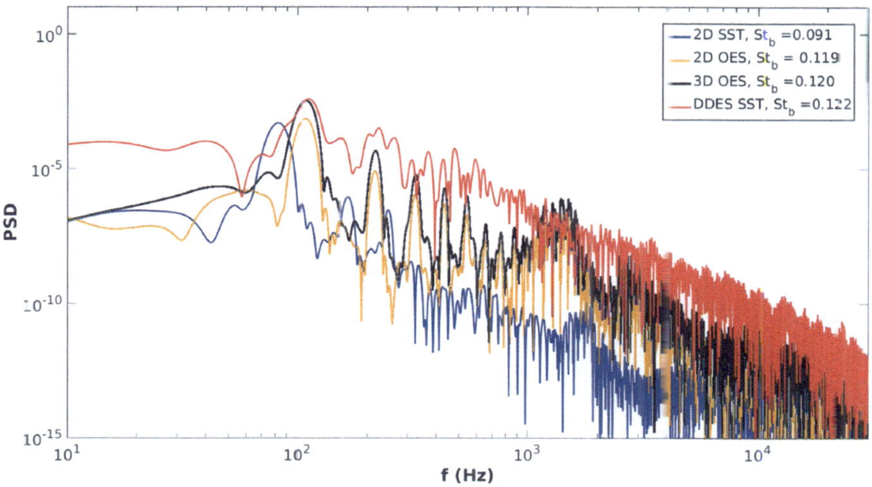

Fig. 178 Spectra of the vertical velocity component

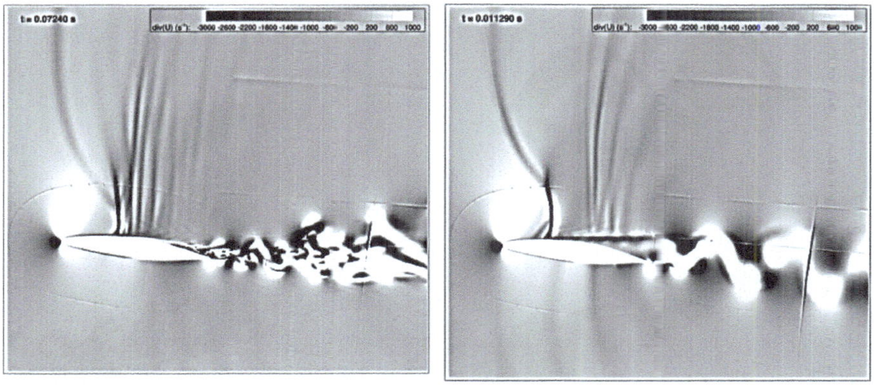

Fig. 179 Instantaneous divergence of the velocity vector illustrating by numerical "Schlieren" visualization the shock wave, the Kutta waves, the SWBLI structure, the shear layer and von Kármán eddies. Left: DDES-SST, right: OES

The higher-order modes are also quite useful in order to be used for a stochastic forcing of the transport equations, able to constrict and thin the shear layers, by considerably reducing the eddy viscosity and improving the forces evaluation, as studied by Szubert et al [27]. Figure 181 shows the spatial view of the flow, intermediate and higher-order modes. In this figure, w designates the vertical velocity component, where v is the spanwise one. Mode 2 is clearly related to the buffet and the shock amplitude formation. Figure 182 shows the spectrum of the second-mode temporal coefficient which displays the spectral bump around the buffet frequency. It will be remembered that the fact of bump formation and not of a sharp peak indicates that the

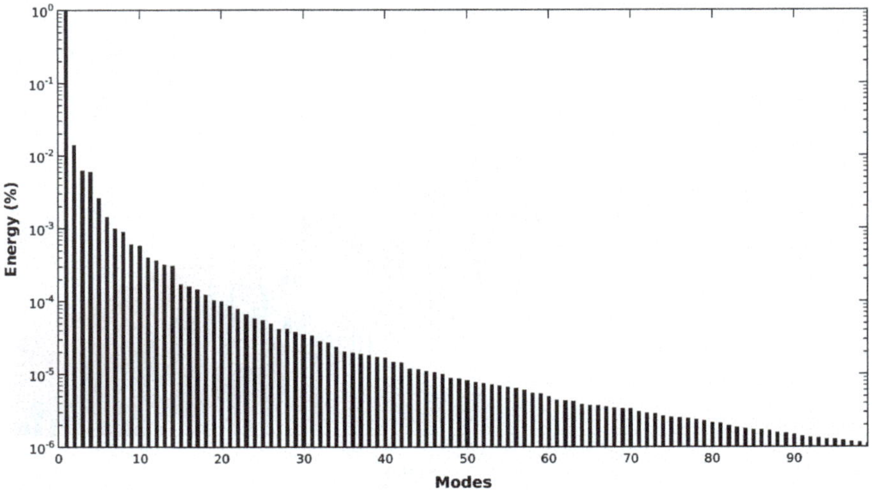

Fig. 180 Energy distribution of the POD modes

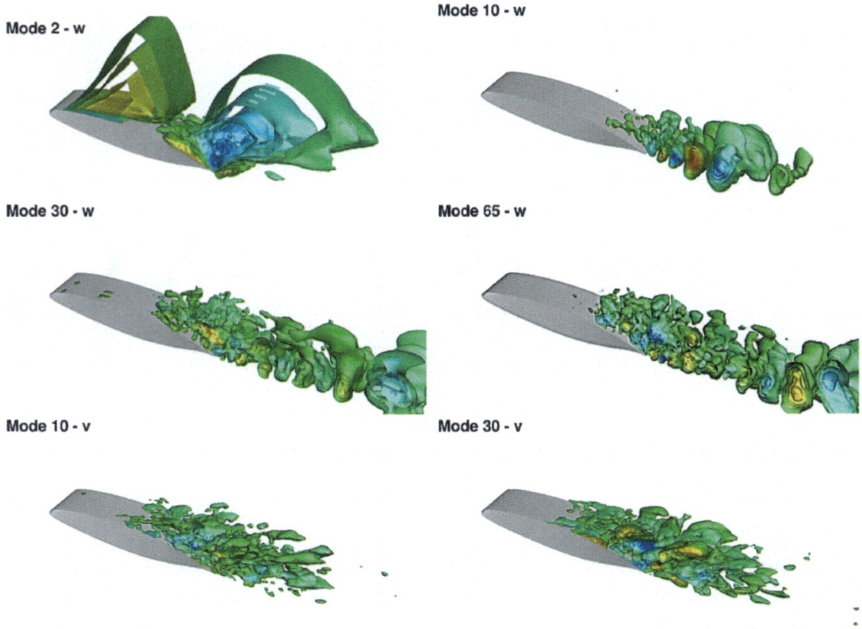

Fig. 181 Spacial view of the POD modes: here w is the vertical velocity and v is the spanwise one

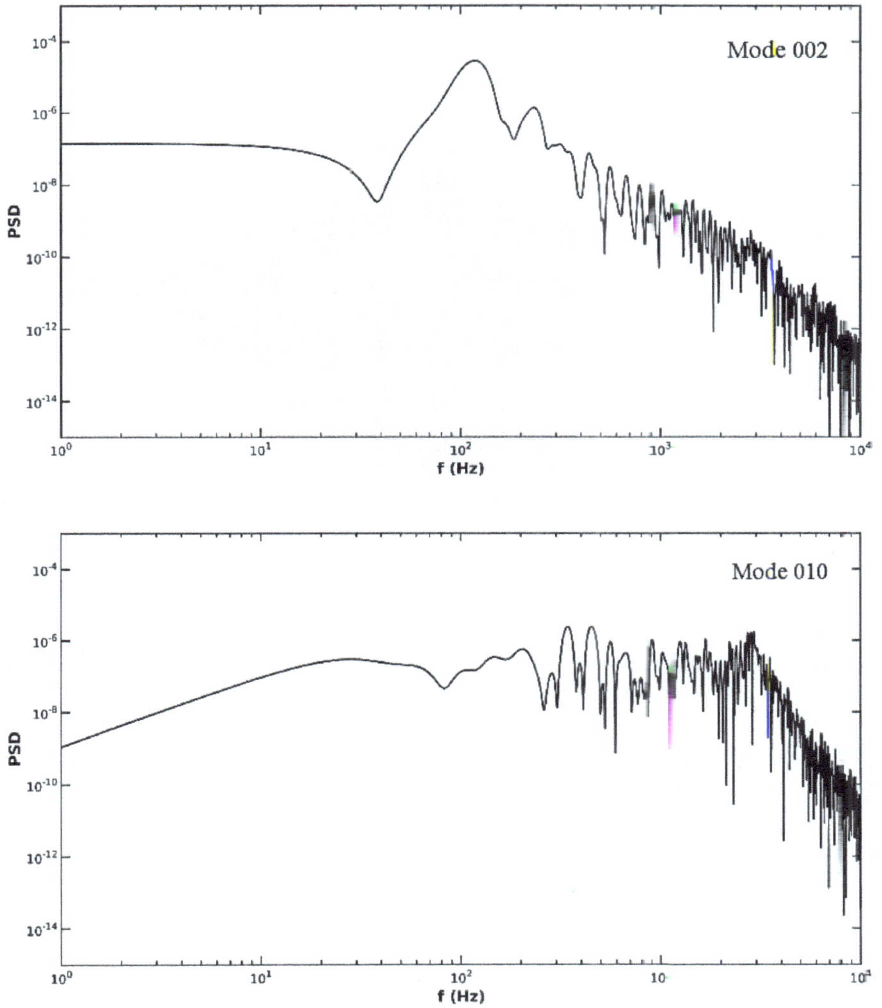

Fig. 182 Spectra of the second and 10th POD temporal coefficients

buffet in the present case is strongly affected by chaotic turbulence motion, enhanced by the strong suction. Figure 181 shows the spatial view of the 10th mode, associated to the shear-layer dynamics. This is shown on the temporal coefficient spectrum of this mode (Fig. 182) where the buffet bump has been attenuated on the profit of the shear layer frequency, near 3×10^3 Hz. The 30th and 55th spatial modes are enriched from the shear layer dynamics (Fig. 181). Concerning the spanwise velocity related to 3D effects, the 10th and 30th POD modes illustrate a chaotic formation of a multitude of spanwise wavelengths (Fig. 181).

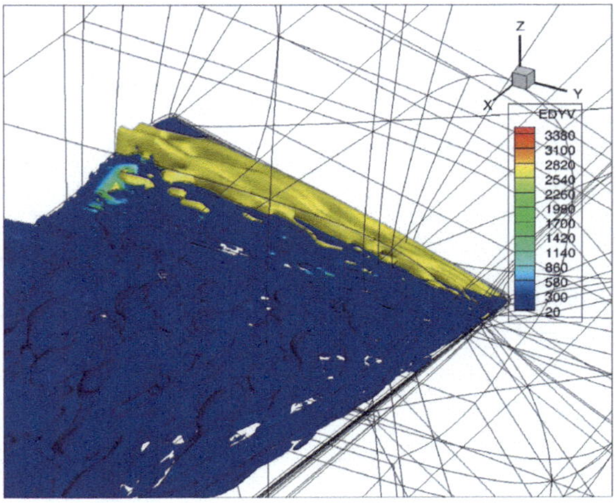

Fig. 183 Instantaneous 3D vorticity coloured by iso-pressure contours, k-ω-OES, fully turbulent case, illustrating the undulation of the coherent vortex rows along the span, according to predominant wavelengths

3.4.4 The Swept Wing

Figure 183 illustrates the instantaneous eddy-viscosity field past the swept wing (fully turbulent case), according to the OES-k-ε modelling. The eddy viscosity displays an organization according to spatial smaller-scale wavelengths due to the above formation of coherent structures. A weak unsteadiness has been maintained at higher time values but is was found that for the present grid and turbulence model, a practically steady state has been reached concerning the forces and near-wall parameters. The effect of the transition position fixed by numerical tripping at 55 and 30% of the chord has been studied at 5° of incidence, because of the TFAST's main goal to provide 'laminarity' in respect of drag reduction.

The choice of 5° is because for the present Mach number, this incidence is in the frontier of buffet onset. In case of a swept wing, the local Reynolds number based on the chord varies along the span. Therefore, only several spanwise sections are over-critical regarding buffet. The interest of the present study is to depict the regions subjected to buffet onset.

3.4.5 Cases of 55 and 30% of Fixed Transition at 5° of Incidence

Concerning the 55% case, Fig. 171 shows the dimensional, non-normalised lift force versus time, displaying buffet oscillations among other smaller-scale fluctuations. This figure also shows the pressure coefficient at different spanwise positions, the value y/c = 0.1 corresponding to the large chord spanwise sections side. It can be seen that towards the small chord side reaching y/c = 0.9, the Cp evolution in the shock area progressively displays an inclination indicating buffet formation, as also shown on the lift coefficient oscillations. Figure 186 shows the iso-Mach contours for the 55% transition case for a mid-span section and for the y/s = 0.9 one. A more detailed investigation of the buffet existence is necessary by using a finer grid, in order to provide a more detailed discussion (Fig. 184).

Concerning the case of 30% fixed transition at 5° of incidence, this reached steady state, with mean drag coefficient 0.0423 and mean lift 0.796. The corresponding Cp coefficients at different spanwise positions is shown on Fig. 184, where only normal shocks appear as expected in case of no buffet spatial amplitude. An onset of weak unsteadiness has been obtained in the near wake (Fig. 185).

The case of 7° and fixed transition at 30% has been finally studied. Figure 174 shows pressure oscillations versus time indicating formation of buffet mode (Fig. 186).

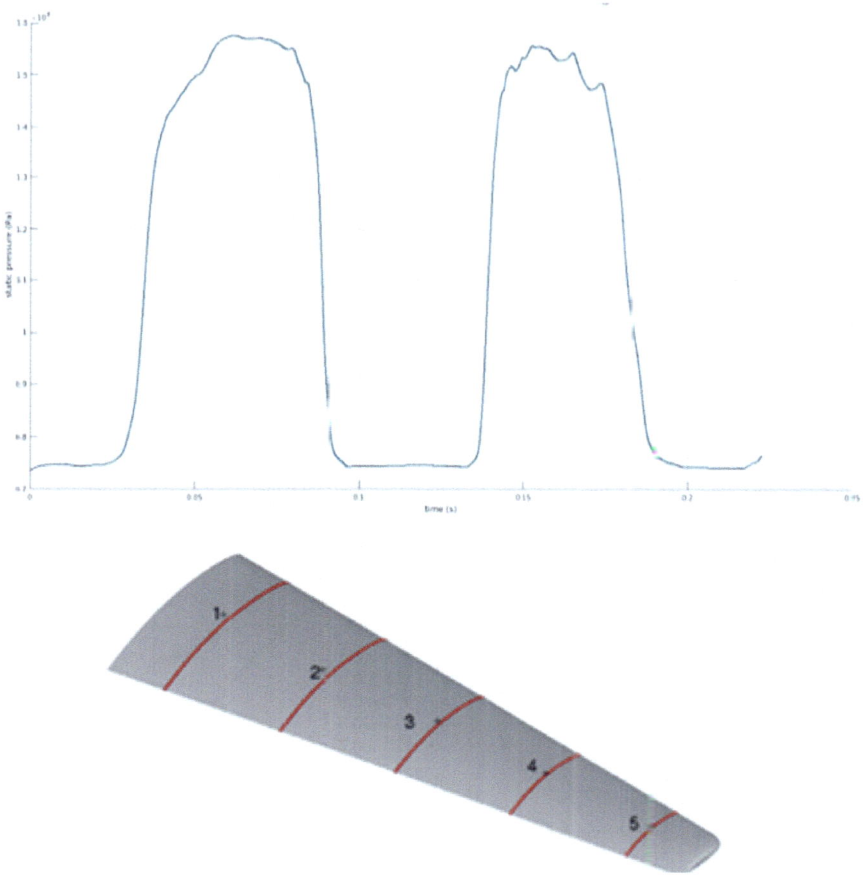

Fig. 184 Top: Wall pressure coefficient versus time indicating the onset of buffet around the swept wing at 7° of incidence; k-ω-OES model. Strouhal number 0.078; bottom: spanwise sections for plane investigations

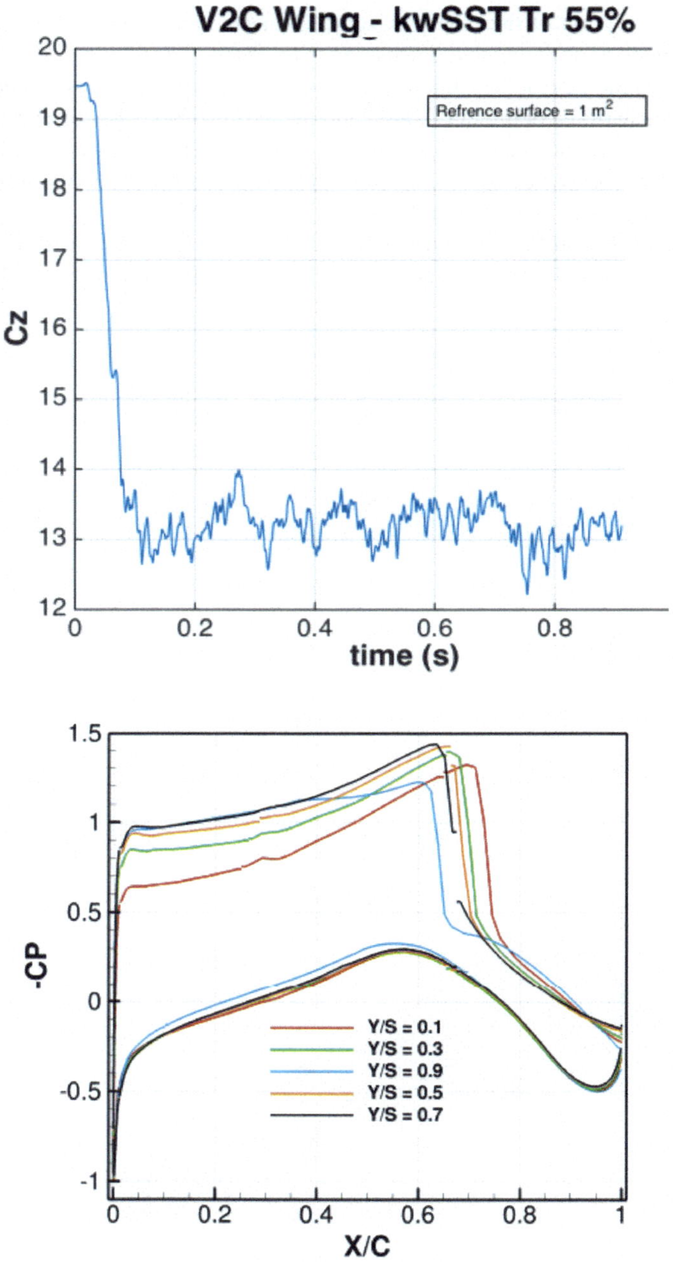

Fig. 185 Top: time-dependent drag coefficient (not normalised by the reference surface): bottom: Pressure coefficient at different sections according to the previous figure

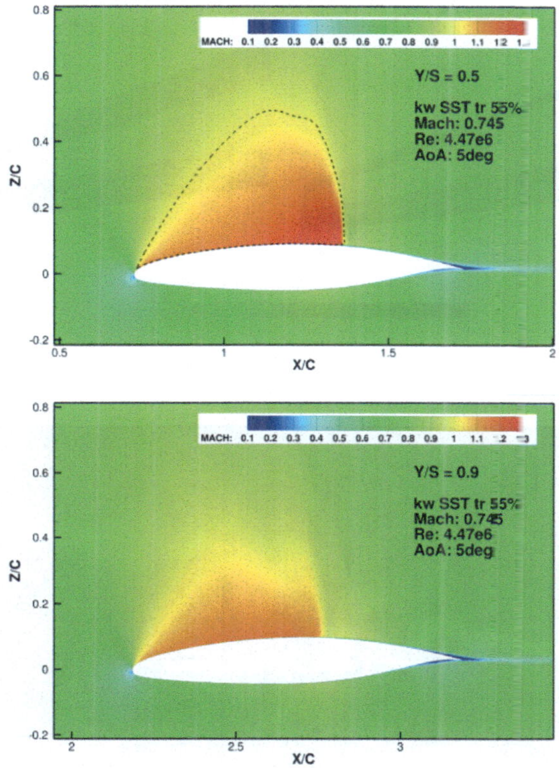

Fig. 186 Iso-Mach contours at two spanwise positions. y/s = 0.9 is located towards the small chord sections

3.4.6 Case of 30% of Fixed Transition and 7° of Incidence

This case displays an onset of buffet at a section located near the 3/4 of the span towards the small chord free edge, as indicated by the pressure versus time at the wall point P4 shown in Fig. 174. The iso-pressure contours of Fig. 175 within a period show the shock motion, the thickening of the boundary layer, separation and formation of the von Kármán vortex street, all these being characteristics of buffet onset. Figure 176 shows the shock location along the span at t = 0.1632.

Figure 187 shows the averaged skin-friction coefficient over the suction side along the span indicating the transition location as well as the mean shock location. An homothetic location of the shock along the span is obtained this indicates that this spanwise structure is not subjected to strong 3D modifications. Figure 190 shows the 3D supersonic region and shock formation along the span.

Because of the chord length variation along the span, the local Reynolds number varies and part of the spanwise sections are under-critical in respect of buffet appearance, whereas the other are over-critical. According to the buffet appearance, Fig. 189

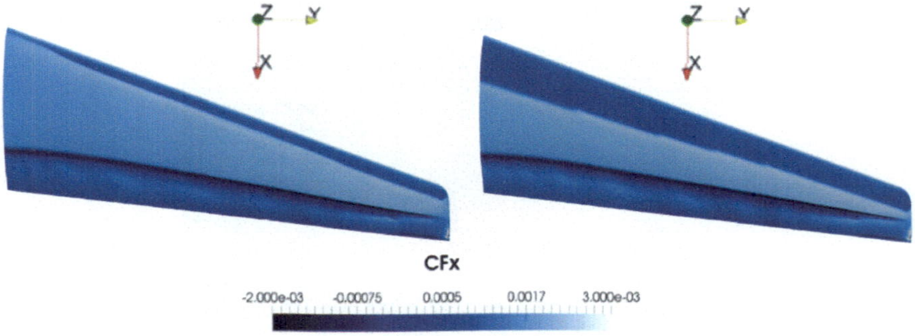

Fig. 187 Time-averaged iso-skin-friction coefficient contours along the span

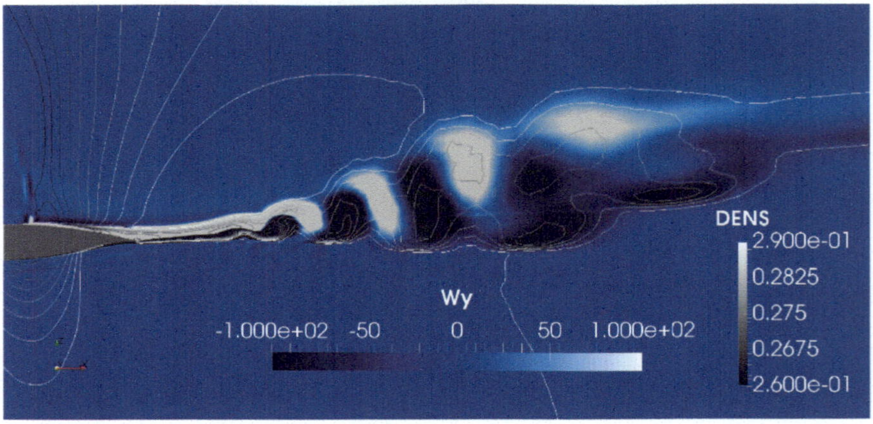

Fig. 188 Density iso-surfaces and iso-pressure contours illustrating the shear—layer and von Kármán vortex dynamics in relation with the lamda-shock region

shows the iso-pressure contours at the spanwise section y/s = 0.8 in association with the thickening of the boundary layer and separated shear layer towards the undulated wake structure. In all the cases, the near trailing-edge structure is found to display von Kármán eddies (Fig. 188), formed past the shear layer dynamics downstream of the lamda-shock region (Fig. 190).

This interaction develops feedback effects which can be used for further control of the buffet dynamics as studied by Tô et al. [34] thanks to trailing edge morphing.

3.4.7 Conclusion

A numerical study by means of statistical and hybrid turbulence modelling has been carried out to analyse the buffet phenomenon around a transonic wing of constant section and a swept wing at high Reynolds number. The predictions have shown the

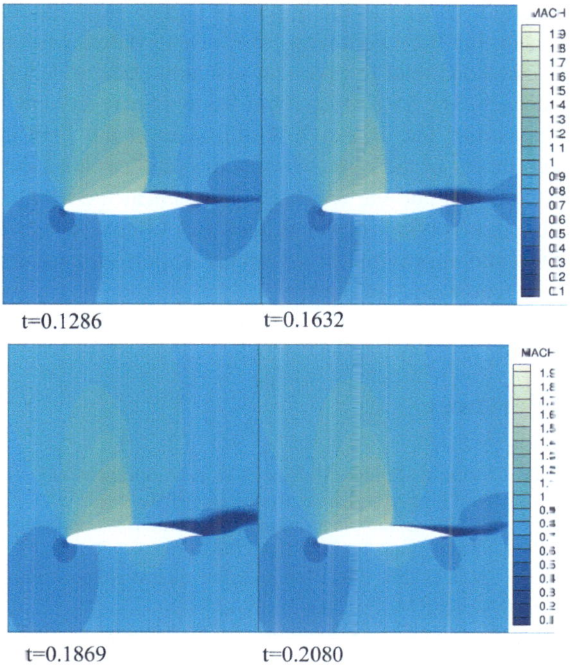

Fig. 189 Instantaneous iso-Cp at y/s = 0.8 spanwise section showing formation of the buffet motion, associated with the boundary layer thickness, separation and the von Kármán mode in the near wake

Fig. 190 Instantaneous iso-pressure surfaces along the span at t = 0.1632 illustrating the shock formation along the span

ability of all the methods in capturing the buffet's frequency bump, but at different spectral amplitudes and width of the bump. The DDES-SST simulations indicated a too strong suction effect and enhancement of the separation and of the shock's amplitude. The OES simulations provided a moderate shock amplitude and a good comparison of the mean forces coefficients with the experiments carried out by the Institute of Aviation (IoA). A detailed POD study using the DDES-SST simulations shows the role of the most energetic modes in respect of the buffet spatial oscillation amplitude as well as the impact of higher-order modes in the shear layer dynamics. Concerning the swept wing, the fully turbulent case and the fixed transition at 30% of the chord lead to practically steady state at 5° of incidence by using the OES approach. The 55% of fixed transition displays a weak buffet towards the small chord dimensions. The case at 30% of transition and 7° of incidence displays buffet onset near the 3/4 of the span, towards the smaller chord dimensions. A strong interaction of the buffet and the vortex dynamics near the trailing edge and the wake has been noticed, involving important feedback effects. These can be used for controlling the buffet and improving the aerodynamic performance through morphing, as is the case in current studies of ours.

References

1. F. Chalot, M. Mallet, M. Ravachol, A comprehensive finite element Navier-Stokes solver for low- and high-speed air-craft design (1994)
2. D. Arnal, Boundary layer transition: prediction, application to drag reduction. In Skin Friction Drag Reduction. AGARD Rep. **786** (1992)
3. B.C.A.D. Courty, J.-C., Laminar flow investigation: computations and flight tests at Dassault Aviation (1993)
4. T.S.C. Davidson, H. Babinsky, An investigation of interactions between normal shocks and transitional boundary layers—Control ID 1889354, in *44th AIAA Fluid Dynamics Conference* (2014), pp. 1–16
5. S.P. Colliss, Vortical structures on three-dimensional shock control bumps. AIAA J. (2014)
6. N. Titchener, S. Colliss, H. Babinsky, On the calculation of boundary-layer parameters from discrete data. Exp. Fluids **56**(8), 1–18 (2015)
7. H. Babinsky, J.K. Harvey, *Shock Wave-Boundary-Layer Interactions* (2011)
8. D. Hue, O. Vermeersch, D. Bailly, V. Brunet, M. Forte, Experimental and numerical methods for transition and drag predictions of laminar airfoils. AIAA J. **53**(9), 2694–2712 (2015)
9. W. Stryczniewicz, R. Placek, R. Szczepaniak, PIV measurements of flow separation over laminar airfoil at transonic speeds. J. KONES (2016)
10. W. Stryczniewicz, R. Placek, P. Ruchała, Investigation of unsteady transonic flow field above laminar airfoil by PIV method (2016)
11. R. Placek, W. Stryczniewicz, Identification of the boundary layer shock wave interaction type in transonic flow regime. J. KONES. Powertrain Transp. **23**, 285–290 (2016)
12. R. Placek, M. Miller, P. Ruchala, The roughness position influence on laminar aerofoil aerodynamic characteristics in transonic flow regime (2016)
13. J. Perraud, D. Arnal, G. Casalis, J.-P. Archambaud, R. Donelli, Automatic transition predictions using simplified methods. AIAA J. **47**(11), 2676–2684 (2009)
14. R.E. Mayle, The 1991 IGTI scholar lecture: the role of laminar-turbulent transition in gas turbine engines. J. Turbomach. **113**(4), 509 (1991)
15. S. Dhawan, R. Narasimha, Some properties of boundary-layer flow during the transition from laminar to turbulent motion. J. Fluid Mech. **3**(1951), 418–453 (1958)
16. A. Probst, R. Radespiel, U. Rist, Linear-stability-based transition modeling for aerodynamic flow simulations with a near-wall Reynolds-stress model. AIAA J. **50**(2), 416–428 (2012)
17. G.A. Gerolymos, I. Vallet, Wall-normal-free Reynolds-stress closure for three-dimensional compressible separated flows. AIAA J. **39**(10), 1833–1842 (2001)
18. M. Ol, B. McCauliffe, E. Hanff, U. Scholz, C. Kaehler, Comparison of laminar separation bubble measurements on a low Reynolds number airfoil in three facilities, in *35th AIAA Fluid Dynamics Conference and Exhibit*, June, 2005

19. M.C. Galbraith, M.R. Visbal, Implicit large Eddy simulation of low-Reynolds-number transitional flow past the SD7003 airfoil. AIAA J. 225 (2008)
20. S. Deck, Numerical simulation of transonic buffet over a supercritical airfoil. AIAA J. **43**(7), 1556–1566 (2005)
21. J. Crouch, A. Garbaruk, D. Magidov, L. Jacquin, *Global Structure of Buffeting Flow on Transonic Airfoils* (2009), pp. 297–306
22. F.R. Menter, Zonal two equation k-w, turbulence models for aerodynamic flows, in *24th Fluid Dynamics Conference July 6–9, 1993* (Orlando, Florida, 1993)
23. R.B. Langtry, F.R. Menter, Correlation-based transition modeling for unstructured parallelized computational fluid dynamics codes. AIAA J. **47**(12), 2894–2906 (2009)
24. F.R. Menter, Y. Egorov, The scale-adaptive simulation method for unsteady turbulent flow predictions. Part 1: Theory and model description. Flow, Turbul. Combust. **85**(1), 113–138 (2010)
25. J. Dudek, Modeling vortex generators in the wind-US code, in *48th AIAA Aerospace Sciences Meeting Including the New Horizons Forum and Aerospace Exposition.* (American Institute of Aeronautics and Astronautics, 2010)
26. G. Berkooz, P. Holmes, J. L. Lumley, The proper orthogonal decomposition in the analysis of turbulent flows. Annu. Rev. Fluid Mech. **25**(1), 539–575 (1993)
27. D. Szubert, F. Grossi, A. Jimenez Garcia, Y. Hoarau, J.C.R. Hunt, M. Braza, Shock-vortex shear-layer interaction in the transonic flow around a supercritical airfoil at high Reynolds number in buffet conditions. J. Fluids Struct. **55**, 276–302 (2015)
28. B.H.K. Lee, Self-sustained shock oscillations on airfoils at transonic speeds. Prog. Aerosp. Sci. **37**(2), 147–196 (2001)
29. D.G. Mabey, Oscillatory flows from shock-induced separations on biconvex aerofoils, in *AGARD Conferences Proceedings* (Vol. 296, No. 11, 1980)
30. R.B. Langtry, A correlation-based transition model using local variables for unstructured parallelized. CFD codes (2016)
31. B.J. Abu-Ghannam, R. Shaw, Natural transition of boundary layers—the effects of turbulence, pressure gradient, and flow history. J. Mech. Eng. Sci. **22**(5), 213–228 (1980)
32. P. Malan, K. Suluksna, E. Juntasaro, Calibrating the gamma-Re_theta transition model for commercial CFD, in *47th AIAA Aerospace Sciences Meeting Including The New Horizons Forum and Aerospace Exposition* (2019). p. 1142
33. N.N. Sorensen, CFD modelling of laminar-turbulent transition for airfoils and rotors using the $\gamma-$model. Wind Energy: Int. J. Prog. Appl. Wind Power Convers. Technol. **12**(8), 715–733 (2019)
34. J-B. Tô, et al, Effects of vibrating and deformed trailing edge of a morphing supercritical airfoil in transonic regime by numerical simulation at high Reynolds number. J. Fluids Struct. **91**(2019), 102595 (2019)
35. M.E. Kelterer, R. Pecnik, W. Sanz, Computation of laminar-turbulent transition in turbomachinery using the correlation based γ-Reθ transition model. *ASME Turbo Expo 2010: Power for Land, Sea, and Air* (American Society of Mechanical Engineers Digital Collection, 2010)
36. A. Toyoda, T. Misaka, S. Obayashi, An application of local correlation-based transition model to JAXA high-lift configuration model, in *25th AIAA Applied Aerodynamics Conference* (2007). p. 4286
37. F.R. Menter, Review of the shear-stress transport turbulence model experience from an industrial perspective. Int. J. Comput. Fluid Dyn. **23**(4), 305–316 (2009)
38. F.R. Menter, Two-equation eddy-viscosity turbulence models for engineering applications. AIAA J. **32**(8), 1598–1605 (1994)
39. O.Z. Mehdizadeh, L. Temmerman, B. Tartinville, C. Hirsch, Applications of earsm turbulence models to internal flows, in *ASME Turbo Expo 2012: Turbine Technical Conference and Exposition* (American Society of Mechanical Engineers Digital Collection, 2012). pp. 2079–2086
40. Y. Hoarau, D. Pena, J.B. Vos, D. Charbonier, A. Gehri, M. Braza ... E. Laurendeau, Recent developments of the navier stokes multi block (nsmb) cfd solver, in *54th AIAA Aerospace Sciences Meeting* (2016). p. 2056

41. J. Vos, A. Rizzi, A. Corjon, E. Chaput, E. Soinne, Recent advances in aerodynamics inside the NSMB (Navier Stokes multi block) consortium, in *36th AIAA Aerospace Sciences Meeting and Exhibit* (1998). p. 225
42. P.L. Roe, Approximate Riemann solvers, parameter vectors, and difference schemes. J. Comput. Phys. **43**(2), 357–372 (1981)
43. B. Van Leer, Towards the ultimate conservative difference scheme. V. A second-order sequel to Godunov's method. J. Comput. Phys. **32**(1), 101–136 (1979)
44. F. Grossi, Physics and modeling of unsteady shock wave/boundary layer interactions over transonic airfoils by numerical simulation (Doctoral dissertation, 2014)
45. D. Szubert, Physics and modelling of unsteady turbulent flows around aerodynamic and hydrodynamic structures at high Reynold number by numerical simulation (Doctoral dissertation, 2015)
46. R. Bourguet, M. Braza, G. Harran, R. El Akoury, Anisotropic Organised Eddy Simulation for the prediction of non-equilibrium turbulent flows around bodies. J. Fluids Struct. **24**(8), 1240–1251 (2008)
47. P.R. Spalart, S. Deck, M.L. Shur, K.D. Squires, M.K. Strelets, A. Travin, A new version of detached-eddy simulation, resistant to ambiguous grid densities. Theor. Comput. Fluid Dyn. **20**(3), 181 (2006)
48. P.R. Spalart, C.L. Rumsey, Effective inflow conditions for turbulence models in aerodynamic calculations. AIAA J. **45**(10), 2544–2553 (2007)
49. H. Persillon, M. Braza, Physical analysis of the transition to turbulence in the wake of a circular cylinder by three-dimensional Navier–Stokes simulation. J. Fluid Mech. **365**, 23–88 (1998)
50. M. Braza, D. Faghani, H. Persillon, Successive stages and the role of natural vortex dislocations in three-dimensional wake transition. J. Fluid Mech. **439**, 1–41 (2001)
51. M. Braza, R. Perrin, Y. Hoarau, Turbulence properties in the cylinder wake at high Reynolds numbers. J. Fluids Struct. **22**(6–7), 757–771 (2006)

Summary, Conclusions and Lessons Learned

Closing Remarks

Pawel Flaszynski, Piotr Doerffer, Sergio Pirozzoli, Jean-Paul Dussauge, Pierre Dupont, Lionel Larchevêque, Holger Babinsky, Patrick Grothe, Anna Petersen, and Flavien Billard

Abstract The TFAST project covered very wide range of knowledge starting from the basic research to more applied flow configurations. New knowledge generated in this project casts new light on the possibilities of flow control methods in reducing negative effects of shock wave interaction with laminar boundary layer. On the other hand application of flow control devices in the laminar boundary layer, just to inspire transition, is a valuable result. As usually in such complex flow structure the results are strongly case dependent, but the complexity of the presented approach shows great potential for improvements and provides a good basis for future research.

1 Conclusions on Reference Cases of Laminar and Turbulent Interactions

Two main configurations have been studied: the Oblique Shock Wave Reflection and the Normal Shock Wave. Mach numbers ranging from 1.45 to 1.7 were considered

P. Flaszynski (✉) · P. Doerffer
Institute of Fluid Flow Machinery, Polish Academy of Sciences, IMP PAN, Gdansk, Poland
e-mail: pflaszyn@imp.gda.pl

S. Pirozzoli
University of Rome "La Sapienza", Rome, Italy

J.-P. Dussauge · P. Dupont · L. Larchevêque
Aix-Marseille University, Marseille, France

H. Babinsky
University of Cambridge, Cambridge, United Kingdom

P. Grothe
Rolls-Royce Deutschland, Blankenfelde-Mahlow, Germany

A. Petersen
German Aerospace Center (DLR), Institute of Propulsion Technology, Goettingen, Germany

F. Billard
Dassault Aviation, Biard, France

P. Doerffer et al. (eds.), *Transition Location Effect on Shock Wave Boundary Layer Interaction*, Notes on Numerical Fluid Mechanics and Multidisciplinary Design 144,
https://doi.org/10.1007/978-3-030-47461-4_7

Table 1 WP-1 partners and investigated cases

Partner	Location	Flow	Mach	Contribution
ITAM	Novosibirsk, Russia	OSWR	1.45	EXP
IUSTI	Marseille, France	OSWR	1.6	EXP/CFD
ONERA	Meudon, France	OSWR/NSW	1.6/1.2	EXP/CFD
TUD	Delft, Netherlands	OSWR	1.7	EXP
UCAM	Cambridge, UK	NSW	1.2	EXP
IMFT	Toulouse, France	OSWR	1.7	CFF
SOTON	Southampton, UK	OSWR	1.5	CFD
URMLS	Rome, Italy	NSW	1.2–1.3	CFD

for the OSWR while a single Mach number of 1.2 was considered for the NSW. The different configurations have been distributed among the eight partners involved in this work package (Table 1). The normal shock wave/boundary layer interaction (NSBLI) flow case of the TFAST project was aimed at establishing the effect of the incoming boundary layer state (laminar/turbulent/transitional) upon impingement of a normal (or quasi-normal) shock wave, either generated externally through a shock generator, or by adverse pressure gradient, such as on the suction side of airfoils under transonic flw conditions. Complementary experimental and numerical campaigns were carried out. Dedicated facilities were designed and successfully built at ONERA and at the University of Cambridge (UCAM). However, achievement of the desired flow conditions, especially as far as the laminar regime is concerned, proved to be all but a simple task. In particular, severe unstart problems were encountered by ONERA owing to lack of power of the set of pumps, which prevented achievement of sustained transonic flow. Greater success was achieved by the UCAM team, who used a clever set-up exploiting the favorable pressure gradient of a suitably designed flat plate with elliptical or modified super-elliptical leading edge to accelerate the flow to mildly supersonic conditions, together with outlet pressure control to keep the recompression shock to a nearly fixed position. Measurements included flow visualizations and quantitative streawise velocity measurements through LDV. Several tests were carried out at Reynolds number of about 1.5×10^6, based on the distance of the shock from the leading edge of the flat plate, and corresponding to transitional flow conditions. Clean flow cases yielded a laminar type interaction, starting with a weak precursor compression wave with upstream Mach number $M = 1.16$, and nominal pressure rise of about 5%, and yielding no sensible associated growth of the boundary layer thickness. As claimed by UCAM, a reasonable interpretation is that the boundary layer separates while remaining laminar. This extended region (many tens of times the boundary layer thickness) starts with a weak oblique compression shock, and it is terminated by a normal shock with significantly greater strength (the relative pressure rise is about 23%). Past the terminating shock the boundary layer is observed to grow abruptly, however without any evidence of massive flow separation. This is a convincing evidence that, under the flow conditions under scrutiny the boundary layer undergoes sudden transition to turbulence, perhaps triggered by undulations of the separated laminar shear layer. The effect of two-dimensional and three-dimensional tripping devices was also explored by UCAM, showing that laminar separation is suppressed along with the separation shock, and the interaction becomes very similar to a fully turbulent interaction, with very nearly wall-normal shock foot and no lambda structure. Remarkably, it is found that the eventual downstream structure of the boundary layer is the same, regardless of the upstream state.

Numerical simulations of (nominally) laminar and turbulent interactions have been carried out by URMLS using a in-house solver tailored for DNS. The flow conditions under study were similar to those considered in the UCAM experiments, with free-stream Mach number $M = 1.2$, and Reynolds number at the shock location of about 1.7×10^6. As experiments, numerical simulations were also hampered by substantial technical difficulties, associated with the need of anchoring a normal shock on a flat wall, and of using extremely large computational domains to fit the interaction zone with strong scale disparity between the boundary layer thickness and the interaction length scale. Numerical simulations were carried out both in constrained two-dimensional configurations, and in three-dimensional configurations, with some initial random seeding to stimulate the onset of flow three-dimensionality. The results generally support and strengthen the experimental findings, in that the interaction is initiated by a weak separation shock, followed by an extended narrow region with laminar reversed flow which extends for about 100 upstream boundary layer thicknesses. Interaction with the terminal shock foot is responsible for sudden boundary layer transition, marked by an abrupt increase of magnitude of the skin friction, corresponding to a region with turbulent reversed flow. Reattachment occurs on a short length scale past the shock, and the friction coefficient then becomes large and positive. The streamwise velocity fluctuations are found to be enhanced since the very beginning of the separated flow region, whereas all other components of the Reynolds stress tensor are only excited past the terminal shock. Numerical simulations allowed close insight into the flow physics, and to more directly appreciate the unsteady features of the interaction. In this respect, a strongly intermittent behavior of the interaction zone was found, whereby intervals with relatively disorganized flow are interspersed with intervals characterized by coherent shedding of large spanwise-oriented rollers. However, the typical time scales of this intermittency are extremely long, making precise spectral quantification difficult to achieve. Fully turbulent interactions have been also computed, using vortex generators as tripping devices. In this case, the Mach number ($M = 1.3$) is sufficient to cause marginal separation of the turbulent boundary layer, with the corresponding formation of a quasi-steady lambda shock pattern with small size. Comparison with available experimental data showed excellent quantitative agreement of the wall pressure and skin friction distributions.

Overall, experiments and numerics consistently point to a rather 'reassuring' scenario, whereby nominally laminar interactions do not feature the expected extensive separated flow regions, nor the associated strong buffet. The likely reason is that, under the explored range of Reynolds numbers the boundary layer is very sensitive to upstream perturbations, and prone to undergo transition at the terminal shock foot. Transition has the effect of stabilizing and anchoring the shock system, hence large-scale excursions are not observed, whereas the precursor shock is too weak to cause massive flow separation. It is noteworthy that a very similar scenario is also observed in incompressible separation bubbles under transitional flow conditions, hence it appears that genuine compressibility effects in NSBLI are weak, and the shock is mainly responsible for establishing the required pressure gradient, but it does not participate the dynamics directly. Hence, the effects of triggering the transition are mainly negative, as friction is increased upstream of the shock without any significant benefit in terms of reduction of the boundary layer thickness past the interaction zone. This important result is also confirmed by experiments and simulations of transonic buffet on airfoils, which have also been the subject of the TFAST project. In that case it was found that buffet is much weaker in nominally laminar conditions, as sudden flow transition has the apparent effect of anchoring the shockwave.

Perhaps not surprisingly, the NSBLI flow case was the least populated in the TFAST project, nevertheless some important, and probably unexpected conclusions can be drawn. Given the available amount of experimental and numerical data, the main conclusions from the TFAST project is that laminar shock/boundary layer interactions, at least in the normal shock case, are quite benign, hence stimulating boundary layer transition to counteract possible detrimental unsteady effects is of little if any use.

Open issues however remain. The main limitation of the studies carried out within the TFAST project is the limited range of Reynolds numbers under scrutiny. Owing to technical difficulties with both experiments and numerical simulations, typical Reynolds number were in the natural transition regime between one and two millions. Hence, strong response of the boundary layer to imposed disturbances is expected, as in fact observed, which is the likely reason for the counter-intuitive suppression of large-scale unsteadiness. While this range of Reynolds number is relevant for aeronautic applications, it would be very interesting to extend the analysis to lower Reynolds numbers of interest in turbomachinery, and in which the boundary layer would be less prone to transition. Recent DNS of transonic buffet at $Re_c = 300{,}000$ carried out by the URLMS group has indeed highlighted a very different behavior, whereby nominally laminar interactions yield stronger buffet, and boundary layer tripping is effective in suppressing the unsteadiness. Additional numerical simulation including the effect of tripping devices on the structure of the NSBLI would also be highly desirable, to compare with the database developed by UCAM. Finally, access to wall properties (pressure and friction) in experiments would be of major use to allow quantitative validation of numerical results.

The technical specificities of the different wind tunnels cover a wide range of unit Reynolds number (5×10^6–35×10^6) and of initial perturbations (external pressure perturbations from 0.4 to 2%).

All experimental partners succeeded in developing operational set-ups for the OSWR. It should be noted that highly stable and repeatable experimental conditions have been obtained, allowing cross comparisons between several metrologies and experiments. Laminar to turbulent upstream conditions have been obtained, documented and used for numerical simulations.

Partners involved in the NSW configuration have experienced serious experimental worries. None was able to obtain stable laminar NSW interactions for such low Mach number, with upstream laminar flow with a plate sharp leading edge. UCAM succeeded to obtain a stable configuration using an elliptical leading edge.

The experimental partners used a wide range of methods: schlieren (in some cases time resolved), hot wire, Pitot probes, high spatial resolution LDA and PIV, Pressure and Temperature Sensitive Paints, unsteady wall pressure sensors.

In parallel, a wide panel of numerical simulations has been used: 2D URANS, DDESs, 3D LES and DNS. LES and DNS results confirm the experimental observations: the transitional interactions are unsteady (low frequencies) and separated, and the turbulent one are attached and steady. SOTON develops multimodal LST analysis, with forcing at the inflow, based on the ONERA and ITAM experimental configurations.

All partners obtained upstream laminar conditions following theoretical laminar boundary layer properties. Transition of the boundary layer was obtained for Reynolds number around 2×10^6 (IUSTI did not obtain natural transition along the plate). For the highest possible total pressure in IUSTI facility, transition begins close to the trailing edge of the plate, which corresponds to a Reynolds number of 2×10^6).

A particular effort has been made to evaluate the onset of separation of the laminar boundary layer. The direct evaluation was difficult due to blockage effects for very low flow deviations. TU-Delft observed marginal flow separation for imposed flow deviation as low as 1°. The onset of separation has been studied from the IUSTI Large Eddy Simulations. These simulations have shown that a flow deviation as small as 1° was sufficient to obtain an incipient separation. This is consistent with empirical relations based upon the free-interaction theory: this validation, provided comparisons between the considered OSWI. When turbulent upstream conditions are considered, no separation is observed, whatever the flow deviation. Beyond 5° (or less for lower Mach numbers) turbulent OSWI induces Mach stems.

The length scales of the interaction have been derived for all configurations. In case of the OSWR, dependences on the imposed flow deviation, unit Reynolds number and upstream level of perturbations have been documented. Large differences between experimental results have been obtained concerning the resulting aspect ratio (L/δ_0, with L the interaction length and δ_0 the displacement thickness of the natural boundary layer at the onset of the interaction): they can differ by more than 100% between the different partners. The influence on the mean field of the upstream unsteadiness has been described in details from IUSTI's experiments and LES and from ITAM's experiments. Combining natural transition and tripped upstream boundary layer (keeping a laminar upstream flow, see WP2 results for more details), these large differences on the aspect ratio of the interaction have been explained by the upstream level perturbations, which may differ significantly between the different experimental facilities. Therefore, only qualitative comparisons were achieved with the transitional experiments. Quantitative comparisons between simulations and experiments would require adjusting the amplitude of disturbances in the upstream boundary layer in order to obtain the same length of interaction. It has to be pointed out that in this case the simulations are no more predictive. The capability to predict accurately the length of interaction if the input perturbation are known (size, amplitude, space-time properties) remains an open question. For the turbulent cases, LES and DNS numerical simulations compared well with experiments from TU-Delft, IUSTI and UCAM if main characteristic properties of the downstream flow are considered (displacement and momentum thicknesses, shape factor, friction coefficient C_f).

The geometrical properties of the interaction (length of interaction, reattachment location, transition location) have been documented from all partners, using different methods. In general, no fully laminar interaction is experimentally observed. Transition occurs always inside the interaction. ONERA and TU-Delft showed that for small flow deviation and/or reduced unit Reynolds number, transition is observed slightly downstream from the impingement location. When flow deviation and/or unit Reynolds number are increased, transition occurs just downstream of the shock impingement and the flow is reattaching. The reversed region shares some characteristic features with the *Free Interaction Theory*. IUSTI's experiments and LES and TU-Delft's experiments have shown that the initial part of the separation region does not depend on the imposed deviation. Constant pressure rises as well as flow deviation (of about 2°) have been observed for all cases and the separation bubbles have been found homothetic for a wide range of shock intensities.

Whatever the geometry (OSWR or NSW), the downstream conditions are always turbulent (or nearly turbulent): the interaction can be considered as an effective tripping system for the transition to turbulence for the upstream laminar boundary layer. The downstream mean field relaxes quickly to fully turbulent conditions. As for turbulent interactions, the turbulent fields relax significantly more slowly than the mean fields.

Unfortunately, the turbulent fields have been less documented, because of the extreme experimental conditions. ITAM (micro-PIV) and IUSTI (38 μm LDA) provided turbulent fields along the interaction. The transitional evolution of the mixing layer, which is developing over the separated region, has been documented. Experimental and numerical IUSTI results showed that the transition of the mixing layer, characterized by a saturation level of the velocity fluctuations amplitudes, defines the reattachment of the flow.

Significant efforts have been carried out to obtain experimental descriptions of the unsteadiness along the interaction. This point was not a centrepiece of the TFAST program. Nevertheless, a large dependence of the mean field on the amplitude of the upstream unsteadiness has been demonstrated during the project which required to document this point in more details as initially planned.

Experiments from IUSTI, ITAM and ONERA have shown that the considered transitional OSWR were unsteady, with evidences of low frequencies, which about one order of magnitude are lower than the characteristic unsteadiness observed in the laminar boundary layer from unsteady hot wire measurements. IUSTI provided extensive results concerning the longitudinal evolution of these unsteadiness for various upstream conditions and flow deviation angles. The unsteadiness developing inside the interaction involves lower frequencies than those observed in the upstream laminar boundary. Non-dimensional frequencies (Strouhal numbers) have been estimated: some controversy remains between the different results. IUSTI and SOTON simulations suggest Strouhal numbers about 0.04 and 0.5 for the low frequencies and mixing-layer frequencies, respectively, as in separated turbulent interaction. On the opposite, IUSTI experiments suggest values of respectively 0.12 and 0.5, when ONERA simulations propose intermediate values of 0.09. If all partners observed evidences of low frequency unsteadiness in transitional interactions, the accurate estimation (and of course the origin) of the unsteadiness remains an open question. Numerical results (IUSTI, ONERA, SOTON) showed that the low frequencies are associated with negative convection velocity: a Fourier analysis would suggest that they are generated in vicinity of the reattachment region and propagate in the upstream direction. Unfortunately, no phase information was available from the experiments during the project. Anticipating the WP2 outcome, a surprising result came from IUSTI experiments and numerical simulations: for a fixed flow deviation, the amplitude of the low frequency unsteadiness has been found to vary as the aspect ratio L/hof the interaction. Shorter is the interaction (i.e. larger the upstream perturbations) higher are their amplitudes.

Finally, as already mentioned, transition to turbulence is observed at reattachment, and large energetic structures are shed downstream of the interaction as in turbulent separated interaction. Therefore, the turbulent field remains out of equilibrium downstream from the reattachment, being possibly an issue for the modelling of this region.

These results of course have strong consequences for the aeronautical applications. Transitional interactions have been proved to be an efficient way to promote transition to turbulence downstream from the interaction. Nevertheless, special care must be taken regarding the low and high frequency unsteadiness developed in separated interactions. These unsteadiness have been shown to have a large influence on the mean field of the interaction. Now, the elements and the expertise for a more detailed analysis of the unsteadiness of such flows, their origin and their precise characterization are available.

It is recommended also to analyse the "turbulent" field formed downstream of the interactions, and to focus on such conditions in the future, to test current modelling. Turbulence generated in such conditions is very particular, and has probably not yet reached asymptotic properties. This has to be described experimentally, and their modelling will require probably non-standard adaptations.

In aeronautical situations, the consequence is that the effort should be put on the definition of the perturbations upstream of the interactions. This depends of course on each particular case, but advanced engineering methods can investigate this point: this is the key issue for the prediction of such flows. A last recommendation can be made: it would be probably useful to use advanced modelling like LES or even DNS in some cases, to calibrate simpler and cheaper models of handy use for current design.

2 Conclusions on Flow Control Devices and Transition Induction

Table 2 lists the partners involved in WP2 and the flow control devices investigated. It can be seen that the research covered a wide range of devices from traditional 'mechanical' transition trips to more complex active controls as plasma actuators. In each case the partners used the same facilities and flow configurations as described in WP1. Therefore the methodologies will not be repeated here.

Typically, transition control devices were placed at Reynolds numbers (based on streamwise distance from the origin of the boundary layer) of 1.5×10^5–7×10^6. Mostly, devices were located at Reynolds numbers in the range of 5×10^5–1.5×10^6. Each device was tested in a variety of sizes or strengths, as shown in Table 3.

In the task 2.1 (*Study of operation of transition control devices*), the effectiveness of control devices to trigger transition ahead of the shock wave was investigated. Broadly speaking, it was observed that all mechanical devices triggered transition successfully. The only exceptions were very small devices applied in facilities with very low free-stream Reynolds numbers. Any device with a device-height Reynolds number of 1000 or greater was found to triggered transition. In fact, the relatively small device height required for transition tripping resulted in most devices under

Table 2 WP-2 partners and flow control devices investigated

	2D mechanical		3D		VG		Plasma
	Roughness	Step	Zig-zag	Dots	Rod-VG	Air-jet	
IUSTI		X					
ONERA			X	X		X	
ITAM		X					X
UCAM		X		X			
TUD	X	X	X				
SOTON							X
URMLS					X		
IoA						X	

Table 3 Control devices and relative height (expressed in local boundary-layer thickness) or 'strength' (expressed in local flow momentum)

Device type	Size or 'strength'
Roughness	3–10% of δ
Step, zig-zag tape	10–130% of δ
Dots	10% of δ to $\gg\delta$
Vortex generators (mechanical)	δ to several δ
Air-jet vortex generator	20–60% of U_∞
Plasma	<5% of U_∞

study being considerably larger than necessary. Indeed, many partners reported evidence of 'over-tripping'. The main symptom of this was a rapid boundary-layer growth across the device, leading to the flow downstream of such devices featuring a thicker boundary layer than that observed in natural (free) transition.

For the 'momentum' devices it was observed that a minimum of around 20% of 'free-stream momentum is necessary to successfully trip the boundary layer to turbulence. Consequently, plasma devices were unable to promote turbulence (as seen in Table 3, their strength was well below this level). A 'hypothetical' numerical study at SOTON investigated (as yet unfeasible) plasma devices at greater strength and here transition was observed when the plasma-induced jet velocity exceeded approximately 20% of the local free stream.

In the task 2.2 (*Interaction sensitivity to transition location*) the effect of artificially tripping the boundary layer ahead of a shock on the shock wave/ boundary layer interaction was studied.

The majority of research groups noted no fundamental difference in the overall flow field between the tripped and untripped flows. While laminar flow ahead of the shock wave generally led to the existence of a laminar separation, the common observation was that this very quickly caused transition and turbulent re-attachment in the vicinity of the main shock wave as illustrated schematically in Fig. 1a. When the flow was tripped ahead of the shock (or transitioned naturally to turbulence) no laminar separation was seen, as expected (see Fig. 1b). In all cases the flow downstream of the SBLI appeared relatively similar, whether it had transitioned inside a laminar separation bubble or ahead of the interaction, suggesting that there were no adverse effects caused by the existence of a laminar separation. In fact, the most common observation was that tripped flows exhibited thicker boundary layers (due to over-tripping) across and downstream of the interaction, suggesting that in such cases the effects of tripping were predominantly unfavourable.

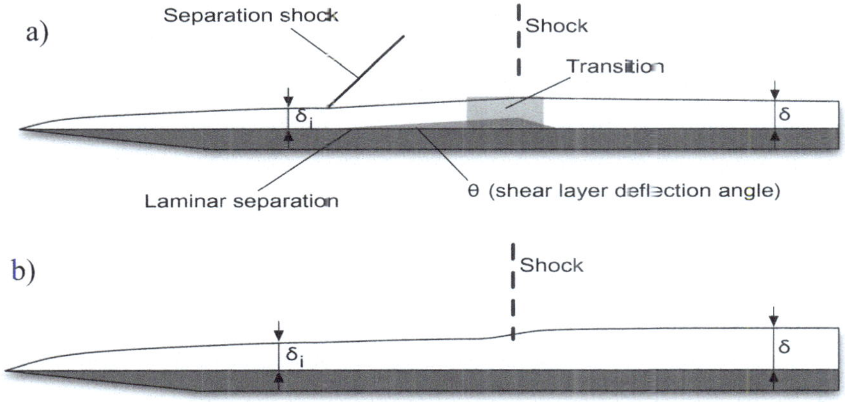

Fig. 1 Schematic view of **a** laminar separation bubble caused by shock-induced pressure rise and, **b** turbulent or tripped interaction (no separation)

This raises the question why a laminar separation bubble is seen as a relatively benign event. The answer to this lies in the fact that the pressure rise at separation is very small for laminar boundary layers. Thus, the pressure jump at the separation point (and thus the strength of the separation shock wave) is very small and the flow deflection angle θ (formed by the separating shear layer and the wall) is also very small (of the order of 1°). Hence, although laminar separations are typically quite long (the separation location is many boundary layer thicknesses upstream of the shock wave) the height of such separations remain quite small. Therefore, the existence of a laminar separation causes negligible effects on the outer flow because the associated shock wave is vanishingly weak and the displacement of the boundary layer edge is very small. Following separation, the instability of the separated shear layer makes this very prone to transition. Thus, transition is observed either upstream of the main shock or, at the latest, underneath the main shock wave where the additional pressure pulse leads to a very quick transition and subsequent flow re-attachment as seen in Fig. 1a.

Consequently, the boundary layer thickness δ downstream of the interaction is often smaller than that observed in a turbulent or tripped interaction where the incoming boundary layer δ_i may be significantly thicker than in the laminar case.

This finding is illustrated by Fig. 2 which shows the development of boundary layer thickness and shape factor across and downstream of a normal SBLI, comparing 'clean' cases (laminar inflow) with various tripped cases. It can be seen that, despite the fact that the laminar cases featured a separated interaction, there are no fundamental differences in the post-interaction boundary layer development. In fact,

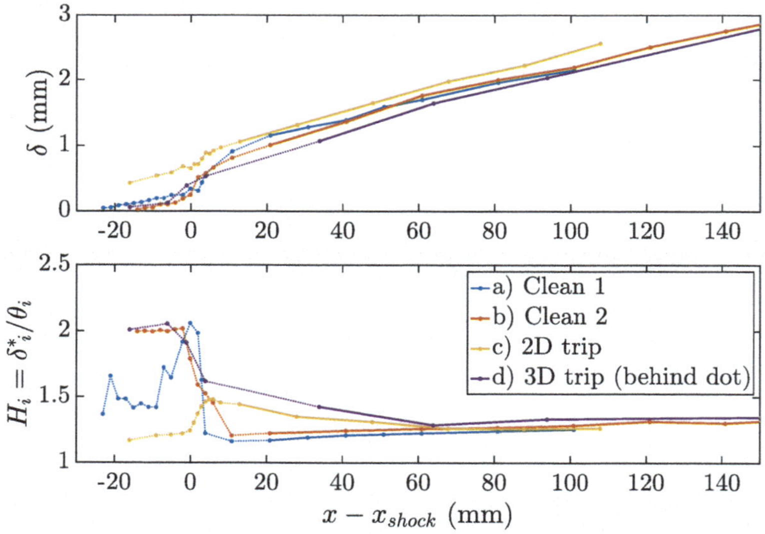

Fig. 2 Boundary layer properties across a normal SBLI, comparing 'Clean' (natural transition in SBLI region) and tripped cases. Despite being separated, 'clean' cases exhibit faster recovery and thinner boundary layers downstream of the interaction (UCAM)

it can be seen that the 'clean' flow cases experienced a faster recovery to typical equilibrium shape factors and a slightly reduced boundary layer thickness compared to tripped cases.

These findings were common to all partners of WP-2. Most groups concluded that for the flow configurations under study, namely flat plate interactions of normal and oblique shock waves, there was no benefit in tripping the flow ahead of the shock wave. Given that most trips exhibited evidence of over-tripping the recommendation is thus to avoid tripping altogether for such geometries.

Flow unsteadiness was rarely found to be affected by the presence of laminar separation. The only flow scenario where unsteadiness was enhanced was reported by IUSTI for the case of tripped flows. Thus, it was once again concluded that allowing the interaction to trigger turbulence is not detrimental.

The main conclusions of this work package can be summarised as follows:

- Most mechanical tripping devices were successful at promoting turbulent flow, except for very small devices (below $Re_H = 1000$ at low Re).
- 'Active devices' (air-jet VGs or plasma) needed a minimum 'strength' to successfully promote transition. This is roughly of the order of 20% of free stream flow momentum. Plasma devices are currently unable to reach this level of 'strength' in high-speed flows.
- Most devices caused over-tripping, leading to thicker (and less full) boundary-layers post-shock compared to natural transition.

When the flow was left to transitional naturally in the SBLI region, the most commonly observed flowfield featured a laminar separation followed by transition in the vicinity of the shock and turbulent re-attachment. Despite the fact that clear differences in the flow were observed at a local level, the global flowfield was not fundamentally different from an attached turbulent interaction. This included flowfield unsteadiness. Thus, it can be concluded that in many cases a laminar separation does not significantly alter the outer flow or cause any long-term damage to the boundary-layer flow across and downstream of the interaction

3 Conclusions on Internal Flow Case—Compressor

In order to have deeper insight to the compressor cascade application two test facilities were used. One for basic studies and one which allows resolving more detailed flow quantities. First test section is a single passage test section representative for the flow conditions in real cascade. The experimental and numerical investigations were carried out at IMP PAN within Task 3.1 (Single passage test section). The second test section is a linear cascade, built and investigated at DLR in Cologne. The experimental and numerical investigations for both cascade configuration were split into two tasks: Task 3.2 *Compressor cascade basic flow* and Task 3.3 *Compressor cascade flow with transition control.* It is necessary to highlight few aspects of transonic compressor test section that make investigations very demanding in single passage test section. The profile defined by industrial partner RRD is representative for an in-service fan blade section. It appeared a very challenging aerodynamic duty to fulfil inlet Mach number = 1.22, flow deflection angle = 15° and AVDR = 1.2 (axial Velocity Density Ratio) in a single passage configuration. The influence of stream tube contraction, typically expressed with AVDR parameter, and particularly short spans to chord ratios of the test section, both promoting flow separation and strong main and end-wall flow interactions, that resulted in unexpected increase in effort, which was required to get the test section designed. Finally, the system of suction slots at wind tunnel walls was designed and the required conditions were obtained. Test section was designed by means of numerical simulations at IMP PAN and CFD predictions by NUMECA.

In case of cascade configuration (DLR), an extensive numerical investigation of the bleed slot location effect on AVDR and the sensitivity of the sucked mass flow was done by RRD and NUMECA (task. 3.2). Due to the complexity of the end wall flows, uncertainties of URANS predictability of those flows and possible negative interactions in respect to flow visualisation in the experiment, the proposed suction slot design of DLR was chosen finally.

In task 3.1, for reference (laminar interaction) and transition flow cases the following measurements are carried out: static pressure along the lower wall and blade, schlieren and oil visualization, PSP and TSP (Pressure and Temperature Sensitive Paint), shock oscillations measurements, velocity traverses (LDA measurements) of the upstream profile and in the wake. Transition was induced by three methods:

- strip: 3 locations: at leading edge and x/c = 0.16; 0.22,
- 3 step heights: 50, 100, 400 μm
- roughness patches
- Air Jet Vortex Generators: hole diameter D 0.3 mm (~boundary layer thickness), pitch angle $\phi = 30°$ and skew angle $\theta = 90°$, spanwise distance L/D = 15.

Based on the results of investigations in single passage test section (task 3.1), the transition control cases were defined and investigated in DLR cascade test facility (task 3.3). The measurements were carried out for two transition control cases. The first one is roughness patch Ra = 10 μm located on suction side at x/c = 0.11 and 10 mm width. The second one are AJVGs with a hole diameter of D = 0.4 mm, pitch angle $\phi = 30°$ and skew angle $\theta = 75°$, spanwise distance (pitch) L/D = 15. The AJVG configuration was optimised for the boundary layer as on the investigated profile. The detailed analysis of the steady and unsteady shock behaviour as well as the shock wave boundary layer interaction is based on:

- schlieren photographs—characterization of the shock structure
- PIV measurements—detection of the shock position within the cascade and measurement of the shock wave boundary layer interaction region
- High Speed Shadowgraphy (HSS)—characterization of unsteady behaviour of the shock fluctuation.

Numerical simulations for DLR cascade configuration were performed by 3 partners: RRD, Numeca and UAN. UAN performed simulations for basic flow configurations without transition control (then UAN left consortium due to research group disintegration as the international conflict consequence). RRD investigated reference and transitional configurations with HYDRA code and k-ω Menter-SST model (including Langtry and Menter correlation if transition is taken into account). The roughness model which is applied in RRD flow solver is based on the ideas from Aupoix and Eca and Hoekstra. In case of AJVG configuration, two different approaches are applied. For the first, the aerofoil and the real AJVG geometries tested in DLR cascade are meshed using the unstructured grid generator BOXER. For the second, source terms are built on the suction side of the baseline structured mesh.

NUMECA investigated the capabilities of state-of-the-art turbulence models such as SST and of new promising model such as Explicit-Algebraic Reynolds Stress (EARSM) model in predicting shock-boundary layer interaction in laminar, turbulent, and transitional state. Using both models simulations were performed in fully turbulent mode and also coupled to a transport model for laminar-to-turbulent intermittency as originally described in Langtry (2006). NUMECA activity aimed at computing (Fine/Turbo) the DLR test section with roughness patch and investigating the possibilities of the turbulence and transitional models. The application of the S-BSL-EARSM turbulence model coupled to a laminar to turbulent transition module has been investigated with a roughness patch. The model has been adapted in order to be able to detect laminar to turbulent transition promoted by roughness elements. The effect of roughness patches onto the shock-boundary layer interaction was investigated by roughness elements included in the turbulence and the transition models following the methodology developed in WP-2. Therefore, a modified solid wall boundary condition has been applied to ω equation, and a new correlation for laminar to turbulent transition promoted by roughness elements has been implemented.

All measurements for reference flow case in the single passage test section show very good agreement to the design requirements thus the flow structure on the transonic compressor profile (on suction side) is well reproduced. Reference case measurements prove that laminar shock wave boundary layer interaction leads to separation upstream of the main shock, whereas separation due to turbulent interaction just appears at the λ-foot. Experimental investigations of different boundary layer transition control devices show their influence on the shock in transonic conditions and possibility of separation bubble and shock wave oscillations reduction. However, the interaction is very sensitive to the upstream boundary layer and the aerodynamic performance can be easily decreased if tripping device is not properly applied. If the boundary layer was tripped just downstream of the leading edge (strip heights 0.1 and 0.4 mm) a thicker boundary layer is obtained, what increases λ-foot size and the shock wave pressure jump moves more upstream. Too thick boundary layer leads to the large separation zone downstream of shock wave, which in selected cases reaches even the trailing edge. As a consequence the wake width is much higher than without tripping device (Fig. 3).

Fig. 3 Velocity (wake)

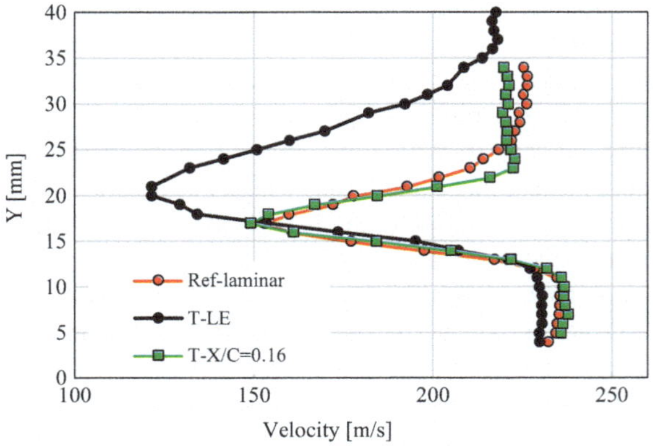

Fig. 4 Velocity (wake)—the best cases

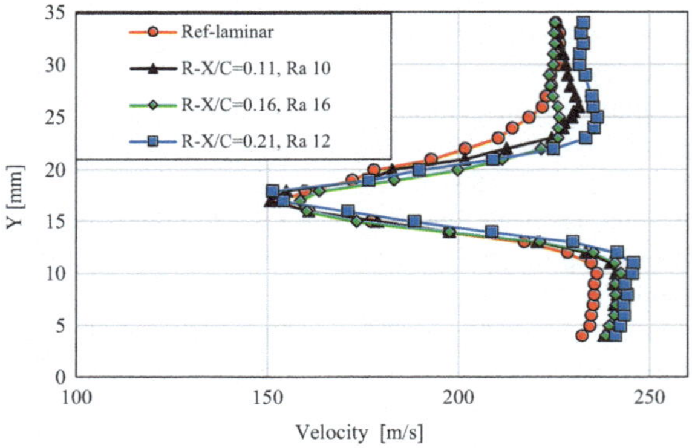

Positive effect was obtained if roughness patch is properly located (Fig. 4). Within the investigated cases, the improvement of the wake parameters is the best for the roughness patch of Ra = 12 μm located 21% of chord length. Oil flow visualization on the other hand shows that the boundary layer is not tripped upstream the shock along the whole span. The shock wave unsteadiness decreases for moderate treatment of boundary layer. The highest reduction of average shock oscillations was obtained for the roughness patch of Ra = 10 μm at 11% of chord length.

Application of AJVGs as tripping device influences on boundary layer upstream of the shock wave and shows also promising potential for shock oscillation and pressure loss reduction.

Based on the experimental investigations at DLR cascade it is observed that lower losses appear for the fully turbulent SWBLI than for the laminar SWBLI. From the results of AJVGs as tripping device, one can state that AJVGs lead to a reduction of viscous losses on the blade suction side which is reflected by a significant reduction of the wake losses. This means, that only a small amount of additional losses is added to the viscous losses within the boundary layer by this control method. However, the shock structure is changed which leads to additional shock losses at the suction side region. Therefore, the sum of viscous and shock losses which is mainly influenced by these both effects is lower than for the laminar case but higher than for the fully turbulent case, where the transition is forced by the high turbulence intensity of the inflow. The measurements for the blade with roughness patch show a different impact of this boundary layer control method on the loss behaviour. There is only a small reduction of the viscous losses, in spite of a turbulent shock boundary layer interaction. Probably, it arises from the additional losses which are caused by the roughness patch. The measured boundary layer thickness at 30% of the chord length is 50% higher than at measurements of the fully turbulent and AJVG case. In the roughness patch case a higher amount of the loss reduction is caused by the changed shock structure in the passage which leads to a reduction of the shock losses compared to the laminar case.

Both experiments show that pressure losses can be reduced when tripping the boundary layer upstream the shock, however over tripping easily appears when incoming boundary layer is increased too much and causes massive increase in pressure losses.

URANS simulations done by RRD indicate that the roughness model which is implemented in Hydra solver is capable to achieve transition in a physical manor. However the range of different roughness heights in-between which successful boundary layer tripping appears is rather small and over tripping appears quickly. Compared to the reference, transitional simulations only small reduction in pressure losses were observed for simulations including roughness patch where all post processed parameters are showing that transition is not yet appearing upstream the shock. The reduction of pressure losses for the cases with roughness patch but not yet tripped boundary layer can be explained with slightly lower viscous losses due to a laminar shock boundary layer interaction with a laminar boundary layer with slightly more momentum compared to the reference laminar boundary layer. For the roughness height which is leading to a reasonable tripped boundary layer, the pressure loss is increased compared to the reference transitional simulation. Although the shock induced separation is considerably reduced for the tripped case, the pressure loss is lining up to the loss of the fully turbulent simulation.

Sensitivity study to the roughness height has been conducted by NUMECA. Five different roughness heights have been investigated ranging from $Ra = 8\ \mu m$ to $Ra = 20\ \mu m$. They all correspond to a transitionally roughness surface. Numerical results show that the pressure losses are a direct function of the height of the roughness element. Furthermore, a too big roughness height over-tripped the boundary layer upstream of the interaction and leads to bigger wake losses. All investigated roughness heights lead to increased pressure losses compared to the reference laminar interaction. Above a certain threshold value of about $Ra = 9\ \mu m$, the pressure losses are higher than the losses computed by the fully turbulent simulation.

Numerical simulations for both test configurations with AJVGs as tripping device are done by IMP PAN (single passage test section) and RRD (DLR cascade configuration). General conclusion is that the boundary layer tripping was achieved, but an increase of pressure losses is observed compared to the reference laminar SWBLIs.

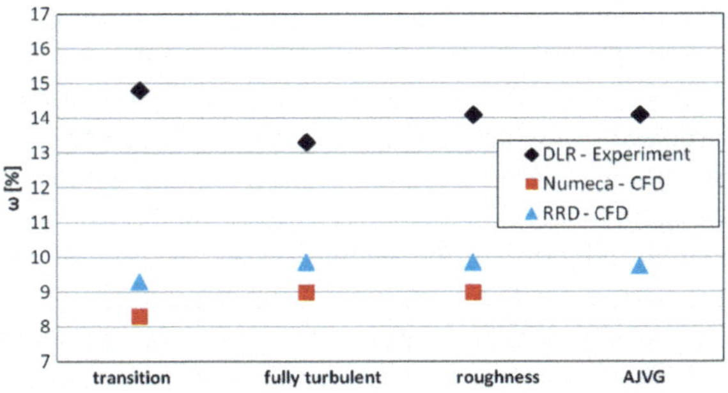

Fig. 5 Pressure losses in compressor cascade

Investigated cases and obtained experimental results allow to draw general conclusions. However, qualitative comparison of numerical results and experimental data for design condition show different effect on pressure losses when tripping the boundary layer (Fig. 5). Both measurements at IMP PAN in the single passage test section and at DLR in the cascade facility show that pressure losses can be reduced when proper tripping of the boundary layer is achieved. However over-tripping easily appears, for e.g. only a small range of roughness heights are capable to reduce pressure losses compared to the reference laminar SWBLI.

In all performed numerical simulations at IMP PAN, RRD and NUMECA tripping the boundary layer upstream the shock lead to increased pressure losses compared to the reference laminar cases. Those contradictory results between experiments and U/RANS simulations need further investigations and leaves the opened question: if U/RANS approach enables to obtain correct prediction of transition location and shock wave boundary layer interaction.

In respect to the unsteady behaviour due to SWBLI conclusions can be only derived from experimental investigations, since typical low frequency oscillations were not observed in URANS simulations. Experiments at IMP PAN did show that for certain cases including tripping devices the amplitude and the shock movement can be reduced compared to the laminar SWBLI.

Generally, improvement of flow parameters downstream of shock wave boundary layer interaction requires moderate treatment of the boundary layer upstream of the interaction. Overtripping (thicker boundary layer generated) appears in step manner and causes the deterioration of wake parameters and also sometimes increases the shock unsteadiness.

4 Conclusions on Internal Flow Case—Turbine

The strong acceleration of the flow leads to a relaminarisation of the boundary layer and the interaction with shocks may cause separation. Investigations of the flow development on the blade suction side were carried out in task 4.1. As described in the project DoW, there was planned to use the existing test section which was investigated in FP6 AITEB-2 project. However, RRD has decided to change in WP-4 the reference. The new single passage test section was designed by means of numerical simulations at IMP PAN. The partners NUMECA and RRD were involved with computational simulations to support the test section design.

Originally, it was planned that the boundary layer tripping should be realised by means of the AJVG located downstream of the cooling holes. Due to the serious technological problem with the manufacturing of the appropriate configuration of the jet holes (skewed, inclined and small diameter 0.3 mm) this method of boundary layer tripping was not investigated. The boundary layer transition control was realised by means of tripping step (height h = 50 μm, h/δ = 0.1). The following experimental techniques were applied at IMP PAN: static pressure along the lower wall and blade, schlieren visualization, oil visualization, shock oscillations measurements, velocity traverses measurements by means of LDA. Numerical investigations were carried out by means of Fine/Turbo NUMECA with EARSM turbulence model and transition model $\gamma - Re_\theta$. AJVG location on the separation downstream of the shock wave were investigated numerically only (for single passage test section). Numerical simulations were also done by NUMECA with the local source terms for the coolant flow.

AJVG configuration for cascade measurements (DLR) were defined according to IMP PAN experience from AITEB-2 project and numerical simulations results. Manufacturing reasons had to be taken into account. A compromise of the location and inclination to the blade surface angle θ had to be found due to a strong profile curvature. Under consideration of all manufacturing reasons the diameter of cooling holes was set to 0.3 mm. This was also the criteria to choose the AJVG exit location. The boundary layer thickness was taken from numerical results from a cooled blade to determine the location. The AJVG location was set to $x/ca_x = 54.4\%$, angles $\alpha = 65°$ and $\theta = 35°$.

The turbine blade was designed by RRD. To achieve relaminarization on the blade with film cooling, the two criteria from literature were applied. The first one is the acceleration parameter:

$$AC = \frac{\nu}{u_e^2} \cdot \frac{du_e}{ds}.$$

The acceleration should be higher than 3.5×10^6. The second one is the Buri parameter, which takes into account the acceleration and the Reynolds number based on the momentum thickness:

$$Bu = \mathrm{Re}_\theta^{\frac{5}{4}} \cdot AC$$

This parameter should be higher than 0.005. One coolant row was put into the Buri-region (stagnation flow) and the other one in the acceleration region. The pitch/c_{ax} ratio of the cascade is approximately 2. The Mach number at the outlet Ma_{exit} was set to 1.05 and the Reynolds number was $Re_{exit} \approx 1.1 \times 10^6$. The design was performed by means of a 2D cascade simulations with Mises and the TAU code.

The experimental investigations for cascade configuration within task 4.2 and task 4.3 were carried out in DLR test section in Gottingen. Following measurement techniques were applied: schlieren, pressure taps and total pressure wake traverses and hot wire measurements. To detect the transition location infrared thermography and hot film measurements on a coolant blade were conducted. The coolant cavity pressure was set 5% higher than the inlet pressure. The AJVG cavity pressure was the same as the main flow inlet pressure.

Numerical simulations were carried out by RRD, NUMECA and UAN. RRD applied two solvers: Fluent solver with k-ω SST turbulence model including the γ-Re_θ transition transport equation and CFX 14 with SAS-SST turbulence model and γ-Re_θ (unsteady simulations). NUMECA investigated the impact of AJVGs on the interaction between the oblique shock wave and the blade boundary layer along the DLR turbine cascade. The applicability of the S-BSL-EARSM turbulence model coupled to a laminar to turbulent transition module has been investigated with AJVGs. Furthermore, sensitivity study to AJVG mass flow rate and location should also be conducted. UAN numerical simulations (k-ω SST model) were focused on the influence of mesh refinement, turbulence intensity and turbulent length scale at the inlet, as well as values of transitional model constants on the flow structure, separation and kinetic energy losses. The work was performed for the basic cascade (without film cooling) by UAN and afterwards the remaining work was taken over by NUMECA.

NUMECA carried out additional investigations, the influence of an uncertainty on the blade geometry of the DLR cascade, where the geometrical variation is achieved by modifying the trailing edge angle.

Investigations of the shock wave boundary layer interaction in the single passage investigations show that the boundary layer is laminar up to the shock if no film cooling is applied. Laminar type of interaction is indicated by the fact that at such relative low Mach number upstream of the shock ($M = 1.26$) a very large λ-foot appears. Shock waves topology changes if cooling is applied and it is dependent on the blowing ratio.

Boundary layer separation is reduced in relation to the clean flow case (blade without film cooling). The separation size at such low Mach number upstream of the shock wave is very small and there is no particular difference in terms of its size between the investigated cooling blowing ratio.

Application of tripping device (strip) decreases the amplitude and RMS of shock oscillations. In the investigated cases the RMS is about 10% lower and the amplitude is about 20–30% lower in relation to the flow with laminar interaction. Separation size is significantly reduced (nearly fully attached) in relation to the case without tripping device. In case of turbulent interaction at low Mach number upstream of the shock the λ-foot does not appears in the flow. Normal shock wave reaches the wall boundary layer. The disturbances upstream of the normal shock are week Mach waves induced by the tripping strip.

In case of DLR (cascade) configuration, the losses are calculated by the pressure ratios without considering the influence of coolant flow. It is assumed that since the same temperature (no differences between main flow and coolant) at the inlet is

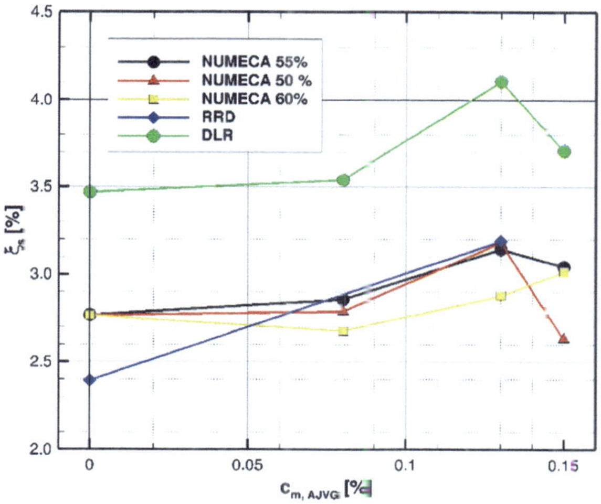

Fig. 6 Loss coefficient (cooling and AJVG influence)

applied, it does not have any effect on the enthalpy and can therefore be excluded from the evaluation. The results for the loss coefficients differ due to the different shock boundary layer interaction effect. In case of measurements, the coolant induces a slight decrease of losses with increasing mass flow ratio.

DLR measurements show an increase of losses in most of the investigated cases if boundary layer is tripped by Air Jet Vortex Generators (Fig. 6) it was found that the separation downstream of the shock wave is reduced.

Numerically predicted trend of loss coefficient are comparable to the experimental results. Noticeable is the loss of peak at $c_m = 0.13\%$. The same effect is obtained by sensitivity study to AJVG location done by NUMECA. It is shown that AJVG can have a positive effect on the shock boundary layer interaction if they are located further upstream. Numerical results show that by reducing this interaction the downstream boundary layer remains partially laminar. The maximum extension of the recirculation zone has been reduced and the maximum thickness of the separation zone, has also been reduced. It has a direct impact on the viscous forces acting on the blade surface and potentially on the heat transfer and the life-time of such a high pressure turbine blade. RRD unsteady simulations results show decrease of shock wave oscillations, but higher total pressure losses.

Experimental investigations lead to conclusion that the laminar boundary layer is very sensitive due to disturbances on the blade surface, especially due to surface roughness and due to the cooling/AJVG holes (even if there is no coolant or no jet flow).

Generally, improvement of flow parameters downstream of shock wave boundary layer interaction requires moderate treatment of the boundary layer upstream of the interaction. Overtripping may appear in step manner and cause the deterioration of boundary layer parameters downstream of the interaction region.

5 Conclusions on External Flow Case—Wing

Due to the high cost of specialized large-scale transonic wind tunnels, existing facilities were used in the consortium. One of these is a transonic test section at UCAM where laminar and transitional profiles had been studied previously at Reynolds numbers up to 2 million (based on chord length). In the first task 5.1, a 2-D laminar airfoil was specifically designed by DAAV for transonic tests, and tested in the UCAM wind tunnel, with the aim to provide basic investigations of the transition location effects on shock-induced separation and unsteadiness for a relatively large number of parameters.

Then, tests in the larger wind tunnel of IoA were performed in task 5.2 which enabled investigation of a much larger aspect ratio profile. In this facility, it was possible to measure a whole force polar up to and including the buffet boundary. The experiment was conducted for natural boundary layer transition location as well as different methods of tripping. The task 5.2 included validation and improvement of various steady and scale resolving numerical methods, without and with transition control devices.

Finally, a purely CFD task 5.3 consisted of simulating a 3-D laminar wing, based on the experience gathered by CFD from simulations in tasks 5.2. The goal was to understand transition control effect on a real wing with a leading edge sweep and a tip vortex.

DAAV has designed a laminar profile (termed V2C) used for task 5.1 and 5.2 and a laminar wing used for task 5.3. The wing is based on the V2C profile with sweep and twist for a closer representation of a real geometry. In both geometries, parabola method coupled with a boundary layer code were used to assess that laminarity was effective up to the shock (amplification factor and wing root contamination below a given limit). The laminarity analysis has also provided guidelines for experimentalists in terms of maximum inlet turbulence intensity.

A wide range of measurement techniques were employed by IoA and UCAM, as summarized in the Table 4, to characterize turbulent, transitional and laminar SBLI.

Table 4 WP-5 experiments

Steady SBLI	[UCAM/IoA/both]
SBLI flow topology	Schlieren, PSP, pressure taps
Aerodynamic forces	Drag (LDV in wake), PSP CL (pressure taps), CD (aero. rake)
Buffet onset limit	Strain gauge
Unsteady (buffet)	
Shock structure	Oil flow, senflex sensors, PIV
Buffeting frequency, fluctuating pressure	High speed video Schlieren Kulites, Pressure taps
Buffeting amplitude	High-speed vid. Schlieren Schlieren, Kulites

Prior to buffet onset (steady regime), the flow topology at SBLI, in particular the presence of a laminar or transitional separation bubble at shock foot, was evidenced by Schlieren visualization as well as surface pressure coefficient obtained with pressure taps or pressure sensitive painting. Pressure transducers signal integration on the surface and the wake allowed lift and drag coefficient to be measured, as well as their dependence on the tripping location. Buffeting occurring at unsteady SBLI was observed using high-speed video Schlieren visualization by UCAM, and IoA employed Kulites, unsteady pressure taps, strain gauge and senflex sensors. Oil flow visualizations were provided by both partners to determine transitional flow pattern and shock structure. Finally transition control effects were studied by UCAM using 2-D and 3-D trip dots as well as painting, and by IoA with carborundum of three different heights (0.1, 0.15, 0.26 mm). The chord length of the model used at UCAM is c = 100 mm, with an aspect ratio of L/c = 1 and test section height H = 2c, whereas for IoA c = 200 mm, L/c = 3 and H = 3c. Due to smaller model and higher blockage, UCAM focused on moderate values of angle of attack, whereas IoA performed full polar measurements, but the complementarity of metrology employed, as shown in the above table, and the earlier availability of UCAM data has justified the importance of both test campaigns in task 5.1 and task 5.2 for the work-package achievements.

The Table 5 summarizes numerical contributions for WP5.2 and WP5.3 for the three types of interactions. This includes implementation, calibration and application of the k-w-γ-R_θ transition model (IoA, LIV, IMFT, NUMECA) for laminar and

Table 5 WP-5 numerical study

	Laminar	Transitional (trip location)	Turbulent
WP5.2 steady		[IoA/URMLS/DAAV/NUMECA/IMFT/LIV/all]	
Flow topology	k-w-γ-R_θ DNS (4°)		RANS
Aerodynamic forces		k-w-γ-R_θ/RANS + optimisation study + drag decomposition study	
Buffet onset limit		26 k-w-γ-R_θ URANS	40 DDES + 35 RANS/SST
WP5.2 unsteady			
Shock structure Buffeting frequency Pressure fluctuations Buffet amplitude	URANS-k-w-γ-R_θ DNS (7°)	DNS (Trip 10%) Tripped ZDES k-w-γ-R_θ DDES, K-w-γ-R_θ URANS SST	OES k-ε, DDES Fully turbulent ZDES
WP5.3 laminar wing			
Application of WP5.2 and WP5.3 findings to a «real life» configuration		EARSM LCTM K-w-γ-R_θ Tripped IDDES Tripped OES k-ε, K-w-γ-R_θ	

transitional SBLI, for which three direct simulations (DNS) were also performed by URMLS thus providing additional reference data for models validation. The study of transition location effect was covered by IoA and IMFT to determine the optimal tripping location which yields best aerodynamic performances (in terms of lift-over-drag ratio). Sensitivity to the nature of tripping was also investigated by IoA (using vortex generators of heights similar to carborundum height used in IoA experiment and located at 0, 6, 15, 30 and 60% chord) and LIV (using CoC1 as well as air jet vortex generators). The determination of buffet onset limit (in terms of critical Mach M_∞ and flow incidence α) was carried out by URMLS (URANS and DDES), IMFT and LIV (URANS) who performed parametric study in the (M_∞, α) domain. Finally, several partners (DAAV, URMLS, IMFT, LIV) performed scale-resolving simulations ((I)DDES, OES) of buffet using various mesh resolutions and underlying RANS turbulence models.

Steady SBLI for $M = 0.7$, $\alpha = 4°$ (IoA)

The presence of a separation bubble at shock foot for laminar or transitional SBLI yields the presence of a plateau in pressure coefficient distribution, with laminar (resp. turbulent) separation bubble characterized by a strong rear (resp. front) foot of the λ-shock, seen on experimental and numerical results shown in Fig. 7. Plots in black (resp. red) are data for the open-air (resp. in-tunnel) configurations. The analysis enabled to distinguish laminar interaction from mere boundary layer (BL) thickness effects of turbulent interaction, even for BL tripped just before shock. The importance of transition modeling (approaches providing smooth evolution of intermittency γ rather than simple Heaviside-function used for numerical tripping) proved essential for this type of flows.

Dependency of Aerodynamic Coefficients on Tripping Location

It was generally observed that increased laminarity (i.e. tripping closer to shock) expectedly yielded lower friction drag, but resulted in strong shock, thus higher wave drag. The maximum lift-over-drag ratio therefore is obtained for a tripping location

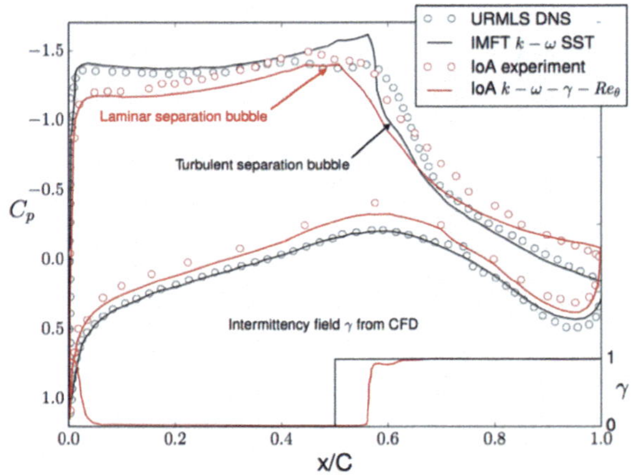

Fig. 7 Pressure coefficient distribution

Fig. 8 Buffet map

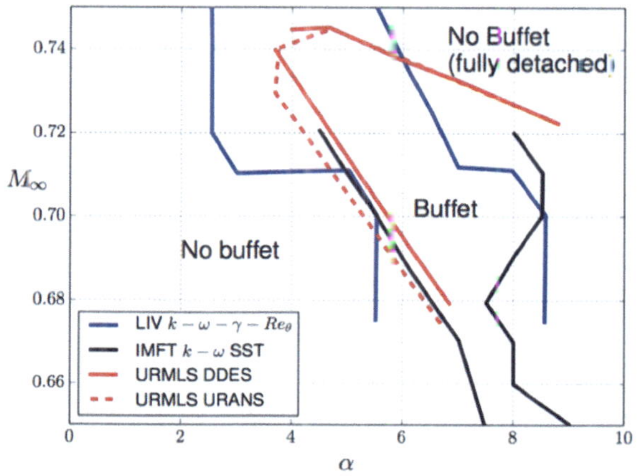

somewhere between leading edge and shock position, commonly found at around 30% chord length for the present case.

Buffet Limit

The experimental detection of buffet onset when varying the Mach number or the flow incidence was performed by IoA using strain gauge. The study was also numerically carried out using several URANS models as well as DDES by LIV, IMFT and URMLS. Buffet map was determined by Mach/Incidence parametric study requiring between 25 and 40 simulations to be carried out for each approach. The study (Fig. 8) reveals a good agreement between SST URANS and DDES methods, whereas the k-w-γ-R_θ predicts an earlier onset with respect to incidence.

Unsteady SBLI with Buffet for Mach = 0.7 and α = 7°

The full buffet regime was the object of several DDES and DNS at IMFT, URMLS, LIV and DAAV. The investigation focused on shock amplitude, lift and drag oscillations amplitude and frequency, and pressure fluctuations downstream of the shock. Only transitional or turbulent interactions were covered by DDES, with a strong dependency of shock oscillation amplitude and extreme shock position on turbulence model and grid, although a Strouhal number associated to buffeting instability was commonly found to be around St = 0.1 based on chord length. The plot in Fig. 9 presents the drag coefficient history, provided by the DDES of UMLRS and DAAV and the k-ε OES of IMFT. Laminar buffeting study was covered by IoA experiment and URMLS DNS that both reported very low oscillation amplitude of shock, lift and drag coefficients, suggesting an opposite trend to turbulent buffet.

Natural and Transitional SBLI on the Laminar Wing

Starting from DAAV guidelines (M_∞, α) values for buffet onset, partners involved performed URANS with transition modeling (k-w-γ-R_θ) (IMFT, LIV, NUMECA) and scale resolving simulations (LIV: IDDES, IMFT: k-ωOES), with fine-tuning of the tripping position to account for models influence and/or the Mach number to correct the sweep effect, with the aim to detect buffet. Simulations from LIV and IMFT reveals milder shock oscillation than for the previous V2C case, and

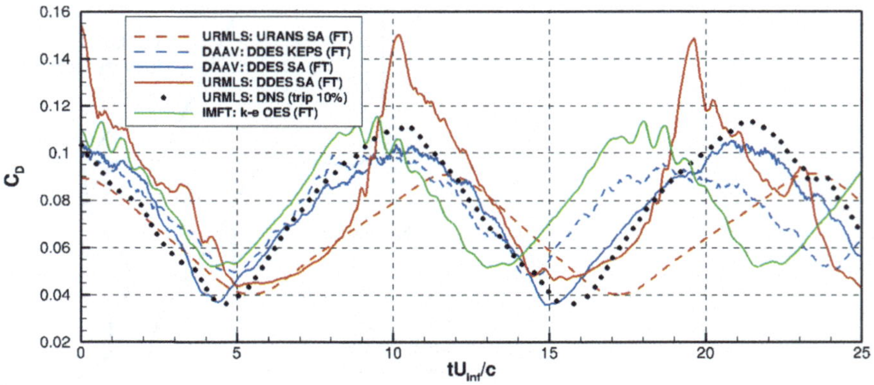

Fig. 9 Drag coefficient history

IMFT investigated the span-wise dependency of buffeting frequency (due to the local chord-length based Reynolds number being different). LIV performed additional simulations to verify the effectiveness of VGs, AJVGs and flaps to suppress buffet.

6 Main Achievements of TFAST Project

It is known that designing an experiment in transonic and transitional conditions is a difficult task, which required a number of adjustments by UCAM and IoA. This took time, but was finally successful with configurations consistent with the qualitative descriptions of results obtained in the past. Methods like PSP and TSP were successfully used to give a rapid and global view of the flows at the wall. For more detailed information, a particular effort had to be put on particle methods. Despite severe limitations related to seeding and space resolution problems, they bring new quantitative data, useful for the validation of the numerical simulations and for the physical description of the flow. Fluctuating fields could be obtained only with particular flow and optical settings. It is clear that such methods, although of unique potential, have still to be complemented for unsteadiness measurements by classical methods like hot-wire anemometry which in this case remains a simple, accurate and reliable tool.

Most of the work of WP-1 was done in 2-D situations, but the mechanisms put in evidence are believed to be general enough to be used as a guide for more complex situations. Some points have been clarified. During the project, the different partners have used several Mach and unit Reynolds numbers. Possible ways to compare the different results and their mutual consistency are now clearer. Similarly, the point of the onset of (2-D) separation has been documented.

Conclusions can be drawn about the numerical methods. Methods based on stability analysis proved to be insightful. Moreover, the low Reynolds number flows studied by TFAST are attractive for methods like DNS and LES which proved to be efficient. Their efficiency has been shown during the project, including wall perturbations. Of course their cost is still such that it is difficult to use them in the design routines for industry, but their potential is real and present.

A major conclusion is probably that these transitional flows are very sensitive to the level of external background fluctuations, so that their organization (size of the separated zone, spreading of the pressure distribution, etc.) depends critically on the nature of the perturbation: the mean flow is directly influenced by them. This means that the external level of background turbulence, or the fluctuations introduced by wall perturbations are major parameters for the determination of the mean field and of the low Reynolds number turbulence formed in and downstream of the interaction. These points have been documented in the TFAST experimental results database.

A second conclusion is relative to the low frequency unstead ness which is found in the interaction. It seems directly related to the transitional mechanism along the interaction. There is still some controversy on their dimensionless values between numerical and experimental results, but similar space-time properties and amplitudes have been observed.

The main achievement of WP-2 is to produce a comprehensive database for transitional SBLIs under the influence of transition trips and flow control. Thus, the existing knowledge of transition tripping in high-speed flows (at practical Reynolds numbers) was extended comprehensively. A comparison of tripped and un-tripped configurations also revealed that in many cases it is not detrimental (even beneficial) to allow the shock wave 'to do the work' and promote transition in the interaction zone. These findings are only valid for interactions on flat surfaces which was the scope of this work package. Nevertheless, this unexpected result is a valuable contribution to the knowledge of compressible flow.

Investigated turbomachinery configurations (WP-3 and WP-4) show that improvement of flow parameters downstream of the shock wave boundary layer interaction, separation and shock oscillations reduction, requires moderate treatment of the boundary layer upstream of the interaction. Overtripping (thicker boundary layer generated) appears in rapid manner and causes the deterioration of wake parameters and also sometimes increases the shock unsteadiness.

A considerable amount of experimental and numerical data were produced to characterize tripping location influence on SBLI prior, at, and beyond buffeting onset, although no unique conclusion could be drawn due to (1) heterogeneity of results and flow conditions, (2) blockage effect in both tunnels that influenced buffet onset and was not (as intended) always been accounted for in simulations, (3) challenges inherent to experiments with laminarity (influence of model roughness and manufacturing, intrusive metrology), (4) prevalence of fully turbulent methods in buffet simulations and (5) grid and models dependency, mostly for transitional buffet simulations.

Nonetheless, the following lessons could be formulated:

- The laminar flow benefit (lower friction drag and enhanced lift due to the shock being pushed backwards) could be reduced when wave drag becomes comparatively significant, as evidenced both experimentally and numerically, showing wave drag could increase for laminar or transitional SBLI.
- Experimental studies could not characterize a strong buffet onset dependency on tripping. At full buffet regime, experimental observations (IoA) and DNS results (URMLS) tend to suggest buffeting frequency and aerodynamic fluctuations are considerably reduced for transitional or laminar SBLI, which would tend towards the conclusion that unsteady laminar interaction is not detrimental and does not need to be avoided. However these conclusions are only based on two-dimensional cases and the influence of sweep angle and finite span is still an open question.

Physical models for laminar and transitional SBLI are upgraded. Several partners have implemented transition models (such as the k-w-γ-R_θ model) with calibration of the correlation function and testing on both 2-D and 3-D reference cases. The intermittency provided by these methods proved essential to predict laminar separation bubble at shock foot, as well as a realistic flow tripping when tripping devices are numerically represented or modeled, therefore demonstrating readiness of such methods for industrial use. Moreover, unsteady configurations were the object of a large number of scale resolving simulations ((I)DDES and OES), that, in the framework of the TFAST project, proved very useful in providing information on coupling mechanisms at the origin of the buffeting phenomenon.

The manufacturer's authorised representative in the EU is Springer Nature Customer Service Centre GmbH, Europaplatz 3, 69115 Heidelberg, Germany. If you have any concerns regarding our products, please contact ProductSafety@springernature.com

Printed and bound by CPI Group (UK) Ltd, Croydon, CR0 4YY

30/09/2026

02225648-0018